Inorganic Biochemistry
Volume 1

A Specialist Periodical Report

Inorganic Biochemistry
Volume 1

A Review of the Recent Literature Published up to late 1977

Senior Reporter

H. A. O. Hill *Inorganic Chemistry Laboratory, University of Oxford*

Reporters

J. V. Bannister *University of Malta*
D. Barber *University of East Anglia*
N. J. Birch *University of Leeds*
M. Brunori *University of Rome*
A. Galdes *University of Oxford*
B. T. Golding *University of Warwick*
C. Greenwood *University of East Anglia*
B. Giardina *University of Rome*
P. M. Harrison *University of Sheffield*
M. N. Hughes *Queen Elizabeth College, London*
P. F. Knowles *University of Leeds*
G. J. Leigh *University of Sussex*
C. A. McAuliffe *UMIST, Manchester*
P. J. Sadler *Birkbeck College, London*
B. E. Smith *University of Sussex*
A. Treffry *University of Sheffield*

The Chemical Society
Burlington House, London W1V 0BN

British Library Cataloguing in Publication Data

Inorganic biochemistry.–
(Chemical Society. Specialist periodical reports).
Vol. 1
1. Biological chemistry   2. Chemistry, Inorganic
I. Hill, Hugh Allen Oliver   II. Series
574.1'921   QP531

ISBN 0-85186-570-4
ISSN 0142-9698

Printed in Great Britain by
Adlard & Son Ltd., Bartholomew Press,
Dorking

Foreword

Inorganic Biochemistry is concerned with the function of all metallic and most non-metallic elements in biology. It can be defined as the biochemistry of those elements whose chemistry normally consititutes the province of inorganic chemists. Its net is cast widely, stretching from chemical physics to clinical medicine. Its influence is all-pervasive. Thus 'there probably does not exist a single enzyme-catalyzed reaction in which either enzyme, substrate, product or a combination of these is not influenced in a very direct and highly specific manner by the precise nature of the inorganic ions which surround and modify it'.[1] The subject has developed dramatically during this decade. This is reflected in the number of publications[2] and even a journal[3] devoted to it. The upsurge of interest coincides with improved analytical methods, less time-consuming preparative techniques, the successful application of spectroscopy and diffraction, the improved synthesis of simple inorganic complexes used to model or mimic various aspects of biological molecules, the increased concern about the environmental hazards caused by some metal ions, the use of metal ions or complexes as therapeutic agents, and the recognition of the importance of an increasing number of trace elements in plant, animal, and human nutrition. All have contributed towards the growth of a subject which previously attracted only a few lone devotees. I hope that this volume and those to follow will ease the burden of inorganic biochemists as they seek to relate their work to fields far beyond the confines of the traditional disciplines.

[1] H. R. Mahler, in 'Mineral Metabolism', Vol 1, Part B, ed. C. L. Comas and F. Bronner, Academic Press, New York, 1961, pp. 743–879.

[2] (a) B. L. Vallee and W. E. C. Wacker, in 'The Proteins', ed. H. Neurath, 2nd. edn., Vol.5, Academic Press, New York, 1970; (b) M. N. Hughes, 'The Inorganic Chemistry of Biological Processes', Wiley, London, 1972; (c) 'Inorganic Biochemistry', ed. G. L. Eichhorn, Elsevier, Amsterdam, 1973; (d) H. Sigel, 'Metal Ions in Biological Systems', Vols. 1–7, Marcel Dekker A. G., Basel; (e) 'Techniques and Topics in Bioinorganic Chemistry', ed. C. A. McAuliffe, Macmillan, London, 1975; (f) E. J. Underwood, 'Trace Elements in Human and Animal Nutrition', Academic Press, New York, 4th edn., 1977; (g) 'Trace Elements in Human Health and Disease', ed. A. S. Prasad, Academic Press, New York, 1976; (h) E. J. Hewitt and T. A. Smith, 'Plant Mineral Nutrition', English Universities Press, London, 1975; (i) 'An Introduction to Bio-inorganic Chemistry', ed. D. R. Williams, C. C. Thomas, Springfield, Illinois, 1976; (j) D. A. Phipps, 'Metals and Metabolism', Oxford University Press, 1976; (k) E. I. Ochiai, 'Bioinorganic Chemistry: An Introduction', Allyn & Bacon Inc., J. Wiley & Sons, Rockleigh, New Jersey, 1977; (l) R. P. Hanzlik, 'Inorganic Aspects of Biological and Organic Chemistry', Academic Press, New York, 1976; (m) 'Bioinorganic Chemistry – II', ed. K. N. Raymond, ACS Advances in Chemistry Series No. 162, 1977; (n) 'Biological Aspects of Inorganic Chemistry', ed. A. W. Addison, W. R. Cullen, D. Dolphin, and B. R. James, John Wiley & Sons, New York, 1977; (o) A. S. Brill, 'Transition Metals in Biochemistry', in 'Molecular Biology, Biochemistry and Biophysics', Vol. 26, ed. A. Kleinzeller, G. F. Springer, and H. E. Wittmann, Springer-Verlag, Berlin, 1977; (p) R. J. P. Williams and J. R. R. F. da Silva, 'New Trends in Bio-inorganic Chemistry', Academic Press, London, 1978.

[3] *Journal of Inorganic Biochemistry*, Elsevier, New York.

The Reporters have striven valiantly to present, in a palatable form, all relevant work published in 1977. Some subjectivity in selection is inevitable. This is particularly true of work coming from the extremes of the discipline. The criterion used has been that the work should be concerned, at least in passing, with the molecular aspects of the subject, thereby excluding papers concerned solely with, for example, the assay of metalloenzymes in clinical chemistry. I know that all the Reporters would welcome comments and criticisms; I would be glad to have readers' opinions on the coverage and content. Subsequent volumes will be produced somewhat closer in time to the year of interest; the division of the subject matter will not always remain the same.

I would like to thank the Reporters for their labours, the committee of the Inorganic Biochemistry Subject Group of the Chemical Society for advice and encouragement, and Mr. B. Starkey and other members of the staff of the Society, who have attended with diligence and care the production of this volume.

H. A. O. HILL

Contents

Chapter 8 Metalloenzymes   317
By A. Galdes and H. A. O. Hill

1
Inorganic Analogues of Biological Molecules

C. A. McAULIFFE

1 Complexes of Amino-acids and Peptides

Of all the possible model studies of metal ions with biologically important ligands, those relating to metal–amino-acid interactions have been the longest studied; the justification being, of course, that proteins contain the same types of donor groups and the study of simple models is thus worthwhile. With the availability of peptides of increasing complexity, it has recently become possible to study much more complicated model systems. Of course, it should be borne in mind that amino-acids are interesting ligands in their own right. In this section the area is divided into three parts, namely metal complexes of (i) simple amino-acids and peptides; (ii) amino-acids containing sulphur donors; and (iii) amino-acids containing heterocyclic nitrogen donors.

Simple Amino-acids and Peptides.—It is not entirely clear what biological justification there is for studying chromium complexes of amino-acids, though there may be some relevance to the 'glucose tolerance factor'. Two lengthy studies have been reported.[1, 2] In one, the rather novel matrix method was used. This involves mixing hexamminechromium(III) nitrate with the amino-acid in 1:3 ratio in a mortar and then heating to 150 °C in an oven, or merely mixing the constituents in warm water. A large number of compounds were isolated with L-α-amino-acids (alanine, aminobutyric acid, norvaline, norleucine, valine, isoleucine, and leucine), *viz.* [Cr(ala)₃], (+)-[Cr(am-but)₃], [Cr(OH)(am-but)₂]₂, (+)- and (−)-[Cr(norval)₃], (−)-[Cr(norleu)₃], [Cr(isoleu)₂(isoleu-O)(NH₃)], (+)- and (−)-[Cr(leu)₃] were prepared by the matrix method, and differences were noted[1] in the formation of the tris-type and the hydroxo-dimer complexes by the two methods. All of the tris-type complexes were of the *fac* structure. A number of complexes with L- and DL-asparagine, L- and DL-aspartic acid, L- and DL-glutamine, L- and DL-glutamic acid, L- and DL-lysine, and L-ornithine were also isolated, including a new complex of chromium(III) containing terdentate DL-aspartic acid and a species containing three different ligands (L-lysine, ammine, and water).[2] A mixed salt of the (glycinato)bis(oxalato)chromate(III) complex has been prepared and characterized with the use of ion-exchange chromatography. In acidic media this complex has been shown to aquate to the *cis*-diaquabis(oxalato)chromate(III) ion.[3] Formation constants of molybdenum(VI)

[1] H. Oki, *Bull. Chem. Soc. Japan*, 1977, **50**, 680.
[2] H. Oki and Y. Takahashi, *Bull. Chem. Soc. Japan*, 1977, **50**, 2288.
[3] T. W. Kallen and R. E. Hamm, *Inorg. Chem.*, 1977, **16**, 1147

and tungsten(VI) with aspartic and glutamic acids have been determined by the potentiometric method,[4] and indicate that previous values[5] are incorrect.

Very few iron–amino-acid complexes have been synthesized and studied. Holt and co-workers[6] have determined the crystal structure of tri-μ_3-oxo-triaqua-hexakis(glycine)tri-iron(III) perchlorate, $[Fe_3O(glyH)_6(H_2O)_3](ClO_4)_7$, and have shown it to be similar to that of basic iron acetate, $[Fe_3O(CH_3CO_2)_6(H_2O)_3]ClO_4$, with a trimeric unit having an oxygen atom at the centre. The remaining co-ordination sites of each iron are occupied by four carboxylate oxygens from bridging zwitterionic glycines and a single water molecule. This general area needs more study, especially by X-ray techniques. (See, for example the vast discrepancy in something as essential as calculated/found elemental analyses by Holt's group[7]).

The synthesis, resolution of enantiomers, and rates of H exchange of Λ,Δ-α-(R,R,S,S)-[Co(trien)(glyOEt)Cl]$^{2+}$ have been described, and structural assignments have been confirmed by an X-ray study of Λ,Δ-α-(R,R,S,S)-[Co(trien)-(glyO)]I$_2$·3H$_2$O.[8] A series of mixed ligand cobalt(III)–Schiff base complexes with the general formula [Co{sal$_2$-(S,S)-chxn}(aa)] (aa = glyH, L- and D-alaH, L- and D-valH, L- and D-leuH, L- and D-thrH, L- and D-trpH) are readily obtained from [Co{sal$_2$-(S,S)-chxn}] and amino-acids by oxidation with air. All of these complexes adopt, stereoselectively, the Λ-cis-$\beta_1(fac)$-structure, irrespective of the configuration of the amino-acids. On the other hand, the reactions of [Co-{sal$_2$-(S,S)-chxn}] with an excess of DL-amino-acids in open air gave [Co-{sal$_2$-(S,S)-chxn}(aa)], in which all L-amino-acids except proline are selectively co-ordinated.[9] Higa and co-workers[10] have published details of the separation and optical resolution of the isomers of bis(β-alaninato)(oxalato)cobalt(III) and (β-alaninato)(glycinato)(oxalato)cobalt(III) complexes. The synthesis of mixed ligand cobalt(III) complexes with (S)-aspartic-N-monoacetic acid, (S)-AMA, and different amino-acids Na[Co{(S)-AMA}(aa)] {aa = glyH, (R)- and (S)-alaH, valH, pheH, serH, and leuH} leads to a mixture of cis-N- and $trans$-N-isomers.[11] The cis-N/$trans$-N interconversion of these compounds containing glycine and (R)- and (S)-leucine is further reported,[12] and it has been shown that the chirality of the bidentate amino-acidato species has a pronounced influence on the cis–$trans$ equilibrium of these systems.

The oxidative rearrangement of dinuclear cobalt–dioxygen complexes to mononuclear cobalt(III) chelates in aqueous solution has been studied for the series of ligands glycylglycine, glycyl-L-alanine, glycyl-L-serine, L-serylglycine, L-alanyl-L-alanine, L-alanylglycine, and glycyl-L-tyrosine. The reaction proceeds in two first-order steps; the first is the more rapid ($t_{1/2} \simeq$ minutes), and it probably

4 D. L. Rabinstein, M. S. Greenberg, and R. Saltre, *Inorg. Chem.*, 1977, **16**, 1241.
5 M. K. Singh and M. N. Srivastava, *J. Inorg. Nuclear Chem.*, 1972, **34**, 2081.
6 R. V. Thundathil, E. M. Holt, S. L. Holt, and K. J. Watson, *J. Amer. Chem. Soc.*, 1977, **99**, 1818.
7 W. F. Tucker, R. O. Asplund, and S. L. Holt, *Arch. Biochem. Biophys.*, 1975, **166**, 433.
8 B. F. Anderson, J. D. Bell, D. A. Buckingham, P. J. Cresswell, G. J. Gainsford, L. G. Marzilli, G. B. Robertson, and A. M. Sargeson, *Inorg. Chem.*, 1977, **16**, 3233.
9 Y. Fujii, M. Sano, and Y. Nakano, *Bull. Chem. Soc. Japan*, 1977, **50**, 2609.
10 T. Ama, M. Higa, N. Kione, and T. Yasui, *Bull. Chem. Soc. Japan*, 1977, **50**, 2632.
11 G. Colomb and K. Bernauer, *Helv. Chim. Acta*, 1977, **60**, 459.
12 G. Colomb and K. Bernauer, *Helv. Chim. Acta*, 1977, **60**, 468.

involves the oxidation of one ligand molecule per two cobalt atoms and the conversion of the bridging dioxygen moiety into water. The second step is slower ($t_{1/2} \simeq$ hours), and probably involves displacement of the oxidized ligand by the excess dipeptide present in solution.[13]

Circular dichroism (c.d.) and proton magnetic resonance (^{1}H n.m.r.) spectra of nickel(II) complexes containing a series of ethylenediamine-NN'-diacetic acids (H_2edda) (see Figure 1), which are optically active quadridentate ligands, have

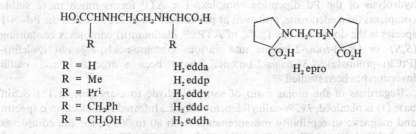

$$
\begin{array}{cc}
\text{HO}_2\text{CCHNHCH}_2\text{CH}_2\text{NHCHCO}_2\text{H} & \\
| \qquad\qquad\qquad | & \\
\text{R} \qquad\qquad\qquad \text{R} &
\end{array}
$$

R = H   H$_2$edda

R = Me   H$_2$eddp

R = Pr^i H$_2$eddv

R = CH$_2$Ph   H$_2$eddc

R = CH$_2$OH   H$_2$eddh

H$_2$epro

Figure 1 *The edda-type polyamino-carboxylic acids. All except H$_2$edda possess two asymmetric carbon atoms with the S configuration*

been measured in aqueous solution. The epro complex, which contains L-proline residues on the ligand, is found to exhibit the same c.d. spectrum as complexes of other edda-type ligands containing L-alanine, L-valine, L-phenylalanine, and L-serine residues. These complexes stereospecifically take the Δ-S-cis-form in solution, and the large contact shift observed for the α-protons of the amino-acid moieties indicates that the substituent groups become axial to the chelate plane.[14] Both ^{1}H n.m.r. and e.p.r. spectroscopy have been used to deduce the structures of the complexes that are formed between nickel(II) or copper(II) and Thr-Lys-Ala-Ala in aqueous solution over a broad pH range. The binding sites in this tetrapeptide are $-$NH$_2$ of threonine and three deprotonated nitrogens of the peptide linkages. A co-operative interaction was observed in the case of nickel(II) ions.[15] A compound of the type [Ni(Bz-β-ala)$_2$]·2H$_2$O (Bz-β-ala = benzoyl-β-alaninate) and its amine adducts of the type [Ni(Bz-β-ala)$_2$B$_n$]·xH$_2$O (B = nitrogen base) have been reported. All of the complexes are hexaco-ordinate, and it is concluded that NiO$_6$ and NiN$_6$ chromophores exist for [Ni(Bz-β-ala)$_2$]·2H$_2$O and [Ni(en)$_2$(Bz-β-ala)$_2$], respectively.[16]

Results of potentiometric titrations[17] of [Pd(en)(H$_2$O)$_2$]$^{2+}$ with glycylglycine (glyglyO) and glycinamide (glyNH$_2$) are consistent with equilibria (1) and (2), and similar conclusions have been drawn about this system with L-asparagine

$$
\begin{array}{ccc}
\text{[Pd(en)(H}_2\text{O)}_2]^{2+} & & \text{[Pd(en)(glyglyO)]} \\
+ & \rightleftharpoons \text{[Pd(en)(glyglyO)]}^+ \rightleftharpoons & + \qquad\qquad (1) \\
\text{glyglyO} & & \text{H}^+
\end{array}
$$

[13] W. R. Harris, R. C. Bess, A. E. Martell, and T. H. Ridgway, *J. Amer. Chem. Soc.*, 1977, **99**, 2958.

[14] T. Murakami, I. Hirako, and M. Hatano, *Bull. Chem. Soc. Japan*, 1977, **50**, 164.

[15] G. Formicka-Kozlowska, H. Koslowski, and B. Jezowska-Trzebiatowska, *Inorg. Chim. Acta*, 1977, **24**, 1.

[16] G. Marcotrigiano, L. Menabue, and G. C. Pellacani, *Bull. Chem. Soc. Japan*, 1977, **50**, 742.

[17] M. C. Lim, *J.C.S. Dalton*, 1977, 15.

$$[Pd(en)(H_2O)_2]^{2+} \qquad\qquad\qquad [Pd(en)(glyNH'_2)]$$
$$+ \qquad \rightleftharpoons [Pd(en)(glyNH_2)]^{2+} \rightleftharpoons \qquad + \qquad\qquad (2)$$
$$glyNH_2 \qquad\qquad\qquad\qquad\qquad\qquad H^+$$

and L-glutamine.[18] The palladium(II) complex of glycyl-L-aspartic acid forms monomeric and dimeric species with adenosine and ATP. At higher pH the palladium atom favours the N-1 over the N-7 co-ordination in nucleoside and nucleotide molecules. At pH > 10 adenosine is unbound, and promotes a 'double' hydrolysis of the Pd–dipeptide complex. The ATP forms much more stable complexes than adenosine, and even at pH values > 10 in 2:1 solution the Pd–N-1 species is the dominant one (80% of ATP).[19] Platinum(II) complexes containing (S,S)- or (R,R)-*trans*-2-butene and various L-amino-acids, *e.g.* *cis*(N, olefin)-[PtCl(L-prolinate){(S,S)-*trans*-2-butene}], have been synthesized and olefin inversion has been studied.[20]

Regardless of the molar ratio of salicylaldehyde to L-arginine, a 1:1 Schiff base (1) is obtained, *i.e.* N-salicylidenearginine,L. Infrared and electronic spectra and magnetic susceptibility measurements from 80 to 300 K for the complexes [Cu(L)NO$_3$], [Cu(L)Cl]·2H$_2$O, [Ni(L)NO$_3$], and [Ni(L)Cl]·H$_2$O indicate that the copper nitrate complexes are antiferromagnetic (T_N = 250 K), and a tetrameric structure has been proposed. The chloride appears to be a mixture of the same tetrameric species and a magnetically dilute form which may be dimeric. Low magnetic moments suggest that the nickel complexes are mixtures of octahedral and planar ones.[21] Hatfield and co-workers[22] have reported electronic structures of [CuL$_2$] complexes (L = anion of L-asparagine or DL-α-amino-n-butyric acid).

(1)

Gergely and Nagypál report the stoicheiometries, stability constants, and the enthalpies and entropies of formation of the complexes formed in the systems of copper(II) with glycylglycine, glycyl-DL-α-alanine, DL-α-alanylglycine, and DL-α-alanyl-DL-α-alanine, and conclude that there is no possibility of the formation of [CuL$_2$] and [Cu(L$_2$H$_{-2}$)]$^{2-}$ (L = dipeptide anion, or H$_2$NCHR^{1}CONHCHR^{2}CO$_2^-$).[23] Further studies are reported for thirty-nine equilibrium systems involving mixed dipeptide–amino-acid complexes of copper(II), and from the high relative stabilities of the mixed-ligand complexes containing α-alanine and aspartic acid it seems that the amino-acids occupy one

[18] M. C. Lim, *J.C.S. Dalton*, 1977, 1398.
[19] H. Kozlowski, *Inorg. Chim. Acta*, 1977, **24**, 215.
[20] Y. Terai, H. Kido, K. Kashiwabara, J. Fujita, and K. Saito, *Bull. Chem. Soc. Japan*, 1977, **50**, 150.
[21] S. T. Chow, D. M. Johns, C. A. McAuliffe, and D. J. Machin, *Inorg. Chim. Acta*, 1977, **22**, 1.
[22] H. W. Richardson, J. R. Wasson, W. E. Estes, and W. E. Hatfield, *Inorg. Chim. Acta*, 1977, **23**, 205.
[23] A. Gergely and I. Nagypál, *J.C.S. Dalton*, 1977, 1104.

equatorial and one axial site around the metal.[24] May *et al.*[25] have computed the distribution of Cu^{2+}, and of divalent Ca, Mg, Mn, Zn, Pb, and Fe^{3+}, amongst 5000 complexes formed with 40 ligands in a simulation of metal ion equilibria in 'biofluids'. Correlation of the stereochemical differences of the copper(II) and nickel(II) complexes of the diastereoisomeric dipeptides L-alanyl-L-alanine, D-alanyl-L-alanine, L-leucyl-L-tyrosine, and D-leucyl-L-tyrosine with their different aqueous equilibrium constants indicates that the hydrophobic nature of the side-chains plays an important part in both conformation and equilibria.[26] pH titration data have afforded equilibrium constants for divalent copper and zinc complexes of glycylglycyl-L-histidine, Gly-Gly-His-Gly-Gly, their alkyl esters, and their benzyloxycarbonyl derivatives. The copper–peptide complexes are reasonable models for the binding of Cu^{2+} by albumin, but the disparity between constants for zinc–peptide and –albumin complexes indicates that a different binding site is involved.[27]

The therapeutic implications of the study of metal–amino-acid complexes must be mentioned. Thus, the perfusion of intact cat skin by a saline solution of bis(glycinato)copper(II) that is labelled with ^{64}Cu has been studied in a diffusion cell, and such work is relevant to the solubility of metallic copper in human sweat and to the possible therapeutic value of the 'copper bracelet'.[28] Equilibrium-dialysis studies of the ternary systems albumin–copper(II)–ligand (ligand = 3,6-diazaoctane-1,8-diamine or D-penicillamine) have been carried out in dilute NaCl solution to measure the amounts of low-molecular-weight membrane-diffusable copper(II) species formed. There is a substantial difference between the two ligands in this respect which correlates with their different clinical behaviour when used in the treatment of patients with Wilson's disease.[29] Other ternary systems have been examined, *viz.* formation constants of [Cu(DL-histidinate)(L-amino-acidate)],[30] and the optical resolution of DL-aspartic acid and of DL-glutamic acid has been achieved *via* the formation of ternary complexes of cupric complexes of these ligands and L-arginine, L-lysine, or L-ornithine.[31] Detailed investigations of the influence of glycine, alanine, histidine, glycylglycine, imidazole, and 1,10-penanthroline on the Cu^{2+}-catalysed rates of decarboxylation and enolization of oxaloacetate have been performed. Calculations show that many of the differences observed between the metal-ion-catalysed enzymatic and the non-enzymatic decarboxylation of $oxac^{2-}$ can be accounted for almost entirely by increases in the stabilities of the known complexes in the environment presented by the enzyme (the low-dielectric region at the active site of the enzyme?).[32] Other metal-ion-promoted (*e.g.* Cu^{2+}, Ni^{2+}, Zn^{2+}, Co^{2+}) hydrolyses of amino-acid esters have been studied.[33]

[24] A. Gergely and I. Nagypál, *J.C.S. Dalton*, 1977, 1109.
[25] P. M. May, P. W. Linder, and D. R. Williams, *J.C.S. Dalton*, 1977, 588.
[26] A. Kaneda and A. E. Martell. *J. Amer. Chem. Soc.*, 1977, **99**, 1586,
[27] R. P. Agarwal and D. D. Perrin, *J.C.S. Dalton*, 1977, 53.
[28] W. R. Walker, R. R. Reeves, M. Brosnan, and G. D. Coleman, *Bioinorg. Chem.*, 1977, **7**, 271.
[29] S. H. Laurie and B. Sarkar, *J.C.S. Dalton*, 1977, 1822.
[30] G. Brookes and L. D. Pettit, *J.C.S. Dalton*, 1977, 1918.
[31] O. Yamauchi, T. Sakurai, and A. Nakahara, *Bull. Chem. Soc. Japan*, 1977, **50**, 1776.
[32] N. V. Raghavan and D. L. L. Leussing, *J. Amer. Chem. Soc.*, 1977, **99**, 2188,
[33] S. A. Bedell and R. Nakon, *Inorg. Chem.*, 1977, **16**, 3055,

Numerous copper-containing proteins utilize molecular oxygen in respiratory and biosynthetic functions, but there are few well-characterized copper complexes bound to small molecules. The four-co-ordinate copper(I) complex difluoro-3,3'-(trimethylenedinitrilo)-bis(2-butanone oximato)borate copper(I) can be produced by electrochemical reduction of the corresponding copper(II) species. The copper(I) derivative binds unidentate ligands (*e.g.* CO, 1-methylimidazole, or MeCN) to yield pentaco-ordinate adducts. The carbonyl derivative is square-pyramidal, with Cu displaced 96 pm out of the basal nitrogen plane; Cu–CO = 178 pm.[34] Electrode potentials have been measured for forty $Cu^{III, II}$–peptide couples (including peptide amides) in aqueous solution. These potentials (1.02—0.45 V) are very sensitive to changes in the nature of the ligand, and decrease with an increase in the number of deprotonated peptide groups. The triply deprotonated peptide and the highly *C*-substituted tripeptide complexes have effective potentials at physiological pH such that oxidation by O_2 to Cu^{III} is possible.[35] Similar studies for thirty $Ni^{III, II}$–peptide couples have been reported.[36]

The crystal and molecular structures of L-prolinatodiphenylboron[37] and L-prolinatodimethylgallium[38] have been obtained, and Zuckerman[39] has synthesized diglycinatotin(II), dimethyltin(IV) diglycinate, and dimethyltin(IV) di-α-alaninate. Circular polarized emission (CPE) and total emission (TE) spectra for Eu^{3+} and Tb^{3+} complexes of L-aspartic acid, L-serine, L-threonine, and L-histidine in D_2O solution have been measured, and spectral parameters can be correlated with ligand–lanthanide ion binding characteristics.[40]

The free radical 4-amino-2,2,6,6-tetramethylpiperidinyl-1-oxyl (ATMPO) forms *cis*-[Pt(ATMPO)$_2$(NO$_3$)$_2$], and the latter has been used to label poly(L-glutamate), and poly(L-aspartate) and poly(L-lysine); labelling occurs by the displacement of nitrate by polymer side-chains. E.s.r. spectra of labelled (Glu)$_n$ are anisotropic, and monitor the helix–coil transition and polymer aggregation. The Pt is probably bifunctionally anchored to adjacent carboxylate groups.[41] Crystals of [*cyclo*(-L-Pro-Gly-)$_4$RbSCN·3H$_2$O]$_2$ have been obtained from rubidium thiocyanate and the cyclic octapeptide *cyclo*(L-prolylglycyl)$_4$ in H$_2$O–Me$_2$CO. The rubidium cation has a distorted octahedral environment consisting of four glycyl carbonyl oxygens from one cyclic peptide of the dimer, one glycyl carbonyl oxygen from the other cyclic peptide of the dimer, and one oxygen from a water molecule.[42] Such studies are of relevance to ion transport across membranes, and cyclic peptides themselves possess potent biological activities as antibiotics, toxins, and hormones. This same *cyclo*(L-prolylglycyl)$_4$ forms complexes of 1:2 and 1:1 cation–peptide stoicheiometries with a variety of alkali-metal and alkaline-earth cations. The larger cations, Cs$^+$ and Ba^{2+},

[34] R. R. Gagne, J. L. Allison, R. S. Gall, and C. A. Koval, *J. Amer. Chem. Soc.*, 1977, **99**, 7170.
[35] F. P. Bossu, K. L. Chellappa, and D. W. Margerum, *J. Amer. Chem. Soc.*, 1977, **99**, 2195.
[36] F. P. Bossu and D. W. Margerum, *Inorg. Chem.*, 1977, **16**, 1210.
[37] S. J. Rettig and J. Trotter, *Canad. J. Chem.*, 1977, **55**, 958.
[38] K. R. Breakell, S. J. Rettig, A. Storr, and J. Trotter, *Canad. J. Chem.*, 1977, **55**, 4174.
[39] W. T. Hall and J. J. Zuckerman, *Inorg. Chem.*, 1977, **16**, 1239.
[40] H. G. Brittain and F. S. Richardson, *Bioinorg. Chem.*, 1977, **7**, 233.
[41] Y. Y. Chao, A. Holtzer, and S. H. Mastin, *J. Amer. Chem. Soc.*, 1977, **99**, 8024.
[42] Y. H. Chin, L. D. Brown, and W. D. Lipscomb, *J. Amer. Chem. Soc.*, 1977, **99**, 4799.

have binding constants comparable to those with naturally occurring cyclic peptides.[43] The synthesis of *cyclo*(-L-Val-Gly-L-Pro-)$_3$ (2) has been achieved. This homodetic cyclic dodecapeptide contains only naturally occurring amino-acids and is a model of an ion carrier that is related to valinomycin. The stabilities of some 1:1 complexes can be correlated with the diameter of the cation, *viz.* Mg^{2+} ≪ Ca^{2+} ≪ Ba^{2+}.[44] The binding of copper(II) to some poly(α-amino-acids) has been reported.[45]

(2)

Sulphur-containing Amino-acids.—The use of such ligands as penicillamine in chelation therapy is currently used to justify an upsurge in work on this and similar sulphur-containing amino-acids. Thus Hodgson, Freeman, and co-workers report[46] the crystal structure of Na[Cr(L-cys)$_2$]·2H$_2$O, shown in Figure 2, in which the central Cr atom is in a slightly distorted octahedral co-ordination, with two Cr—S (241.6 pm), two Cr—O (198.1 pm), and two Cr—N (206.2 pm) bonds; the two sulphido donors are mutually *trans*. The stereospecific synthesis of a sulphenamide–cobalt(III) complex derived from (R)-cysteine has been achieved; Figure 3 shows the crystal structure of Δ-(S)-{ethylenediamine-(R)-cysteinesulphenamide}(ethylenediamine)cobalt(III), isolated as the ZnCl$_4$$^{2-}$ salt.[47] Chromous reductions of [Co(en)$_2$(met)]$^{2+}$, [(Co(en)$_2$(Mecys)]$^{2+}$, and [Co(en)$_2$(cys)]$^{2+}$ (met = methionine, Mecys = anion of methylcysteine) have been studied.[48] The cobalt(III) complexes of methionine and methylcysteine contain *O,N*-bonded ligands. Attack of Cr^{2+} occurs at the O atom in each of these complexes to produce the *O*-bonded unidentate chromium(III) product, which slowly aquates.

In an attempt to determine the effect, if any, of an interfacial environment on transition-metal complexes, a number of copper(II) species have been allowed

[43] V. Madison, C. M. Deber, and E. R. Blout, *J. Amer. Chem. Soc.*, 1977, **99**, 4788.

[44] D. Baron, L. G. Pease, and E. R. Blout, *J. Amer. Chem. Soc.*, 1977, **99**, 8299.

[45] M. Palumbo, A. Cosani, M. Terbojevich, and E. Peggion, *J. Amer. Chem. Soc.*, 1977, **99**, 939.

[46] P. de Meester, D. J. Hodgson, H. C. Freeman, and C. J. Moore, *Inorg. Chem.*, 1977, **16**, 1494.

[47] G. J. Gainsford, W. G. Jackson, and A. M. Sargeson, *J. Amer. Chem. Soc.*, 1977, **99**, 2383.

[48] R. J. Balahura and N. A. Lewis. *Inorg. Chem.*, 1977, **16**, 2213.

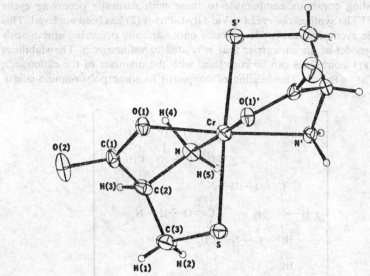

Figure 2 *The structure of the anion in sodium bis(L-cysteinato)chromate(III) dihydrate*
(Reproduced by permission from *Inorg. Chem.*, 1977, **16**, 1494)

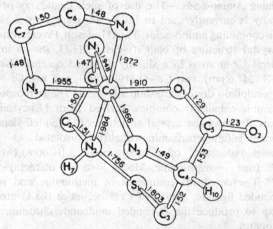

Figure 3 *The crystal structure of Δ-(S)-{ethylenediamine-(R)-cysteinesulphenamide}(ethylenediamine)cobalt(III) tetrachlorozincate*
(Reproduced by permission from *J. Amer. Chem. Soc.*, 1977, **99**, 2383)

to react with the side-chains of *N*-dodecanoyl-L-lysinol, N^{α}-dodecanoyl-L-glutaminol, and *N*-dodecanoyl-L-methioninol at the interfacial domain of an oil-continuous microemulsion.[49] Sigel *et al.*[50] have examined the influence of

[49] G. D. Smith, B. B. Garrett, S. L. Holt, and R. E. Barden, *Inorg. Chem.*, 1977, **16**, 558.
[50] H. Sigel, C. F. Naumann, B. Prijis, D. B. McCormick, and M. C. Falk, *Inorg. Chem.*, 1977, **16**, 790.

alkyl side-chains with hydroxy-groups or thioether groups on the stability of binary and ternary copper(II)-dipeptide complexes.[50] A number of chelates of bivalent metals (Cu, Ni, Zn, Pb, and Co) with L-cystine have been prepared which apparently contain intact —S—S— bonds, but little structural information is presented.[51]

Dinuclear molybdenum(v) complexes containing di-μ-oxo ($Mo_2O_4{}^{2+}$), μ-oxo-μ-sulphido ($Mo_2O_3S^{2+}$), and di-μ-sulphido ($Mo_2O_2S_2{}^{2+}$) centres co-ordinated to EDTA, cysteine, and ethylcysteinate ligands have been obtained.[52] The EDTA and cysteine complexes undergo electrochemical reduction in a single four-electron step to Mo^{III} dimers, and, though the ease of reduction and the electrochemical reversibility of the $Mo^V{}_2/Mo^{III}{}_2$ couple increase with insertion of S into the bridge system, the Mo^{III} dimers become increasingly unstable upon bridge-S insertion.[52] Biological systems show a marked preference for molybdenum over tungsten. Studies with methyliminodiacetic acid and L-cysteine show that formation constants for Mo^{VI} and W^{VI} are very similar, implying that these elements would be bound approximately equally strongly to an apoenzyme or to a carrier, whether or not these proteins contain a ligating sulphydryl group.[53]

Gold(I) complexes containing S or P donors have been studied as anti-arthritic agents; hence the study by Smith and co-workers of some new complexes of type [Au(PR$_3$)X] (R = Et or Ph; X = Cl, L-cysteinate, D-penicillaminate, or thiomalate).[54]

Carty and co-workers have embarked on a programme involving structural studies of bivalent mercury and cadmium with S-containing amino-acids.[55-58] In MeHg(DL-Met) the amino-acid is bound *via* the amino-group;[55] in [MeHg(pen)]·H$_2$O the ligand is co-ordinated *via* the sulphido atom; and in [Me$_2$Hg$_2$(pen)] *via* amino- and sulphido-groups.[57] In [Cd(pen)Br(H$_2$O)]·2H$_2$O the crystal consists of infinite chains of alternating cadmium atoms and bridging terdentate pen⁻ moieties, the latter being co-ordinated *via* S⁻ and bidentate carboxylate (the amine group is present as —NH$_3{}^+$).[58]

Amino-acids containing Heterocyclic N-Donors.—Meester and Hodgson report two *X*-ray crystallographic studies of mixed histidine/penicillamine complexes. In [Cr(L-his)(D-pen)]·H$_2$O the distorted octahedral co-ordination contains penicillamine bound as the dianion, and terdentate, whereas the terdentate histidinate is monoanionic; Cr—N(amino) = 206.3 pm, Cr—N(imidazole) = 205.7 pm (see Figure 4).[59] The co-ordination entity in [Co(L-his)(D-pen)]·H$_2$O is almost identical.[60] Larkworthy and Tabatabai have isolated [Cr^{II}(his)Cl]·H$_2$O; this compound is antiferromagnetic and is believed to be chloride-bridged.[61]

51 R. J. Gale and C. A. Winkler, *Inorg. Chim. Acta*, 1977, **21**, 151.
52 V. R. Ott, D. S. Swieter, and F. A. Schultz, *Inorg. Chem.*, 1977, **16**, 2538.
53 G. E. Callis and R. A. D. Wentworth, *Bioinorg. Chem.*, 1977, **7**, 57.
54 D. H. Brown, G. McKinlay, and W. E. Smith, *J.C.S. Dalton*, 1977, 1874.
55 Y. S. Wong, A. J. Carty, and P. C. Chieh, *J.C.S. Dalton*, 1977, 1157.
56 N. J. Taylor and A. J. Carty, *J. Amer. Chem. Soc.*, 1977, **99**, 6143.
57 Y. S. Wong, A. J. Carty, and P. C. Chieh, *Inorg. Chem.*, 1977, **16**, 1801.
58 A. J. Carty and N. J. Taylor, *Inorg. Chem.*, 1977, **16**, 177.
59 P. de Meester and D. J. Hodgson, *J.C.S. Dalton*, 1977, 1604.
60 P. de Meester and D. J. Hodgson, *J. Amer. Chem. Soc.*, 1977, **99**, 101.
61 L. F. Larkworthy and J. M. Tabatabai, *Inorg. Chim. Acta*, 1977, **21**, 265.

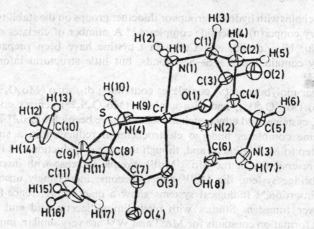

Figure 4 *The co-ordination around chromium in* [Cr(L-his)(D-pen)]·H$_2$O. *Hydrogen atoms are represented as spheres of arbitrary size*
(Reproduced from *J.C.S. Dalton*, 1977, 1604)

Proton n.m.r. chemical shifts and nuclear relaxation rates have been measured for the protons attached to C-2 and C-4 of poly-L-histidine (PLH) (3) as a function of pH. N.m.r. titration curves indicate a pK_a value of 6.1 for the imidazole groups. In the presence of copper(II) the relaxation rates show a pronounced maximum at *ca.* pH 3.5; above this pH, temperature dependence studies indicate that slow exchange is occurring between Cu^{II} and PLH.[62] Angelici and co-workers[63] have bound *N*-carboxymethyl-β-(2-pyridyl)-L-α-alanine (4) to nickel(II) and copper(II) complexes and studied the stereoselective binding of optically active amino-acids.

2 The Binding of Small Molecules by Transition-metal Complexes

The critical role played by transition metals in both biochemical and industrial processes involving molecular oxygen makes the chemistry of metal–dioxygen

(3)   (4)

[62] R. E. Wasylishen and J. S. Cohen, *J. Amer. Chem. Soc.*, 1977, **99**, 2480.
[63] S. A. Bedell, P. R. Rechani, R. J. Angelici, and R. Nakon, *Inorg. Chem.*, 1977, **16**, 972.

complexes especially important. The binding of other small molecules, *e.g.* SO_2, CO, CO_2, and N_2, is also included in this section.

Dioxygen.—Among the three co-ordination modes known for dioxygen, namely unidentate (5), bridging (6), and chelating (7), the first has particular significance as it almost certainly occurs in oxygenated haem proteins (with M = Fe).[64, 65] However, the great majority of known dioxygen complexes contain chelated O_2; a recent review[66] lists 43 *X*-ray structures of type (7), compared with only 16 of types (5) and (6) combined. The results of SCF-X_α-SW calculations on the model chelated dioxygen complex $[Pt(PH_3)_2(O_2)]$ and its ethylene analogue

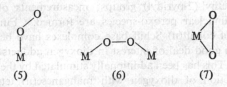

$[Pt(PH_3)_2(C_2H_4)]$ have been compared; the covalent bonding of both of these small molecules is due to mixing of metal $5d_{x^2-y^2}$ and $5d_{z^2}$ orbitals with ligand bonding orbitals of both σ- and π-type.[67] Other studies of a relatively theoretical type include the note by Muetterties *et al.* on a co-ordination chemistry guide to structural studies of chemisorbed molecules,[68] a comparison of the electronic spectra and structure of mononuclear and dinuclear dioxygen complexes,[69] and the energy transfer from luminescent transition-metal complexes to dioxygen.[70]

Various dioxygen adducts of polymer-supported $TiCl_3$ complexes have been prepared in pyridine or THF from $TiCl_3$ and copolymers composed of 4-vinylpyridine, divinylbenzene, and styrene. The e.s.r. spectra indicate three different types of dioxygen adducts, characterized by *g* values of 2.027—2.028, 2.022—2.024, and 2.017—2.018.[71]

Chelate formation, amide deprotonation, and dioxygenation reactions of 1:1 cobaltous complexes of glycyl-L-histidine, glycylhistamine, glycyl-L-aspartic acid, L-histidyl-L-histidine, L-histidylglycine, and L-aspartylglycine have been studied, and thermodynamic equilibrium constants have been reported.[72] The reaction of cyanocobaltate(II) and cyanomolybdate(V) ions with molecular oxygen results in the formation of the bimetallic dioxygen adduct (8). Infrared evidence for the presence of peroxide (band at 893 cm^{-1}) has been presented.[73] Oxygenation of $[Co(tren)(H_2O)_2]^{2+}$ in 6M aqueous ammonia or by ligand exchange,

[64] J. R. Collman, J. I. Brauman, and K. S. Suslick, *J. Amer. Chem. Soc.*, 1975, **97**, 1427.
[65] J. H. Dawson, R. H. Holm, J. R. Trudell, G. Barth, R. E. Linder, E. Bunnenberg, C. Djerassi, and S. C. Tang. *J. Amer. Chem. Soc.*, 1976, **98**, 3707.
[66] L. Vaska, *Accounts Chem. Res.*, 1976, **9**, 175.
[67] J. G. Norman, *Inorg. Chem.*, 1977, **16**, 1328.
[68] E. L. Muetterties, J. C. Hemminger, and G. A. Somarjai, *Inorg. Chem.*, 1977, **16**, 3381.
[69] G. McLendon, S. R. Pickens, and A. E. Martell, *Inorg. Chem.*, 1977, **16**, 1551.
[70] J. N. Demas, E. W. Harris, and R. P. McBride, *J. Amer. Chem. Soc.*, 1977, **99**, 3547.
[71] Y. Chimura, M. Beppu, S. Yoshida, and K. Tarama, *Bull. Chem. Soc. Japan*, 1977, **50**, 691.
[72] W. R. Harris and A. E. Martell, *J. Amer. Chem. Soc.*, 1977, **99**, 6746.
[73] H. Arzoumanian, R. C. Alvarez, A. D. Kowalak, and J. Metzger, *J. Amer. Chem. Soc.*, 1977, **99**, 5175.

$$K_6 \left[(CN)_5 Co^{III} O_2 Mo^{VI} (CN)_5 \right] \cdot 3H_2O$$

with the structure showing Cl above and O double-bonded below the Mo.

(8)

starting with $[(NH_3)_5Co(O_2)Co(NH_3)_5](NO_3)_4$, yields $[(tren)(NH_3)Co(O_2)Co-(NH_3)(tren)]^{4+}$, in which the Co–O–O–Co unit is planar and the Co–O–O angle is 111.5°.[74] Cobaltous complexes have been isolated that contain polyamines with 2-pyridyl and 6-methyl-(2-pyridyl) groups; measurements of oxygen uptake indicate that the dinuclear peroxo-species are formed.[75] Further examples of dioxygen adducts of cobalt(II)–Schiff base complexes have been obtained.[76, 77]

There has been a good deal of interest in dioxygen adducts of manganese(II) complexes recently. This has been additionally stimulated by the work of Basolo's group on the reaction of dioxygen with manganese(II) tetraphenylporphin, Mn^{II}(TPP). From the results of an optical and an e.s.r. study, an extensive $Mn \rightarrow O_2$ charge transfer has been found, and this adduct has been formulated as $Mn^{IV}(O_2^{2-})$ $(S=\frac{3}{2})$, in which the Mn^{IV} is in the (t_2^3) ground state.[78,79] An *ab initio* calculation is at odds with this formulation, however.[80] Chiswell has studied Mn^{II} complexes of the ligand (9) (N_4ligand), and observed O_2 uptake in pyridine solution to be equivalent to 1.3 oxygen atoms per Mn; this could be explained by the isolation from the oxygenation reactions of four products: (i) MnO_2, (ii) $[Mn(N_4ligand)(py)(OH)]$, (iii) $[Mn(N_4ligand)(py)O_2]$, (iv) $[Mn(N_4ligand)(py)O]$.[81] Oxygen-uptake measurements on [Mn(salen)] indicate that each Mn atom reacts with one O atom.[82] However, manganese(II) complexes containing linear quinquedentate O_2N_3 ligands take up O_2 to give irreversible oxidation of Mn^{II} to Mn^{III} along with oxidation of the ligand.[83]

The synthetic molecular oxygen carrier (10) has an FeN_4 chromophore and exhibits essentially spin-only magnetic behaviour; $\mu_{eff} = 2.94$ μ_B at 300 K., and the value remains as such to as low as 25 K.[84] There is a report of steric factors

$$X \underset{N}{\underbrace{}} N-N=C-C=N-N \underset{N}{\underbrace{}} X$$

with H, H on the terminal N's and R^1, R^2 on the central C's

(9)

[74] U. Thewalt, M. Zehnder, and S. Fallab, *Helv. Chim. Acta*, 1977, **60**, 867.
[75] I. Exnar and H. Mäcke, *Helv. Chim. Acta*, 1977, **60**, 2504.
[76] E. Cesarotti, M. Gullotti, A. Pasini, and R. Ugo, *J.C.S. Dalton*, 1977, 757.
[77] M. Corrigan, K. S. Murray, B. O. West, P. R. Hicks, and J. R. Pilbrow, *J.C.S.Dalton*, 1977, 1478.
[78] C. J. Weschler, B. M. Hoffman, and F. Basolo, *J. Amer. Chem. Soc.*, 1975, **97**, 5278.
[79] B. M. Hoffman, C. J. Weschler, and F. Basolo, *J. Amer. Chem. Soc.*, 1976, **98**, 5473.
[80] A. Dedieu and M. M. Rohmer, *J. Amer. Chem. Soc.*, 1977, **99**, 8050.
[81] B. Chiswell, *Inorg. Chim. Acta*, 1977, **23**, 77.
[82] C. J. Boreham and B. Chiswell, *Inorg. Chim. Acta*, 1977, **24**, 77.
[83] W. M. Coleman and L. T. Taylor, *Inorg. Chem.*, 1977, **16**, 1114.
[84] W. M. Reiff, H. Wong, J. E. Baldwin, and J. Huff, *Inorg. Chim. Acta*, 1977, **25**, 91.

(10)

which control one- or two-electron reduction of O_2 by cuprous complexes of substituted imidazoles.[85]

Further work on the reaction of $[RhCl(PPh_3)_2]_2$ and $[RhCl(PPh_3)_3]$ with molecular oxygen has been reported.[86] The crystal and molecular structure of $[RhCl(O_2)(PPh_3)_3] \cdot 2CH_2Cl_2$ shows a trigonal-bipyramidal configuration, with phosphines at the apices and at one equatorial position. The O_2 molecule is π-bonded, and a slight asymmetry in the Rh—O bond lengths (208, 201 pm) is observed.[87] The dimeric $[RhCl(O_2)(PPh_3)_2 \cdot CH_2Cl_2]_2$ contains two trigonal-bipyramidal subunits, the bridge being formed by one oxygen atom of each subunit; the O_2 molecules thus have features similar to those of both π-bonded ligands and chelating peroxo-groups.[88] Side-on-bonded N_2 and O_2 exist in $[RhCl(N_2)(PPr^i_3)_2]$ and $[RhCl(O_2)(PPr^i_3)_2]$, respectively.[89] Hanlon and Ozin, using matrix-isolation techniques, have found evidence for $[Rh(O_2)_2]$ and $[Rh(O_2)]^{90a}$ as well as $[Rh_2(O_2)n]$ and $[Rh_3(O_2)m]$.[90b] Finally, reactions of the two-co-ordinate $[Pd(PR_3)_2]$ and $[Pt(PR_3)_2]$ with dioxygen to produce $[MO_2L_2]$ have been reported, co-ordination of dioxygen being reversible in $[PdO_2(PPhBu^t_2)_2]$ but not in the corresponding Pt compound.[91]

Carbon Monoxide and Carbon Dioxide.—Both carbon monoxide and benzyl isocyanide, L, are substantially more inert in $[L_2Fe(dmgH)_2]$ (dmgH = dimethylglyoxime) complexes than in the corresponding porphyrin and phthalocyanine systems.[92] Photolabilization studies of ligands, including carbon monoxide from low-spin d^6 iron(II) macrocyclic complexes, have been described.[93]

[85] M. Guntensperger and A. D. Zuberbühler, *Helv. Chim. Acta*, 1977, **60**, 2584.
[86] G. L. Geoffroy and M. E. Keeney, *Inorg. Chem.*, 1977, **16**, 205.
[87] M. J. Bennett and P. B. Donaldson, *Inorg. Chem.*, 1977, **16**, 1581.
[88] M. J. Bennett and P. B. Donaldson, *Inorg. Chem.*, 1977, **16**, 1585.
[89] C. Busetto, A. D'Alfonso, F. Maspero, G. Perego, and A. Zazzetta, *J.C.S. Dalton*, 1977, 1828.
[90] (a) A. J. L. Hanlan and G. A. Ozin, *Inorg. Chem.*, 1977, **16**, 2848. (b) A. J. L. Hanlan and G. A. Ozin, *Inorg. Chem.*, 1977, **16**, 2857.
[91] T. Yoshida and S. Otsuka, *J. Amer. Chem. Soc.*, 1977, **99**, 2134.
[92] I. W. Pang and D. V. Stynes, *Inorg. Chem.*, 1977, **16**, 590.
[93] M. J. Incorvia and J. I. Zink, *Inorg. Chem.*, 1977, **16**, 3161.

Uptake of carbon dioxide and the chelation of intramolecular carbonato ligands in aqueous solutions of *cis-* and *trans*-diaquo(1,4,8,11-tetra-azacyclotetradecane)cobalt(III) cations have been studied, and comparisons drawn between this system and carbonic anhydrase.[94] Some rhodium(I)–CO_2 complexes have been isolated,[95] and the transformation (3) has been studied

$$Ir(dmpe)_2Cl \cdot CO_2 \xrightarrow{\Delta} [Ir(H)\{Me(CH_2CO_2)PC_2H_4PMe_2\}(dmpe)]Cl \qquad (3)$$

(dmpe = $Me_2PCH_2CH_2PMe_2$).[96] *trans*-[PtH_2(PCy_3)_2] reacts with CO_2 to form both the formato compound *trans*-[PtH_2(O_2CH)(PCy_3)_2] and the monomethyl-carbonato compound *trans*-[PtH(O_2COMe)(PCy_3)_2].[97] The reaction of NiL_4 (L = PEt_3 or PBu^n_3) with CO_2 in toluene affords [Ni(CO_2)L_2] *via* the [Ni(CO_2)L_3] species, and [Ni(CO_2)(PCy_3)_2] reacts with O_2 to give peroxo-carbonatobis(tricyclohexylphosphine)nickel(II).[98]

Sulphur Dioxide.—Ryan, Moody, and Eller have published an impressive body of results during 1977, including the results of some extended Hückel M.O. calculations on $L_nM–SO_2$ species,[99] descriptions of the structures of [(C_5H_5)Rh(C_2H_4)(SO_2)] (a d^8 complex with a planar $RhSO_2$ group),[100] [Pt_3(SO_2)_3(PPh_3)_3] \cdot C_7H_8 \cdot SO_2 (which contains a triangle of Pt atoms, each of which is bound to two bridging SO_2 groups and a single PPh_3),[101] [Pt(SO_2)_2-(PPh_3)_2] \cdot C_7H_8,[102] [Pt(PPh_3)_3(SO_2)] \cdot 0.7SO_2,[103] [Rh(NO)(SO_2)(PPh_3)_2],[104] and details of the preparation of some sulphur dioxide adducts of organophos-phinecopper(I) mercaptide complexes [Cu(PR^1_3)_n(SR^2)(SO_2)], which are rather sluggishly reversible SO_2-bound complexes.[105] Blum and Meek have prepared [Rh(triphosphine)X(SO_2)] complexes.[106]

Dinitrogen.—Although co-ordination complexes containing dinitrogen as a ligand are now quite commonplace, the facile conversion of dinitrogen into ammonia still seems to be a long way from reality. Significant steps continue to be made, however. For instance, Chatt and co-workers[107] have found that, whilst treatment of *trans*-[M(N_2)(dpe)_2] (dpe = $Ph_2PCH_2CH_2PPh_2$; M = Mo or W) with H_2SO_4 gives only [M(HSO_4)(NNH_2)(dpe)_2](HSO_4) and no ammonia or hydrazine, the complexes *cis*-[M(N_2)_2(PPhMe_2)_4] and *trans*-[M(N_2)_2(PPh_2Me)_4] (M = M or W) react with H_2SO_4 in methanol at 20 °C to give ammonia (*ca.* 1.9 NH_3 per W atom and *ca.* 0.7 NH_3 per Mo atom). Schrauzer continues with his studies, which he titles 'The Chemical Evolution of a Nitrogenase Model', and has recently concluded that the reducing site of nitrogenase more probably

[94] T. P. Dasgupta and G. M. Harris, *J. Amer. Chem. Soc.*, 1977, **99**, 2490.
[95] M. Aresta and C. F. Nobile, *Inorg. Chim. Acta*, 1977, **24**, L49.
[96] T. Merskovitz, *J. Amer. Chem. Soc.*, 1977, **99**, 2391.
[97] A. Immirzi and A. Musco, *Inorg. Chim. Acta*, 1977, **22**, L35.
[98] M. Aresta and C. F. Nobile, *J.C.S. Dalton*, 1977, 708.
[99] R. R. Ryan and P. G. Eller, *Inorg. Chem.*, 1977, **16**, 494.
[100] R. R. Ryan, P. G. Eller, and G. J. Kubas, *Inorg. Chem.*, 1977, **16**, 797.
[101] D. C. Moody and R. R. Ryan, *Inorg. Chem.*, 1977, **16**, 1052.
[102] D. C. Moody and R. R. Ryan, *Inorg. Chem.*, 1977, **16**, 1823.
[103] P. G. Eller, R. R. Ryan, and D. C. Moody, *Inorg. Chem.*, 1977, **16**, 2442.
[104] D. C. Moody and R. R. Ryan, *Inorg. Chem.*, 1977, **16**, 2473.
[105] P. G. Eller and G. J. Kubas, *J. Amer. Chem. Soc.*, 1977, **99**, 4346.
[106] P. R. Blum and D. W. Meek, *Inorg. Chim. Acta*, 1977, **24**, L75.
[107] J. Chatt, A. J. Pearman, and R. L. Richards, *J.C.S. Dalton*, 1977, 1852.

contains a mononuclear Mo^{IV} rather than a Mo^{III} species in the active reduced form.[108] The reduction of co-ordinated isocyanide in cationic complexes of oxomolybdate(IV), *i.e.* $[Mo(O)X(CNR)_4]^+$, produces methylamine and various hydrocarbons – the same products as have previously been observed on reduction of isocyanides with functional nitrogenase. In the presence of ATP, molecular nitrogen partially inhibits the production of hydrocarbons from co-ordinated MeNC and is itself reduced to ammonia *via* di-imide and hydrazine intermediates.[109]

Treatment of $[W(N_2)_2(PPh_2Me)_4]$ with HCl in dichloromethane gives N_2 (1 mol) and $[WCl_3(NHNH_2)(PPh_2Me)_2]$, which with H_2SO_4 in methanol gives hydrazine (0.28 mol) and ammonia (0.23 mol), but KOH gives mainly ammonia.[110] Other reactions of similar complexes have been reported by the Sussex group.[111, 112]

3 Non-haem Iron

Iron–Sulphur Compounds.—Non-haem iron–sulphur proteins are found in many bacterial, plant, and animal cells, and act as electron carriers in these systems. Many such proteins contain an Fe_4S_4 cluster shaped approximately as a distorted cube, with iron and sulphur atoms at alternate corners.[113]

To complete the set of synthetic analogues corresponding to the three known types of active sites in iron–sulphur redox proteins, Holm and co-workers[114] have prepared bis(o-xylyl-α,α'-dithiolato)ferrate(II,III) anions $[Fe(S_2-o-xyl)_2]^{2-,-}$. These complexes are shown to be related to the active sites of the 1-Fe rubredoxin (Rd) proteins. Previous work by Holm's group[115, 116] has demonstrated that the complexes $[Fe_4S_4(SR)_4]^{2-}$ and $[Fe_2S_2(SR)_4]^{2-}$, which are synthetic analogues of the $[Fe_4S_4^*(S\text{-cys})_4]$ and $[Fe_2S^*_2(S\text{-cys})_4]$ active sites of oxidized ferredoxin proteins, undergo facile ligand-substitution reactions with added thiols at ambient temperature. These reactions have been applied to the extrusion of intact $Fe_4S_4^*$ and $Fe_2S_2^*$ cores of protein sites in the form of their spectrally characteristic R = Ph analogues.[117] Further recent work has demonstrated the chemical and electrochemical inter-relationships of the 1-Fe, 2-Fe, and 4-Fe analogues of the active sites in these proteins.[118] A new synthetic tetranuclear

[108] P. R. Robinson, E. L. Moorehead, B. J. Weathers, E. A. Ufkes, T. M. Vickrey, and G. N. Schrauzer. *J. Amer. Chem. Soc.*, 1977, **99**, 3657.

[109] E. L. Moorehead, B. J. Weathers, E. A. Ufkes, P. R. Robinson, and G. N. Schrauzer, *J. Amer. Chem. Soc.*, 1977, **99**, 6089.

[110] J. Chatt, A. J. Pearman, and R. L. Richards, *J.C.S. Dalton*, 1977, 2139.

[111] J. Chatt, A. A. Diamantis, G. A. Heath, N. E. Hooper, and G. J. Leigh, *J.C.S. Dalton*, 1977, 688.

[112] P. C. Bevan, J. Chatt, A. A. Diamantis, R. A. Head, G. A. Heath, and G. J. Leigh, *J.C.S. Dalton*, 1977, 1711.

[113] W. H. Orme-Johnson, *Ann. Rev. Biochem.*, 1973, **42**, 159.

[114] R. W. Lane, J. A. Ibers, R. B. Frankel, G. C. Papaefthymiou, and R. H. Holm, *J. Amer. Chem. Soc.*, 1977, **99**, 84.

[115] R. H. Holm, *Endeavour*, 1975, **34**, 38.

[116] R. W. Lane, J. A. Ibers, R. B. Frankel, and R. H. Holm, *Proc. Nat. Acad. Sci. U.S.A.*, 1975, **72**, 2868.

[117] W. O. Gillum, L. E. Mortenson, J. S. Chen, and R. H. Holm, *J. Amer. Chem. Soc.*, 1977, **99**, 584.

[118] J. Cambray, R. W. Lane, A. G. Wedd, R. W. Johnson, and R. H. Holm, *Inorg. Chem.*, 1977, **16**, 2565.

iron–sulphur complex has been the subject of a crystallographic study; the hexanegative anion $[Fe_4S_4\{S(CH_2)_2CO_2\}_4]^{6-}$, shown in Figure 5, is balanced by six cations (one NBu_4^+ and five Na^+).[119]

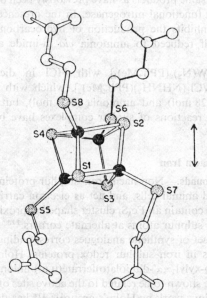

Figure 5 *The structure of* $[Fe_4S_4\{S(CH_2)_2CO_2\}_4]^{6-}$. *Black circles are iron atoms and speckled ones are sulphur*
(Reproduced by permission from *J. Amer. Chem. Soc.*, 1977, **99**, 3683)

Schwartz and van Tamelen[120] have shown that $Fe_4S_4L_4$–isocyanide adducts markedly promote the α,α-addition of mercaptans to isocyanides, which is the first non-enzymic catalytic reaction of this cluster type to be observed. Gray and co-workers have examined the oxidation of reduced spinach ferredoxin by $[Fe(edta)]^-$, $[Fe(Hedta)]$, horse heart ferricytochrome *c*, and horse metmyoglobin; each reaction follows second-order kinetics (rate $= k_{12}[$ferredoxin$]$ $[$oxidant$]$).[121]

Iron Transport.—Among the chelating functional groups found in the microbial iron-transport agents called siderophores are hydroxamate, catecholate, and thiohydroxamate. Model systems containing the anion or dianion of thiobenzohydroxamic acid, $PhC(=S)N(OH)H$, have been prepared, but little is known of their chemical properties.[122] The resolution of tris(hydroxamato)- and tris(thiohydroxamato)-complexes of high-spin iron(III) has been achieved.[123]

[119] H. L. Carrell, J. P. Glusker, R. Job, and R. C. Bruice, *J. Amer. Chem. Soc.*, 1977, **99**, 3683.
[120] A Schwartz and E. E. van Tamelen, *J. Amer. Chem. Soc.*, 1977, **99**, 3189.
[121] J. Rawlings, S. Wherland, and H. B. Gray, *J. Amer. Chem. Soc.*, 1977, **99**, 1968.
[122] K. Abu-Dari and K. N. Raymond, *Inorg. Chem.*, 1977, **16**, 807.
[123] K. Abu-Dari and K. N. Raymond, *J. Amer. Chem. Soc.*, 1977, **99**, 2003.

4 Copper Proteins

'Blue' copper proteins, which occur widely in Nature as electron carriers, have attracted particular attention because of their unique properties. The distinctive features of 'blue' copper centres are usually high extinction coefficients near 600 nm ($\varepsilon = 1000$—5000), anomalously small copper hyperfine coupling constants ($A_\parallel = 30$—100 G), and markedly positive copper redox potentials.[124] Recent resonance Raman[125] and X-ray P.E.[126] spectral studies have suggested the presence of Cu^{II}–S(Cys) co-ordination for blue copper sites and S $\rightarrow$ Cu^{II} charge transfer to account for the electronic absorption band at 600 nm. On the other hand, the ^{1}H n.m.r. spectra of plastocyanins suggest that imidazole groups of histidine residues are bound to copper.[127]

The copper(II) complexes of new sulphydryl- and imidazole-containing peptides such as *N*-mercaptoacetyl-L-histidine (MAH) and 2-mercaptopropionyl-L-cysteine have been characterized. The 1:1 MAH–Cu^{II} complex, which shows an intense absorption near 600 nm ($\varepsilon = 830$) and a small copper hyperfine coupling constant ($A_\parallel = 93$ G), has similar characteristics to the blue proteins, and it is postulated that such proteins involve cysteine sulphydryl and histidine imidazole co-ordination.[128] On the other hand, McAuliffe and co-workers[129] have examined copper(II) perchlorate complexes of some linear quadridentate thioethers, $[Cu(S_4ligand)ClO_4]ClO_4$, and relate that the electronic spectra are similar to that of azurin from *Pseudomonas aeruginosa*; other results also suggest the predominance of Cu–S binding.[130] Bosnich and co-workers[131] have suggested the co-ordination sites shown in Figure 6(a) and Figure 6(b) for the blue type I copper proteins and oxyhaemocyanin (type III copper, they say, may have a similar environment, though there is no evidence for the peroxide ligand in its resting state). Of possible peripheral interest is the synthesis of the ten-copper aggregate $Cu_{15}\{S_2CCH(CO_2Bu^t)_2\}_6\{S_2CC(CO_2Bu^t)_2\}_2$.[132] Some e.p.r. results are consistent with tetrahedral distortion at the metal-binding sites of the blue proteins,[133] and Gray and co-workers have reported the kinetics of the oxidation of *Pseudomonas aeruginosa* azurin, bean plastocyanin, and *Rhus vernicifera* stellacyanin by $[Co(phen)_3]^{3+}$.[134]

[124] R. Malkin and B. G. Malmstrom, *Adv. Enzymol.*, 1970, **33**, 177.
[125] V. Miskowski, S. W. P. Tang, T. G. Spiro, E. Shapiro, and T. H. Moss, *Biochemistry*, 1975, **14**, 1244.
[126] E. I. Solomon, P. J. Clendening, H. B. Gray, and F. J. Grunthaner, *J. Amer. Chem. Soc.*, 1975, **97**, 3878.
[127] J. L. Markley, E. L. Ulrich, S. P. Berg, and D. W. Krogmann, *Biochemistry*, 1975, **14**, 4428.
[128] Y. Sugiura and Y. Hirayama, *J. Amer. Chem. Soc.*, 1977, **99**, 1581.
[129] M. H. Jones, W. Levason, C. A. McAuliffe, S. G. Murray, and D. M. Johns *Bioinorg. Chem.*, 1978, **8**, 267.
[130] E. N. Baker and G. E. Norris, *J.C.S. Dalton*, 1977, 877.
[131] A. R. Amundsen, J. Whelan, and B. Bosnich, *J. Amer. Chem. Soc.*, 1977, **99**, 6730.
[132] D. Coucouvanis, D. Swenson, N. C. Baenziger, R. Pedelty, and M. L. Caffery, *J. Amer. Chem. Soc.*, 1977, **99**, 8097.
[133] H. Yoko and A. W. Addison, *J.C.S. Dalton*, 1977, 1341.
[134] J. V. McArdle, C. L. Coyle, H. B. Gray, G. S. Yoneda, and R. A. Holwerda, *J. Amer. Chem. Soc.*, 1977, **99**, 2483.

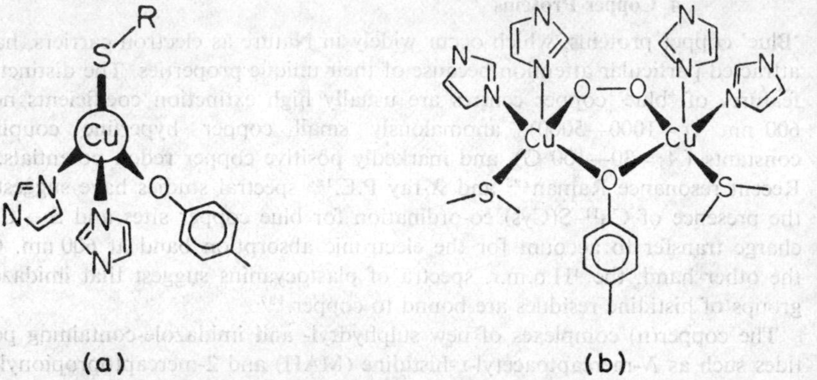

(a)   (b)

Figure 6 (a) *The proposed structure and the ligand co-ordination in the blue type I*
copper proteins. (b) *The proposed structure and ligand environment of*
the copper site in oxyhaemocyanin. The phenolate oxygen atom, one
oxygen of the perioxide ligand, the sulphur of a thioether ligand, and a
nitrogen atom of an imidazole molecule form approximately square-
planar arrays around each copper atom. Two imidazole molecules are in
quasi-axial positions
(Reproduced by permission from *J. Amer. Chem. Soc.*, 1977, **99**, 6730)

5 Complexes of Constituents of Nucleic Acids

The study of the interaction of metal complexes with biological moieties has
recently received added impetus because of the importance of the methylation
of heavy metals in aqueous environments,[135] the potential use of heavy metals
to sequence nucleic acids using electron microscopy,[136] and recent attempts to
understand the chemotherapeutic action of inorganic species.[137, 138]

The extent of interaction of *cis*-[Pt(NH₃)₂] with the ribonucleosides guanosine,
adenosine, and cytidine at pH 6.5 has been examined, and in all cases the data
are consistent with the nucleoside functioning as a unidentate ligand, with no
net deprotonation involved in the binding process. Metal binding at multiple
sites on the nucleoside may occur at high *cis*-Pt^{II} to nucleoside mole ratios, and
elevated temperatures appear to promote secondary reactions.[139] In some complexes of the type [Pd(adenosine)₂X₂] it appears that nucleoside is bound by the
N-1 atom.[140] Adenine nucleosides, but not cytosine, uracil, or guanine nucleosides, readily form isolable complexes of type [Co(acac)₂(NO₂)(nucleoside)]; the
deoxyadenosine compound is pseudo-square-bipyramidal, with two acac

135 J. M. Wood, M. W. Penley, and R. E. DeSimone, 'Mercury Pollution of the Environment',
    International Atomic Energy Commission, Vienna, 1972.
136 M. Beer, D. W. Gibson, and T. Koller, in 'Effects of Metals on Cells, Subcellular Elements
    and Macromolecules', ed. J. Maneloff, J. R. Coleman, and M. W. Miller, Charles C.
    Thomas, Springfield, Illinois, 1970, p. 131.
137 R. H. Freyberg in 'Arthritis and Allied Conditions', 6th edn., ed. J. L. Hollander, Lea and
    Febiger, 1960, Ch. 19.
138 T. A. Connors and J. J. Roberts, in 'Platinum Complexes in Cancer Chemotherapy',
    Springer-Verlag, New York, 1974.
139 W. M. Scovell and T. O'Connor, *J. Amer. Chem. Soc.*, 1977, **99**, 120.
140 R. Ettore, *Inorg. Chim. Acta*, 1977, **25**, L9.

ligands defining the equatorial plane and the N-bonded nitro and the N-7-bonded nucleoside ligands in axial positions.[141] Pneumatikakis *et al.*[142] have suggested N-7, O-6 chelation by guanosine in a palladium complex, and McAuliffe and co-workers[143] postulate N-7 co-ordination only in [Pt(guanosine)$_2$I$_2$]. The latter workers also suggest both NH$_2$ + N-3 and N-1 + O binding in [Pt(cytosine)I$_2$].[143] Solid 1:1 silver(I) complexes with cytidine and guanosine have been isolated from aqueous solutions at pH 10. Binding sites for cytidine are NH$_2$ and N-3; guanosine chelates *via* N-7 and O-6.[144] Marzilli and co-workers have isolated an unusual polymeric silver(I) complex with 1-methylcytosine which may provide a partial model for the cross-linking of two strands of a DNA helix.[145] In mixed-ligand complexes of cobalt(II) and copper(II) with glycyl-L-tyrosine and cytidine, the latter is bonded through N-3.[146]

N.m.r. spectroscopy has been used to look at the self-association of cAMP and ATP, the association of cAMP with ATP, and the interaction of the nucleo-tide dimers with Mn^{II}. In [Mn(cAMP)$_2$] the metal is bound to the phosphate of one cAMP and to the adenine ring of the other; base stacking between the bases also occurs. A similar structure is found for [Mn(cAMP)(ATP)]$^{3-}$; the Mn^{2+} is co-ordinated to the triphosphate chain of the ATP and to the adenine ring of the cAMP.[147] The formation of a stacked adduct between 2,2'-bipyridyl and the purine moiety of adenosine or ATP, and inosine or ITP, causes a change in absorption spectra. Comparison of the molar absorptivities of all the systems indicates that the stacked isomer dominates in the equilibrium between an open and stacked form of [M(bipy)(NTP)]$^{2-}$ (NTP = nucleoside; M = Co, Ni, Cu, or Zn).[148] The monometallic Co^{2+}, Ni^{2+}, Mg^{2+}, and Ca^{2+} complexes with ATP have been studied by ^{1}H n.m.r. spectroscopy at various pD's from the point of view of the effect of ring protonation or the secondary phosphate hydrogen ionization on the metal–ring and metal–chain interactions.[149] Direct platinum–phosphate bonding (Pt–O = 197 pm) has been observed in the 5'-cytidine monophosphate complex [Pt(en)(5'-CMP)]$_2 \cdot$ 2H$_2$O.[150] The inter-actions of Cu^{2+} with the ribose moiety of ATP exhibit characteristic c.d. spectra, and two distinct complexes are observed for which the 2'- and 3'-hydroxy-groups are ionized, respectively.[151]

Because of the general importance of metal–imidazole interactions, it is proper to mention briefly here some pertinent studies relating to the ambidentate nature of the imidazole ring system,[152] some vanadium(II) complexes of substituted

[141] T. Sorrell, L. A. Epps, T. J. Kistenmacher, and L. G. Marzilli, *J. Amer. Chem. Soc.*, 1977, **99**, 2173.
[142] G. Pneumatikakis, N. Hadjiliadis, and T. Theophanides, *Inorg. Chim. Acta*, 1977, **22**, L1.
[143] K. P. Beaumont, C. A. McAuliffe, and M. E. Friedman, *Inorg. Chim. Acta*, 1977, **25**, 241.
[144] R. Cini, P. Colamarino, and P. L. Orioli, *Bioinorg. Chem.*, 1977, **7**, 345.
[145] L. G. Marzilli, T. J. Kistenmacher, and M. Rossi, *J. Amer. Chem. Soc.*, 1977, **99**, 2197.
[146] J. Dehand, J. Jordanov, and F. Keck, *Inorg. Chim. Acta*, 1977, **21**, L13.
[147] S. Fan, A. C. Storer, and G. G. Hammes, *J. Amer. Chem. Soc.*, 1977, **99**, 8293.
[148] P. Chandhuri and H. Siegel, *J. Amer. Chem. Soc.*, 1977, **99**, 3142.
[149] J. Granot and D. Fiat, *J. Amer. Chem. Soc.*, 1977, **99**, 70.
[150] S. Louie and R. Bau, *J. Amer. Chem. Soc.*, 1977, **99**, 3874.
[151] M. Gabriel, D. Larcher, C. Thirion, J. Torreilles, and A. Crastes de Paulet, *Inorg. Chim. Acta*, 1977, **24**, 187.
[152] B. S. Tovrog and R. S. Drago, *J. Amer. Chem. Soc.*, 1977, **99**, 2203.

imidazoles,[153] Co^{II} and Zn^{II} complexes of imidazole and N-methylimidazole with regard to the activity-related ionization in carbonic anhydrase,[154] equilibria in the methylmercury(II)–imidazole system,[155] and some ternary complexes involving imidazole donor systems.[156]

Molybdenum and Tungsten Complexes.—Research on the co-ordination chemistry of molybdenum–sulphur compounds has been stimulated by evidence that oxidation–reduction reactions which are catalysed by molybdenoenzymes occur at sites where molybdenum is co-ordinated by one or more sulphur atoms.[157] Two isomeric forms of the redox-active dinuclear Mo^V anion $[Mo_2S_4(S_2C_2H_7)_2]^{2-}$ have been isolated, both of which undergo a reversible one-electron reduction at -1.87 V and also exhibit two irreversible oxidation waves, each of which appears to involve two electrons per dimer.[158] On the other hand, although the major evidence for Mo–S co-ordination in flavoenzymes is the e.p.r. g values of 1.97—1.98, McAuliffe and Sayle have observed an e.p.r. signal ($g = 1.979$) in the molybdenum(V)–8-hydroxyquinoline system, where there are no sulphur donors.[159] The crystal and molecular structures of cis-dioxodichloro(9,10-phenanthrenequinone)molybdenum(VI) are six-co-ordinate; $Mo-O_q = 231$ pm.[160] The entire series of eight-co-ordinate complexes $[W(mpic)_n(dcq)_{4-n}]$ (mpic = 5-methylpicolinic acid; dcq = 5,7-dichloro-8-quinolinol) have been prepared and characterized.[161]

6 Photolysis of Water

There are now a large number of groups examining model transition-metal complexes as catalysts for the photodecomposition of water. The potential importance of this work as a means of storing solar energy as molecular hydrogen need not be emphasized.[162]

Calvin has previously claimed[163] that solutions of $[(bipy)_2MnO_2Mn(bipy)_2]^{3+}$ aid the photodecomposition of water. Though subsequent analysis of his results did not substantiate this conclusion, the recent crystal structure elucidation[164] of this mixed Mn^{III}, Mn^{IV} complex is of interest because there is much to be said for Calvin's claim[163] that such dimeric high-valent manganese complexes are models for natural O_2-evolving systems in the chloroplast. Sawyer et al.[165] continue their work on the electrochemical and spectroscopic characterization

[153] M. Ciampolini and F. Mani, *Inorg. Chim. Acta*, 1977, **24**, 91.
[154] D. W. Appleton and B. Sarkar, *Bioinorg. Chem.*, 1977, **7**, 211.
[155] C. A. Evans, D. L. Rabenstein, G. Geier, and I. W. Erni, *J. Amer. Chem. Soc.*, 1977, **99**, 8106.
[156] H. Sigel, B. E. Fischer, and B. Prijs, *J. Amer. Chem. Soc.*, 1977, **99**, 4489.
[157] R. C. Bray, in 'The Enzymes', Vol. 12, 3rd end., ed. P. D. Boyer, Academic Press, New York, 1975, p. 299.
[158] G. Bunzey, J. H. Enemark, J. K. Howie, and D. T. Sawyer, *J. Amer. Chem. Soc.*, 1977, **99**, 4168.
[159] C. A. McAuliffe and B. J. Sayle, *Bioinorg. Chem.*, 1978, **8**, 331.
[160] C. G. Pierpont and H. H. Downs, *Inorg. Chem.*, 1977, **16**, 2970.
[161] C. J. Donahue and R. D. Archer, *J. Amer. Chem. Soc.*, 1977, **99**, 6613.
[162] C. A. McAuliffe, *Chem. and Ind.*, 1975, 725.
[163] M. Calvin, *Science*, 1974, **184**, 375.
[164] S. R. Cooper and M. Calvin, *J. Amer. Chem. Soc.*, 1977, **99**, 6623.
[165] M. E. Bodini, L. A. Willis, T. L. Reichel, and D. T. Sawyer, *Inorg. Chem.*, 1976, **15**, 1538.

of manganese-(II), -(III), and -(IV) gluconate complexes, which may be models for both mitochondrial superoxide dismutase and photosystem II. A range of metal oxides (SiO_2, Al_2O_3, ZnO, CuO, NiO, CoO, Fe_2O_3, MnO, Cr_2O_3, V_2O_5, and TiO_2) have been examined as catalysts for the oxygen-evolving step in the thermal splitting of sulphuric acid.[166]

The claim by Whitten and co-workers[167] that surface-bound $[Ru(bipy)_3]^{2+}$ complexes catalyse the photodecomposition of water remains unsubstantiated, but work on the photochemical reactivity of surfactant ruthenium(II) complexes in monolayer assemblies and at water/solid interfaces[168, 169] goes on apace. Gray and co-workers have examined the novel dinuclear rhodium(I) species $[Rh_2(bridge)_4]^{2+}$ (bridge = 1,3-di-isocyanopropane) and claim that the reaction (4) is of relevance to solar energy storage.[170] The recent arguments about conversion are of interest to anyone concerned with this area.[171, 172]

$$[Rh_2(bridge)_4H]^{3+} Cl^- + H^+ + Cl^- \xrightarrow[\text{12-MH-Cl}]{\text{546 nm}} [Rh_2(bridge)_4Cl_2]^{2+} + H_2 \quad (4)$$
$$\text{(blue)} \hspace{8cm} \text{(yellow)}$$

7 Complexes of Metals with Cyclic Ligands

Porphyrins and Metalloporphyrins.—Semi-empirical CNDO/2 M.O. calculations[173] have been performed for the ground states of porphyrin, 2,4-divinylporphyrin (DVP), and $\alpha\beta\gamma\delta$-tetraphenylporphyrin (TPP). Results for TPP indicate that the conformation with all phenyl groups perpendicular to the ring is 108 kJ mol^{-1} more stable than the coplanar conformation. The lowest singlet states are closely similar in energy and composition for all three molecules except for an extra state and more complex compositions in DVP above 3 eV. The lowest triplet states of porphyrin and DVP are very similar, whilst those of TPP are comparable in energy or composition, but not both. Further calculations for dilithium and disodium porphyrin indicate that little mixing of metal and porphyrin orbitals takes place; the two lowest unoccupied and the two highest occupied M.O.s hardly differ from those in porphyrin, but lower M.O.s are considerably rearranged.[174] The crystal and molecular structures of the free base porphyrin protoporphyrin-IX dimethyl ester have been reported by Caughey and Ibers.[175]

Although TPP is undoubtedly the easiest porphyrin to handle, and it readily forms complexes with metal ions, in many cases it is not a close enough model to natural porphyrins. Deuteroporphyrin-IX dimethyl ester is a better model for

166 M. Dokiya, T. Kameyama, K. Fukuda, and Y. Kotera, *Bull. Chem. Soc. Japan*, 1977, **50**, 2657.
167 See references cited in ref. 168.
168 G. Sprintschnik, H. W. Sprintschnik, P. P. Kirsch, and D. G. Whitten, *J. Amer. Chem. Soc.*, 1977, **99**, 4947.
169 K. P. Seefeld, D. Mobius, and H. Kuhn, *Helv. Chim. Acta*, 1977, **60**, 2608.
170 K. R. Mann, N. S. Lewis, V. M. Miskowski, D. K. Erwin, G. S. Hammond, and H. B. Gray, *J. Amer. Chem. Soc.*, 1977, **99**, 5525.
171 L. M. Fetterman, L. Galloway, N. Winograd, and F. K. Fong, *J. Amer. Chem. Soc.*, 1977, **99**, 654.
172 Govindjee and J. T. Warden, *J. Amer. Chem. Soc.*, 1977, **99**, 6089.
173 S. J. Chantrell, C. A. McAuliffe, R. W. Munn, A. C. Pratt, and R. F. Weaver, *Bioinorg. Chem.*, 1977, **7**, 283.
174 S. J. Chantrell, C. A. McAuliffe, R. W. Munn, A. C. Pratt, and R. F. Weaver, *Bioinorg. Chem.*, 1977, **7**, 297.
175 W. S. Caughey and J. A. Ibers, *J. Amer. Chem. Soc.*, 1977, **99**, 6639.

metallo-derivative study, and a new synthesis of this ligand is thus welcome.[176] The effect of substituents on π-radical reactions of *para*-substituted TPP's has been investigated by cyclic voltammetry in methylene chloride.[177]

In recent years thallium salts have become important in organic syntheses, and interest in porphyrin derivatives has grown. The crystal structure of 2,3,7,8,12,13,17,18-octaethylporphinatochlorothallium(III) shows it to be square-pyramidal, with an axial chlorine; Tl—Cl = 244.9, Tl—N = 221.1 (av.) pm, and the metal is located 69 pm out of the mean plane of the four pyrrole nitrogen atoms. The porphinato core is highly expanded, with an average radius of 210 pm, and there is some deviation from planarity.[178] The structure of methyl-5,10,15,20-tetraphenylporphinatothallium(III), [MeTl(TPP)], is similarly square-pyramidal; Tl—C = 97.9, Tl—N = 229(av.) pm.[179]

Adducts of chloro-*meso*-tetraphenylporphinatochromium(III) [Cr(TPP)Cl] and neutral oxygen-, sulphur-, and nitrogen-donor ligands of the type [Cr(TPP)Cl(L)] have been prepared, and it appears that N-donor ligands are the most strongly bound.[180] As mentioned earlier, the observation that certain Mn–porphyrin systems bind dioxygen has led to new work on these systems, and Anderson and Lavallee have reported the crystal structure of chloro-*N*-methyl-$\alpha\beta\gamma\delta$-tetraphenylporphinatomanganese(II), which is a distorted square pyramid; three Mn—N = 211.8—215.5 pm, whilst the alkylated N atom forms a longer bond (Mn—N = 236.8 pm) and Mn—Cl = 229.5 pm. The N-alkylated pyrrole ring deviates from the main plane and blocks access to the sixth co-ordination position.[181] These workers find essentially similar behaviour in chloro-*N*-methyl-$\alpha\beta\gamma\delta$-tetraphenylporphinatocobalt(II).[182]

The binding of dioxygen to cobalt(II) meso-, deutero-, and proto-porphyrin-IX dimethyl esters complexed with pyridine or 2-methylimidazole between -10 and $-60\,°C$ in toluene or DMF is greater by a factor of 1.4—2.0 for the meso-porphyrin complex than for the other complexes.[183] The equilibrium constant for the first thiocyanate addition to cobalt(III) tetra(3-*N*-methylpyridyl)porphine is a factor of 2 smaller than for the less basic 4-*N*-methylporphine isomer, due primarily to a larger Co—SCN dissociation rate constant for the 3-isomer.[184]

Electron nuclear double resonance (ENDOR) permits the mapping out of the odd-electron distribution within the Ag^{II}– and Cu^{II}–TPP species, and there is σ-delocalization of the odd electron of the metal ion through four bonds onto the protons.[185] The synthesis of *meso*-tetraferrocenylporphyrin, H$_2$TFcP (11), has been achieved, and this black substance appears to have properties similar

176 A. C. Adler, D. L. Ostfeld, and E. H. Abbott, *Bioinorg. Chem.*, 1977, **7**, 187.
177 K. M. Kadish and M. M. Morrison, *Bioinorg. Chem.*, 1977, **7**, 107.
178 D. L. Cullen, E. F. Mayer, and K. M. Smith, *Inorg. Chem.*, 1977, **16**, 1179.
179 K. Henrick, R. W. Matthews, and P. A. Tasker, *Inorg. Chem.*, 1977, **16**, 3293.
180 D. A. Summerville, R. D. Jones, B. M. Hoffman, and F. Basolo, *J. Amer. Chem. Soc.*, 1977, **99**, 8195.
181 O. P. Anderson and D. K. Lavallee, *Inorg. Chem.*, 1977, **16**, 1634.
182 O. P. Anderson and D. K. Lavallee, *Inorg. Chem.*, 1977, **16**, 1404.
183 H. Yamamoto, T. Takayanagi, T. Kwan, and T. Yonetani, *Bioinorg. Chem.*, 1977, **7**, 189.
184 G. N. Williams and P. Hambright, *Bioinorg. Chem.*, 1977, **7**, 267.
185 T. G. Brown, J. L. Petersen, G. P. Lozos, J. R. Anderson, and B. M. Hoffman, *Inorg. Chem.*, 1977, **16**, 1563.

(11)

to those of other *meso*-substituted porphyrins. [Cu(TFcP)] has been isolated, and the shifting of the Soret band from the normal position at 400 nm to 350 nm is probably due to a π-electron interaction between the porphyrin and ferrocenyl groups.[186] Resonance Raman spectral results provide direct evidence for π-delocalization linking *meso*-aryl groups to the porphyrin ring system in porphyrin dications, neutral porphyrins, and copper(II) porphyrins; even dihedral angles greater than 80 ° do not preclude π-delocalization.[187] Both chloro-*N*-methyltetraphenylporphinatonickel(II) and the corresponding copper(II) complex dimethylate by a two-path mechanism in which each path involves removal of $-CH_3^+$ by a nucleophile, whereas the reactions of the corresponding Zn^{II} and Mn^{II} complexes involve nucleophilic attack of the methyl group by a single-path mechanism.[188] Equilibrium constants for the metallation of zinc porphyrins have been measured.[189] Some thermodynamics of the formation of adducts of [Zn(TPP)] with *N*-, *O*-, *S*-donors in cyclohexane have been reported.[190]

There have been a number of recent studies of porphyrin complexes of heavy 2nd- and 3rd-row transition-metal ions. These include the report of the remarkably different structures of two metalloporphyrins containing $M_2O_3^{4+}$ units, the complexes $[O_3M_2(TPP)_2]$ ($M = Mo^V$ or Nb^V), shown in Figures 7 and 8. The average Nb—N distance is 224.6 pm, each niobium being displaced 101 pm out of the mean porphine plane; for the Mo derivative, Mo—N = 209.4 pm, but the Mo atom is displaced only 9 pm out of the plane towards the terminal oxo ligand.[191] Definitive evidence for the 'sitting-atop' complex for the intermediate in metalloporphyrin formation is supplied by Fleischer and Dixon for the interaction of *meso*-TPP with $[Rh(CO)_2Cl]_2$.[192] Some new results for rhodium octaethylporphyrin complexes include the synthesis of the hydridorhodium

[186] R. G. Wollmann and D. N. Hendrickson, *Inorg. Chem.*, 1977, **16**, 3079.
[187] W. H. Fuchsman, Q. R. Smith, and M. M. Stein, *J. Amer. Chem. Soc.*, 1977, **99**, 4190.
[188] D. K. Lavallee, *Inorg. Chem.*, 1977, **16**, 955.
[189] P. Hambright, *Inorg. Chem.*, 1977, **16**, 2987.
[190] G. C. Vogel and J. R. Stahlbush, *Inorg. Chem.*, 1977, **16**, 950.
[191] J. J. Johnson and W. R. Scheidt, *J. Amer. Chem. Soc.*, 1977, **99**, 294.
[192] E. B. Fleischer and F. Dixon, *Bioinorg. Chem.*, 1977, **7**, 129.

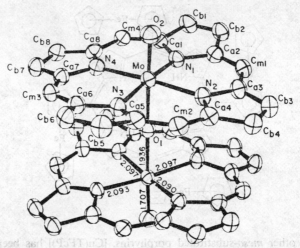

Figure 7 *A model, drawn in perspective, of the central unit of the molecule*
[O₃Mo₂(TPP)₂]. *The phenyl rings of the ligand have been omitted*
(Reproduced by permission from *J. Amer. Chem. Soc.*, 1977, **99**, 294)

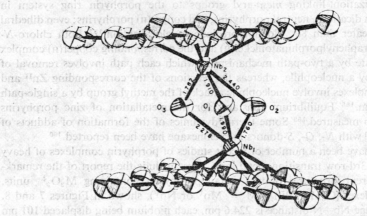

Figure 8 *A model of the central unit of the molecule* [O₃Nb₂(TPP)₂], *drawn in
perspective and with the phenyl groups omitted for clarity*
(Reproduced by permission from *J. Amer. Chem. Soc.*, 1977, **99**, 294)

complex [(OEP)Rh^{III}H] and the dimeric complex formulated as [(OEP)Rh^{II}]₂,
with Rh–Rh bonds.[193] Rates of rotation of the phenyl ring in *para*-substituted
TPP complexes with rhuthenium 4-t-butylpyridine, indium chloro, and titanyl
ions increase in the order Ru < In < Ti.[194] The crystal structure of [Au(TPP)Cl]
shows distorted square-pyramidal geometry, with quite short Au—N(≈ 200 pm),
probably because the Au atom is essentially in the plane of the ring.[195]

[193] H. Ogoshi, J. Setsune, and Z. Yoshida, *J. Amer. Chem. Soc.*, 1977, **99**, 3869.
[194] S. S. Eaton and G. R. Eaton, *J. Amer. Chem. Soc.*, 1977, **99**, 6594.
[195] R. Timkovich and A. Tulinsky, *Inorg. Chem.* 1977 **16**, 962.

As expected, the majority of published work has been on iron complexes with porphyrins. Dioxygen-bridged di-iron adducts have frequently been proposed as important, though unstable, intermediates in the autoxidation of ferrous complexes.[196] Such bridged species have also been proposed to represent the oxygenated state of the respiratory pigment haemerythrin.[197] The detection and characterization of this long-postulated Fe—OO—Fe intermediate in the oxidation of iron(II) porphyrins has been achieved by LaMar and co-workers,[198] who suggest the mechanism shown in Scheme 1. High-spin iron(II) porphyrins in

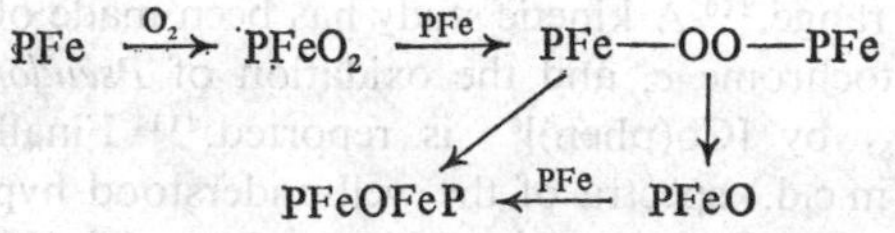

$$\text{PFe} \xrightarrow{O_2} \text{PFeO}_2 \xrightarrow{\text{PFe}} \text{PFe—OO—PFe}$$

$$\text{PFeOFeP} \xleftarrow{\text{PFe}} \text{PFeO}$$

Scheme 1

N-methylpyrrolidone–acetic acid, methanol, or benzene are very rapidly oxidized by quinones at room temperature; the corresponding iron(III) porphyrins and hydroquinones are the only products, and the reaction is not reversible.[199] Nitro-aromatics also oxidize high-spin iron(II) porphyrins.[200] Proton n.m.r. linewidth analysis has been used to characterize the axial lability of nitrogenous bases in bis-ligated low-spin iron(III) complexes of a range of synthetic porphyrins. In the presence of excess ligand, exchange proceeds by a dissociative mechanism with a five-co-ordinate transition state.[201] Proton n.m.r. studies of model high-spin iron(II) complexes of porphyrins have also been reported by LaMar's group.[202] Karplus and co-workers claim that the nature of the iron–oxygen bond in oxyhaemoglobin is more like that in ozone than as a formal Fe^{3+}—O_2^- formulation.[203] Sams and co-workers[204] have re-prepared octaethylporphyrinatoiron(III) perchlorate and shown that it has an intermediate spin state $(S = \frac{3}{2})$. Espenson and Christensen[205] have shown that, in the demetallation of iron(III) porphyrins by HCl in acetic acid catalysed by iron(II), iron(II) porphyrin is an intermediate.[205] Not surprisingly, it has been discovered that there are substituent effects on the formation constants of iron(III) and iron(II) tetraphenylporphyrin–pyridine complexes.[206]

The six-co-ordinate complexes [L₂Fe(OMBP)] (OMBP = dianion of octamethylbenzoporphyrin; L = 1-methylimidazole, pyridine, or piperidine) reversibly bind carbon monoxide in toluene solution.[207] The kinetics of the

[196] G. S. Hammond and C. H. S. Wu, *Adv. Chem. Series*, 1968, **77**, 186.
[197] M. Y. Okamura and I. M. Klotz, in 'Inorganic Biochemistry', Vol. 1, ed. G. L. Eichorn, Elsevier, New York, 1973, p. 320.
[198] D. H. Chin, D. Del Gaudio, G. N. LaMar, and A. L. Balch, *J. Amer. Chem. Soc.*, 1977, **99**, 5486.
[199] C. E. Castro, G. M. Hathaway, and R. Havlin, *J. Amer. Chem., Soc.*, 1977, **99**, 8032.
[200] J. H. Ong and C. E. Castro, *J. Amer. Chem. Soc.*, 1977, **99**, 6740.
[201] J. D. Saterlee, G. N. LaMar, and T. J. Bold, *J. Amer. Chem. Soc.*, 1977, **99**, 1088.
[202] H. Goff and G. N. LaMar, *J. Amer. Chem. Soc.*, 1977, **99**, 6599.
[203] B. H. Huynh, D. A. Case, and M. Karplus, *J. Amer. Chem. Soc.*, 1977, **99**, 6103.
[204] D. H. Dolphin, R. H. Sams, and T. B. Tsin, *Inorg. Chem.*, 1977, **16**, 711.
[205] J. H. Espenson and R. J. Christensen, *Inorg. Chem.*, 1977, **16**, 2561.
[206] K. M. Kadish and L. A. Bottomley, *J. Amer. Chem. Soc.*, 1977, **16**, 2380.
[207] B. R. James, K. J. Reimer, and T. C. T. Wong, *J. Amer. Chem. Soc.*, 1977, **99**, 4815.

reaction of ferroprotoporphyrin-IX with carbon monoxide in glycerol–water mixtures have been reported.[208] Several new haemochromes have been complexed with phosphines and phosphites.[209]

Deuteroferrihaem peroxide compounds have been formed by the reaction of deuteroferrihaem with H_2O_2, Bu^tOOH, and ten perbenzoic acids. The rate constant is independent of pH in the range 7.75—10.0, but shows a 3-fold increase as the pH decreases in the range 7.75—6.0. In contrast, the rate constants for the oxidation of iodide by compounds I and II of the ferrihaem hydroperoxidase enzyme horseradish peroxidase (E.C.1.11.1.7) are directly proportional to $[H^+]$ over a wide range.[210] A kinetic study has been made of the oxidation of horse heart ferrocytochrome c, and the oxidation of *Pseudomonas aeruginosa* ferrocytochrome c_{551} by $[Co(phen)]^{3+}$ is reported.[211] Finally, the magnetic circular dichroism (m.c.d.) spectra of the well-understood hyperporphyrins are remarkably similar to that of 'reduced + CO' cytochrome P-450.[212]

Cobalamins and Cobaloximes.—The best known biological function of cobalt is in its intimate involvement in the coenzymes related to vitamin B_{12}. The macrocyclic ring involved is the *corrin* system; it is not unlike the porphyrin system except that one methine bridge is absent between one pair of pyrrole rings. There are a number of model systems, of which the best known is cobalt(II) dimethylglyoxime, *i.e.* cobaloxime.

The ^{13}C n.m.r. spectra of 51 cobaloxime complexes have been measured when complexed with phosphines or phosphites and an X group (X = Cl, Br, NO_3, NO_2, or N_3).[213] Two main types of reaction take place when alkyl-cobaloximes react with cobaloxime(II) complexes in methanol or methylene chloride. First, homolytic displacement of cobaloxime(II) from the alkyl-cobaloxime takes place by attack of the cobaloxime(II) reagent on the alkyl group. Secondly, after the onset of alkyl exchange, additional exchange of equatorial ligands between reagent cobaloxime(II) and displaced cobaloxime(II) complexes takes place rapidly.[214] A deuterium-exchange study has been made of the equatorial groups of methylbis(dimethylglyoxime)cobalt(III) complexes.[215]

The complexes of adenosine, methylcobalamin, adenosylcobalamin, and the derivatives of adenosylcobalamin with $PdCl_4^{2-}$ are of three types: with methylcobalamin the Pd is co-ordinated at the 'lower' benzimidazole ligand; in adenosylcobalamin the Pd bonds to the upper adenosyl ligand; whereas the N-7 position in adenosine and the N-3 position in 5,6-dimethylbenzimidazolyltribotide are co-ordinated.[216] Using stopped-flow techniques, the Hg^{II}-induced dealkylation of methylcobalamin was studied in three solvents: (*a*) chloride, unbuffered, (*b*) acetate, buffered, and (*c*) water, unbuffered. In solvents (*a*) and (*b*) the transfer

[208] B. B. Hasinoff, *Canad. J. Chem.*, 1977, **55**, 3955.
[209] W. N. Connor and D. K. Straub, *Inorg. Chem.*, 1977, **16**, 491.
[210] P. Jones and D. Mantle, *J.C.S. Dalton*, 1977, 1849.
[211] J. V. McArdle, K. Yocum, and H. B. Gray, *J. Amer. Chem. Soc.*, 1977, **99**, 4141.
[212] J. H. Dawson, J. R. Trudell, G. Barth, R. E. Linder, E. Bunnenberg, C. Djerassi, M. Goutermann, C. R. Connell, and P. Sayer, *J. Amer. Chem. Soc.*, 1977, **99**, 641.
[213] R. C. Stewart and L. G. Marzilli, *Inorg. Chem.*, 1977, **16**, 424.
[214] D. Dodd, M. D. Johnson, and B. L. Lockman, *J. Amer. Chem. Soc.*, 1977, **99**, 3664.
[215] A. V. Cartano and L. L. Ingraham, *Bioinorg. Chem.*, 1977, **7**, 351.
[216] A. M. Yurkevich, E. G. Chauser, and I. R. Rudokova, *Bioinorg. Chem.*, 1977, **7**, 315.

of the methyl group from the central cobalt atom to Hg^{II} was found to occur without cleavage of the 5,6-dimethylbenzimidazolyl (DMBz) moiety of the coenzyme, thereby dispelling the notion that displacement of DMBz from methylcobalamin in acetate occurs as a prerequisite to methyl transfer.[217] The emission Mössbauer spectra of 'base on' and 'base off' forms of the cyanocob(III)-alamin (B$_{12}$), the cyanocob(II)alamin (B$_{12r}$), and the cyanocob(I)alamin (B$_{12s}$) have been measured with and without the co-ordination of DMBz.[218] The kinetics of BMDz dissociation in methylcobalamin in methanol have been evaluated, using variable-temperature ^{1}H or ^{13}C n.m.r. spectra, and signals due to the axial methyl group were observed. Milton and Brown[219] have concluded that the comparative strengths of interaction of nitrogen bases, as mediated by steric effects, may play an important role in determining the reactivity of the cobalt–carbon bond in methylcobalamin or coenzyme B$_{12}$ that is bound to proteins.

Phthalocyanine Complexes.—The phthalocyanines form a special type of macro-cyclic ligand, more nearly analogous to the porphyrins than to any other types of synthetic N$_4$ ring system.

Detailed analysis of the e.s.r. signal observed in some diamagnetic phthalo-cyanines and in metal-free TPP suggests that the cause of this signal is various charge-transfer interactions between phthalocyanine or porphyrin cations and dioxygen anions that are stabilized by the surface of the diamagnetic host.[220] E.s.r. spectroscopy has also been used to identify the products formed by the reduction of the low-spin Mn^{II} chelates of tetrasodium 3,10,17,24-tetra-sulphonatophthalocyanine, [Na$_4$(tspc)], on addition of various reducing agents; *e.g.*, addition of hydrazine to an aqueous solution containing [Mn^{II}(tspc)]$^{4-}$ leads to the formation of low-spin [Mn^0(tspc)]$^{6-}$. When either pyridine or imidazole is present in an aqueous solution of [Mn^{II}(tspc)]$^{4-}$, the addition of NaBH$_4$ leads to a free-radical species which is thought to involve the phthalocyanine ring system.[221]

Six-co-ordinate low-spin iron(II) phthalocyanine complexes [L$_2$FePc] (L = 1-methylimidazole, piperidine, or pyridine) reversibly bind benzyl iso-cyanide (RNC) in toluene solution *via* a dissociative mechanism. The formation constants for the [LFe(Pc)(RNC)] complexes are a thousand times greater than those for [Fe(Pc)(RNC)$_2$]; RNCS is bound to FePc in preference to CO by a factor of 1000, and it is a factor of 1000 times more inert than CO.[222] A re-investigation of the reaction of dioxygen with phthalocyaninetetrasulphonate-iron(II) has been made.[223] The molecular stereochemistry of phthalocyanato-zinc(II), [ZnPc], shows that the Pc ligand constrains the Zn atom to an effectively square-planar co-ordination. Surprisingly, the Zn atom has contracted sufficiently to fit into the central hole of the Pc ligand; the Zn atom is very nearly centred

[217] V. C. W. Chu and D. W. Gruenwedel, *Bioinorg. Chem.*, 1977, **7**, 169.
[218] K. Inone and A. Nath, *Bioinorg. Chem.*, 1977, **7**, 159.
[219] P. A. Milton and T. L. Brown, *J. Amer. Chem. Soc.*, 1977, **99**, 1390.
[220] J. B. Raynor and A. S. M. Torrens-Burton, *J.C.S. Dalton*, 1977, 2360.
[221] D. J. Cookson, T. D. Smith, J. F. Boas, P. R. Hicks, and J. R. Pilbrow, *J.C.S. Dalton*, 1977, 211.
[222] D. V. Stynes, *Inorg. Chem.*, 1977, **16**, 1170.
[223] G. McLendon and A. E. Martell, *Inorg. Chem.*, 1977, **16**, 1812.

in the plane of the four nitrogen atoms of Pc. The core of the Pc ligand is somewhat expanded; the average Zn—N bond is 198.0 pm long.[224]

Synthetic Macrocyclic Complexes.—The subject of macrocyclic complexes continues to expand. Most of it offers little insight into what factors determine structure–reactivity relationships in natural macrocyclic systems. In this section are included a few of the many studies reported which may be of interest to the readers of this volume.

The photochemistry of Fe^{II} and Fe^{III} complexes has itself been the subject of intensive study for several decades, and it has recently been shown that when 10^{-3}—10^{-4} M solutions of an iron(II) complex of 2,3,9,10-tetramethyl-1,4,8,11-tetra-azacyclotetradeca-1,3,8,10-tetraene (TIM) in methanol are exposed to sunlight, oxidation of methanol to formaldehyde occurs:

$$CH_3OH + \tfrac{1}{2}O_2 \longrightarrow CH_2O + H_2O \qquad (5)$$

In the absence of dioxygen, the oxidation of methanol is accompanied by stoicheiometric reduction of $[Fe^{III}(TIM)(CH_3OH)(OCH_3)]^{2+}$ to $[Fe^{II}(TIM)(CH_3-OH)_2]^{2+}$.[225] Busch and co-workers have recently reported[226] a series of 14-, 15-, and 16-membered tetra-aza-tetraene macrocyclic ligands (12a—e) that are devoid of functional substituents. In the dicationic iron(II) complexes it was shown that, for the 14-membered ligand (12a), the bis-β-di-imine complex (13) is obtained, whereas under similar conditions the larger ligands (12b—e) produce the novel hexaco-ordinate (14). The two types of protonation products have been obtained, including the complexes of quadridentate bis-β-di-imine macrocyclic ligands and two novel derivatives in which acetonitrile molecules have been adjacent to the apical γ-carbon of the charged six-membered chelate rings.[227]

(12) a; $X = Y = (CH_2)_2$, $R^1 = Me$, $R^2 = H$
    b; $X = (CH_2)_2$, $Y = (CH_2)_3$, $R^1 = Me$, $R^2 = H$
    c; $X = (CMe_2)_2$, $Y = (CH_2)_3$, $R^1 = Me$, $R^2 = H$
    d; $X = Y = (CH_2)_3$, $R^1 = Me$, $R^2 = H$
    e; $X = Y = (CH_2)_3$, $R^1 = R^2 = Me$

[224] W. R. Scheidt and W. Dow, *J. Amer. Chem. Soc.*, 1977, **99**, 1101.
[225] D. W. Reichgott and N. J. Rose, *J. Amer. Chem. Soc.*, 1977, **99**, 1813.
[226] D. P. Riley, J. A. Stone, and D. H. Busch, *J. Amer. Chem. Soc.*, 1976, **98**, 1752; and references therein.
[227] D. P. Riley, J. A. Stone, and D. H. Busch, *J. Amer. Chem. Soc.*, 1977, **99**, 767.

$$\left[cis\text{-} \quad X \underset{\underset{\displaystyle H_\gamma}{\displaystyle Me}}{\overset{\displaystyle H_\gamma}{\underset{\displaystyle R^2 \quad R^1}{\overset{\displaystyle R^2 \quad R^1}{Fe^{2+}}}}} \right]^{2+} \quad 2PF_6^-$$

(14)   a; X = $(CH_2)_2$, R^1 = Me, R^2 = H_δ
       b; X = $(CMe_2)_2$, R^1 = Me, R^2 = H_δ
       c; X = $(CH_2)_3$, R^1 = Me, R^2 = H_δ
       d; X = $(CH_2)_3$, R^1 = R^2 = Me

The reactions of $\cdot O_2^-$ with $[Co^{II}(4,11\text{-dieneN}_4)]$ and with $[Co^{II}(1,3,8,10\text{-tetraeneN}_4)]$ have been studied by pulse radiolysis methods.[228] and the general observation has been made that, since $\cdot O_2^-$ adds irreversibly, this reaction must be included when consideration is given to the role of $\cdot O_2^-$ in the damage of biological systems. Observations of the growth of obligate anaerobes, which do not contain superoxide dismutase (SOD), under aerobic conditions[229] in the presence of added Co^{2+} appear to suggest that there is interaction between $\cdot O_2^-$ and cellular Co^{II} complexes. Busch and co-workers[230] have studied the influence of ring size on the synthesis, stereochemistry, and electrochemistry of cobalt(III) complexes with the unsubstituted saturated tetra-aza-macrocycles shown in Figure 9[230] and the spectrochemical properties of tetragonal complexes of high-spin nickel(II) with these ligands.[231] Diaquo(5,7-dimethyl-1,4,8,11-tetra-azacyclotetradeca-4,7-diene)cobalt(II) hexafluorophosphate, $[Co([14]4,7\text{-dieneN}_4)(OH_2)](PF_6)_2$, where [14]4,7-dieneN$_4$ is (15), reacts with O_2 to form a new macrocyclic ligand ([14]4,7-dieneN$_4$-one) (16) which has a ketone oxygen

(15)   (16)

[228] M. G. Simic and M. Z. Hoffman, *J. Amer. Chem. Soc.*, 1977, **99**, 2370.
[229] G. A. Dedic and O. G. Koch, *J. Bacteriol.*, 1956, **71**, 126.
[230] Y. Hung, L. Y. Martin, S. C. Jackels, A. M. Tait, and D. H. Busch, *J. Amer. Chem. Soc.*, 1977, **99**, 4029.
[231] L. Y. Martin, C. R. Sperati, and D. H. Busch, *J. Amer. Chem. Soc.*, 1977, **99**, 2968.

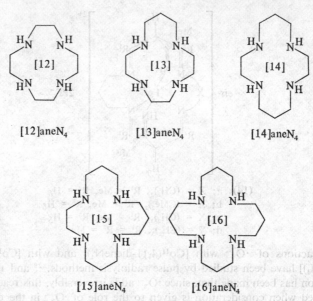

Figure 9 *Some homologous fully saturated tetra-aza-macrocycles with which cobalt (III) forms complexes*

on the central carbon of the 2,4-pentanedi-iminato moiety. The cobalt(II) centre appears to be necessary for this reaction, yet is not oxidized in the process.[232]

Of only peripheral interest to the biologist, but of great interest to the inorganic chemist, are the two reports of the syntheses of polyphosphino macrocyclic systems.[233, 234]

Crown Ether Complexes and Cryptates. (see also Chapter 3)—The ability of crown ethers to associate with a variety of charged and uncharged substrates[235] bears some resemblance to the initial step in reactions catalysed by enzymes. In developing enzyme mimics, it may be hoped that, by the use of correct crown ethers, both a high degree of substrate selectivity and high reactivity may be attained. The feasibility of this idea has recently been illustrated by Chao and Cram[236] with a crown ether model that is capable of mimicking reactions catalysed by the enzyme trypsin. Van Bergen and Kellogg have recently synthesized a crown ether mimic of NAD(P)H.[237] Macrocyclic polyether molecules such as 18-crown-6 (17) may both be compared to and contrasted with valinomycin (Figure 10) and other macrocyclic antibiotics in terms of structure and reactivity.

[232] B. Durham, T. J. Anderson, J. A. Switzer, J. F. Endicott, and M. D. Glick, *Inorg. Chem.*, 1977, **16**, 271.

[233] E. P. Kyba, C. W. Hudson, M. J. McPhaul, and A. M. John, *J. Amer. Chem. Soc.*, 1977, **99**, 8053.

[234] T. A. DelDonno and W. Rosen, *J. Amer. Chem. Soc.*, 1977, **99**, 8051.

[235] J. M. Lehn, *Structure and Bonding*, 1973, **16**, 1 is a recent review.

[236] Y. Chao and D. J. Cram, *J. Amer. Chem. Soc.*, 1976, **98**, 1015.

[237] T. J. Van Bergen and R. M. Kellogg, *J. Amer. Chem. Soc.*, 1977, **99**, 3882.

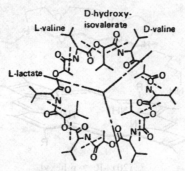

Figure 10 *The structure of valinomycin*
(Reproduced by permission from *J. Amer. Chem. Soc.*, 1977, **99**, 3882)

(17)   (18)   (19)

Both types of molecules selectively bind certain alkali-metal cations.[238] Two novel analogues of 18-crown-6, *i.e.* 2,6-dioxo-18-crown-6 (18) and 2,4-dioxo-19-crown-6 (19), have recently been obtained.[239] Some unusual stability characteristics in methanol of the complexes of a new pyridine-substituted polyether-ester compound with Na⁺, K⁺, Ag⁺, and Ba²⁺ have been observed,[240] and a stacked double-macrocyclic ligand (20), a 'crowned' porphyrin, has also been synthesized.[241] Some non-cyclic complexes, *e.g.* of CaCl₂ with *NNN′N′*-tetrapropyl-3,6-dioxaoctane diamide, are also of peripheral interest.[242]

Lehn and co-workers continue their excellent studies on cryptates. For example, the cylindrical macrotricyclic ligands (21)—(23) yield inclusion complexes, [3]-cryptates, with various metal cations,[243] and Lehn[244] has suggested that his

[238] E. Grell, T. Funck, and F. Eggers, in 'Membranes', ed. G. Eisenman, Vol. III, Marcel Dekker, New York, 1974.
[239] R. M. Izatt, J. D. Lamb, G. E. Maas, R. E. Asay, J. S. Bradshaw, and J. J. Christensen, *J. Amer. Chem. Soc.*, 1977, **99**, 2365.
[240] R. M. Izatt, J. D. Lamb, R. E. Assay, G. E. Maas, J. S. Bradshaw, and S. S. Moore, *J. Amer. Chem. Soc.*, 1977, **99**, 6134.
[241] C. K. Chang, *J. Amer. Chem. Soc.*, 1977, **99**, 2819.
[242] K. Neuport-Laves and M. Dobler, *Helv. Chim. Acta*, 1977, **60**, 1861.
[243] J. M. Lehn and J. Simon, *Helv. Chim. Acta*, 1977, **60**, 141.
[244] J. M. Lehn, S. H. Pine, E. Watanabe, and A. K. Willard, *J. Amer. Chem. Soc.*, 1977, **99**, 6766.

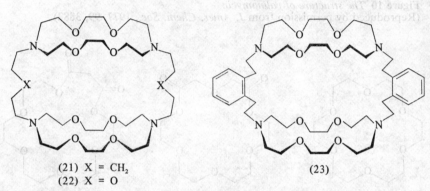

(20) R = n-hexyl

(21) X = CH$_2$
(22) X = O

(23)

bis-tren macrobicyclic ligands may be capable of forming a bis(metal)–substrate complex, as shown in Figure 11. The acid-catalysed dissociation of metal–cryptate complexes,[245] some dinuclear copper(I) and copper(II) inclusion complexes of cylindrical polythia-macrotricyclic ligands,[246] and even a sepulchrate (a macrobicyclic nitrogen cage for metal ions) have been reported.[247]

8 Miscellaneous Items

Some interesting reports must be included here as they do not readily fit into other sections. Mention will be very brief.

Reports have appeared of the crystal and molecular structures of *cis*-dichloro[octa(methylamino)cyclotetraphosphazene-*NN''*]platinum(II)[248] and of di-μ-hydroxo-bis(diammine)platinum(II) nitrate;[249] there has been a kinetic study of the hydrolysis of acetonitrile co-ordinated to platinum, leading to platinblau,[250] and on the interaction of thiamine and its phosphate ester with Pt[II] and Pd[II].[251]

[245] J. M. Lehn, J. Simon, and J. Wagner, *Nouveau J. de Chimie*, 1977, **1**, 77.
[246] A. H. Alberts, R. Annunziata, and J. M. Lehn, *J. Amer. Chem. Soc.*, 1977, **99**, 8502.
[247] J. M. Lehn, *Pure Appl. Chem.*, 1977, **49**, 857.
[248] R. W. Allen, J. P. O'Brien, and H. R. Allcock, *J. Amer. Chem. Soc.*, 1977, **99**, 3987.
[249] R. Faggiani, B. Lippert, C. J. L. Lock, and B. Rosenberg, *J. Amer. Chem. Soc.*, 1977, **99**, 777.
[250] A. K. Johnson and J. D. Miller, *Inorg. Chim. Acta*, 1977, **22**, 219.
[251] N. Hadjiliadis, J. Markouplar, G. Pneumatikakis, D. Katakis, and T. Theophanides, *Inorg. Chim. Acta*, 1977, **25**, 21.

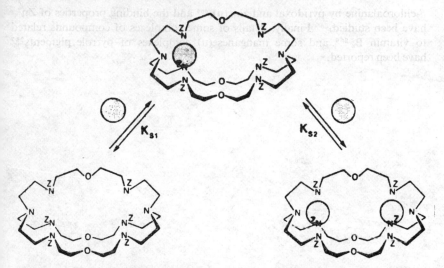

Figure 11 *A representation of the monuclear and dinuclear cryptate complexes formed between metal (shown as speckled circles) and bis-tren-type ligands*
(Reproduced by permission from *J. Amer. Chem. Soc.*, 1977, **99**, 6766)

Relating to the possible mechanism of the anti-tumour activity of *cis*-Pt[II] are the studies of the interactions of platinum complexes, peptides, methionine, and dehydrogenases,[252] and of the interaction of platinum complexes with dehydrogenase enzymes in the presence of different mono- and poly-nucleotides; there is evidence for a ternary complex.[253] Work on selenium in biochemistry continues, including such things as the statistical association of mortality from cancer with dietary selenium intakes[254, 255] and selenium-77 relaxation-time studies that are of relevance to biological systems.[256] Farago and co-workers report on plants which accumulate metals (*e.g.* Ni, Zn).[257–259] Some cobalt(II) complexes containing L-sparteine have been synthesized,[260] as have some mercury– and cadmium–thio-steroid complexes.[261] Interactions of DNA with the copper–thiosemicarbazide system are reported.[262] The cobalt(III)-promoted hydrolysis of phosphate ester,[263] the catalysis of the β-elimination of O-phosphoserine and

[252] P. Melius, C. A. McAuliffe, I. Photaki, and M. Sakerellou-Daitsioutu, *Bioinorg. Chem.*, 1977, **7**, 203.
[253] M. E. Friedman, P. Melins, J. Teggins, and C. A. McAuliffe, *Bioinorg. Chem.*, 1977, **7**, 211.
[254] G. N. Schrauzer, D. A. White, and C. J. Schneider, *Bioinorg. Chem.*, 1977, **7**, 23.
[255] G. N. Schrauzer, D. A. White, and C. J. Schneider, *Bioinorg. Chem.*, 1977, **7**, 35.
[256] W. H. Dawson and J. D. Odom, *J. Amer. Chem. Soc.*, 1977, **99**, 8352.
[257] M. E. Farago, A. J. Clark, and M. J. Pitt, *Inorg. Chim. Acta*, 1977, **24**, 53.
[258] M. E. Farago and M. J. Pitt, *Inorg. Chim. Acta*, 1977, **24**, 127.
[259] M. E. Farago and M. J. Pitt, *Inorg. Chim. Acta*, 1977, **24**, 211.
[260] J. T. Wrobleski and G. J. Long, *Inorg. Chem.*, 1977, **16**, 704.
[261] G. Pouskouleli, P. Kourounakis, and T. Theophanides, *Inorg. Chim. Acta*, 1977, **24**, 45.
[262] C. K. S. Pillai, U. S. Nandi, and W. Levinson, *Bioinorg. Chem.*, 1977, **7**, 151.
[263] B. Anderson, R. M. Milburn, J. M. Marrowfield, G. B. Robertson, and A. M. Sargeson, *J. Amer. Chem. Soc.*, 1977, **99**, 2652.

β-chloroalanine by pyridoxal and zinc(II),[264] and the binding properties of Zn^{2+} have been studied.[265] Finally, details of some complexes of compounds related to vitamin B_6[266] and some manganese(II) complexes of pyrrole pigments[267] have been reported.

[264] K. Tatsumoto and A. E. Martell, *J. Amer. Chem. Soc.*, 1977, **99**, 6082.
[265] D. Demoulin, A. Pullman, and B. Sarkar, *J. Amer. Chem. Soc.*, 1977, **99**, 8498.
[266] J. T. Wrobleski and G. J. Long, *Inorg. Chem.*, 1977, **16**, 2752.
[267] Y. Matsuda and Y. Murakami, *Bull. Chem. Soc. Japan*, 1977, **50**, 2321.

2
Inorganic Analogues of Biological Processes

B. T. GOLDING & G. J. LEIGH

1 Introduction

This chapter is about functional models for enzymes, the catalytic mechanism of which is dependent on the participation of one or more metal ions. The objective of model studies is to simulate in a non-enzymatic system those apparent features of an enzymatic reaction which cannot be studied readily with the enzyme.[1] For example, the first step of the condensation of reaction (1), catalysed by prenyltransferase, is thought to be ionization of substrate (1) to pyrophosphate and a carbocation. Prenyltransferase binds two metal ions per catalytic site, and one of their functions may be to assist the formation of pyrophosphate from (1).

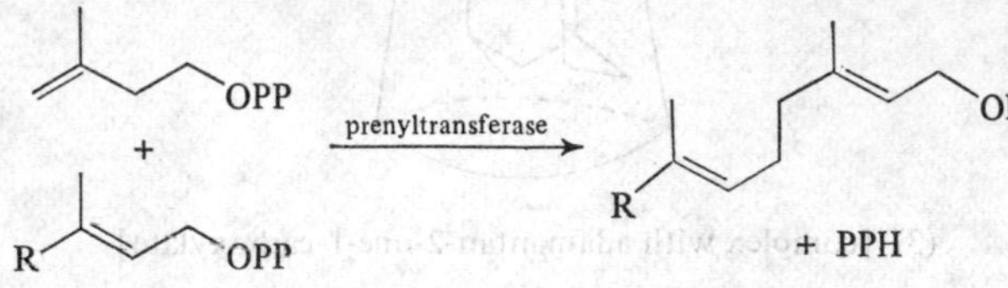

$$\text{(1)}$$

(PP = pyrophosphate, R = e.g. Me)

Brems and Rilling[2] therefore studied the influence of Mg^{II} and Mn^{II} on the hydrolysis of geranyl pyrophosphate (2) [cf. reaction (2)]. Up to 1 mmol l^{-1} of Mg^{II} or Mn^{II} did not significantly affect the hydrolysis of (2), but raising the concentration of metal ion from 1 to 10 mmol l^{-1} caused a gradual increase in

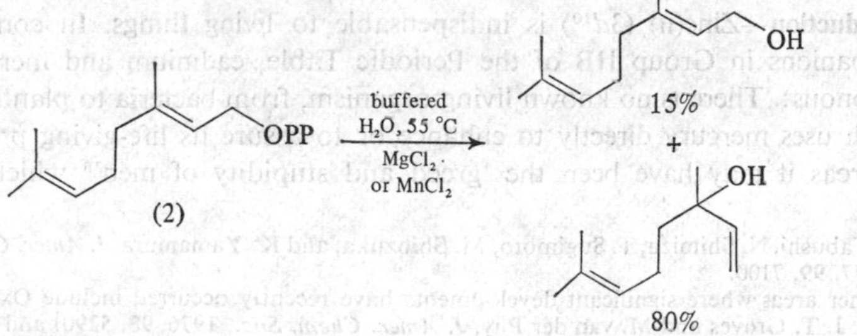

$$\text{(2)}$$

---

[1] For a comprehensive introduction to metal-dependent enzymes and their models, see A. S. Mildvan in 'The Enzymes' ed. P. D. Boyer, Academic Press, New York, 1970, Vol. II, Ch. 9, p. 446.

[2] D. N. Brems and H. C. Rilling, J. Amer. Chem. Soc., 1977, 99, 8351.

the rate of hydrolysis. These results indicate that faster hydrolysis proceeds from species M_2–(2) [M = Mg^{II} or Mn^{II}] and denote a role for the two metal ions at the active site of prenyltransferase.

The structural sophistication of models varies from a pinch of metal salt, which effects the catalysis of a reaction by reversibly binding to and activating a substrate (as in the above example), to elaborate small molecules which hold a metal ion, a substrate group, and an attacking group in proximity (see, *e.g.*, models for carboxypeptidase A described in Section 2). Capped cyclodextrins [*e.g.* (3)] which bind a metal ion at the bottom of their cavity and an organic molecule within the cavity are as yet only impressive structural models for metal-dependent enzymes.[3] If catalytic chemistry could be devised with such model compounds, they might be designated the first metal-dependent 'mini-enzymes'.

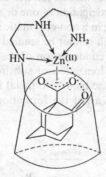

(3) [complex with adamantan-2-one-1-carboxylate]

= β-cyclodextrin

This year we have selected three topics to review in detail from those areas where substantial progress in the development of model systems has recently been made.[4]

2 Models for Zinc-dependent Enzymes

Introduction—Zinc(II) ($3d^{10}$) is indispensable to living things. In contrast, its companions in Group IIB of the Periodic Table, cadmium and mercury, are poisonous: 'There is no known living organism, from bacteria to plants to man, which uses mercury directly to enhance or to assure its life-giving processes'.[5] Whereas it may have been the 'greed and stupidity of men'[6] which created

[3] I. Tabushi, N. Shimizu, T. Sugimoto, M. Shiozuka, and K. Yamamura, *J. Amer. Chem. Soc.*, 1977, **99**, 7100.

[4] Other areas where significant developments have recently occurred include Oxidases (see *e.g.* J. T. Groves and M. van der Puy, *J. Amer. Chem. Soc.*, 1976, **98**, 5290) and Phosphate hydrolysis and phosphoryl transfer (*cf.* C-M. Hsu and B. S. Cooperman, *ibid*, pp. 5652 and 5657 and H. Sigel and P. E. Amsler, *ibid*. p. 7390).

[5] Anon, 'Mercury and the Environment,' OECD, Paris, 1974, p. 15.

[6] D. Hunter, 'The Diseases of Occupations', English Universities Press, London, 1975, p. 320.

methylmercury after unlocking mercury from cinnabar, primitive biological systems inevitably acquired Zn^{II} from their environment.[7] Some well-studied Zn^{II}-dependent enzymes are alcohol dehydrogenases (yeast and horse liver), δ-aminolaevulinate dehydratase (*e.g.* from bovine liver), carbonic anhydrases, certain carboxypeptidases, thermolysin, aldolases (from yeast and *Bacillus stearothermophilus*), alkaline phosphatase, and DNA and RNA polymerases. In those enzymes (*e.g.* bovine carboxypeptidase B) where Zn^{II} has been replaced by Cd^{II} or Hg^{II}, catalytic activity towards natural substrates may be completely lost. We will review recent studies which contribute to understanding the role of Zn^{II} in the mechanisms of action of the above enzymes, excluding aldolases, alkaline phosphatase, and the DNA and RNA polymerases. For recent investigations of these enzymes, *cf.* refs. 8 (aldolases), 9 (alkaline phosphatase), and 10 (polymerases).

The function of Zn^{II} has been elegantly discussed by Dunn[11] and amplified by Cheh and Neilands.[12] In common with other metal ions, Zn^{II} [normally 4-, 5-, or 6-co-ordinate] may act as a meeting place for reactants (template effect). In addition, the properties of groups ligated to Zn^{II} may be appreciably changed; *e.g.*, water, alcohols, aldehydes, and ketones are more acidic owing to the stabilization of OH^-, RO^-, or enolate ion; carbonyl groups are more susceptible to nucleophilic attack owing to enhanced polarization [$C\overset{\delta-}{=}O\cdots Zn^{II}$]. Because other oxidation states of Zn in aqueous media are unknown, it is unlikely that Zn^{II} can engage in electron-transfer processes requiring the formation of Zn^I or Zn^{III}. Evidence[13] that Zn^{II} forms π-complexes with arenes suggests that such interactions are conceivable between Zn^{II} and an aromatic ring of a substrate or a structural component (*e.g.* phenylalanine) of an enzyme.

Alcohol Dehydrogenases.—Alcohol dehydrogenases (ADH's) from liver [horse (HLADH), human, rat], yeast (YADH), and *Bacillus stearothermophilus* have been intensively studied.[14] These enzymes catalyse the overall reaction:

$$R^1R^2CHOH + \underset{NAD^+}{\text{[structure]}} \longrightarrow R^1R^2C{=}O + \underset{NADH}{\text{[structure]}} + H^+ \quad (3)$$

[7] J. J. R. Frausto de Silva and R. J. P. Williams, *Structure and Bonding*, 1976, **29**, 67.
[8] H. A. O. Hill, R. R. Lobb, S. L. Sharp, A. M. Stokes, J. I. Harris, and R. S. Jack, *Biochem. J.*, 1976, **153**, 551.
[9] W. E. Hill and B. D. Sykes, *Biochemistry*, 1976, **15**, 1535; R. A. Anderson and B. L. Vallee, *ibid.*, 1977, **16**, 4388.
[10] H. Lattke and U. Weser, *F.E.B.S. Letters*, 1977, **83**, 297; D. C. Speckhard, F. Y.-H. Wu, and C.-W. Wu, *Biochemistry*, 1977, **16**, 5228.
[11] M. F. Dunn, *Structure and Bonding*, 1975, **23**, 61.
[12] A. M. Cheh and J. B. Neilands, *Structure and Bonding*, 1976, **29**, 123.
[13] L. C. Damude and P. A. W. Dean, *J. Organometallic Chem.*, 1977, **125**, C1.
[14] C-I. Bränden, H. Jornvall, H. Eklund, and B. Furugren in 'The Enzymes' ed. P. D. Boyer, Academic Press, New York, 1975, Vol. XI, Ch. 3, p. 104.

Apart from their metabolic importance, alcohol dehydrogenases are valuable as analytical aids (*e.g.* in the determination of steroids) and as catalysts for stereo-selective reductions of ketones (see below). Sequencing and crystal structure analysis (2.4 Å resolution) of HLADH show that its molecules (mol. wt. 80 000) consist of two identical subunits, each of which contains two Zn^{II} within a chain of 374 amino-acid residues.[15] The subunits are composed of two domains, one of which (residues 176—318) binds the coenzyme (NAD^+ or NADH) and has been conserved during evolution. Similar coenzyme-binding domains are found in lactate dehydrogenase and glyceraldehyde 3-phosphate dehydrogenase, and even in functionally quite different enzymes such as the ferredoxins. The other domain (residues 1—175, 319—374) binds the substrate and is structurally unrelated to the catalytic domains of other types of enzyme. Each catalytic domain contains the two Zn^{II}, but only one of these is directly engaged in reactions catalysed by HLADH. This Zn is tetrahedrally co-ordinated [Cys-46, Cys-174, His-67, and H_2O (OH at higher pH)] and is deeply embedded (~ 20 Å from the surface of the protein). However, it can be approached through two channels, one of which binds and is blocked by the coenzyme (NAD^+ or NADH). The pyridinium (dihydropyridine) ring of the coenzyme is close to the catalytic Zn. The other channel is lined with hydrophobic groups and is very probably the substrate-binding pocket. The water ligand to the catalytic Zn is hydrogen-bonded to Ser-48, which is H-bonded to His-51. These interactions may provide a means of proton release following binding of NAD^+ (or proton gain following binding of NADH) or may help to activate the substrate (see below).

The second Zn has a zinc-blende-like structure, being co-ordinated to Cys-97, -100, -103, and -111 (distorted tetrahedron) within a lobe near the surface of the protein (residues 95—113). Its function is unknown. This region of the enzyme has been called[15] a molecular fossil because it looks like a catalytic centre reminiscent of the Fe_4S_4 proteins.

The various alcohol dehydrogenases have distinctive substrate specificities. LADH is catholic compared with YADH, which will not oxidize cyclohexanol or secondary alcohols R^1CHOHR^2 when R^1 and R^2 are larger than methyl. These differences are due to variations in the groups lining the substrate-binding pocket, *e.g.* YADH has tryptophans in place of Leu-57 and Phe-93 in LADH and also has two extra prolines flanking Leu-58. The substrate-binding pocket of LADH can be represented by a model hewn from a diamond lattice. This enables its substrate specificity and the chirality of alcoholic products to be predicted.[16] LADH is a useful enzyme for the synthesis of chiral organic molecules, *e.g.* optically pure (*S*)-1-deuterio- or 1-tritio-alkan-1-ols.[17]

According to ref. 14, the sequence of events during conversion of an alcohol into a carbonyl compound by ADH's is as follows: the enzyme binds NAD^+, which induces a conformational change and causes one proton to be released

[15] H. Eklund, N. Nordström, E. Seppezauer, G. Söderlund, I. Ohlsson, T. Boiwe, B.-O. Söderberg, O. Tapia, C.-I. Brändén, and A. Akeson, *J. Mol. Biol.*, 1976, **102**, 27; T. Boiwe and C.-I. Brändén, *European J. Biochem.*, 1977, **77**, 173 and references cited therein.
[16] J. B. Jones and H. B. Goodbrand, *Canad. J. Chem.*, 1977, **55**, 2685.
[17] *cf.* B. T. Golding, P. V. Ioannou, and I. Eckhard, *J.C.S. Perkin I*, 1978, 774.

(Zn—OH$_2$→Zn—OH); alcohol binds near the catalytic Zn and is converted into an alkoxide ion. Loss of H$_s$ from C-1 of the alkoxide to the *re* face (C-4) of the pyridinium ring of NAD$^+$ produces aldehyde (or ketone), which dissociates from the enzyme, and NADH (for stereochemical details, see ref. 18). After a conformational change, NADH is released. Klinman[19] made a steady-state kinetic study of interconversions between benzyl alcohol and benzaldehyde, and *para*-substituted benzyl alcohols and the corresponding aldehydes, catalysed by YADH + NADH/NAD$^+$. Substantial isotope effects were associated with k_{cat} for oxidation of alcohol to aldehyde (*e.g.* $k_H/k_D = 4.2$ for *p*-chlorobenzyl alcohol), indicative of a rate-limiting transfer of hydrogen (hydride ion) from alcohol to NAD$^+$. It was found that k_{cat} for oxidation of the alcohols was independent of the *para*-substituent, indicating that there is little or no alteration of charge at C-1 of alcohol substrate between its ground state and the transition state for oxidation. Observations of the pH-dependence of k_{cat} identified a single group (pK_a 8.25; Zn-bound water or amino-acid side-chain) at the active site, which functions as an acid-base catalyst in the hydrogen-transfer step.

In discussions of the role of the Zn^{II} in the above sequence, it is customary to consider the reaction in the direction aldehyde (or ketone)→alcohol, and nearly all the model systems to be described effect a conversion of a C=O group into CHOH. Three modes of binding and activation of substrate by Zn have been proposed:

(i) Replacement of Zn-bound water by a substrate molecule occurs, resulting in enhanced polarization of substrate carbonyl by Zn:

$$H_2O \overset{\displaystyle |}{\underset{\displaystyle \diagdown}{Zn}} \cdots \quad \overset{\delta+}{\diagup}\!C\!\!=\!\!O^{\delta-} \longrightarrow \overset{\Delta+}{\diagup}\!C\!\!=\!\!O^{\Delta-} \cdots \overset{\displaystyle |}{\underset{\displaystyle \diagdown}{Zn}} \quad (4)$$

NADH

(ii)[20] Zn-bound water activates a substrate molecule by hydrogen bonding:

$$H_2O \overset{\displaystyle |}{\underset{\displaystyle \diagdown}{Zn}} \cdots \quad \overset{\delta+}{\diagup}\!C\!\!=\!\!O^{\delta-} \longrightarrow \overset{\Delta+}{\diagup}\!C\!\!=\!\!O^{\Delta-} \cdots H \underset{H}{\overset{}{\diagdown}} O \overset{\displaystyle |}{\underset{\displaystyle \diagdown}{Zn}} \quad (5)$$

NADH

In this case, Zn has an indirect influence on substrate carbonyl.

(iii)[21] There is an intermediate five-co-ordinate complex of Zn containing a substrate molecule and water; substrate is activated through co-ordination to

[18] W. L. Alworth, 'Stereochemistry and its Application in Biochemistry', Wiley-Interscience, New York, 1972, pp. 240–244.

[19] J. P. Klinman, *J. Biol. Chem.*, 1975, **250**, 2569; *Biochemistry*, 1976, **15**, 2018.

[20] D. J. Sloan, J. M. Young, and A. S. Mildvan, *Biochemistry*, 1975, **14**, 1998.

[21] R. T. Dworschack, and B. V. Plapp, *Biochemistry*, 1977, **16**, 2716.

Zn and by proton transfer from the adjacent water, which is coupled to Ser-48 and His-51:

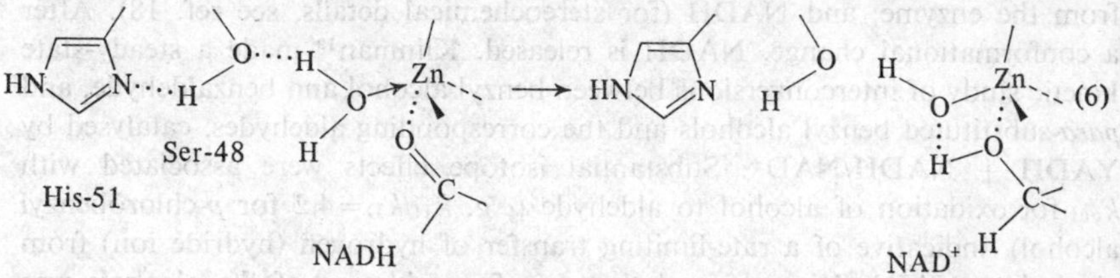

(6)

In each mechanism, Zn functions essentially as a Lewis acid, enhancing the susceptibility of carbonyl to either nucleophilic attack by hydride ion or electron transfer from NADH.

Mechanism (i) was recently supported[22] by experiments with p-dimethylamino-cinnamaldehyde (λ_{max} = 398 nm). On incubating this aldehyde with LADH + NADH, an intermediate (λ_{max} = 464 nm) is formed, which decays to expected products. It was suggested to have structure (4), a proposal supported by model studies in which the effect of Zn^{II} and other Lewis acids on the visible spectrum of p-dimethylaminocinnamaldehyde was monitored.

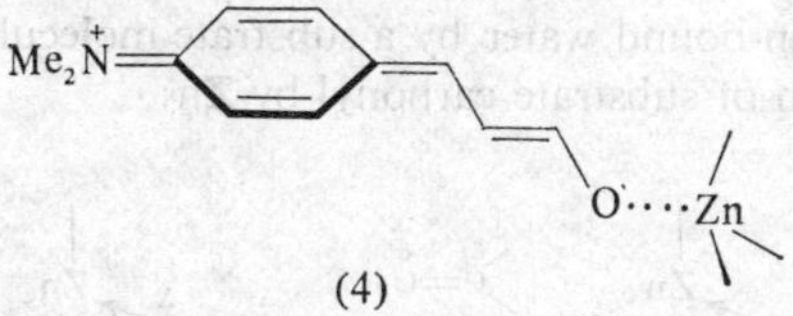

(4)

Mechanism (ii) is supported by n.m.r. studies of the relaxation of protons in, *e.g.*, ethanol in an abortive ternary complex of ethanol and NADH with Co^{II}-enzyme (LADH), from which Mildvan and co-workers[20] calculated Co^{II}-ethanol distances of 6.2 ± 1.3 Å (Me protons) and 6.3 ± 1.3 Å (CH_2 protons). They suggested that a water molecule interposed between Zn^{II} and a substrate molecule, occupying the same site as ethanol in the Co^{II}-enzyme, could act as a general acid catalyst for substrate reduction. The group of pK_a 8.25 that Klinman[19] identified in her studies of YADH (see above) could be the water molecule of this proposal.

Mechanism (iii) was deduced[21] from studies of hydroxybutyrimidylated HLADH, on which steady-state kinetic parameters can be measured, complementary to Klinman's studies[19] of YADH. Precedent for five-co-ordinate Zn in ADH's exists in the o-phenanthroline complex of ADH.[15] From studies of isotope effects (*e.g.* oxidation of PhC^2H_2OH) and considering ρ^+ values for oxidations of *para*-substituted benzyl alcohols ($\rho^+ = -0.2$) and reductions of *para*-substituted benzaldehydes ($\rho^+ = 1.1$), it was concluded that there is a concerted transfer of hydride ion from NADH to substrate carbonyl with little charge development at the transition state. Oxidation of benzyl alcohol was shown

[22] C. T. Angelis, M. F. Dunn, D. C. Muchmore, and R. M. Wing, *Biochemistry*, 1977, **16**, 2922.

to be mediated by a group of pK_a 8.4 (assigned to Zn-bound water), which should be unprotonated for maximal activity in the direction alcohol → aldehyde.

To study the influence of Zn^{II} on the reactivity of C=O and CHOH groups towards NADH/NAD$^+$, it has proved valuable to examine the effect of metal ions on the rate of reduction of carbonyl groups by model compounds for NADH. These models are 1,4-dihydronicotinamides such as 'Hantzsch ester' (5) and 1-benzyl-1,4-dihydronicotinamide (6a), which are cheap and easy to prepare [*e.g.*, for (6a), by reducing the corresponding pyridinium compound (7a) with

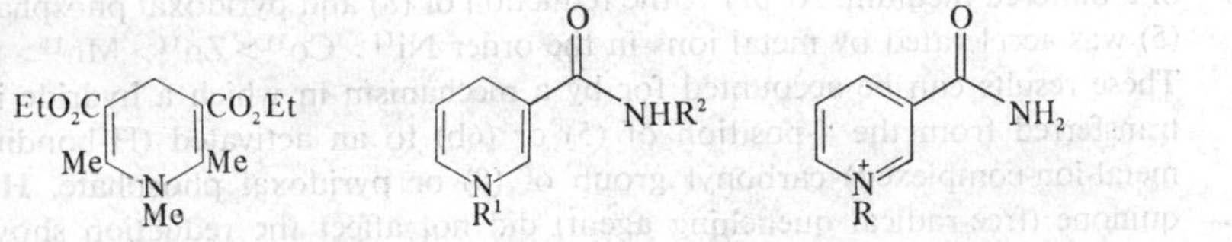

(5)

(6a) R^1 = CH$_2$Ph, R^2 = H (7a) R = CH$_2$Ph
(6b) R^1 = Pr^n, R^2 = H   (7b) R = Pr^n

(6c) R^1 = Pr^n, R^2 = —C—Me (with Ph and H attached)

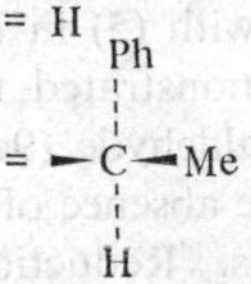

dithionite], and are soluble in a wider range of solvents than NADH. 1,4-Dihydronicotinamides are thermodynamically unstable with respect to their 1,2-isomers, into which they are converted under the catalytic influence of the corresponding pyridinium salt, as shown in reaction (7). However, 1,2-dihydronicotinamides are feeble reducing agents compared with the 1,4-isomers.

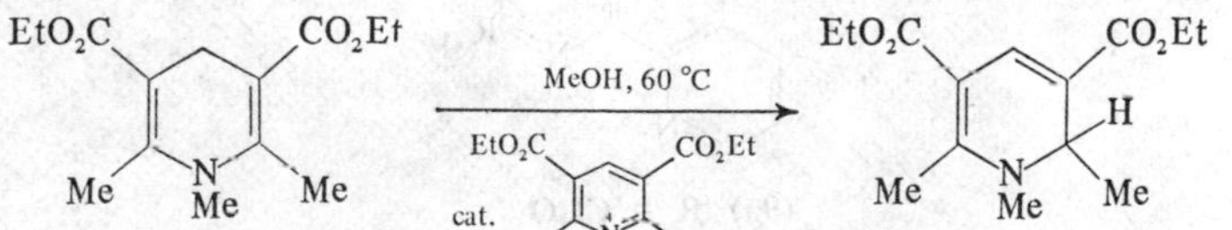

(7)

In 1973, T. C. Bruice wrote 'searches have been made, to no avail, for aldehydes which are reducible in water at ambient temperatures by dihydronicotinamides'.[23, 24] It had already been shown that hydrogen at C-4 of dihydronicotinamides can be transferred to the electrophilic carbon atom of thioketones, malachite green [(p-NMe$_2$C$_6$H$_4$)$_3$C̈], and activated ketones (*e.g.* PhCOCF$_3$). Shinkai and Bruice[23] have described reductions of 4-formylpyridines by dihydronicotinamides which take place under mild conditions and which can be accelerated by metal ions. 4-Formylpyridine was not reduced under these conditions (see Scheme 1), showing that the hydroxy-group of (8) activates the carbonyl group by intramolecular hydrogen-bonding (*i.e.* general acid catalysis). [*N.B.* reduction of PhCOCF$_3$ by (6b) goes faster in protic solvents[25]]. As the reaction of Scheme 1

[23] S. Shinkai and T. C. Bruice, *Biochemistry*, 1973, **12**, 1750.
[24] U. K. Pandit and F. R. Mas Cabre, *J.C.S. Chem. Comm.*, 1971, 552.
[25] P. Van Eikeren and D. L. Grier, *J. Amer. Chem. Soc.*, 1976, **98**, 4655.

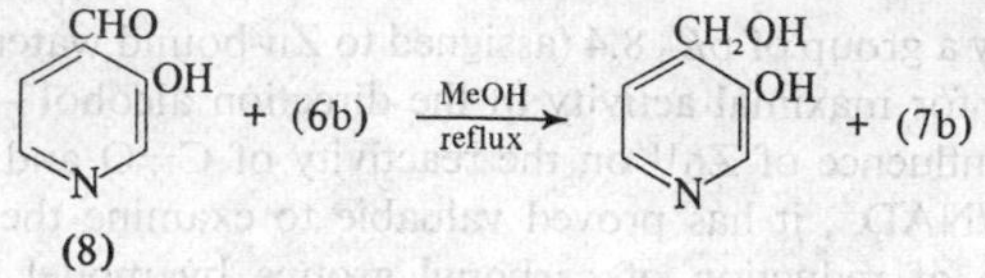

Scheme 1

progresses, methoxide ion is produced, and deprotonates this hydroxy-group, giving an unreactive monoanion of (8). This problem can be avoided by the use of a buffered medium. At pH 7, the reduction of (8) and pyridoxal phosphate by (5) was accelerated by metal ions in the order $Ni^{II} > Co^{II} > Zn^{II} > Mn^{II} > Mg^{II}$. These results can be accounted for by a mechanism in which a hydride ion is transferred from the 4-position of (5) or (6b) to an activated (H-bonding or metal-ion-complexed) carbonyl group of (8) or pyridoxal phosphate. Hydroquinone (free-radical quenching agent) did not affect the reduction shown in reaction (6), and no deuterium was incorporated into products from the reaction of pyridoxal phosphate with (5) in MeO^2H.

Creighton *et al.*[26] demonstrated that Zn^{II} catalyses the reduction of 1,10-phenanthroline-2-carboxaldehyde (9a) to the corresponding carbinol (9b) by (6b) in acetonitrile. In the absence of metal ions, no reduction of (9a) to (9b) was detected during 4 days. Reduction of (9a) by $[4-^2H_2]$-(6b) was slower ($k_H/k_D = 1.74 \pm 0.6$) and gave (9c). These reactions were insensitive to hydroquinone. Clearly, Zn^{II} co-ordinates the nitrogen atoms of (9a) and exerts a polarizing influence on the nearby carbonyl group, activating it to nucleophilic attack.

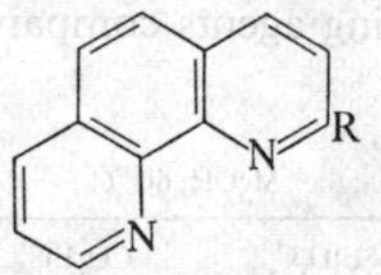

(9a) R = CHO
(9b) R = CH_2OH
(9c) R = CH^2HOH

Shirai *et al.*[27] studied reactions of 2-, 3-, and 4-formylpyridine with (6a) and concluded that: (*a*) 2-formylpyridine reacts with (6a) in the presence of methanolic $Zn(OAc)_2$ (2 days, 40 °C) to give 2-hydroxymethylpyridine; (*b*) 3-formylpyridine does not react under these conditions, 4-formylpyridine reacts slowly; (*c*) other bivalent metal ions catalyse the reduction of 2-formylpyridine [Cu > Zn > Pb > Cd]; (*d*) rate constants for the reduction of 2-formylpyridine in the presence of ZnX_2 depended on X ($NO_3^- > Br^- > Cl^- > AcO^-$). In this report the yield of 2-hydroxymethylpyridine was not given, and it was identified only by the melting point of a picrate and a retention time on g.l.c. The use of methanolic solvent with a metal salt may diminish the concentration of (6a) [1,4-dihydronicotinamides undergo

[26] D. J. Creighton, J. Hajdu, and D. S. Sigman, *J. Amer. Chem. Soc.*, 1976, **98**, 4619.
[27] M. Shirai, T. Chishina, and M. Tanaka, *Bull. Soc. Chem. Japan*, 1975, **48**, 1079

acid-catalysed additions to their 5,6-double bond[28]] and so the rate constants determined from the fall of absorbance of (6a) at 355 nm might be inaccurate. Also, acetal formation from aldehydes is likely to be catalysed by metal salts in methanol. As these possibilities were not considered in ref. 27, at least the quantitative aspects of this report have to be viewed with caution.

Pandit and co-workers[29] showed that 2-benzoylpyridine (but not the 3- and 4-isomers) was reduced to 2-(1'-phenyl)hydroxymethylpyridine ($>70\%$) on treatment with (6a) in acetonitrile containing Mg^{II} or Zn^{II}. N.m.r. studies indicated that Mg^{II} could bind to both 2-benzoylpyridine and (6a), but more tightly to the latter. The results of kinetic measurements were said to favour a mechanism whereby (6a) reduced 2-benzoylpyridine bound to Zn^{II} or Mg^{II}. The rate of reduction increased on addition of Mg^{II} until $[Mg^{II}]/[(6a)] = 0.4$. Alcohol complexes of metal salts were recommended as catalysts on the grounds that the 'use of metal halides or perchlorates suffers from serious disadvantages, ... the former are incompletely dissociated, while the latter have been observed to give insoluble complexes which complicate spectroscopic or kinetic studies'. However, alcohol in the reaction mixture provides a further potential reactant (see above).

Hughes and Prince[30] studied the reaction between 2-formylpyridine and (6a) in the presence of metal salts (MX_2) in acetonitrile and found that the rate of reduction passed through a maximum as the salt concentration was increased. There was a dependence on the square root of the concentration of the metal ion, indicating the importance of MX^+ species in catalysis. It was stated that the kinetic data could only be interpreted in terms of a reactive ternary complex composed of Zn^{II}, substrate, and (6a). ^1H n.m.r. spectra showed that the interaction between Zn^{II} and (6a) occurred *via* C=O, whereas only the nitrogen of 2-formylpyridine is co-ordinated to Zn^{II}. Although 2-formylpyridine prefers the *trans*-conformation in solution, provided that the *cis*-conformer is accessible *via* a low energy barrier, Zn^{II} already co-ordinated to N could activate the formyl group to nucleophilic attack by direct co-ordination to O. A significant point made in ref. 30 was that 3- and 4-formylpyridine are not reduced by (6a)/Zn^{II} because in the ternary complexes (6a) is incorrectly orientated to deliver H (from C-4) to the formyl group.

Ohno and co-workers have reported numerous reductions of 1,2-diketones, α-keto-esters, and other carbonyl compounds (*e.g.* refs. 31—33) by (6a) and chiral 1,4-dihydronicotinamides. Benzil did not react with (6a) in acetonitrile during 22 h, at 50 °C, in darkness, but it gave 50% benzoin when equimolar $Mg(ClO_4)_2$ was added.[31] This reduction was not catalysed by $LiClO_4$ and was inhibited by water (10% water in acetonitrile reduced the yield of benzoin from 47 to 23%). It was unaffected under any conditions by hydroquinone (0.82 mol %) but was catalysed by light ($\lambda \geqslant 350$ nm). α-Keto-esters ($PhCOCO_2Et$ and $MeCOCO_2Et$) were quantitatively reduced to the corresponding α-hydroxy-esters by (6a) in

28 *cf.* S. L. Johnson and P. T. Tuazon, *Biochemistry*, 1977, **16**, 1175.
29 R. A. Gase, G. Boxhoom, and U. K. Pandit, *Tetrahedron Letters*, 1976, 2889.
30 M. Hughes and R. H. Prince, *Chem. and Ind.*, 1975, 648.
31 Y. Ohnishi, M. Kagami, and A. Ohno, *Tetrahedron Letters*, 1975, 2437.
32 Y. Ohnishi, M. Kagami, and A. Ohno, *J. Amer. Chem. Soc.*, 1975, **97**, 4766.
33 Y. Ohnishi, T. Numakunau, and A. Ohno, *Tetrahedron Letters*, 1976, 3813.

acetonitrile (17 h, at r.t.) containing either $Mg(ClO_4)_2$ or $Zn(ClO_4)_2$ (but not $LiClO_4$). With the chiral 1,4-dihydronicotinamide (6c) and Mg^{II}, ethyl benzoyl-formate was reduced to ethyl mandelate containing 60% of the (R)-isomer.[32] Reduction of $PhCOCF_3$ with (6c) in acetonitrile gave (after 6 days, at r.t.) 30% of racemic 1-phenyl-2,2,2-trifluoroethanol.[33] In the presence of Mg^{II}, a higher yield of the alcohol was obtained, and it contained an excess of the (S)-isomer (optical purity 16%). It was suggested[33] that the reduction in the presence of Mg^{II} takes place *via* a ternary complex, leading to a tighter transition state and to greater induction of optical activity.

Some reductions by 1,4-dihydronicotinamides [*e.g.* hexachloroacetone + (6a) → 1,1,1,3,3,3-hexachloropropan-2-ol + (7a)] are retarded by Zn^{II}.[34] It is believed that Zn^{II} and dihydronicotinamide form a non-productive complex, which reduces the concentration of free, reactive dihydronicotinamide.

For the reductions of carbonyl compounds discussed so far, a mechanism can be written in which a hydride ion is transferred from C-4 of the dihydro-nicotinamide to the carbon atom of a carbonyl group activated by co-ordination to a metal ion [*cf.* reaction (4)]. However, evidence has been presented that the reaction pathway for some examples is not a single bimolecular step, but rather a stepwise electron transfer, proton transfer, and second electron transfer (or electron transfer followed by hydrogen-atom transfer). Evidence for an electron-transfer pathway was assembled by Creighton *et al.*,[26] following an earlier important contribution of Steffens and Chipman.[35] A discrepancy was noted between the isotope effect for transfer of H from C-4 of (6b) to (9a) that was calculated from kinetic data and that calculated from product analysis; it was accounted for by postulating non-covalent intermediates, as in the following equation[26,35] $(k_{-1} \approx k_2)$:

$$RH_2 \ + \ A \ \underset{}{\overset{rapid}{\rightleftharpoons}} \ Y_1 \ \underset{k_{-1}}{\overset{k_1}{\rightleftharpoons}} \ Y_2 \ \overset{k_2}{\longrightarrow} \ RH + AH \quad (8)$$

$$\underset{\substack{\text{Donor} \\ e.g. \text{ (6a)}}}{} \quad \underset{\text{Acceptor}}{} \qquad \underset{\substack{\text{Charge-transfer} \\ \text{complex}}}{}$$

In the reduction of the N-methylacridinium ion (10) by (6a) in methanol, an intense dark colour immediately appeared and then disappeared at a rate con-sistent with that of the overall reaction.[36] The colour was ascribed to a charge-transfer complex [Y_1 in equation (8)] between (6a) and (10). Charge-transfer complexes are usually formed at diffusion-controlled rates, and so it was neces-sary to propose an additional intermediate on the reaction pathway [Y_2 in equation (8)] in order to explain the non-equality of kinetic isotope effect and isotopic partitioning ratio in products. As to the nature of Y_2, Hajdu and Sigman[36] imply that it is a radical pair which collapses to products.

[34] D. C. Dittmer, A. Lombardo, F. H. Batzold, and C. S. Greene, *J. Org. Chem.*, 1971, **93**, 6694.

[35] J. J. Steffens and D. S. Sigman, *J. Amer. Chem. Soc.*, 1971, **93**, 6694.

[36] J. Hajdu and D. S. Sigman, *J. Amer. Chem. Soc.*, 1976, **98**, 6060.

(10)

Reactions of 1,4-dihydronicotinamides in which radicals participate occur with ferricyanide,[37] sulphite/O_2,[38] and geminal bromo-nitro-alkanes.[39] With sulphite/ O_2, NADH and its model compounds [*e.g.* (6b)] are oxidized to pyridinium salts (*e.g.* NADH → NAD$^+$) by pathways that are inhibited by hydroquinone. The reaction between (6b) and $Fe(CN)_6^{3-}$ has the following stoicheiometry:

$$(6b) + 2 Fe(CN)_6^{3-} \longrightarrow (7b) + 2 Fe(CN)_6^{4-} + H^+ \qquad (9)$$

Kinetic measurements showed a second-order rate law, with rate $= k[(6b)][Fe(CN)_6^{3-}]$, $\Delta H^{\neq} = 2$ kcal mol^{-1}, and $k_H/k_D = 1.68$ {measured with [4-$^2H_2$]-(6b)}. These data led Ohno and co-workers[37] to suggest rate-determining electron transfer from a C-4—H bond, through the π-system of (6b), to $Fe(CN)_6^{3-}$.

Kill and Widdowson[39] found that geminal bromo-nitro-alkanes are reduced to nitro-alkanes by (6a). The reduction of a bromo-nitro-bicycloalkane is shown in reaction (10). Geminal bromo-nitro-alkanes do not undergo ready S_N1 or S_N2 displacement of bromide ion, but do react readily with, *e.g.*, nitronate anions to yield vicinal dinitro-alkanes *via* an electron-transfer mechanism. The reduction of bromo-nitro-alkanes by (6a) therefore involves an initial rate-determining electron transfer, generating the (6a)$^{+\cdot}$ radical cation, Br$^-$, and a nitro-radical [*e.g.* $Me_2\dot{C}NO_2$], which was detected by e.s.r. spectroscopy.

Signals attributed to radical ions have been observed in e.s.r. spectra of reaction mixtures containing (6a) and other reducible partners. In a further study[40] of the Zn^{II}-catalysed reduction of 2-formylpyridine by (6a), it was found that addition of 2-methyl-2-nitrosopropane did not spin-trap any radicals. Exposure of a reaction mixture containing (6a), 2-formylpyridine, and Zn^{II} to light showed up a triplet signal by e.s.r. spectroscopy, but the same signal was observed when the aldehyde was replaced by pyridine, Ph$_3$P, or other additives. Therefore, the radical detected in this system does not lie on the reaction profile for reduction of 2-formylpyridine by (6a).

[37] T. Okamoto, A. Ohno, and S. Oka, *J.C.S. Chem. Comm.*, 1977, 181.
[38] P. T. Tuazon and S. L. Johnson, *Biochemistry*, 1977, **16**, 1183.
[39] R. J. Kill and D. A. Widdowson, *J.C.S. Chem. Comm.*, 1976, 755.
[40] R. A. Hood, R. H. Prince, and K. A. Robinson, *J.C.S. Chem. Comm.*, 1978, 300.

Ohno *et al.*[41] found that the typical absorption at λ_{max} 351 nm in the u.v.—visible spectrum of (6c) shifts to longer wavelength in the presence of Mg^{II} {with [Mg(ClO$_4$)$_2$]/[(6c)] = 8, in acetonitrile, the absorption shifts to 357 nm}. From such tenuous evidence it was proposed that Mg^{II} co-ordinates to the dihydropyridine ring of (6c) rather than to its amide carbonyl group. Kinetic studies of reactions in MeCN were said to reveal that a 1 : 1 complex between (6c) and Mg^{II} is formed which reacts with methyl benzoylformate in the rate-determining step. It was envisaged that Mg^{II} promotes the transfer of an electron from (6c) to methyl benzoylformate. This process is followed by a proton transfer from C-4 of (6c) to the ketonic carbonyl group of methyl benzoylformate. The reaction is completed by another electron transfer. Alternatively, following the initial electron transfer, a hydrogen-atom transfer from (6c)$^{+\cdot}$ to [methyl benzoylformate]$^{-\cdot}$ would yield products directly.

N-Methylacridan (11) is oxidized by powerful π-acceptors to (10). For tetracyanoethylene and four quinones studied,[42] the kinetic isotope effect was equal

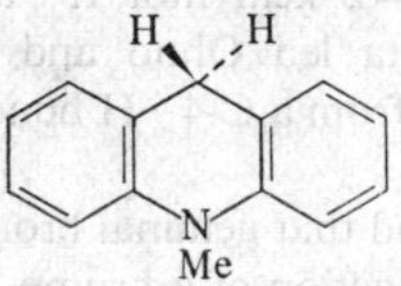

(11)

(within experimental error) to the isotopic partitioning ratio in products. These results demand that breakage of the C—H bond is both product- and rate-determining. Two mechanisms were considered: (i) rapid formation of a charge-transfer complex, followed by slow transfer of hydride ion: (ii) formation of a charge-transfer complex followed by rapid electron transfer, then slow H-atom transfer [*cf.* equation (8) with $k_{-1} \gg k_2$]. Trends in isotope and solvent effects were believed[42] to favour mechanism (i), although it was pointed out that the low ionization potential of *N*-methylacridan and the small gain in aromatic resonance energy on converting *N*-methylacridan into (10) should have favoured mechanism (ii).

Shirra and Suckling[43] demonstrated that lithium alkoxides react with alkylpyridinium salts to give an equilibrium mixture containing the corresponding aldehyde and dihydropyridine [reaction (11)]. Kinetic studies revealed a reaction

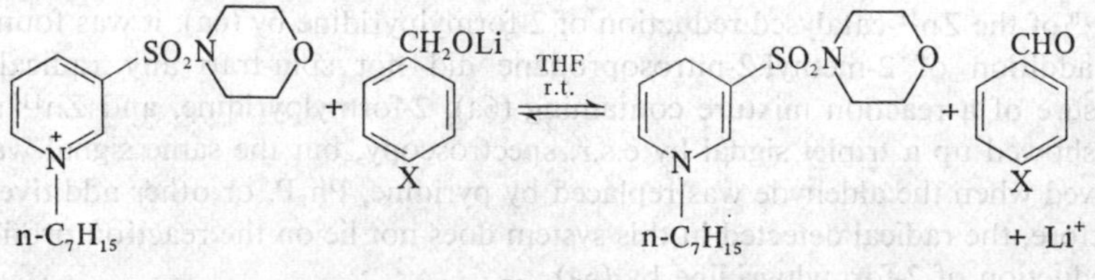

(*e.g.* X = NMe$_2$)

[41] A. Ohno, T. Kimura, H. Yamamoto, S. G. Kim, S. Oka, and Y. Ohnishi, *Bull. Soc. Chem. Japan*, 1977, **50**, 1535; A. Ohno, H. Yamamoto, T. Okamoto, S. Oka, and Y. Ohnishi, *ibid.*, p. 2385.
[42] A. K. Colter, G. Saito, and F. J. Sharom, *Canad. J. Chem.*, 1977, **55**, 2741
[43] A. Shirra and C. J. Suckling, *J.C.S. Perkin I*, 1977, 759

constant of $\rho = 0.23$, indicative of a transition state with little charge development. One of the objectives of this work was to test a mechanism for NAD$^+$ oxidations, whereby a covalent adduct from NAD$^+$ and alcohol rearranges to NADH and aldehyde (ketone) by an intramolecular H-shift, as is shown in Scheme 2. The model compound (12) was synthesized, but was found to be stable at 100 °C in the presence of trapping agents for the aldehyde group in the potential product (13).

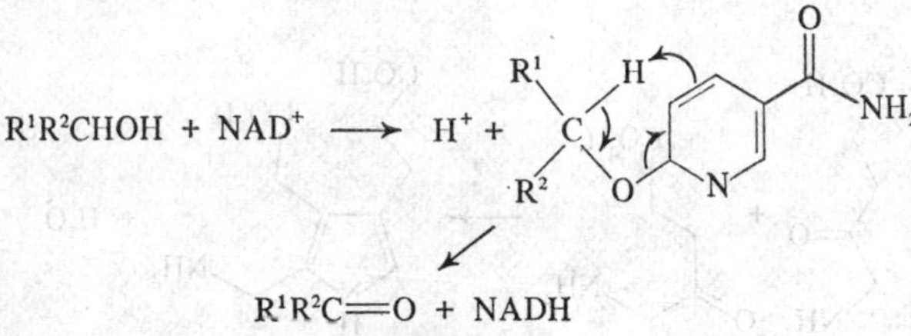

$$R^1R^2CHOH + NAD^+ \longrightarrow H^+ +$$

$$R^1R^2C{=}O + NADH$$

Scheme 2

Model studies that have been described show that reaction (4) is a chemically reasonable description of the role of Zn in ADH's. However, other than the observations that hydrogen bonding (intramolecular with OH, or intermolecular with protic solvent) will activate a carbonyl group to reduction, there are no

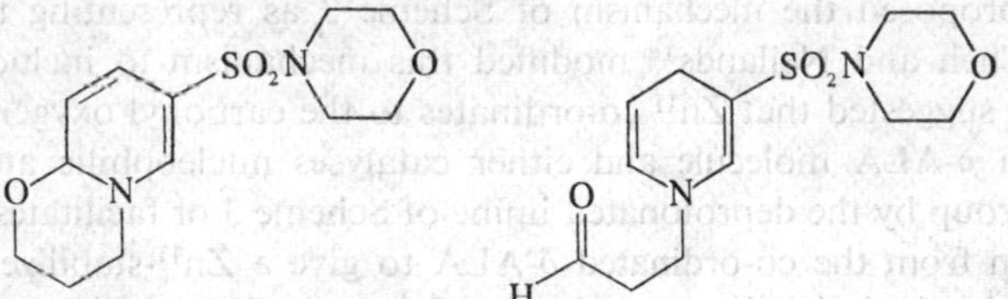

(12)   (13)

model studies which allow reactions (5) and (6) to be evaluated with respect to reaction (4). A further uncertainty, unresolved by model studies, is the nature of the hydrogen transfer in the enzymatic reactions. Undoubtedly, radical ions feature in some of the model reactions. Nevertheless, Bruice[44] concluded that the similarity of the reactivities of Zn^{II} and Co^{II} ADH's argues against the participation of radical species in the enzymatic reactions, and Klinman[19] proposed a direct hydride-ion transfer *via* transition state (14), noting that 'there is no evidence to date implicating kinetic intermediates in the reactions catalysed by dehydrogenases'. However, she was unable to exclude the alternative transition state (15), in which a hydrogen atom is transferred to a protonated radical intermediate.

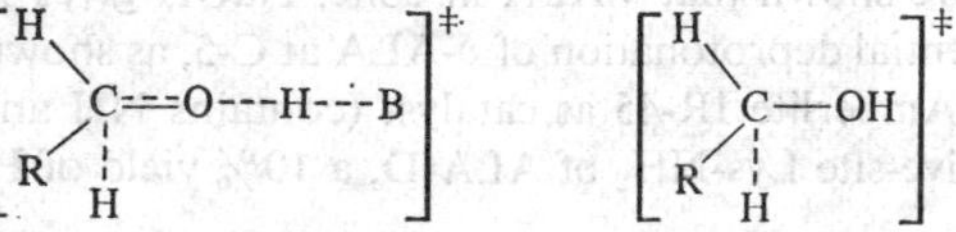

(14)   (15)

[44] T. C. Bruice, *Ann. Rev. Biochem.*, 1976, **45**, 331.

δ-Aminolaevulinate Dehydratases—These ubiquitous enzymes catalyse an early step in the biosynthesis of chlorophylls, corrinoids, and porphyrins, *i.e.* the production of one molecule of porphobilinogen (16) from two molecules of δ-aminolaevulinic acid (δ-ALA). Cheh and Neilands[12] have divided ALA-D's into two types, analogous to the subdivision of aldolases. Only eukaryotic ALA-D's may have a dependence on Zn^{II}. The enzyme from beef liver is probably an octamer, with a binding site for one Zn^{II} on each subunit of mol. wt. *ca.* 35 000. In common with other ALA-D's, the beef liver enzyme is irreversibly

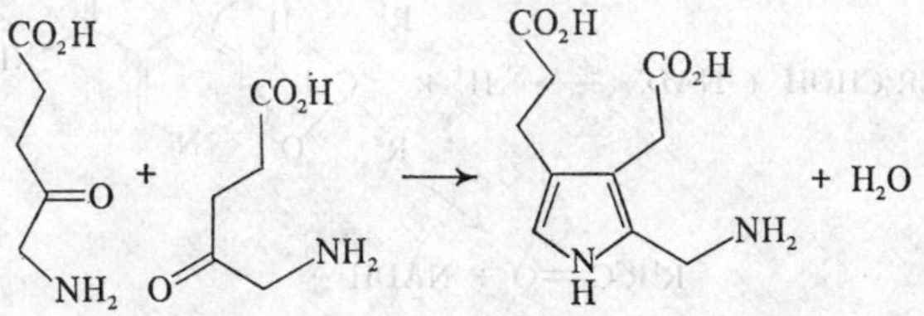

$$+ H_2O \qquad (12a)$$

(16)

inactivated by borohydride ion in the presence of substrate. This suggests the intermediacy of a protonated imine ('Schiff base') in the catalytic mechanism, this imine arising from a substrate molecule and the ε-NH$_2$ of a lysine residue of the enzyme and being reduced by borohydride. With this vital information, Shemin[45] proposed the mechanism of Scheme 3 as representing the action of ALA-D. Cheh and Neilands[12] modified this mechanism to include a role for Zn^{II}. They suggested that Zn^{II} co-ordinates to the carbonyl oxygen and amino-group of a δ-ALA molecule and either catalyses nucleophilic attack on that carbonyl group by the deprotonated imine of Scheme 3 or facilitates the removal of a proton from the co-ordinated δ-ALA to give a Zn^{II}-stabilized carbanion, which attacks the imine from another δ-ALA molecule.

Two molecules of δ-ALA can self-condense in three ways. Besides producing porphobilinogen (PBG), which in principle does not require an auxiliary amino-group (Lys-NH$_2$ of Scheme 3), dihydropyrazine (17) or 'iso-PBG' (18) could be formed. Production of (17) is a reversible process. In the presence of suitable oxidants to convert (17) into a pyrazine, this pathway might predominate.

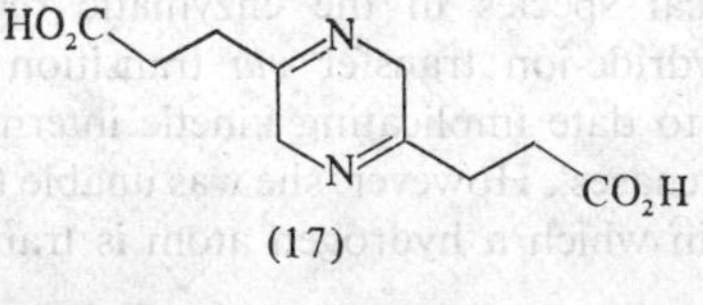

(17)

Scott *et al.*[46] have shown that δ-ALA in conc. NaOH gives 70% iso-PBG (18), owing to preferential deprotonation of δ-ALA at C-5, as shown in reaction (12b). However, with Amberlite IR-45 as catalyst (contains NH and NH$_2$ groups) to simulate the active-site Lys-NH$_2$ of ALA-D, a 10% yield of PBG was obtained

[45] D. Shemin, *Naturwiss.*, 1970, **57**, 185.
[46] A. I. Scott, C. A. Townsend, K. Okada, and M. Kajiwara, *New York Acad. Sci. Trans.*, 1973, **35**, 72.

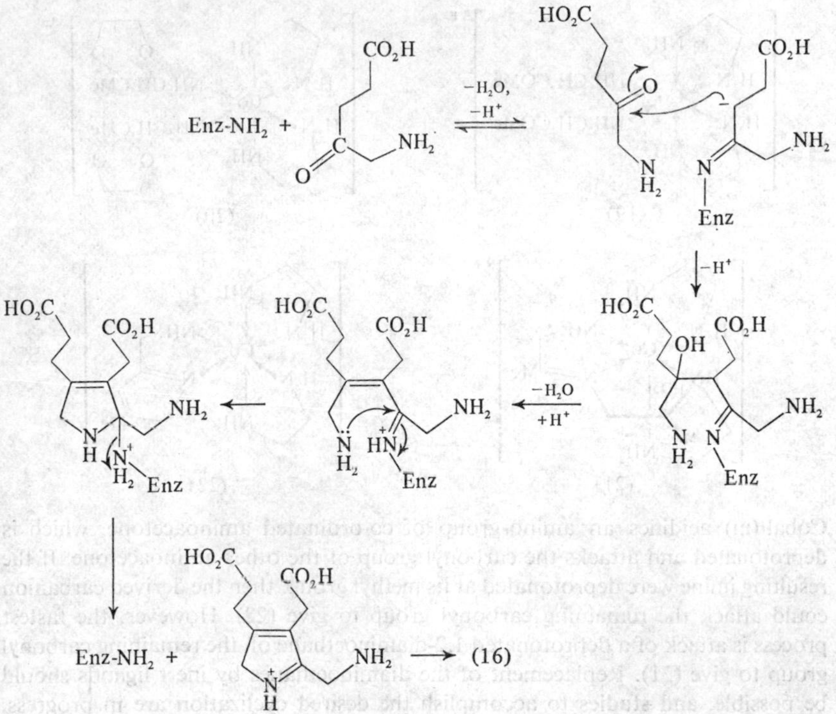

$$\text{Enz-NH}_2 + \underset{\text{O}}{\overset{\text{CO}_2\text{H}}{\bigcup}}\text{NH}_2 \underset{-\text{H}^+}{\overset{-\text{H}_2\text{O,}}{\rightleftharpoons}}$$

Mechanism[45] for the biosynthesis of porphobilinogen (16) from 2 molecules of δ-amino-laevulinic acid (Enz-NH₂ = ALA-D)

Scheme 3

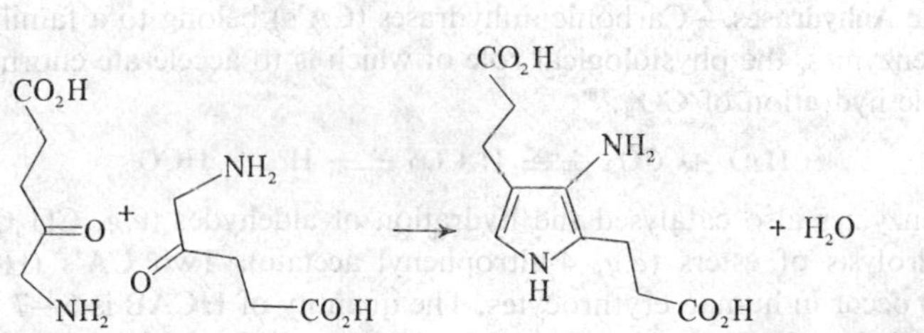

$$+ \text{H}_2\text{O} \quad (12b)$$

from δ-ALA. Bell *et al.*[47] sought a template synthesis of 2-aminomethyl-4-methylpyrrole from two molecules of aminoacetone, as a simple model for a metal-ion-directed condensation of two molecules of δ-ALA to PBG. The *cis*-bis(aminoacetone)bis-(1,2-diaminoethane)cobalt(III) ion (19) was prepared by treating *cis*-bis-(2-aminomethyl-2-methyl-1,3-dioxolan)-bis-(1,2-diaminoethane)-cobalt(III) (20) with acid. At pH 7, (19) underwent, within seconds at room temperature, a stereoselective and regioselective condensation to produce 86% of (21).

[47] J. D. Bell, A. R. Gainsford, B. T. Golding, A. J. Herlt, and A. M. Sargeson, *J.C.S. Chem. Comm.*, 1974, 980

3

$$\left[\begin{array}{c} \text{H}_2\text{N} \xrightarrow{\text{NH}_2} \text{Co} \xleftarrow{} \text{NH}_2\text{CH}_2\text{COMe} \\ \text{H}_2\text{N} \xrightarrow{} \text{NH}_2\text{CH}_2\text{COMe} \\ \text{NH}_2 \end{array} \right]^{3+}$$

(19)

$$\left[\begin{array}{c} \text{H}_2\text{N} \xrightarrow{\text{NH}_2} \text{Co} \xleftarrow{} \text{NH}_2\text{CH}_2\text{CMe} \begin{smallmatrix} \text{O} \ \ \text{O} \end{smallmatrix} \\ \text{H}_2\text{N} \xrightarrow{} \text{NH}_2\text{CH}_2\text{CMe} \begin{smallmatrix} \text{O} \ \ \text{O} \end{smallmatrix} \\ \text{NH}_2 \end{array} \right]^{3+}$$

(20)

$$\left[\begin{array}{c} \text{H}_2\text{N} \xrightarrow{\text{NH}_2} \text{Co} \xleftarrow{} \text{NH}_2 \\ \text{HN} \xrightarrow{} \text{HO} \ \ \text{N} = \text{Me} \\ \text{Me} \ \ \text{NH}_2 \end{array} \right]^{3+}$$

(21)

$$\left[\begin{array}{c} \text{H}_2\text{N} \xrightarrow{\text{NH}_2} \text{Co} \xleftarrow{} \text{NH}_2 \\ \text{H}_2\text{N} \xrightarrow{} \text{N} \\ \text{NH}_2 \ \ \ \ \ \text{Me} \end{array} \right]^{3+}$$

(22)

Cobalt(III) acidifies an amino-group of co-ordinated aminoacetone, which is deprotonated and attacks the carbonyl group of the other aminoacetone. If the resulting imine were deprotonated at its methyl group, then the derived carbanion could attack the remaining carbonyl group to give (22). However, the fastest process is attack of a deprotonated 1,2-diaminoethane on the remaining carbonyl group to give (21). Replacement of the diaminoethanes by inert ligands should be possible, and studies to accomplish the desired cyclization are in progress. The function of Zn^{II} in ALA-D's could be to anchor two molecules of δ-ALA, as in complex (19). A protonated imine derived from δ-ALA and Lys-NH_2 could participate in either or both condensation steps leading to pyrrole.

Carbonic Anhydrases.—Carbonic anhydrases (CA's) belong to a family of Zn^{II}-metalloenzymes, the physiological role of which is to accelerate enormously the reversible hydration of CO_2:[48]

$$H_2O + CO_2 \rightleftharpoons H_2CO_3 \rightleftharpoons H^+ + HCO_3^- \qquad (13)$$

These enzymes also catalysed the hydration of aldehydes (*e.g.* CH_3CHO) and the hydrolysis of esters (*e.g.* 4-nitrophenyl acetate). Two CA's (HCAB and HCAC) occur in human erythrocytes. The quantity of HCAB is 6—7 times that of HCAC, but the latter is appreciably more active for hydration of CO_2, and has one of the highest turnover numbers (2.5×10^5 s^{-1} for HCO_3^- substrate at pH 7.4 and 25 °C) known for enzymes. Each enzyme has been sequenced, and their structures have been determined to 2.0 Å resolution by X-ray crystallography.[49] HCAB consists of a polypeptide chain of 260 residues containing one Zn^{II}. HCAC has 259 residues (mol. wt. *ca.* 29 300) and a similar overall shape to HCAB (ellipsoid, $41 \times 41 \times 47$ Å), although its primary structure differs

[48] S. Lindskog, L. E. Henderson, K. K. Kannan, A. Lijas, P. O. Nyman, and B. Strandberg in 'The Enzymes', ed. P. D. Boyer, Academic Press, New York, 1971, Vol. V, Ch. 21, p. 587.
[49] A. Liljas, K. K. Kannan, P.-C. Bergsten, I. Waara, K. Fridborg, B. Strandberg, U. Carlbom, L. Järup, S. Lövgren, and M. Petef, *Nature New Biology*, 1972, **235**, 131; L. E. Henderson, D. Henriksson, and P. O. Nyman, *J. Biol. Chem.*, 1976, **251**, 5457.

at *ca.* 40% of the amino-acid residues. Neither HCAB nor HCAC possess a disulphide bridge (HCAC contains only one half-cysteine residue). The active site in HCAB and HCAC contains the Zn, ligated (distorted tetrahedron) to three histidine residues and a water molecule (hydroxide ion), and is situated at the base of a deep cleft in the protein. The surroundings of the active sites are similar, with only four residues differing out of nineteen within 8 Å of the Zn. One of the common residues is a fourth histidine (His-64) at the entrance to the inner region of the active site. Sulphonamides (*e.g.* the drug acetazolamide) powerfully inhibit CA's by binding to Zn with displacement of Zn-bound water (hydroxide).[50]

At least six mechanisms have been proposed for CA's (see Table 1 in ref. 51 and Schemes 4 and 5, 7—9 below). To help decide which mechanism is correct, two important questions for experiments to answer are: how near does CO_2 get to Zn^{II}, and what is the identity of the group(s) responsible for the pH profile of catalytic activity? Related to the second question is the function of the histidines near or ligated to Zn^{II}. Scheme 4 shows the 'zinc–hydroxide' mechan-

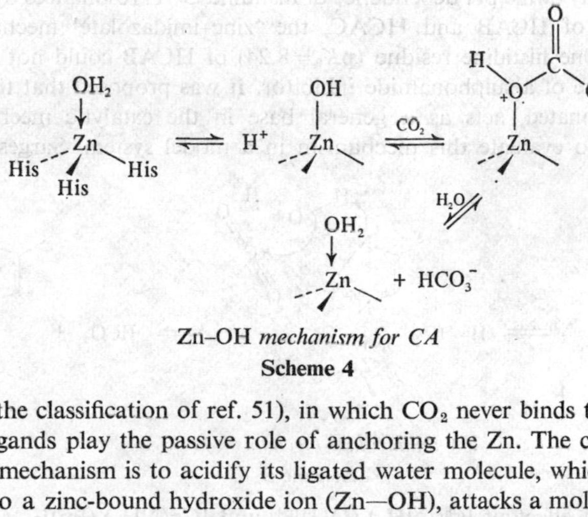

Zn–OH *mechanism for CA*
Scheme 4

ism (Ia in the classification of ref. 51), in which CO_2 never binds to Zn and the histidine ligands play the passive role of anchoring the Zn. The crucial role of Zn in this mechanism is to acidify its ligated water molecule, which, after conversion into a zinc-bound hydroxide ion (Zn—OH), attacks a molecule of CO_2. This water molecule is a candidate for the group of pK_a *ca.* 7 that is known to take part in the catalytic mechanism.[52] *Ab initio* molecular calculations[53] and model studies also show that Zn—OH will catalyse the hydration of CO_2 and CH_3CHO as well as the hydrolysis of 4-nitrophenyl acetate. Woolley[54] prepared Zn^{II} and Co^{II} complexes of the ligands (23a—c). These complexes are five-co-ordinate, with 4 nitrogen donors from (23) and water as the fifth ligand. The pK_a of the water molecule in the Zn^{II} complex of (23a) is 8.69. Cobalt(II) has a similar acidifying effect on its H_2O ligand, but Ni^{II} and Cu^{II} complexes, which

[50] R. L. Petersen, T.-Y. Li, J. T. McFarland, and K. L. Watters, *Biochemistry*, 1977, **16**, 726.
[51] Y. Pocker and D. W. Bjorkquist, *Biochemistry*, 1977, **16**, 5698.
[52] R. G. Khalifah, D. J. Strader, S. H. Bryant, and S. M. Gibson, *Biochemistry*, 1977, **16**, 2241.
[53] D. Demoulin, A. Pullman, and B. Sarkar, *J. Amer. Chem. Soc.*, 1977, **99**, 8498.
[54] P. Woolley, *Nature*, 1975, **258**, 677.

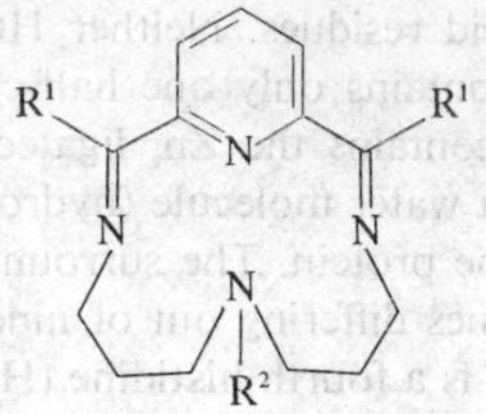

(23a) R^1 = Me, R^2 = H
(23b) R^1 = R^2 = H
(23c) R^1 = R^2 = Me

are six-co-ordinate, with two water ligands, show pK_a's > 11. The Zn^{II} and Co^{II} complexes of (23a) catalyse the hydration of CO_2 and CH_3CHO. The pK_a of $[Co(NH_3)_5OH_2]^{3+}$ is 6.22 (with ClO_4^- as the only anion in solution). The deprotonated complex catalyses hydration of CO_2 with a solvent isotope effect (k^{H_2O}/k^{D_2O}) of 1.0, inferring that there is direct attack on CO_2 by Co—OH.[55,56]

From a study of the pH dependence of histidine C—H resonances by ^{1}H n.m.r. spectroscopy of HCAB and HCAC, the 'zinc–imidazolate' mechanism was suggested.[57] One histidine residue ($pK_a = 8.24$) of HCAB could not be titrated in the presence of a sulphonamide inhibitor. It was proposed that this residue, when deprotonated, acts as a general base in the catalytic mechanism (*cf.* Scheme 5). To evaluate this mechanism in a model system, Sargeson *et al.*[58]

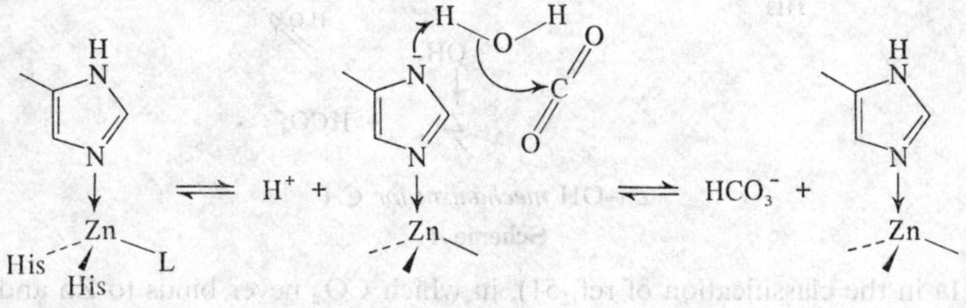

Zn–imidazolate mechanism for CA
Scheme 5

compared the effect of $[Co(NH_3)_5(OH)]^{2+}$ and $[Co(NH_3)_5(Im)]^{2+}$ (Im = imidazolate ion) on 4-nitrophenyl acetate. Both complexes reacted *exclusively via* direct nucleophilic attack on the acetyl group, irrespective of solvent (water or dimethyl sulphoxide). In water, the hydroxide complex is 6×10^3 times less reactive than the imidazolate complex, the co-ordinated imidazolate being of similar nucleophilic reactivity to free OH$^-$ towards 4-nitrophenyl acetate. In DMSO, the reactivities of Co—Im and Co—OH are comparable, owing to enhanced reactivity of the hydroxide complex in DMSO. In water, the nucleophilicity of the co-ordinated OH is attenuated by solvation. It was concluded

55 E. Chaffee, T. P. Dasgupta, and G. M. Harris, *J. Amer. Chem. Soc.*, 1973, **95**, 4169; D. A. Palmer and G. M. Harris, *Inorg. Chem.*, 1974, **13**, 965.
56 Y. Pocker and D. W. Bjorkquist, *J. Amer. Chem. Soc.*, 1977, **99**, 6537.
57 J. M. Pesando, *Biochemistry*, 1975, **14**, 681.
58 I McB. Harrowfield, V. Norris, and A. M. Sargeson, *J. Amer. Chem. Soc.*, 1976, **98**, 7282.

that these results support a mechanism for the esterase activity of CA whereby a zinc imidazolate attacks the carbonyl group of ester substrate, giving an intermediate acylhistidine, which is then cleaved by water. Co-ordination of N-acetylimidazole to Co^{III} was found to enhance the rate of hydrolysis (H_2O or OH^-) by *ca.* 20-fold relative to free N-acetylimidazole. However, direct attack of Zn—Im on CO_2 as a feature of the catalytic mechanism of CA (hydration of CO_2) was recognized as unlikely because the resulting carbamate would probably undergo addition of water to its carbonyl group at too slow a rate. Sargeson suggested that CA might operate *via* two mechanisms: Zn—OH for CO_2, but Zn—Im (acylhistidine intermediate, *cf.* Scheme 6) for esters.

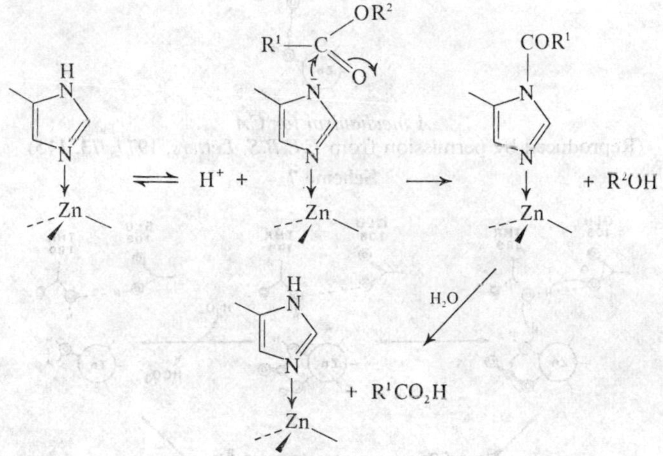

Zn–imidazolate mechanism for ester hydrolysis by CA

Scheme 6

Only one competitive inhibitor, imidazole, is known for the hydration of CO_2 by CA. The crystal structure of the HCAB–imidazole complex has been determined[59] and used to make some provocative suggestions about mechanism. The imidazole appears to occupy a distant fifth co-ordination site on the Zn ion, its nearest nitrogen atom being 2.7 Å from the Zn. It sits in a hydrophobic pocket which, it is believed, is occupied by CO_2 in the catalytic reaction. Two mechanisms (Schemes 7 and 8) for the catalytic reaction are proposed. The distinguishing feature of these mechanisms is the postulated five-co-ordinate Zn, one of its co-ordination sites being occupied by an oxygen atom of CO_2. This Zn—O interaction polarizes the $CO=O$ group ($Zn \leftarrow \overset{\delta-}{O} = C = \overset{\delta+}{O}$) and activates it to nucleophilic attack (*cf.* mechanisms of action of CPA and ADH). The second mechanism (Scheme 8) involving proton transfers accords with the finding of a solvent isotope effect ($k_{cat}^{H_2O}/k_{cat}^{D_2O}$) of 3.3 for the hydration of CO_2 catalysed by bovine CA.[51,56] This result was taken to imply a mechanism for the interconversion of CO_2 and HCO_3^- in which proton transfer is involved in the rate-

[59] K. Kannan, M. Petef, K. Fridborg, H. Cid-Dresdner, and S. Lövgren, *F.E.B.S. Letters* 1977, **73**, 115.

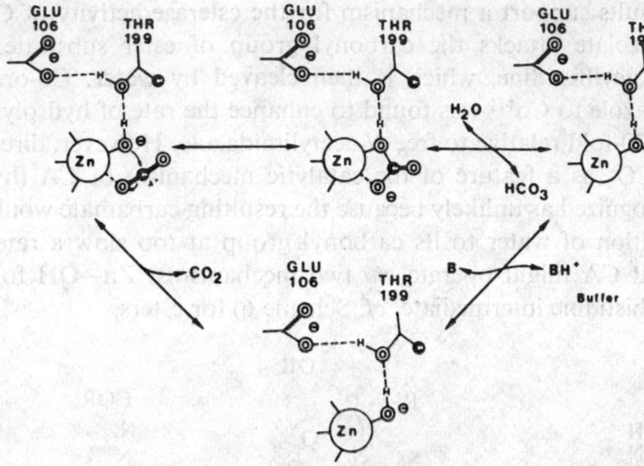

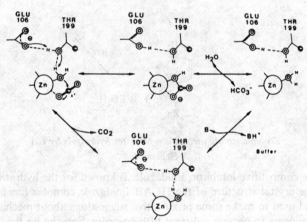

A mechanism for CA
(Reproduced by permission from *F.E.B.S. Letters*, 1977, **73**, 115)

Scheme 7

A mechanism for CA
(Reproduced by permission from *F.E.B.S. Letters*, 1977, **73**, 115)

Scheme 8

determining step. Pocker and Bjorkquist[51] favour either the mechanism of Scheme 9 or that of Scheme 10 (mechanisms Ib and IIa in their classification). Note that, after an i.r. study of bovine CA, it was concluded that CO_2 did not bind to Zn in the catalytic reaction.[60] However, a recent study[61] of relaxation times by ^{13}C n.m.r. spectroscopy, using ^{13}C-enriched CO_2 and Co^{II}-substituted HCAB, found distances between the Co and the carbon atoms of CO_2 and of HCO_3^- of 3.4—3.6 Å (pH 5.5—7.8). These results were interpreted in favour of a mechanism like that in Scheme 8, but rule out mechanisms in which a water molecule is inserted between Zn—OH or Zn—Im and CO_2 (*e.g.* Scheme 5).

[60] M. E. Riepe and J. H. Wang, *J. Biol. Chem.*, 1968, **243**, 2779.
[61] P. J. Stein, S. P. Merrill, and R. W. Henkens, *J. Amer. Chem. Soc.*, 1977, **99**, 3194.

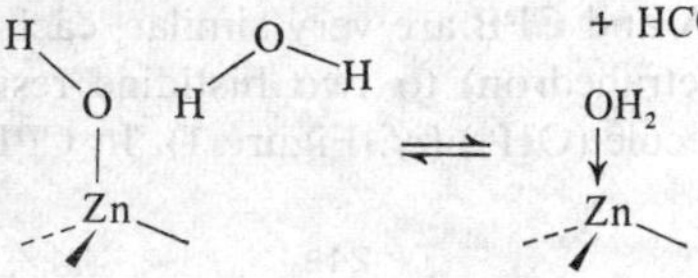

A mechanism for CA (general base-assisted attack by H_2O on CO_2)
Scheme 9

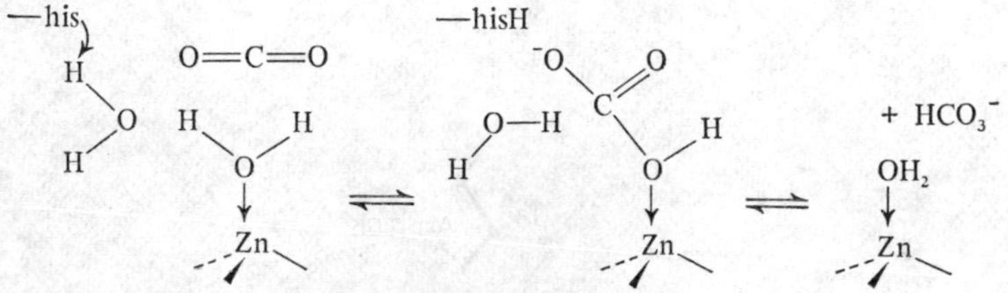

A mechanism for CA (general base-assisted attack of $Zn-OH_2$ on CO_2)
Scheme 10

In spite of intensive studies of an ostensibly simple reaction, the mechanism of action of CA has not been defined. The mechanism of Scheme 8[59] seems to be in best agreement with experimental results.

Carboxypeptidases.—Two classes of carboxypeptidase have been recognized. Enzymes of one class, *e.g.* yeast carboxypeptidase C,[62] are intracellular, showing maximal activity at acidic pH. They are *not* metalloenzymes, and their mechanism of action may be similar to that of the serine proteinases, *e.g.* α-chymotrypsin. The second class of carboxypeptidases[63] contains proteolytic metalloenzymes which are released from their inactive precursors (zymogens) in the pancreatic juice of animals. These enzymes exhibit maximal activity at neutral or slightly alkaline pH, and act extracellularly, aiding protein digestion in the duodenum.

Bovine carboxypeptidases A and B have been sequenced and have had their structures established by X-ray techniques. Carboxypeptidase A (CPA) consists of a single chain of 307 amino-acids embedding one Zn^{II}.[64] Carboxypeptidase B (CPB) has 308 residues, 49% of its sequence being identical to that of CPA.[65] Their three-dimensional structures are similar ellipsoidal shapes ($50 \times 40 \times 38$ Å for CPA). CPA catalyses the hydrolysis of the amide bond at the carboxy-terminus of peptides and proteins, and shows a marked preference for residues which contain aromatic or large aliphatic side-chains:

$$-CONHCHRCO_2^- \longrightarrow -CO_2^- + \overset{+}{N}H_3CHRCO_2^- \qquad (14)$$

CPB also cleaves an amino-acid from the carboxy-terminus of peptides and proteins, but prefers the basic amino-acids lysine and arginine. CPA and CPB also possess esterase activity, hydrolysing, *e.g.*, *O*-acetyl-L-mandelate.

[62] R. W. Kuhn, K. A. Walsh, and H. Neurath, *Biochemistry*, 1976, **15**, 4881.
[63] J. A. Hartsuck and W. N. Lipscomb, in 'The Enzymes', ed. P. D. Boyer, Academic Press, New York, 1971, Vol. III, Ch. 1, p. 1
[64] W. N. Lipscomb, *Chem. Soc. Rev.*, 1972, **1**, 319; *Tetrahedron*, 1974, **30**, 1725.
[65] M. F. Schmid and J. R. Herriott, *J. Mol. Biol.*, 1976, **103**, 175.

The active sites of CPA and CPB are very similar, each enzyme's Zn^{II} being co-ordinated (distorted tetrahedron) to two histidine residues, one glutamate residue, and a water molecule (OH^-) (*cf.* Figure 1). In CPB, Asp-255 lies at the

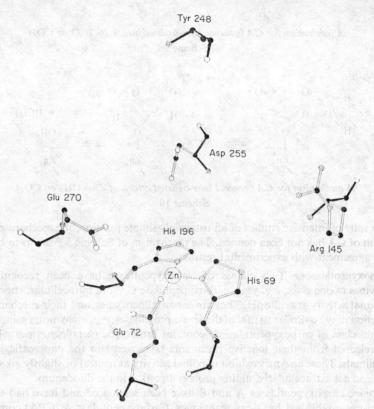

Figure 1 *Amino-acid residues near to the active site of CPB*
(Reproduced by permission from *J. Mol. Biol.*, 1976, **103**, 175)

rear of a hydrophobic pocket which presumably binds the basic side-chain of preferred substrate. In CPA, Asp-255 is replaced by a neutral residue (isoleucine), which explains the difference in substrate specificity between CPA and CPB. Chemical modification studies indicate roles for Glu-270 and Tyr-248 in the catalytic mechanism, whereas Arg-145 helps to bind substrate. Metal-replacement studies denote a vital role for the Zn in CPA and CPB. It is very interesting that replacing Zn^{II} by Cd^{II} abolishes peptidase activity but enhances esterase activity.

Scheiner and Lipscomb[66] have discussed 'zinc–carbonyl' and 'zinc–hydroxide' mechanisms for CPA. One of two favoured 'zinc–carbonyl' mechanisms is shown for a peptide substrate in Scheme 11. A substrate molecule displaces the water ligand of the Zn^{II}, with the amide group to be cleaved co-ordinating to Zn *via*

[66] S. Scheiner and W. N. Lipscomb, *J. Amer. Chem. Soc.*, 1977, **99**, 3466.

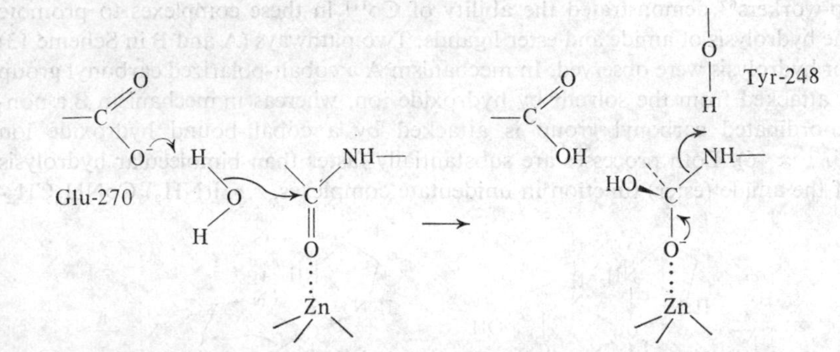

Zn–carbonyl mechanism for CPA (peptide substrate)
Scheme 11

its carbonyl group. The crystal structure of CPA containing glycyltyrosine (a poor substrate) actually shows the amide carbonyl group (of Gly-Tyr) ligated to Zn. The co-ordinated carbonyl group is polarized and attacked by a water molecule activated by Glu-270. The resulting tetrahedral intermediate is protonated on its nitrogen atom by Tyr-248 and can then decompose to amine and carboxylic acid. Scheme 12 shows an alternative 'zinc–carbonyl' mechanism,

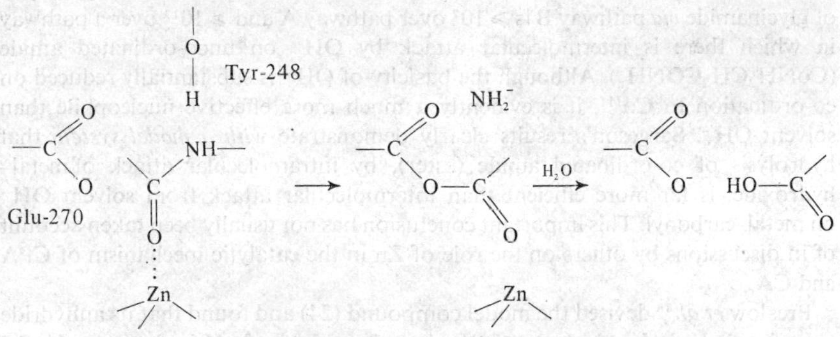

Alternative Zn–carbonyl mechanism for CPA (ester substrate)
Scheme 12

illustrated with an ester substrate. In this case, the carboxylate anion of Gly-270 attacks the Zn-activated carbonyl group of substrate molecule, leading to an intermediate mixed acid anhydride which is hydrolysed, possibly by a Zn-held hydroxide ion. In the 'zinc–hydroxide' mechanism, a Zn-held hydroxide ion (see discussion in the section on carbonic anhydrase regarding the ease of formation of such a species from a water molecule ligated to zinc) attacks the carbonyl group of a substrate molecule, giving a tetrahedral intermediate which breaks down, with assistance from Tyr-248, by a route akin to that shown in Scheme 11.

Model systems pertinent to the mechanism of action of CPA have been devised by Sargeson and Breslow and their co-workers. Using kinetically inert bis(ethylenediamine)cobalt(III) complexes, to allow mechanistic features to be defined without complications from ligand-exchange processes, Sargeson and co-workers[67] demonstrated the ability of Co[III] in these complexes to promote the hydrolysis of amide and ester ligands. Two pathways (A and B in Scheme 13) for hydrolysis were observed. In mechanism A a cobalt-polarized carbonyl group is attacked from the solvent by hydroxide ion, whereas in mechanism B a non-co-ordinated carbonyl group is attacked by a cobalt-bound hydroxide ion ($pK_a \approx 6$). Both processes are substantially faster than bimolecular hydrolysis of the amide (ester) function in unidentate complexes, _e.g._ $[(NH_3)_5CoNH_2CH_2$-

$$\left(X = O— \text{ or } N< \right)$$

Pathways for hydrolysis of amide and ester ligands in Co[III] complexes
Scheme 13

$CONH_2]^{3+}$, which is faster than hydrolysis of free glycinamide, with pathway B being _substantially faster_ than A. At pH 9, the rate enhancement for hydrolysis of glycinamide _via_ pathway B is $\geqslant 10^7$ over pathway A and $\geqslant 10^{11}$ over a pathway in which there is intermolecular attack by OH^- on unco-ordinated amide ($CoNH_2CH_2CONH_2$). Although the basicity of OH^- is substantially reduced on co-ordination to Co[III], it is evidently a much more effective nucleophile than solvent OH^-. Sargeson's results clearly demonstrate _with a model system_ that hydrolysis of co-ordinated amide (ester), by intramolecular attack of metal–hydroxide, is far more efficient than intermolecular attack from solvent OH^- on metal–carbonyl. This important conclusion has not usually been taken account of in discussions by others on the role of Zn in the catalytic mechanism of CPA and CA.

Breslow _et al._[68] devised the model compound (24) and found that its anhydride grouping is hydrolysed at a rate that is independent of pH in the range 1—7.5 (pseudo-first-order $k = 2.7 \times 10^{-3}\ s^{-1}$). With sufficient Zn[II] to saturate (24), forming the complex (25), hydrolysis of the anhydride at pH 7.5 is first-order in $[OH^-]$, and occurs with $k = 3\ s^{-1}$, probably by attack of co-ordinated OH on anhydride carbonyl. The rate constant for this process is comparable to values of k_{cat} at this pH for typical substrates of CPA. In the absence of Zn[II], hydroxylamine is a much better nucleophile than water for the anhydride of (24). When Zn[II] is present, attack on the anhydride by hydroxylamine is _not_ catalysed, and so hydrolysis of the anhydride competes effectively. Breslow and McClure[69] have examined model compounds which contribute to an understanding of the role of Tyr-248 in the catalytic mechanism of CPA. In acidic aqueous acetonitrile,

[67] D. A. Buckingham, D. M. Foster, and A. M. Sargeson, _J. Amer. Chem. Soc._, 1970, **92**, 6151.
[68] R. Breslow, D. E. McClure, R. S. Brown, and J. Eisenach, _J. Amer. Chem. Soc._, 1975, **97**, 194.
[69] R. Breslow and D. McClure, _J. Amer. Chem. Soc._, 1976, **98**, 258.

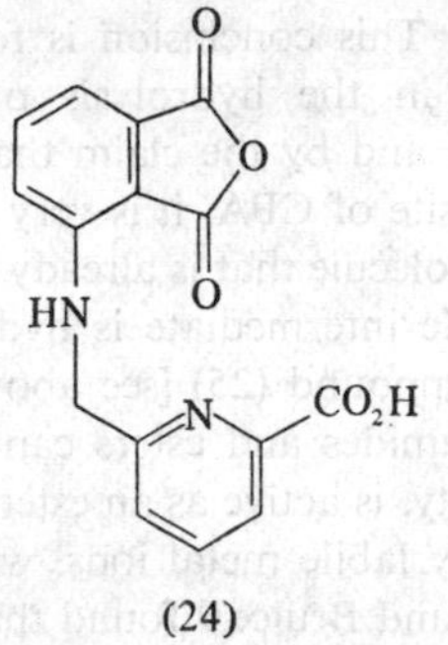

(24)

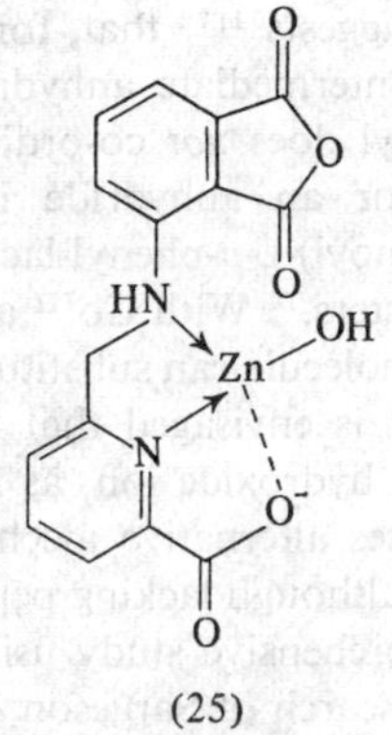

(25)

derivatives (26) of maleamic acid are hydrolysed *via* dimethylmaleic anhydride, as shown in reaction (15). Under these conditions, the rate of hydrolysis of (26a) slightly exceeds that of (26b). The unexpected discovery was made that these derivatives are hydrolysed in neutral aqueous acetonitrile (a medium intended to simulate the interior of an enzyme) without the intermediacy of the anhydride. The rate of hydrolysis of (26b) is now *ca.* 70 times that of (26a). It is proposed that carboxylate anion promotes addition of water to the amide carbonyl group. The phenolic hydroxy-group assists decomposition of the resulting tetrahedral intermediate by protonating NH and then deprotonating OH.

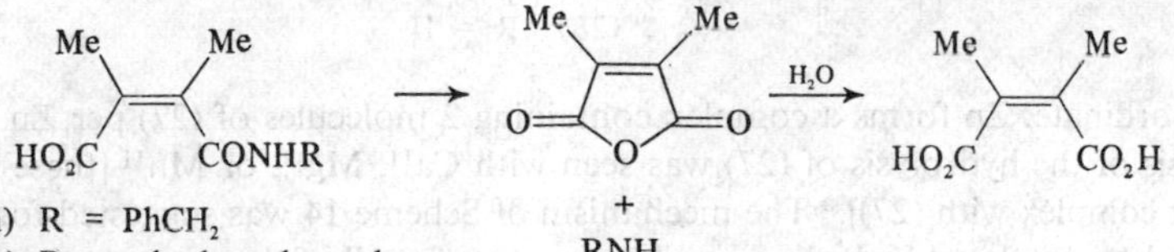

(15)

(26a) R = PhCH₂
(26b) R = o-hydroxybenzyl

Returning to enzyme studies, Breslow and Wernick[70] found that benzoyl-glycine labelled with ¹⁸O in its carboxy-group would not exchange ¹⁸O with water on incubation with CPA unless another component (*e.g.* phenylalanine) is present in the reaction mixture. This experiment indicates that, in the absence of phenyl-alanine, a mixed acid anhydride with Glu-270 is not being formed (*i.e.* the mechanism of Scheme 12 is excluded). In the presence of phenylalanine, ¹⁸O exchange occurs by reversal of the normal catalytic mechanism for which Breslow favours Scheme 11. The function of Tyr-248 in this mechanism is analogous to that of phenolic OH in the hydrolysis of derivatives [(26a) and (26b)] of maleamic acid in neutral aqueous acetonitrile. This conclusion was strengthened by finding that methanol would not substitute for water in the hydrolysis (methanolysis) of benzoylglycylphenylalanine by CPA. This is because the mechanism of Scheme 11 (incorporating proton transfers) involves *both* protons of the attacking water molecule. Note that with some proteolytic enzymes (*e.g.* chymotrypsin) methanol is much preferred to water.

70 R. Breslow and D. Wernick, *J. Amer. Chem. Soc.*, 1976, **98**, 259.

Breslow suggests[71, 72] that, for at least some ester substrates with CPA, there may be an intermediate anhydride (mechanism as in Scheme 12, except that ester carbonyl does *not* co-ordinate to Zn). This conclusion is forced by the evidence[73] for an anhydride intermediate in the hydrolysis of (*O*-*p*-*trans*-chlorocinnamoyl)-L-β-phenyl-lactate by CPA and by the claim that Co^{III}-CPA hydrolyses esters.[74] With Co^{III} at the active site of CPA, it is very unlikely that a substrate molecule can substitute a water molecule that is already bound to the metal ion. It is envisaged that the anhydride intermediate is hydrolysed by a metal-bound hydroxide ion, as in model compound (25) [see above]. The idea that CPA uses alternative mechanisms for amides and esters can explain why Cd^{III}-CPA, although lacking peptidase activity, is active as an esterase.

In a comprehensive study using kinetically labile metal ions, which complements the research of Sargeson *et al.*, Wells and Bruice[75] found that hydrolyses of esters (27a—e) to acid (28) are appreciably catalysed by Co^{II} and Ni^{II} [10^3—10^5 times faster than catalysis by OH$^-$; comparable to the rate of hydrolysis of mandelates by CPA], whereas Cd^{II} and Zn^{II} are poor catalysts. Cd, Co, and Ni form 1 : 1 complexes with (27), and the complex with Co was shown to be

(27a)   R = Me
(27b)   R = CH$_2$CH$_2$Cl
(27c)   R = CH$_2$C≡CH
(27d)   R = CH$_2$CF$_3$
(27e)   R = C$_6$H$_4$Cl-*p*
(28)   R = H

six-co-ordinate. Zn forms a complex containing 2 molecules of (27) per Zn. No catalysis of the hydrolysis of (27) was seen with Ca^{II}, Mg^{II}, or Mn^{II} [these ions do not complex with (27)].[3] The mechanism of Scheme 14 was suggested for the Co- and Ni-catalysed hydrolyses, and it accounts for all of the numerous experimental observations, including the following:

(i) Hydrolyses of (27) are inhibited by the alcohol from which (27) was prepared; in the presence of *p*-chlorophenol, (27a) and (27d) are converted into (27e). These observations are explained by 'metal–carboxyl' anhydride intermediates (T and TH of Scheme 14) which can revert to ester (27).

(ii) When (27) is saturated with metal ion, the pH-dependence of k_{obs} for its hydrolysis suggests that the active form of the M–(27) complex contains M—OH.

(iii) The kinetically measured pK_a of M–(27) complex is smaller than that determined by titration, as expected if there is a tetrahedral intermediate which is slowly formed (on the time scale for titration), but which breaks down rapidly to products.

[71] К. Breslow, in 'Further Perspectives in Organic Chemistry', Elsevier, Amsterdam, 1978, p. 175.
[72] R. Breslow and D. L. Wernick, *Proc. Nat. Acad. Sci. U.S.A.*, 1977, **74**, 1303.
[73] M. Makinen, K. Yamamura, and E. Kaiser, *Proc. Nat. Acad. Sci. U.S.A.*, 1976, **73**, 3882.
[74] E. P. Kang, C. B. Storm, and F. W. Carson, *J. Amer. Chem. Soc.*, 1975, **97**, 6723 (but see H. E. van Wart and B. L. Vallee, *Biochemistry*, 1978, **17**, 3385).
[75] M. A. Wells and T. C. Bruice, *J. Amer. Chem. Soc.*, 1977, **99**, 5341 (*c.f.* T. H. Fife and V. L. Sqillacote, *ibid.*, p. 3762).

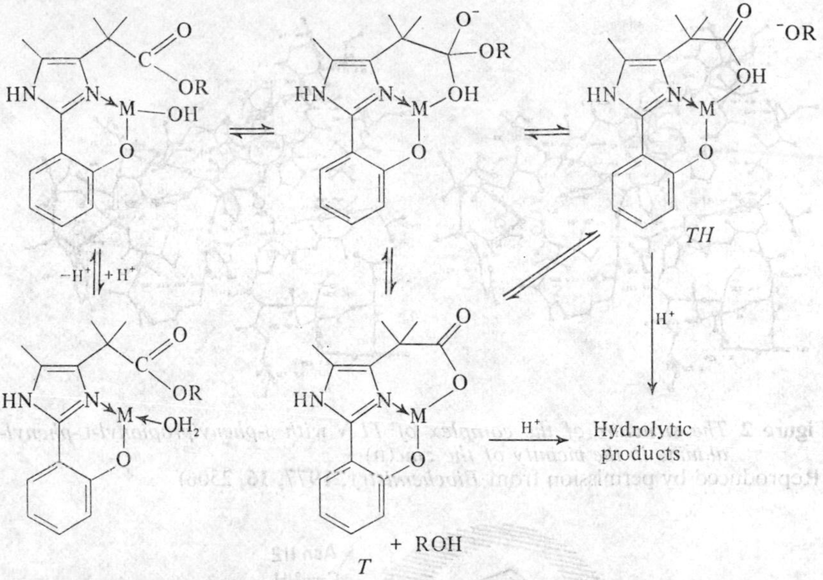

Mechanism for the Co^{II}*- and* Ni^{II}*-catalysed hydrolyses of esters* (27a—e) [75]

Scheme 14

The report of Wells and Bruice[75] reiterates the effectiveness of M—OH for intramolecular hydrolyses of esters and amides. Of course, a mechanism for CPA in which M—OH attacks an amide substrate has to explain satisfactorily Breslow's findings (see above). Bruice questions whether a metal–carboxyl anhydride from CPA would be trapped by methanol.

Thermolysin.—The thermophilic organism *Bacillus thermoproteolyticus* produces an extracellular endopeptidase called thermolysin (TLN). The preferred substrates of this enzyme possess a hydrophobic residue (*e.g.* Leu, Phe) of the imino side of the amide bond, or, better still, hydrophobic residues on both sides of the amide bond. TLN has a polypeptide chain (mol. wt. 34 600) containing one Zn^{II} and four Ca^{II}. The Zn^{II} is engaged in the catalytic mechanism whereas Ca^{II} is responsible for TLN's thermal stability.[76] Matthews[77] has used *X*-ray techniques to examine several enzyme–inhibitor complexes of TLN, and a most detailed picture of this enzyme's active site has emerged. Figure 2 shows the structure of the complex with β-phenylpropionyl-L-phenylalanine in the vicinity of the Zn^{II}. The oxygen atom of this inhibitor's amide carbonyl is 2.1 Å from Zn, having displaced a water ligand. The carbonyl group is 3.5 Å from a water molecule which is probably hydrogen-bonded to the carboxy-group of Glu-143 and the NH of Trp-115. The β-phenyl group sits in a hydrophobic pocket stacked against the phenyl group of Phe-114. From a detailed analysis of this structure and others, Kester and Matthews propose a mode of substrate binding to TLN shown in Figure 3, and favour a catalytic mechanism analogous to that proposed for CPA

[76] F. W. Dahlquist, J. W. Long, and W. L. Bigbee, *Biochemistry*, 1976, **15**, 1103.
[77] W. R. Kester and B. W. Matthews, *Biochemistry*, 1977, **16**, 2506.

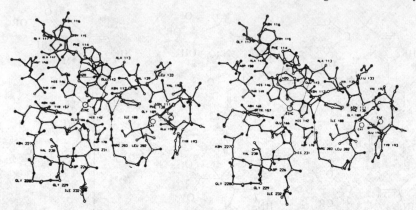

Figure 2 *The structure of the complex of TLN with β-phenylpropionyl-L-phenyl-
alanine in the vicinity of the zinc(II)*
(Reproduced by permission from *Biochemistry*, 1977, **16**, 2506)

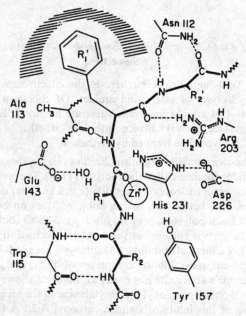

Figure 3 *Binding of substrate to the active site of TLN*
(Reproduced by permission from *J. Biol. Chem.*, 1977, **252**, 7704)

in Scheme 11. The model studies described in the context of CPA are also
pertinent to TLN. Although TLN and CPA have quite different overall shapes,
their active-site regions are similar, and they are examples of metalloenzymes
exhibiting convergent evolution.[78]

[78] W. R. Kester and B. W. Matthews, *J. Biol. Chem.*, 1977, **252**, 7704.

3 Molybdenum-containing Enzymes

Introduction.—Molybdenum is necessary for at least five biological redox reactions, namely reduction of dinitrogen and of nitrate, and also oxidation of purines, aldehydes, and sulphite. The enzymes are called nitrogenase, nitrate reductase, xanthine oxidase or dehydrogenase, aldehyde oxidase, and sulphite oxidase, respectively. Some of these enzymes exist in more than one form, and in every case except nitrogenase, Mo has been shown by e.s.r. studies to take part in the redox reaction.[79] An e.s.r. signal from Mo in nitrogenase has yet to be positively identified. Table 1 [80-84] summarizes the current data concerning these Mo enzymes. It is not yet clear whether any of the enzymes have structural features in common, though there is evidence for an interchangeable Mo cofactor. We will review here studies of nitrogenase and nitrate reductase.

Table 1 *Enzymes which contain molybdenum*

Enzyme	Substrate	Electron donor or acceptor	g atom Mo per mole of enzyme	g atom of other metal per mole of enzyme	Ref.
Nitrogenase	N_2	Ferredoxin	2 (Fraction 1)^a 0 (Fraction 2)	Fe (18—36) Fe(4)	80
Nitrate reductase	NO_3^-	NADPH	?	Fe	81
Xanthine oxidase (dehydrogenase)	Purines	NAD$^+$ or ferredoxin, or O_2	2	Fe (8)	82
Aldehyde oxidase	Aldehydes	O_2	2	Fe (8)	83
Sulphite oxidase	SO_3^{2-}	O_2	2	Fe	84

(a) 'Nitrogenase' consists of two proteins; the larger (mol. wt. *ca.* 200 000) is now generally termed Fraction 1 (F1) and the smaller (mol. wt. *ca.* 60 000) Fraction 2 (F2).

Nitrogenase.—Whatever the source, be it an aerobe, anaerobe, rhizobium, or blue-green alga, nitrogenase seems to possess the same structure.[79] It is isolated as two proteins, one having a molecular weight of about 200 000 and probably containing two molybdenums, 18—36 irons, and an equivalent number of sulphides. In general, this is termed Fraction 1 (F1) and when derived from a specific bacterium the name is amended, so that Fraction 1 from *Klebsiella pneumoniae* is referred to as Kp 1, from *Azotobacter vinelandii* as Av 1, *etc.* The other protein, referred to as Fraction 2 (F2, Kp2, Av2, *etc.*), has a molecular weight of about 60 000, and contains 4 irons and 4 sulphides. Both proteins are oxygen-sensitive.

[79] See *e.g.* R. C. Bray, Proceedings of the International Conference on Chemistry and Uses of Molybdenum, Second, 1976, p. 271.

[80] See *e.g.* (*a*) 'The Chemistry and Biochemistry of Nitrogen Fixation', ed. J. R. Postgate, Plenum, London and New York, 1971; (*b*) R. R. Eady, 'The Evolution of Metalloenzymes, Metalloproteins and Related Materials,' ed. G. J. Leigh, Symposium Press, London, 1977. See also Chapter 7 of this volume.

[81] A. H. Stouthamer, *Adv. Microbial Physiology*, 1976, **14**, 315.

[82] R. C. Bray, Proceedings of the International Conference on Chemistry and Uses of Molybdenum, First, 1974, p. 216; R. C. Bray and T. C. Swann, *Structure and Bonding*, 1972, **11**, 107.

[83] See *e.g.* U. Branzoli and V. Massey, *J. Biol. Chem.*, 1974, **249**, 4347.

[84] See *e.g.* H. J. Cohen, I. Fridovich, and K. V. Rajagopalan, *J. Biol. Chem.*, 1971, **246**, 374.

Fraction 1 can be broken down irreversibly into four subunits, two each of two different kinds, which have not been fully characterized. Fraction 2 can be split irreversibly into two subunits. There have been numerous Mössbauer and e.s.r. studies of F1 and F2. Several states of Fe_4ferredoxins have been identified in F1, but there is no detectable influence of dinitrogen or of Mo on e.s.r. spectra. The e.s.r. spectrum of reduced F2, which may donate electrons to N_2 bound to oxidized F1 (presumably on Mo),[85] has anisotropic line-widths and low integrated intensity,[86] the structural significance of which is unclear.

Brill *et al.*[87,88] have isolated an iron–molybdenum cofactor (Fe–Mo-Co) having a molecular weight of about 1200 and containing 1 Mo, 8 Fe, and 6 S atoms. There are no amino-acids present, which suggests that the 'cofactor' may be an artefact. However, it mediates the conversion of acetylene into ethylene, and more significantly it can confer nitrogen-fixing ability upon mutants unable to incorporate Mo in Fraction 1 due to genetic defects. Fe–Mo-Co is the only material able to do this among a range of compounds tested, including $[MoO_4]^{2-}$ and $[MoS_4]^{2-}$.[89] If the Fe/S ratio in Fe–Mo-Co correctly represents the proportions of these elements around the active site, then not more than one Fe_4S_4 cluster can be present. There is evidence that Fe–Mo-Co and Fraction 1 indeed have Mo in similar environments. Thus, both have the same Mo X-ray absorption edge fine structure (XAFS), which has been interpreted in terms of Mo with 3—5 nearest neighbours at 2.35 Å consisting of S atoms, another S at 2.47 Å, and Fe atoms at 2.78 Å.[90]

Models for Nitrogenase.—The active site of nitrogenase is oxygen-sensitive. Organisms which are aerobic have developed methods to keep oxygen away from their nitrogenase. For example, blue-green algae fix nitrogen in heterocysts, which are separated from the cells in which photosynthesis occurs by an impermeable membrane.[91] Organisms such as *Azotobacter* have enhanced rates of respiration which are related to nitrogenase protection.[92] In model studies, therefore, it does not detract from the validity of the system selected if it is oxygen-sensitive. For example, the ability of the oxygen-sensitive species $[Ru(NH_3)_5(H_2O)]^{2+}$ to form detectable amounts of $[Ru(NH_3)(N_2)]^{2+}$ on exposure to air[93] is a model for the uptake of N_2 by nitrogenase. Whether the active site of nitrogenase is water-sensitive or not has yet to be demonstrated. The active site

[85] See *e.g.* B. E. Smith, R. N. F. Thorneley, M. G. Yates, R. R. Eady, and J. R. Postgate, Proceedings of the International Conference on Nitrogen Fixation, Pullman, 1974, ed. W. E. Newton and C. J. Nyman, Washington State University Press, 1975, Vol. 1, p. 150.

[86] D. J. Lowe, *Biochem. J.*, 1978, **175**, 955.

[87] V. K. Shah and W. J. Brill, *Proc. Nat. Acad. Sci. U.S.A.*, 1977, **74**, 3249; J. Rawlings, V. K. Shah, J. R. Chisnell, W. J. Brill, R. Zimmermann, E. Munck, and W. H. Orme-Johnson, *J. Biol. Chem.*, 1978, **253**, 1001; V. K. Shah, J. R. Chisnell, and W. J. Brill, *Biochem. Biophys. Res. Comm.*, 1978, **81**, 232.

[88] P. T. Pienkos, V. K. Shah, and W. J. Brill, *Proc. Nat. Acad. Sci. U.S.A.*, 1977, **74**, 5468.

[89] R. J. Burt and B. E. Smith, personal communication.

[90] S. P. Cramer, T. K. Eccles, F. W. Kutzler, K. O. Hodgson, and L. E. Mortenson, *J. Amer. Chem. Soc.*, 1976, **98**, 1287; S. P. Cramer and K. O. Hodgson, Stanford Synchrotron Radiation Project, Report 77/07; K. O. Hodgson, personal communication.

[91] W. D. P. Stewart, *Proc. Roy. Soc.*, 1969, **B172**, 367.

[92] See J. R. Postgate in ref. 80 (*a*).

[93] A. D. Allen and F. Bottomley, *Canad. J. Chem.*, 1968, **46**, 469.

may be contained within a hydrophobic pocket of the protein, and so information from anhydrous model systems is admissible.

G. N. Schrauzer and co-workers have produced a series of papers entitled 'Chemical Evolution of a Nitrogenase Model'. This model developed from the observation that among a selection of derivatives (mainly chlorides) of transition metals, those of Mo gave the most efficient reduction of acetylene when mixed with a thiol (*e.g.* thioglycerol) and NaOH in the presence of sodium dithionite.[94] Reduction of acetylene is a characteristic of biological nitrogen-fixing systems, and thioglycerol was used to represent the sulphur-containing principle of nitrogenase. The efficiency of reduction of acetylene depended markedly on the type of thiol used and the molar ratio of thiol to Mo.[95] The reaction pathway of Scheme 15 was suggested for the cysteine-based system.

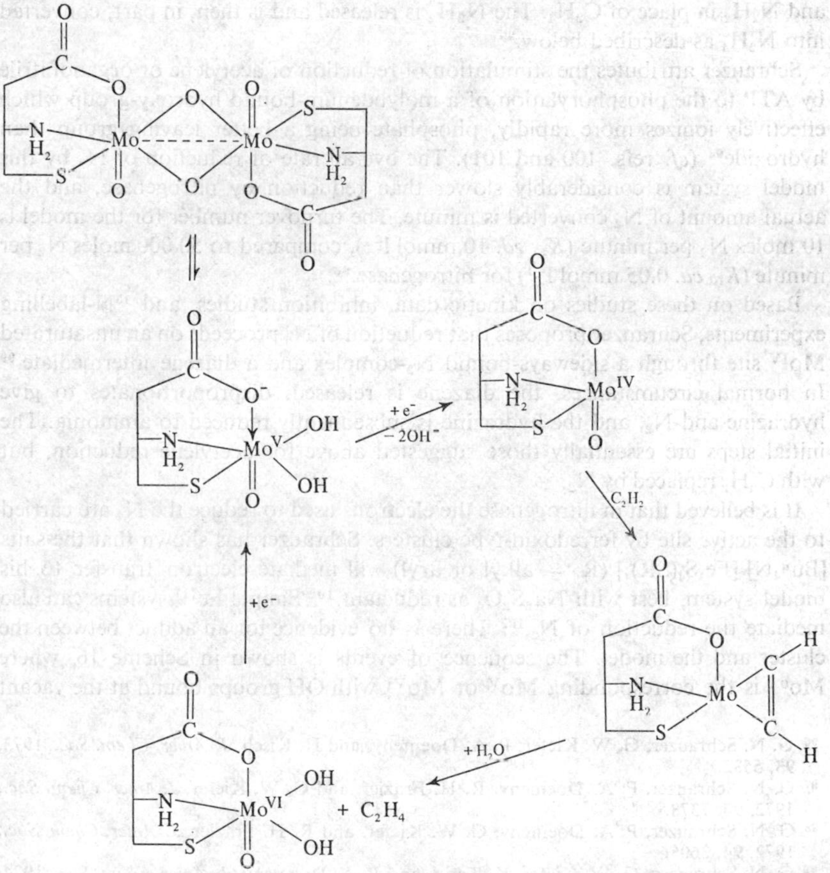

Scheme 15

[94] G. N. Schrauzer and G. Schlesinger, *J. Amer. Chem. Soc.*, 1970, **92**, 1808.
[95] G. N. Schrauzer and P. A. Doemeny, *J. Amer. Chem. Soc.*, 1971, **93**, 1608; G. N. Schrauzer, G. Schlesinger, and P. A. Doemeny, *ibid.*, p. 1803.

Subsequently, the Mo–thiol system was found to catalyse the reduction by borohydride or dithionite of other substrates for nitrogenase, *e.g.* azide,[96] N_2O,[96] cyanide,[96] and organic nitriles[97] and isocyanides.[98] It converted water into hydrogen catalytically, in the absence of any substrate.[95] In addition, the reduction of acetylene was found to be stimulated by ATP.[96–98] Addition of iron to the reducing systems had some effect on the efficacy of reduction, but the iron was believed to be acting as an electron-transfer agent, rather than as a catalyst.[95,98] Reduction of acetylene in a system made up with ^2H_2O rather than ^1H_2O produced dideuterioethylene [$95\%(Z)$-isomer].[95] Finally, Schrauzer's systems were shown to reduce N_2 to ammonia.[99] Since the optimal conditions for reduction of acetylene are also those for reduction of dinitrogen, it is believed that the route of conversion of N_2 into NH_3 is essentially similar to that of the conversion of C_2H_4 into C_2H_2 (*cf.* Scheme 15), but with N_2 in place of C_2H_2 and N_2H_2 in place of C_2H_4. The N_2H_2 is released and is then, in part, converted into N_2H_4 as described below.[99]

Schrauzer attributes the stimulation of reduction of acetylene or organonitrile by ATP to the phosphorylation of a molybdenum-bound hydroxy-group which effectively ionizes more rapidly, phosphate being a better leaving group than hydroxide[95] (*cf.* refs. 100 and 101). The overall rate of reduction of N_2 by this model system is considerably slower than reduction by nitrogenase, and the actual amount of N_2 converted is minute. The turnover number for the model is 10 moles N_2 per minute (K_m *ca.* 10 mmol l^{-1}), compared to 50 000 moles N_2 per minute (K_m *ca.* 0.05 mmol l^{-1}) for nitrogenase.[96]

Based on these studies on kinetic data, inhibition studies, and ^{15}N-labelling experiments, Schrauzer proposes that reduction of N_2 proceeds on an unsaturated Mo^{IV} site through a sideways-bound N_2-complex and a diazene intermediate.[99] In normal circumstances, the diazene is released, disproportionates to give hydrazine and N_2, and the hydrazine is subsequently reduced to ammonia. The initial steps are essentially those suggested above for acetylene reduction, but with C_2H_2 replaced by N_2.

It is believed that in nitrogenase the electrons used to reduce the N_2 are carried to the active site by ferredoxin-type clusters. Schrauzer has shown that the salts $[Bu^n_4N]_2[Fe_4S_4(SR)_4]$ (R = alkyl or aryl) will mediate electron transfer to his model system, best with $Na_2S_2O_4$ as reductant.[102] Simple Fe^{III} systems can also mediate the reduction of N_2.[102] There is no evidence for an adduct between the cluster and the model. The sequence of events is shown in Scheme 16, where Mo^{ox} is the corresponding Mo^V or Mo^{VI} with OH groups bound at the vacant

[96] G. N. Schrauzer, G. W. Kiefer, P. A. Doemeny, and H. Kisch, *J. Amer. Chem. Soc.*, 1973, **95**, 5582.

[97] G. N. Schrauzer, P. A. Doemeny, R. H. Frazier, and G. W. Kiefer, *J. Amer. Chem. Soc.*, 1972, **94**, 7378.

[98] G. N. Schrauzer, P. A. Doemeny, G. W. Kiefer, and R. H. Frazier, *J. Amer. Chem. Soc.*, 1972, **94**, 3605.

[99] G. N. Schrauzer, G. W. Kiefer, K. Tano, and P. A. Doemeny, *J. Amer. Chem. Soc.*, 1974, **96**, 641.

[100] G. W. Parshall, *J. Amer. Chem. Soc.*, 1967, **89**, 1822.

[101] T. A. Vorontsova and A. E. Shilov, *Kinetika i Kataliz*, 1973, **14**, 1326.

[102] K. Tano and G. N. Schrauzer, *J. Amer. Chem. Soc.*, 1975, **97**, 5404; G. N. Schrauzer and T. D. Guth, *J. Amer. Chem. Soc.*, 1976, **98**, 3508.

sites.[102] In assessing the accuracy of this model it should be noted that it is oxidized F1 which is believed to pick up N_2, and not reduced F1, as this model implies. See also comments in ref. 103.

$$[Fe_4S_4(SR)_4]^{2-} + S_2O_4^{2-} + 2H_2O \xrightarrow{\text{slow}} [Fe_4S_4(SR)_4]^{4-} + 2SO_3H^- + 2OH^-$$

$$[Fe_4S_4(SR)_4]^{4-} + Mo^{ox} \xrightarrow[\text{ATP}]{\text{fast}} Mo^{red} + [Fe_4S_4(SR)_4]^{2-}$$

$$N_2 + 2H^+ + Mo^{red} \longrightarrow Mo^{ox} + N_2H_2$$

$$2N_2H_2 \longrightarrow N_2 + N_2H_4$$

$$N_2H_4 + Mo^{red} + 2H^+ \longrightarrow 2NH_3 + Mo^{ox}$$

Mo^{red} is

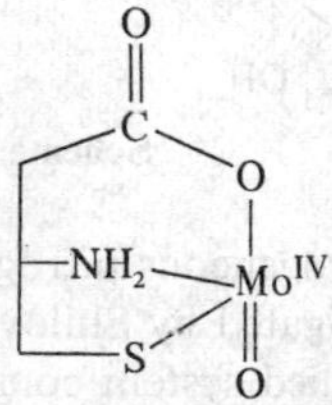

Scheme 16

Schrauzer *et al.* have shown that the compound $[MoO(H_2O)(CN)_4]^{2-}$ gives rise to a catalyst for N_2 conversion when treated with borohydride and ATP, as does MoO_4^{2-} with cyanide.[104] Inhibition and labelling studies like those discussed above lead to the mechanism of Scheme 17.[104-106] Schrauzer and co-workers have also examined the reactions of Mo^V complexes, such as the cysteine derivatives referred to above, and have shown that they disproportionate in alkali to yield a Mo^{IV} species which can react with N_2, and a Mo^{VI} species.[107] In contrast, Mo^{III} species, such as $K_3[MoCl_6]$, react with water to produce hydrogen, and dinitrogen is not reduced. These results were taken to imply that in nitrogenase the active site is a mononuclear Mo^{IV} species which binds N_2 sideways on and reduces it to diazene. The evidence has been reviewed by Schrauzer.[108]

[103] W. E. Newton, J. L. Corbin, P. W. Schneider, and W. A. Bulen, *J. Amer. Chem. Soc.*, 1971, **93**, 268.
[104] G. N. Schrauzer, P. R. Robinson, E. L. Moorehead, and T. M. Vickrey, *J. Amer. Chem. Soc.*, 1976, 2815.
[105] G. N. Schrauzer, P. R. Robinson, E. L. Moorehead, and T. M. Vickrey, *J. Amer. Chem. Soc.*, 1975, **97**, 7069.
[106] E. L. Moorehead, P. R. Robinson, T. M. Vickrey, and G. N. Schrauzer, *J. Amer. Chem. Soc.*, 1976, **98**, 6555.
[107] P. R. Robinson, E. L. Moorehead, B. J. Weathers, E. A. Ufkes, T. M. Vickrey, and G. N. Schrauzer, *J. Amer. Chem. Soc.*, 1977, **99**, 3657.
[108] G. N. Schrauzer, *Angew. Chem. Internat. Edn.*, 1975, **14**, 514.

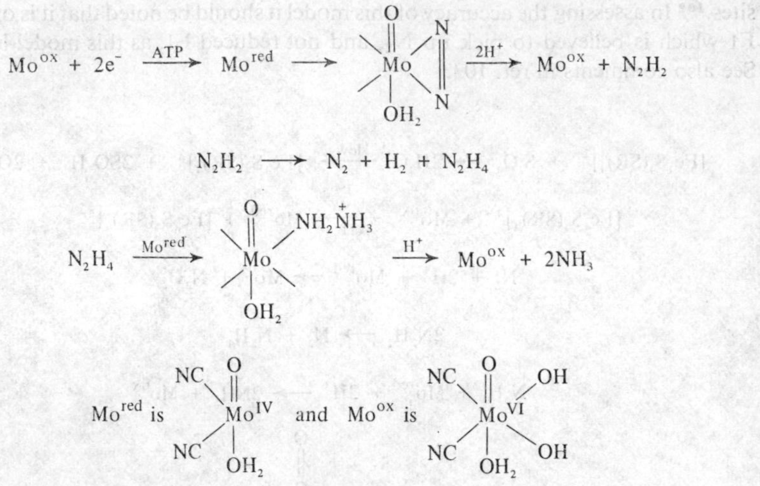

$$Mo^{ox} + 2e^- \xrightarrow{ATP} Mo^{red} \longrightarrow \underset{\underset{OH_2}{|}}{Mo} \overset{\overset{O}{\|}}{\underset{}{}} \overset{N}{\underset{N}{\|}} \xrightarrow{2H^+} Mo^{ox} + N_2H_2$$

$$N_2H_2 \longrightarrow N_2 + H_2 + N_2H_4$$

$$N_2H_4 \xrightarrow{Mo^{red}} \underset{\underset{OH_2}{|}}{Mo} \overset{O}{\underset{}{\|}} \overset{NH_2\overset{+}{N}H_3}{\underset{}{}} \xrightarrow{H^+} Mo^{ox} + 2NH_3$$

$$Mo^{red} \text{ is } \underset{NC}{\overset{NC}{\diagdown}} \underset{\underset{OH_2}{|}}{Mo^{IV}} \overset{O}{\underset{}{\|}} \text{ and } Mo^{ox} \text{ is } \underset{NC}{\overset{NC}{\diagdown}} \underset{\underset{OH_2}{|}}{Mo^{VI}} \overset{O}{\underset{}{\|}} \overset{OH}{\underset{OH}{}}$$

Scheme 17

Homogeneous and heterogeneous nitrogen-reducing systems based on vanadium salts have been investigated by Shilov[109-114] and Schrauzer[115, 116] and their co-workers. The most studied system consists of V^{II} (as sulphate, presumably) with an aromatic diol having two adjacent hydroxy-groups (*e.g.* catechol) plus base in water or methanol.[113] Under these conditions, reduction of N_2 proceeds according to reaction (16). This requires *eight* electrons to be transferred from

$$8V^{2-} + N_2 + 8H_2O \longrightarrow 8V^{3+} + 2NH_3 + H_2 + 8OH^+ \qquad (16)$$

the reducing agent, and not six. The evolution of hydrogen is seen as an integral part of the reaction, and not as an unavoidable waste of electrons. Carbon monoxide inhibits nitrogen fixation but not the evolution of hydrogen.

A kinetic study, by u.v. spectroscopy, of fixation yielded the expression $d[V^{II}]/dt = k_1[V^{II}]^2[N_2] + k_2[V^{II}]^{1/2}$, in which the first term pertains to reduction of N_2 and the second to the parallel and independent evolution of hydrogen arising from reaction (17). The first term is taken to indicate a polynuclear

$$2V^{2+} + 2H_2O \longrightarrow 2V^{3+} + H_2 + 2OH^- \qquad (17)$$

[109] N. T. Denisov, E. I. Rudshtein, N. I. Shuvalova, A. K. Shilova, and A. E. Shilov, *Doklady Akad. Nauk S.S.S.R.*, 1972, **202**, 623.

[110] L. A. Nikonova, A. G. Ovcharenko, O. N. Efimov, V. A. Avilov, and A. E. Shilov, *Kinetika i Kataliz*, 1972, **13**, 1602.

[111] N. T. Denisov, O. N. Efimov, N. I. Shuvalova, A. K. Shilova, and A. E. Shilov, *Zhur. fiz. Khim.*, 1970, **44**, 2694.

[112] A. Shilov, N. Denisov, O. Efimov, N. Shuvalov, N. Shuvalova, and A. Shilova, *Nature*, 1971, **231**, 460.

[113] L. A. Nikonova, S. A. Isaeva, N. I. Pershikova, and A. E. Shilov, *J. Mol. Catalysis*, 1975/76, **1**, 367.

[114] A. E. Shilov, *Russ. Chem. Rev.*, 1974, **43**, 863.

[115] S. I. Zones, T. M. Vickrey, J. G. Palmer, and G. N. Schrauzer, *J. Amer. Chem. Soc.*, 1976, **98**, 7289.

[116] S. I. Zones, M. R. Palmer, J. G. Palmer, J. M. Doemeny, and G. N. Schrauzer, *J. Amer. Chem. Soc.*, 1978, **100**, 2113.

structure for the V complex, probably with four V atoms, forming a hydrazido-complex containing a grouping $V{\diagup}^{\displaystyle N-N}{\diagdown}V$ as a first step in the reduction.

The hydrazido-group does not become free hydrazine, but is subsequently reduced to ammonia by another tetranuclear species, which loses its remaining two electrons to water.[117] This mechanism is very similar to that proposed for heterogeneous systems, *e.g.* $V(OH)_2 + Mg(OH)_2$ under strongly alkaline conditions, where the principal product is hydrazine rather than ammonia.[118,119]

Shilov has also demonstrated that molybdate and a reductant ($TiCl_3$, $SnCl_2$, $CrCl_2$, *etc.*) under strongly alkaline conditions readily reduces N_2 provided that Mg^{II} is present.[111,112,114,120-123] This system, too, is envisaged to fix nitrogen *via* a nitrogen molecule bridging two Mo atoms, with electron transfer to form a hydrazido(4−) complex. The influence of the Mg^{II} is unclear, but addition of $MgCl_2$ to $V(OH)_2$ and strong alkali forms a mixed Mg–V hydroxide which does not reduce water as rapidly as $V(OH)_2$ alone. Furthermore, from Mo systems treated with CO, a tetranuclear[122,124] carbonyl was isolated and shown to contain Mo^{III}. The CO can be removed by, *e.g.*, passing Ar through a solution of the complex, when nitrogen-fixing power is restored. Thus, Shilov's view of the activation of N_2 by his aqueous systems, both heterogeneous and homogeneous, is of N_2 bound between two metal atoms, with Mo^{III} being the significant oxidation state. This contrasts with Schrauzer's postulates (sideways N_2 bound to Mo^{IV}).[7] Both proposals overlook one vital feature of the natural system, namely that the oxidized form of Fl is responsible for binding N_2.

There is evidence that dinuclear complexes (*i.e.* M—N_2—M, where M = metal ion) participate in nitrogen fixation in the non-aqueous systems, originally developed by Volpin.[125,126] The kinetics of fixation in the system $FeCl_3$–PhLi–N_2 are said to suggest intermediates containing groupings of the type $[Fe(I)]_2N_2$.[127] Several groups have isolated or detected dinuclear N_2 adducts in systems containing titanium.[116,128] N_2 in the complex $[(C_5Me_5)_2(N_2)ZrN_2Zr(C_5Me_5)_2N_2]$ can be converted into hydrazine by the addition of acid,[129] although it has not

[117] L. A. Nikonova and A. E. Shilov, Proceedings of the International Conference on Nitrogen Fixation, Salamanca, 1976, ed. W. J. Newton, J. R. Postgate, and C. Rodriguez-Barrueco, Academic Press, London, New York, and San Francisco, 1977, p. 41.

[118] N. T. Denisov, N. I. Shuvalova, I. N. Ivleva, and A. E. Shilov, *Zhur. fiz. Khim.*, 1974, **48**, 2238.

[119] N. T. Denisov, *Kinetika i Kataliz*, 1976, **17**, 1044.

[120] N. T. Denisov, C. G. Terekhina, N. I. Shuvalova, and A. E. Shilov, *Kinetika i Kataliz*, 1973, **14**, 939.

[121] N. T. Denisov, A. E. Shilov, N. I. Shuvalova, and T. P. Panova, *Reaction Kin. Cat. Letters*, 1975, **2**, 237.

[122] A. E. Shilov, A. K. Shilova, and T. A. Vorontsova, *Reaction Kin. Cat. Letters*, 1975, **3**, 143.

[123] G. W. Nikolaeva, O. N. Efimov, Kh. M. Brikenstein, and A. E. Shilov, *Reaction Kin. Cat. Letters*, 1977, **6**, 349.

[124] A. E. Shilov, personal communication.

[125] M. E. Volpin and V. B. Shur, *Organometallic Reactions*, 1970, **1**, 55.

[126] M. E. Volpin, *Pure Appl. Chem.*, 1970, **30**, 607.

[127] G. Le Ny, A. E. Shilov, A. K. Shilova, and B. Tchoubar, *Nouveau J. de Chimie*, 1977, **1**, 397.

[128] R. D. Sanner, D. M. Duggan, T. C. McKenzie, R. E. Marsh, and J. E. Bercaw, *J. Amer. Chem. Soc.*, 1976, **98**, 8358.

[129] J. M. Manriquez, R. D. Sanner, R. E. Marsh, and J. E. Bercaw, *J. Amer. Chem. Soc.*, 1976, **98**, 3402.

been shown whether the intermediate which reacts with H^+ contains bridging
or terminal N_2.

An alternative approach to understanding nitrogenase through models has
been undertaken by members of the A.R.C. Unit of Nitrogen Fixation at
Brighton, who have investigated the reactivity of co-ordinated N_2 in stable, well-
characterized complexes such as cis-$[W(N_2)_2(PMePh_2)_4]$ and trans-$[Mo(N_2)_2$-
$(Ph_2PCH_2CH_2PPh_2)_2]$. The N_2 in these complexes is unequivocally bound to
the metal at one end, and is neither sideways-bound nor bridging. These complexes react with acids to yield stable products. In cases where two diphosphine
ligands are bound to the Mo (or tungsten), the ultimate products are hydrazido-
(2 −)-complexes, which can be deprotonated reversibly to give diazenido-
complexes,[130, 131] as shown in Scheme 18. However, when there are four mono-

$$[M(N_2)_2(Ph_2PCH_2CH_2PPh_2)_2] + HX \longrightarrow [MX(N_2H)(Ph_2PCH_2CH_2PPh_2)_2] + N_2$$

$$HX \big\Updownarrow base$$

$$[MX(N_2H_2)(Ph_2PCH_2CH_2PPh_2)_2]X$$

Scheme 18

tertiary phosphine ligands, the reaction can proceed further. With $[W(N_2)_2$-
$(PMe_2Ph)_4]$, 2 moles of ammonia are produced by the action of sulphuric acid
in tetrahydrofuran.[132] Even acids as weak as methanol[133, 135] are capable of protonating nitrogen in these complexes. The overall change in those examples
producing ammonia may be represented by reaction (18).[56] In general, however,

$$M^0 + N_2 + 6H^+ \longrightarrow M^{VI} + 2NH_3 \qquad (18)$$

the protonated N_2 may appear as N_2H_4 or NH_3, and some N_2 (above 1 mole)
and H_2 may be evolved. The products seem to depend upon acid, solvent, and
starting complex in a complicated fashion that is not yet completely understood.[133−135] So far, complexes such as $[MBr(N_2H)(PMe_2Ph)_4]$,[131] $[MBr_2$-
$(N_2H_2)(PMe_2Ph)_3]$,[134] and $[MBr_3(N_2H_3)(PMePh)_2]$[135] have been isolated.[136] No
complexes containing nitrogen atoms (nitrido-, imido-, or amido-complexes)
have been isolated from these systems, but such complexes should be hydrolysed
smoothly to give ammonia.[137] Scheme 19 shows how the various species may be
interconverted. The relationship of Scheme 19 to the mechanism of nitrogenase
action is, of course, open to question. The species underlined have definitely
been isolated and characterized, and it seems likely that this kind of chemistry
will be used by enzyme systems. In contrast, although the mechanism proposed
by Schrauzer and Shilov produces N_2H_4 or NH_3, no unequivocal characterization of the reactive species has been made. At present, evidence favours end-on

[130] J. Chatt, G. A. Heath, and R. L. Richards, *J.C.S. Dalton*, 1974, 2074.
[131] J. Chatt, A. J. Pearman, and R. L. Richards, *J.C.S. Dalton*, 1976, 1520.
[132] J. Chatt, A. J. Pearman, and R. L. Richards, *Nature*, 1975, **253**, 39.
[133] J. Chatt, A. J. Pearman, and R. L. Richards, *J.C.S. Dalton*, 1977, 1853.
[134] J. Chatt, A. J. Pearman, and R. L. Richards, *J. Organometallic Chem.*, 1975, **101**, C45.
[135] J. Chatt and R. L. Richards, *J. Less-Common Metals*, 1977, **54**, 477.
[136] J. Chatt, A. J. Pearman, and R. L. Richards, *J.C.S. Dalton*, 1977, 2139.
[137] J. Chatt and J. R. Dilworth, *J.C.S. Chem. Comm.*, 1975, 983.

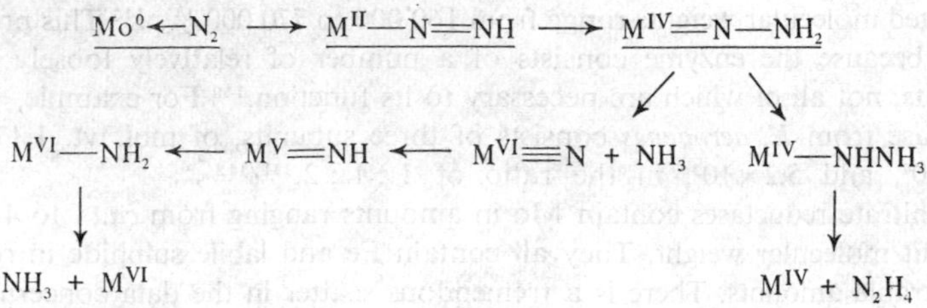

$$Mo^0 \leftarrow N_2 \longrightarrow M^{II}-N{\equiv}NH \longrightarrow M^{IV}{\equiv}N-NH_2$$

$$M^{VI}-NH_2 \longleftarrow M^V{\equiv}NH \longleftarrow M^{VI}{\equiv}N + NH_3 \qquad M^{IV}-NHNH_3$$

$$NH_3 + M^{VI} \qquad\qquad\qquad\qquad\qquad M^{IV} + N_2H_4$$

Scheme 19

binding of N_2 to Mo followed by a flow of electrons from Mo to N_2,[138] with concomitant uptake of protons from solution as a principal feature of nitrogenase action (*cf.* Scheme 20). Evidence that Scheme 20 represents the course of

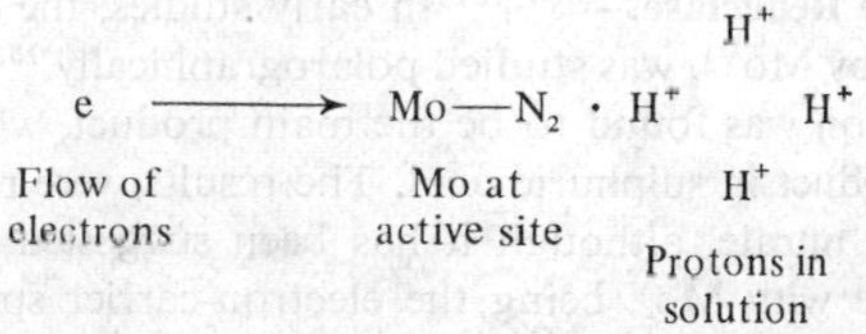

$$e \xrightarrow{} Mo-N_2 \cdot H^+ \qquad H^+$$

Flow of	Mo at	H⁺
electrons	active site	Protons in
		solution

Scheme 20

natural fixation has recently been provided.[139] If a natural system is set up to fix N_2 *in vitro*, but is quenched by acid or alkali after a short time, then hydrazine is produced, reaching a maximum yield about 200 seconds after the start of the reaction, and then falling. Evidently the material which produces hydrazine is an intermediate between N_2 and NH_3. It is unlikely to be a hydrazine complex because hydrazine is a poor substrate for nitrogenase and is reduced much more slowly than ammonia is produced. The evidence is consistent with the source of hydrazine being a species containing $Mo{=}N-NH_2$.

Nitrate Reductase.—Enzymes of this name are found in higher plants and in bacteria. Nitrogen is assimilated into plants as nitrate, which is used for the synthesis of all nitrogen-containing compounds within the cell. Nitrate can also act as a terminal hydrogen acceptor in bacteria under anaerobic conditions, when it is converted into products such as N_2 and N_2O, a process called denitrification.[140] In both cases, the initial step is conversion of nitrate into nitrite.

Two kinds of nitrate reductase are known (termed A and B), one of which (A) can also function as a chlorate reductase. There is also a chlorate reductase which is not a nitrate reductase, but otherwise seems very similar. Nitrate reductase B has different characteristics depending on its method of preparation.

[138] J. Chatt, J. R. Dilworth, and R. L. Richards, *Chem. Rev.*, 1978, **78**, 589.
[139] R. N. F. Thorneley, R. R. Eady, and D. J. Lowe, *Nature*, 1978, **272**, 557.
[140] A. H. Stouthamer, *Adv. Microbial Physiol.*, 1976, **14**, 315.

Reported molecular weights range from 160 000 to 770 000.[141-146] This probably arises because the enzyme consists of a number of relatively loosely bound subunits, not all of which are necessary to its function.[140] For example, nitrate reductase from *K. aerogenes* consists of three subunits, of mol. wt. 1.17×10^5, 5.7×10^4, and 5.2×10^4, in the ratio of 1 : 1 : 2.[141,142]

All nitrate reductases contain Mo in amounts ranging from *ca.* 1 to 4 atoms per unit molecular weight. They all contain Fe and labile sulphide in roughly comparable amounts. There is a tremendous scatter in the data concerning Fe content.[140] Mo is necessary for nitrate reduction, and there is evidence for a Mo cofactor (Mo-Co) common to several Mo-containing enzymes (but not including nitrogenase). For further details, see Chapter 7. The oxidized enzyme shows e.s.r. signals assignable to Mo^V.[147] The e.s.r. spectra also show signals attributable to Fe–S centres.[141,148,149] Reduction of nitrate reductases may produce Mo^{III}.[148,149] Nitrate apparently re-oxidizes the enzyme. In the natural system Mo probably cycles between oxidation states (v) and (vi).

Models for Nitrate Reductase.—[150,151] In early studies, the reduction of nitrate by Sn^{II}, catalysed by Mo^{VI}, was studied polarographically.[152,153] In hydrochloric acid, ammonium ion was found to be the main product, whereas nitrous oxide was the major product in sulphuric acid. The results were rationalized in terms of Mo^{IV} reducing nitrate, although it has been suggested that the results are equally compatible with Mo^V being the electron-carrier species. Similarly, the Mo^{VI}-catalysed reduction of nitrate at a dropping mercury electrode has been said to proceed *via* a Mo^{IV} species,[154] although a monomeric Mo^V species has also been proposed[155] as being consistent with the experimental data. The electrochemical reduction of nitrate has been held to follow the sequence: $NO_3^- \rightarrow NO_2 \rightarrow HNO_2 \rightarrow NO \rightarrow N_2O_2^{2-} \rightarrow NH_2OH \rightarrow NH_3$, with N_2O and N_2 being formed by side-reactions.[156]

Molybdenum oxo-species are prone to polymerization, and the presence of a monomeric rather than a polymeric Mo^V species has been held to be a vital part of nitrate reduction.[157] The possibility that mononuclear Mo^V species are present

141 J. van't Riet, J. H. van Ee, R. Wever, B. F. van Gelder, and R. J. Planta, *Biochim. Biophys. Acta*, 1975, **405**, 306.
142 J. van't Riet and R. J. Planta, *F.E.B.S. Letters*, 1969, **5**, 249.
143 C. H. Macgregor, C. A. Schnaitman, D. E. Normansell, and M. G. Hodgens, *J. Biol. Chem.*, 1974, **249**, 5321.
144 P. Forget, *European J. Biochem.*, 1974, **42**, 325.
145 P. Forget, *European J. Biochem.*, 1971, **18**, 442.
146 J. P. Rosso, P. Forget, and F. Pichinoty, *Biochim. Biophys. Acta*, 1973, **321**, 443.
147 R. C. Bray, Proceedings of the International Conference on Chemistry and Uses of Molybdenum, 1976, p. 271.
148 P. Forget and D. V. Der Vartanian, *Biochim. Biophys. Acta*, 1972, **256**, 600.
149 D. V. Der Vartanian and P. Forget, *Biochim. Biophys. Acta*, 1975, **397**, 1374.
150 E. I. Stiefel, *Progr. Inorg. Chem.*, 1977, **22**, 1.
151 J. T. Spence, in 'Metal Ions in Biological Systems,' ed. H. Sigel, Dekker, New York and Basle, Vol. 5, p. 279.
152 G. P. Haight, P. Mohilner, and A. Katz, *Acta Chem. Scand.*, 1962, **16**, 221.
153 G. P. Haight and A. Katz, *Acta Chem. Scand.*, 1962, **16**, 659.
154 G. P. Haight, *Acta Chem. Scand.*, 1961, **15**, 2012.
155 J. M. Kolthoff and I. Hodara, *J. Electroanalyt. Chem.*, 1963, **5**, 2.
156 R. K. Kvaratskheliya, *Izvest. Akad. Nauk. Gruz. S.S.R.*, Ser. khim., 1976, **2**, 140.
157 C. A. Fewson and D. J. D. Nicholas, *Biochim. Biophys. Acta*, 1961, **49**, 335.

to some degree in solutions containing Mo^V dinuclear species has produced some interesting observations.[158] A Mo^V species prepared from sodium molybdate by electrochemical reduction in tartrate or citrate buffer reduces nitrate to NO. In phosphate buffer, and using the biological electron-transfer agent flavin mononucleotide, the reduction is very slow.[159] An e.s.r. signal from tartrate-buffered solutions was taken to indicate a monomeric Mo^V species, and the mechanism of Scheme 21 was proposed.[158] No e.s.r. signals were observed from

$$Mo^V_2 \rightleftharpoons 2Mo^V$$

$$Mo^V + NO_3^- \longrightarrow Mo^{VI} + NO_2$$

$$2NO_2 \longrightarrow NO^+ + NO_3^-$$

$$NO^+ + Mo^V \longrightarrow NO + Mo^{VI}$$

$$NO^+ + H_2O \rightleftharpoons HNO_2 + H^+$$

Scheme 21

solutions buffered with phosphate. A detailed study of the reduction of Mo^{VI} to Mo^V by flavins, which are probably the biological electron donor to Mo in nitrate reductase, yielded strong evidence for mononuclear Mo^V species of the type $[MoL_3]$ (L = tartrate).[160] This species could also be obtained by adding the Mo^V species $(NH_4)_2[MoOCl_5]$ directly to the buffer solution.[160] Such experiments support the idea that Mo^V mononuclear complexes can be significant components in biological systems.

Most recently, flavin mononucleotide has been shown to reduce nitrate to NO, but a source of Mo^V or Mo^{VI} is also necessary. Both $[MoOCl_5]^{2-}$ and $[MoO_4]^{2-}$ were used.[161] This system was demonstrated to operate through a mononuclear Mo^V tartrate as electron carrier, and to behave essentially as described above in the reduction of nitrate. The product of reduction in these systems is NO, whereas, biologically, nitrite is the reduction product. In other cases (see below) NO_2 is produced, and is a poor ligand for Mo^V. Because nitrite is a competitive inhibitor for nitrate reduction, it is possible that NO_2 is a primary product in the biological system and that it produces nitrite by reaction with water once it leaves the Mo. Since Mo^V reduces nitrite more rapidly than it reduces nitrate in aqueous solution, these considerations suggest that Mo in nitrate reductase may be in a non-aqueous environment.[162,163]

These considerations have led to a study of the reduction of nitrate by Mo^V systems in non-aqueous solvents. The complexes $[MoOCl_3L]$ (L = a bidentate

[158] E. P. Guymon and J. T. Spence, *J. Phys. Chem.*, 1966, **70**, 1964.
[159] J. T. Spence and J. A. Frank, *J. Amer. Chem. Soc.*, 1963, **85**, 116.
[160] J. T. Spence and M. Heydanek, *Inorg. Chem.*, 1967, **8**, 1489.
[161] J. T. Spence, *Arch. Biochem. Biophys.*, 1970, **137**, 288.
[162] J. A. Frank and J. T. Spence, *J. Phys. Chem.*, 1964, **68**, 2131; see also M. R. Hyde and C. D. Garner, *J. C. S. Dalton*, 1975, 1186.
[163] J. T. Spence and R. D. Taylor, Proceedings of the International Conference on Chemistry and Uses of Molybdenum, Second, 1976, p. 230.

ligand such as *o*-phenanthroline or bipyridyl) in anhydrous DMF produce NO_2 and $[MoO_2Cl_2L]$, possibly by the following sequence:

$$[MoOCl_3L] \rightleftharpoons [MoOCl_2L]^+ + Cl^-$$

$$[MoOCl_2L] + NO_3^- \rightleftharpoons [MoO(ONO_2)Cl_2L]$$

$$[MoO(ONO_2)Cl_2L] \longrightarrow [MoO_2Cl_2L] + NO_2$$

This parallels the reaction path inferred for the interaction of nitrate with $[MoOCl_5]^{2-}$,[164] although in the latter case the rate-controlling step is electron transfer, rather than dissociation of chloride as above (the reaction is inhibited by Cl^-)[*cf.* reaction (19)].

$$[MoOCl_5]^{2-} + NO_3^- \longrightarrow [MoO_2Cl_2] + NO_2 + 3Cl^- \qquad (19)$$

The individual steps proposed are:

$$[MoOCl_5]^{2-} \longrightarrow [MoOCl_4]^- + Cl^-$$

$$[MoOCl_4]^- + NO_3^- \longrightarrow [MoOCl_3(NO_3)] + Cl^-$$

$$[MoOCl_3(NO_3)]^- \longrightarrow [MoO_2Cl_2] + Cl^- + NO_2$$

As in the biological system, free NO_2 is suggested to react with water to form nitrite, as shown in reaction (20). The rate-controlling dissociation of chloride

$$H_2O + 2NO_2 \longrightarrow NO_3^- + HNO_2 + H^+ \qquad (20)$$

is believed to take place from *trans* to Mo=O. Co-ordination and reduction of nitrate should yield a *trans*-dioxo-product, but *cis*-dioxo-species are isolated, so rearrangement must occur.[163,164] This picture has been challenged in its details. The substitution reactions of $[MoOCl_3(OPPh_3)_2]$ with chloride and bromide in dichloromethane have been shown to yield $[MoOCl_3X(OPPh_3)]^-$ (X = Cl or Br), with X taking up a final position *cis* to Mo=O, but the reactions are best described as of S_N1 type, with the Ph_3PO *trans* to Mo=O being the leaving group. Furthermore, the reaction of the $[MoOCl_4]^-$, which has a vacant co-ordination site *trans* to Mo=O, with isothiocyanate apparently produces $[MoOCl_4(SCN)]^{2-}$ with thiocyanate *cis* to Mo=O.[165] Subsequently, the reaction of $[MoOCl_3(OPPh_3)_2]$ with nitrate was shown to yield *cis*-$[MoO_2Cl_3(OPPh_3)]^-$ and NO_2.[166,167] Evidence was adduced to show that nitrate binding to Mo^V by a single oxygen is necessary for facile electron transfer. As a consequence of this pattern, it was suggested that in the system $[MoOCl_5]^{2-}$–NO_3^-–DMF, $[MoOCl_4]^-$ is unlikely to be generated as suggested,[163,164] and e.s.r. evidence was presented[168] to support the presence of $[MoOCl_4(DMF)]^-$ in the sequence of Scheme 22. The binding of nitrate is again taken to be *via* a single oxygen.

[164] J. D. Spence and R. D. Taylor, *Inorg. Chem.*, 1975, **14**, 2815.
[165] C. D. Garner, M. R. Hyde, F. E. Mabbs, and V. I. Routledge, *J.C.S. Dalton*, 1975, 1175; P. M. Boorman, C. D. Garner, and F. E. Mabbs, *ibid.*, p. 1899.
[166] R. Durrant, C. D. Garner, M. R. Hyde, F. E. Mabbs, J. R. Parsons, and D. Richens, Proceedings of the International Conference on Chemistry and Uses of Molybdenum, Second, 1976, p. 234.
[167] C. D. Garner, M. R. Hyde, F. E. Mabbs, and V. I. Routledge, *J.C.S. Dalton*, 1975, 1180; *Nature*, 1974, **252**, 579
[168] C. D. Garner, M. R. Hyde, and F. E. Mabbs, *Inorg. Chem.*, 1976, **15**, 2327.

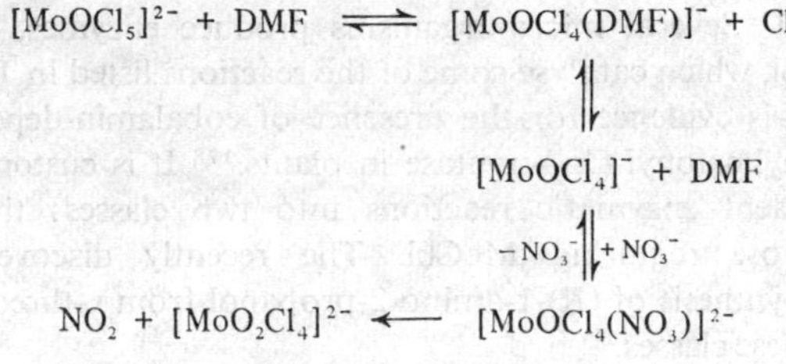

$$[MoOCl_5]^{2-} + DMF \rightleftharpoons [MoOCl_4(DMF)]^- + Cl^-$$

$$[MoOCl_4]^- + DMF$$

$$-NO_3^- \Big\Updownarrow + NO_3^-$$

$$NO_2 + [MoO_2Cl_4]^{2-} \longleftarrow [MoOCl_4(NO_3)]^{2-}$$

Scheme 22

These views have been criticized[169] because this scheme does not necessarily account for inhibition by chloride,[163,164] and the amended mechanism also requires a fast isomerization, which has been observed in other systems. The transformation shown in reaction (21) has been suggested.[169] All participants in the mechanistic discussion agree that the metalloenzyme site may be anhydrous,

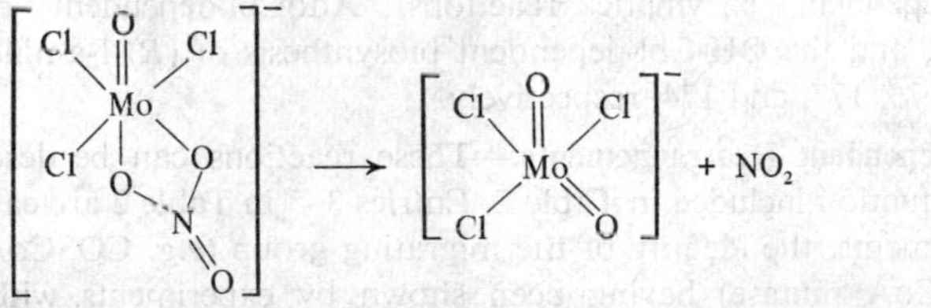

$$\begin{bmatrix} Cl & \overset{O}{\underset{\|}{Mo}} & Cl \\ Cl & & O \\ & O & \\ & & N \\ & & O \end{bmatrix}^- \longrightarrow \begin{bmatrix} Cl & \overset{O}{\underset{\|}{Mo}} & Cl \\ Cl & & O \\ Cl & & O \end{bmatrix}^- + NO_2 \quad (21)$$

that Mo^V and Mo^{VI} are probably involved [note that other oxidation states of Mo such as Mo^{III} also reduce nitrate], and that NO_2 is likely to be the initial product, disproportionating in water. Proton transfer coupled with electron transfer has been suggested as a significant feature of the function of nitrate reductase.[170] Because the best understood models are anhydrous systems, there is a need to develop models in aqueous media.

4 Cobalamin-dependent Enzymatic Reactions

Introduction.—The recognized functions of cobalt in living things derive from its presence in vitamin B_{12} (cyanocobalamin) and other corrinoids.* The main corrinoids in human serum and tissues are hydroxocobalamin (OH-Cbl), methylcobalamin (MeCbl), and adenosylcobalamin (AdoCbl).[171] AdoCbl is a cofactor for the conversion of methylmalonyl-CoA into succinyl-CoA, catalysed by methylmalonyl-CoA mutase (*cf.* Table 2, entry 3). MeCbl participates in the synthesis of methionine from homocysteine, catalysed by N^5-methyltetrahydro-folate–homocysteine methyltransferase (*cf.* Table 2, entry 10). These are the only roles for cobalamins in human metabolism which have been identified at a

---

* For nomenclature of corrinoids, see *Biochemistry*, 1974, **13**, 1555.

169 J. T. Spence and R. D. Taylor, *Inorg. Chem.*, 1977, **16**, 1256.
170 E. I. Stiefel and J. K. Gardner, Proceedings of the International Conference on Chemistry and Uses of Molybdenum, First, 1973, p. 272; E. I. Stiefel, *Proc. Nat. Acad. Sci. U.S.A.*, 1973, **70**, 988.
171 For a review of cobalamin chemistry and biochemistry, B. T. Golding, in 'Comprehensive Organic Chemistry', ed. E. Haslam, Pergamon Press, 1978, Vol. 5, Ch. 24.4

molecular level.[172] Several micro-organisms produce enzymes, dependent on AdoCbl or MeCbl, which catalyse some of the reactions listed in Table 2 (entries 1, 2, 4–6). There is evidence for the presence of cobalamin-dependent leucine mutase and methylmalonyl-CoA mutase in plants.[173] It is customary to divide cobalamin-dependent enzymatic reactions into two classes: those requiring AdoCbl and those requiring MeCbl. The recently discovered OH-Cbl-dependent[174] biosynthesis of (R)-1-amino-2-propanol from L-threonine does not fit into either of these classes.

No cobalamin-dependent enzyme has been sequenced, and only one has been crystallized (methylmalonyl-CoA mutase from *Propionibacterium shermanii*, as a complex with OH-Cbl).[175] Mechanistic discussions have therefore been directed to understanding the function of the cobalamin, whereas the protein has been, of necessity, almost ignored. This year, attention is focussed on AdoCbl-dependent rearrangements because nearly all recent model studies of cobalamin-dependent enzymes have been concerned with these reactions. For model studies on MeCbl-dependent enzymatic reactions, AdoCbl-dependent ribonucleotide reductase, and the OH-Cbl-dependent biosynthesis of (R)-1-amino-2-propanol, see refs. 176, 177, and 174, respectively.

AdoCbl-dependent Rearrangements.—These reactions can be described by the general equation included in Table 2. Entries 3–5 in Table 2 are carbon-skeleton rearrangements, the identity of the migrating group (*e.g.* COSCoA for methylmalonyl-CoA mutase) having been shown by experiments with isotopically labelled substrates in the case of entries 3 and 4. Diol dehydrase catalyses the migration of a hydroxy-group from C-2 of propane-1,2-diol, as proved by ¹⁸O-labelling.[178] With ethanolamine ammonia lyase it is inferred that an amino-group is transferred from C-2 of, *e.g.*, 2-aminoethanol to C-1 of intermediate 1-aminoethanol, which loses ammonia to give ethanal.

Three distinct pathways for AdoCbl-dependent rearrangements have been proposed, and for convenience these are labelled (i), (ii), and (iii).

(i) *Reaction pathway of Abeles and co-workers*[179] (*Scheme* 23). From experiments with AdoCbl and substrates for diol dehydrase specifically labelled with isotopes of hydrogen, Abeles concluded that the reaction is initiated by removal of a hydrogen atom from C-1 of a substrate molecule (*e.g.* MeCHOHCH₂OH = SH) by an adenosyl radical [derived with cob(II)alamin by homolysis of the Co—C bond of AdoCbl], giving a substrate-derived radical (MeCHOHĊHOH = S·) and deoxyadenosine (29). The radical S· rearranges to product-related radical P· [MeĊHCH(OH)₂], which takes a hydrogen atom back from the methyl

172 *cf.* 'Cobalamin, Biochemistry and Pathophysiology', ed. B. M. Babior, Wiley, New York, 1975.
173 J. M. Poston, *Phytochemistry*, 1978, **17**, 401.
174 S. H. Ford and H. C. Friedmann, *Biochim. Biophys. Acta*, 1977, **500**, 217.
175 B. Zagalak, J. Retey, and H. Sund, *European J. Biochem.*, 1974, **44**, 529.
176 G. N. Schrauzer, *Angew. Chem. Internat. Edn.*, 1977, **16**, 233.
177 H. P. C. Hogenkamp and G. N. Sands, *Structure and Bonding*, 1974, **20**, 23; V. I. Borodulina-Shvets, I. P. Rudakova, and A. M. Yurkevich, *J. Gen. Chem.* (*U.S.S.R.*), 1971, **41**, 2832.
178 J. Retey, A. Umani-Ronchi, J. Seibl, and D. Arigoni, *Experientia*, 1966, **22**, 502.
179 T. H. Finlay, J. Valinsky, K. Sato, and R. H. Abeles, *J. Biol. Chem.*, 1972, **247**, 4197; R. H. Abeles and D. Dolphin, *Accounts Chem. Res.*, 1976, **9**, 114.

Table 2 *Enzymatic reactions requiring a corrinoid cofactor*

Ado Cbl-dependent rearrangements

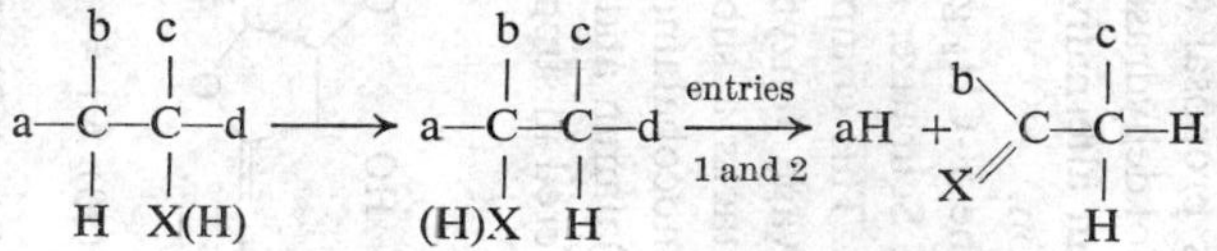

	a	b	c	d	X(H)
1 (a) Diol dehydrase	OH	H	H	Me or H	OH
(b) Glycerol dehydrase	OH	H	H	CH₂OH	OH
2 Ethanolamine ammonia lyase	OH	H	H	Me or H	NH₂
3 (R)-Methylmalonyl-CoA mutase	Me or H	H	CO₂H	H	COSCoA
4 (S)-Glutamate mutase	H	H	CO₂H	H	CHNH₃CO₂⁻
5 α-Methyleneglutarate mutase	H	H	CO₂H	H	(C=CH₂)CO₂⁻ (probable migrating group)
6 Aminomutases, *e.g.* leucine aminomutase	Me₂CH	H	CO₂H	H	NH₂

7 Ribonucleotide reductase [catalyses: Ribonucleotide (*e.g.* ATP) ⟶ Deoxyribonucleotide (*e.g.* dATP)]

MeCbl-dependent reactions:

8 Synthesis of methane from CO₂
9 Synthesis of ethanoic acid from 2CO₂
10 Synthesis of methicnine (*N*-methyltetrahydrofolate + homocysteine ⟶ tetrahydrofolate + methionine)

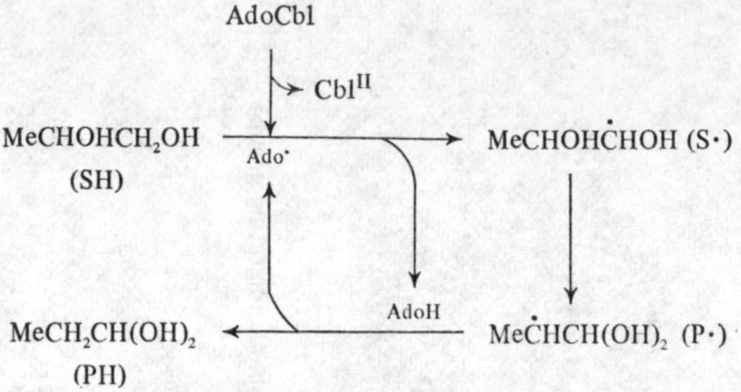

AdoCbl

Mechanism for diol dehydrase (ref. 179)
Scheme 23

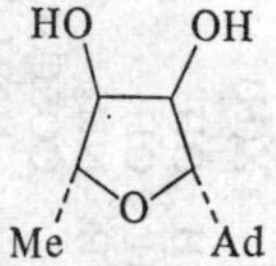

(29) Ad = adenine

group of (29), regenerating adenosyl radical and producing PH [MeCH$_2$-CH(OH)$_2 \to$ MeCH$_2$CHO + H$_2$O]. Experimental evidence supporting this Scheme will be cited below. A similar Scheme can be written for other AdoCbl-dependent rearrangements.

(ii) *Reaction pathway of Schrauzer.*[176, 180] This proposal (Schemes 6 and 7 in ref. 176) was applied to reactions catalysed by diol dehydrase, glycerol dehydrase, and ethanolamine ammonia lyase, and offers an alternative explanation to the labelling results referred to in the previous section.

To explain the connection between label in the 5'-CH$_2$ group of AdoCbl and at C-2 of a product molecule such as propanal, Schrauzer suggests an exchange reaction involving 4',5'-anhydroadenosine (30). This compound is supposed to arise, with cob(i)alamin, by heterolytic cleavage (enzyme-catalysed) of the Co—C bond of AdoCbl. The cob(i)alamin attacks a substrate molecule (*e.g.* propane-1,2-diol) to give an intermediate organocobalamin which decomposes by an intramolecular 1,2-hydride shift to cob(i)alamin and a product molecule. The model experiments (a), (b), and (c) were offered in support of this scheme.

Model experiment (a):

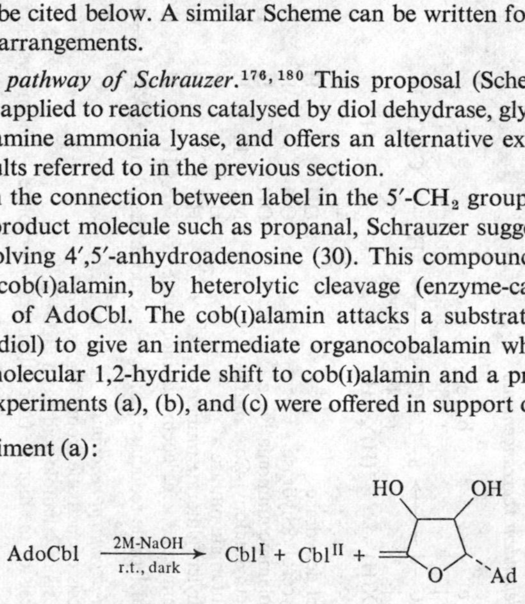

$$\text{AdoCbl} \xrightarrow[\text{r.t., dark}]{\text{2M-NaOH}} \text{Cbl}^{\text{I}} + \text{Cbl}^{\text{II}} + \quad\quad\quad\quad (22)$$

(30) Ad = adenine

[180] G. N. Schrauzer and J. W. Sibert, *J. Amer. Chem. Soc.*, 1970, **92**, 1022.

(30) was not identified from reaction (22). Its 2′,3′-isopropylidene derivative was 'definitely identified'[180] (comparison with an authentic sample in two t.l.c. systems) from treating the 2′,3′-isopropylidene derivative of AdoCbl with alkali. This reaction is likely to proceed by an $E2$ elimination mechanism rather than $E1$ (5′-deoxyadenosyl cation insufficiently stable) or $E1cB$ (4′-H insufficiently acidic). For an $E2$ mechanism in the structural fragment $H-C_\alpha-C_\beta-X$ (X = leaving group), the $H-C_\alpha$ and $C_\beta-X$ bonds must be in the same plane (cisoid or transoid). With AdoCbl (X = Cbl), only one of the permissible

$$\text{AdoCbl} \xrightarrow[?]{\text{OH}^-} \text{Cbl}^{\text{I}} + \ (23)$$

conformations (transoid) for $E2$ elimination allows access of hydroxide ion or of enzymatic base to 4′-H. However, this conformation is probably unfavourable, owing to close contact of the ribose ring with the corrin. A lower energy reaction for AdoCbl with base might be the fragmentation shown in reaction (23), which was not considered in ref. 180 as an alternative to reaction (22). It is concluded that the model reaction (22) requires verification.

Model experiment (b):*

$$[\text{HOCH}_2\text{CH}_2\text{Co(dmgH)}_2(\text{py})] \xrightarrow[\text{r.t.}]{0.1\text{M-NaOH}} \text{cobaloxime(I)} + \text{MeCHO} \quad (24)$$

$$[\text{MeCHOHCH}_2\text{Co(dmgH)}_2(\text{py})] \xrightarrow[\text{warm}]{\text{M-NaOH}} \text{MeCOMe} \quad (25)$$

$$(60\%, \text{ by g.c./m.s.})$$

Intramolecular 1,2-hydride shifts probably occur in these reactions, *i.e.* as shown in reaction (26). An attempt to prove that there is an intramolecular pathway by allowing $[\text{HOCH}_2\text{CH}_2\text{Co(dmgH)}_2(\text{py})]$ to react with $^2\text{H}_2\text{O}-\text{O}^2\text{H}^-$ was stymied by exchange of the methyl protons of ethanal under the reaction

$$(26)$$

conditions. However, a kinetic isotope effect of 5.5 (k_H/k_D) was observed for the decomposition of $\text{MeC}^2\text{HOHCH}_2\text{Cbl}$ and the corresponding cobaloxime in 2M-NaOH. Further indirect evidence for the proposed mechanism was the finding that 2-alkoxyethyl-cobaloximes are resistant to heating in 50% KOH.

Model experiment (c):

$$\xrightarrow[\substack{\text{or} \\ \text{neutral} \\ \text{Al}_2\text{O}_3, \ 65\ °\text{C}}]{\text{MeCO}_2{}^2\text{H}} \quad (27)$$

* dmgH = dimethylglyoxime, py = pyridine; cobaloximes are bis(dimethylglyoximato) cobalt complexes used as models for cobalamins (*cf.* ref. 171 and G. N. Schrauzer, *Angew. Chem. Internat. Edn.*, 1976, **15**, 417).

Presumably ethanoic acid catalyses the exchange of deuterium between propanone and 2-methylenetetrahydrofuran by reversibly protonating each of these
substances. Alumina probably deprotonates propanone, and uses the proton
acquired to protonate 2-methylenetetrahydrofuran. An enzymatic acid or base
could conceivably shuttle protons between (30) and C-2 of propanal in similar
ways.

Schrauzer has described[181] experiments with diol dehydrase and ethanolamine
ammonia lyase in support of his mechanistic scheme. He reported that N_2O,
believed to be a specific oxidant for cob(I)alamin, inhibits these enzymes. He has
also claimed to have isolated (30) from incubating either enzyme with [5'-³H]Ado-
Cbl. The incubations were interrupted by lyophilization and unlabelled (30)
was added. Paper chromatograms of re-isolated (30) showed low radioactivities
(*ca.* 0.5% of the activity of recovered AdoCbl) under spots corresponding to (30).

(*iii*) *Reaction pathway of Corey et al.*[182] This idea (*cf.* Figure 2 in ref. 182) uses
the organometallic process of oxidative addition/reductive elimination as a model
system for AdoCbl-dependent enzymatic reactions. Such a process with AdoCbl
requires that two of the six ligands to Co temporarily depart, so as to provide two
co-ordination sites on Co to which a substrate can oxidatively add. Reversible
detachment of the benzimidazole is well known. To create another co-ordination
site, Corey *et al.*[182] suggest an electrocyclic reaction which is essentially the
reverse of a reaction by which Eschenmoser synthesized corrins. Eventually, this
allows one nitrogen atom (ring A of the corrin) to leave Co. Rearrangement of
substrate to product is accomplished *via* a metal–carbene intermediate.[183] For
the carbon-skeleton rearrangements of Table 2 (entries 3—5), the transformation
of reaction (28) is cited as a model. This paper does not account for butane-2,3-
diol being a substrate for diol dehydrase, and too readily dismisses the evidence
that radicals are intermediates in AdoCbl-dependent rearrangements.

$$(CO)_5W{=}C\underset{Ph}{\overset{CH_2R}{\big<}} \longrightarrow (CO)_5W\underset{CHPh}{\overset{CHR}{\big|}} \qquad (28)$$

$$(R = H,\ Pr^n,\ or\ CH{=}CH_2)$$

Further Evidence for Scheme 23 from Enzymological Experiments.—According to
ref. 182, 'there is no convincing direct evidence for the intermediacy of carbon
radicals' in AdoCbl-dependent reactions. Prior to this statement, four
groups[183–186] had reported on experiments wherein holoenzyme (diol dehydrase,

[181] R. N. Katz, T. M. Vickrey, and G. N. Schrauzer, *Angew. Chem. Internat. Edn.*, 1976, **15**, 542.
[182] E. J. Corey, N. J. Cooper, and M. L. H. Green, *Proc. Nat. Acad. Sci. U.S.A.*, 1977, **74**, 811.
[183] K. W. Shepler, W. R. Dunham, R. H. Sands, J. A. Fee, and R. H. Abeles, *Biochim. Biophys. Acta*, 1975, **397**, 510; J. E. Valinsky, R. H. Abeles, and A. S. Mildvan, *J. Biol. Chem.*, 1974, **249**, 2751; J. E. Valinsky, R. H. Abeles, and J. A. Fee, *J. Amer. Chem. Soc.*, 1974, **96**, 4709.
[184] S. A. Cockle, H. A. O. Hill, R. J. P. Williams, S. P. Davies, and M. A. Foster, *J. Amer. Chem. Soc.*, 1972, **94**, 275.
[185] B. M. Babior, T. H. Moss, W. H. Orme-Johnson, and H. Beinert, *J. Biol. Chem.*, 1974, **249**, 4537.
[186] W. H. Orme-Johnson, H. Beinert, and R. L. Blakley, *J. Biol. Chem.*, 1974, **249**, 2338.

glycerol dehydrase, ethanolamine ammonia lyase, or ribonucleotide reductase +
AdoCbl) in reaction mixtures containing substrate are quenched in liquid N_2
and then observed by e.s.r. spectroscopy. The resulting spectra show a broad
signal near $g = 2.3$ and a doublet around $g = 2.0$. 'In every system studied,
the appearance of the characteristic EPR spectrum occurs at a rate which is
faster than or comparable to the turnover rate of the enzyme'.[183] Schrauzer[176]
has commented that in the case of diol dehydrase 'the signals are not very
characteristic and could not be identified unambiguously'. Actually, the signal
at $g = 2.3$ has been assigned to cob(II)alamin, because this species is known
to give a strong signal at this position. The signal at $g = 2.0$ is believed to
arise from an organic radical derived from a substrate molecule, because spectra
from experiments with isotopically labelled substrates (^2H and ^{13}C) show slight
modifications to the shape of this signal. Attempts[183, 187] have been made to
account theoretically for the doublet character of this signal. Boas *et al.*[187]
interpret the signal using a model which allows for both isotropic exchange
coupling and dipolar interactions between cob(II)alamin and a radical. Spectra
were calculated for various distances (*cf.* Figure 4) between the radical and
cob(II)alamin. The best fit of calculated and observed spectra was obtained with
$r \geqslant 10$ Å ($\xi \approx 0°$). For ethanolamine ammonia lyase, r was found to lie in the
range 10—12 Å. Increasing ξ above 20° caused a 'considerable deterioration' of
the calculated spectrum compared with experiment, and it was concluded that
ξ lies in the range 0—10°.

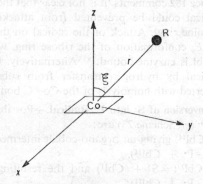

Figure 4 *The parameters involved in calculating spectra for complexes of a radical
(R) and the active centre of cob(II)alamin*

Johnson and co-workers[188] have studied reactions between ethanolamine
ammonia lyase and its substrates ethanolamine and L-2-aminopropan-1-ol,
using a rapid-wavelength-scanning stopped-flow spectrophotometer. With
ethanolamine, it was found that, during the steady state of the reaction, the
cobalamin existed as 58% cob(II)alamin and 42% of a cob(III)alamin (either

[187] J. F. Boas, P. R. Hicks, J. R. Pilbrow, and T. D. Smith, *J.C.S. Faraday II*, 1978, **74**, 417;
G. R. Buettner and R. E. Coffman, *Biochim. Biophys. Acta*, 1977, **480**, 495.
[188] M. R. Hollanay, H. A. White, K. N. Joblin, A. W. Johnson, M. F. Lappert, and O. C.
Wallis, *European J. Biochem.*, 1978, **82**, 143.

AdoCbl or an organocobalamin derived from substrate, the second possibility being favoured). With L-2-aminopropan-1-ol, $>95\%$ of cobalt was identified as cob(II)alamin. No evidence was found for the presence of cob(I)alamin (required for Schrauzer's mechanism; see above), which has a very strong absorption maximum at 386 nm.

Current experimental evidence, therefore, favours Scheme 23 over alternatives, and the remainder of this section is devoted to model studies which are relevant to this Scheme and which seek to define the nature of the conversion of S· into P· radical.

Cleavage of the Co—C Bond of Alkyl-cobalt Compounds.—In the model compounds $[Co([14]tetraeneN_4)(OH_2)Me]^{2+}$ and $[Co([14]aneN_4)(OH_2)Me]^{2+}$, the strength of their Co—Me bond has been estimated as ~ 200 kJ mol^{-1} (covalent contribution ~ 135, ligand-field stabilization energy ~ 65 kJ mol^{-1}).[189] Co—C bond strengths in AdoCbl and other RCbl are unknown. Although homolytic cleavage of the Co—C bond of AdoCbl and other RCbl has been clearly established under conditions of photolysis, this is not so for thermolysis.[190] Anaerobic photolysis of AdoCbl gives CblII and 8′,5′-cyclic-adenosine, presumably *via* an adenosyl radical, but AdoCbl is stable to heating at 94 °C for 5 h. Thermolysis of EtCbl at 80 °C gives CblII and ethylene. This reaction may be a concerted β-elimination or the most favourable pathway for decomposition of a CblII–Et· pair. If a free adenosyl radical is formed in AdoCbl-dependent reactions, the energy needed to cleave the Co—C bond must be recouped from binding energy to the protein. Reference 182 comments 'it is not clear that the extremely reactive 5′-deoxyadenosyl radical could be prevented from attacking nearby groups including its own adenine ring'. Attack of the radical on the adenine ring can only proceed *via* the E_0 conformation of the ribose ring, which might not be available when AdoCbl is enzyme-bound.[191] Alternatively, the formation of a substrate-derived radical by hydrogen transfer from substrate to adenosyl 5′-CH$_2$ could be concerted with homolysis of the Co—C bond of AdoCbl.

The Nature of the Conversion of S· into P· Radical.—Possible pathways for the conversion of S· into P· (*cf.* Scheme 23) are:

(*a*) S· combines with CblII, giving an organo-cobalt intermediate (S-Cbl) which rearranges to P-Cbl (→ P· + CblII).

(*b*) S· is oxidized by CblII(⇌ S$^+$ + CblI) and the resulting S$^+$ rearranges to P$^+$ (which with CblI → P· + CblII).

(*c*) S· is reduced by CblII (⇌ S$^-$ + CblIII) and the resulting S$^-$ rearranges to P$^-$ (which with CblIII → P· + CblII).

(*d*) S· directly rearranges to P· (without intervention of electron-transfer steps or the formation of organo-cobalt intermediates).

The chemistry of carbonium ions and carbanions is relevant to pathways (*b*) and (*c*) respectively. Reasonable mechanisms for diol dehydrase and ethanolamine ammonia lyase based on pathway (*b*) can be written (see, *e.g.*, ref. 192),

[189] C. Y. Mok and J. F. Endicott, *J. Amer. Chem. Soc.*, 1978, **100**, 123.
[190] H. P. C. Hogenkamp, P. J. Vergamini, and N. A. Matwiyoff, *J.C.S. Dalton*, 1975, 2628.
[191] B. T. Golding and L. Radom, *J. Amer. Chem. Soc.*, 1976, **98**, 6331.
[192] J. Halpern, *Ann. New York Acad. Sci.*, 1974, **239**, 2.

whereas for methylmalonyl-CoA mutase, mechanisms based on either (b) or (c) are conceivable [see equations (2) and (3) of ref. 191]. Chemical information relevant to pathways (a) and (d) was lacking, and so model studies have concentrated on trying to simulate these pathways.

Model Studies Pertaining to Pathway (a). 2-Acetoxyalkyl(pyridine)cobaloximes undergo rapid alcoholysis in neutral solution, probably *via* an intermediate π-complex of Co^{III},[193, 194] *e.g.* as shown in reaction (29).

$$[AcOCH_2CH_2Co(dmgH)_2(py)] \xrightarrow[\tau_{1/2} = 40\,h]{EtOH,\ 25\ ^\circ C} [EtOCH_2CH_2Co(dmgH)_2(py)] + AcOH \quad (29)$$

Babior[195] had speculated on the intermediacy of an analogous π-complex in the ethanolamine ammonia lyase reaction. This idea was developed by Dolphin,[196] who was able to generate R-Cbl and alkyl-cobaloximes, ostensibly *via* an intermediate π-complex, by allowing Cbl^{III} to react with nucleophilic alkenes, as shown in reaction (30).

$$OH\text{-}Cbl + EtOCH{=}CH_2 \xrightarrow[Et_3N]{EtOH} (EtO)_2CHCH_2Cbl \quad (30)$$

Ingraham and co-workers[197] demonstrated that 2-hydroxy-1-methylethyl-(pyridine)cobaloxime isomerizes to 2-hydroxypropyl(pyridine)cobaloxime under acidic conditions, and obtained kinetic evidence for the production of an intermediate, suggested to be a π-complex, in the acid-catalysed decomposition of phenacylcobalamin to OH-Cbl and acetophenone. The mechanism of Dolphin and co-workers[196] for diol dehydrase is shown in Scheme 24. A similar mechanism can be written for ethanolamine ammonia lyase, where the leaving group is NH_3, but analogous mechanisms for the AdoCbl-dependent carbon skeleton

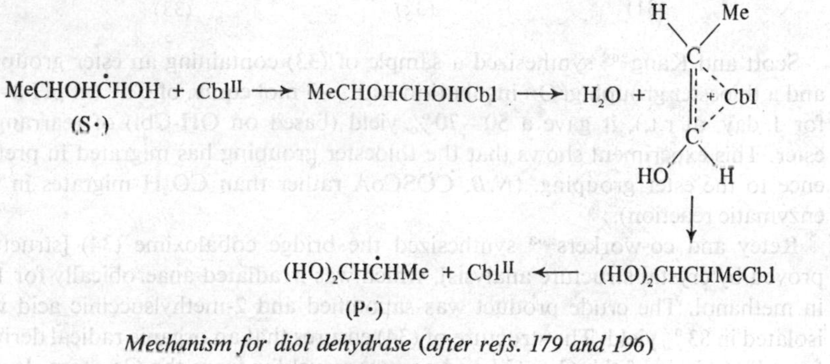

$$MeC\dot{H}OHCHOH + Cbl^{II} \longrightarrow MeCHOHCHOHCbl \longrightarrow H_2O +$$
(S·)

$$(HO)_2CHCHMe + Cbl^{II} \longleftarrow (HO)_2CHCHMeCbl$$
(P·)

Mechanism for diol dehydrase (after refs. 179 and 196)

Scheme 24

[193] B. T. Golding, H. L. Holland, U. Horn, and S. Sakrikar, *Angew. Chem. Internat. Edn.*, 1970, **9**, 959.
[194] E. A. Parfenov, T. G. Chervyakova, M. G. Edelev, and A. M. Yurkevitch. *J. Gen. Chem.* (*U.S.S.R.*) 1974, **44**, 2319.
[195] B. M. Babior, *J. Biol. Chem.*, 1970, **245**, 6125.
[196] R. B. Silverman, D. Dolphin, and B. M. Babior, *J. Amer. Chem. Soc.*, 1972, **94**, 4028; R. B. Silverman and D. Dolphin, *ibid.*, 1976, **98**, 4626, 4633.
[197] K. L. Brown, M. M. L. Chu, and L. L. Ingraham, *Biochemistry*, 1976, **15**, 1402; K. L. Brown and L. L. Ingraham. *J. Amer. Chem. Soc.*, 1974, **96**, 7681.

rearrangements are implausible because the leaving group is a poorly stabilized carbanion. Contrary to a statement in ref. 198, Scheme 24 does account for the stereochemical course of the diol dehydrase reaction.

The intermediate 2,2,-dihydroxyethylcobalamin (*cf.* Scheme 24) could either homolyse to Cbl[II] and a 2,2,-dihydroxyethyl radical or dehydrate to formylmethylcobalamin, which could homolyse to Cbl[II] and formylmethyl radical. However, Schrauzer and co-workers[199] have shown that formylmethylcobalamin shows no tendency to produce Cbl[II]. Formylmethylcobalamin is stable to alkali but decomposes in acid ($\tau_{1/2} \approx 10^3$ h at pH 7, $\tau_{1/2} = 24$ h at pH 5.8) to OH-Cbl and ethanal. Thermolysis (160 °C) or photolysis of formylmethylcobalamin in neutral solution gives OH-Cbl and ethanal, apparently without the intermediacy of Cbl[II].

Within the context of pathway (*a*), (31) is a hypothetical intermediate for the methylmalonyl-CoA mutase reaction, and it appeared in a plausible mechanism for methylmalonyl-CoA mutase suggested by Ingraham.[200] To prepare a model compound for (31), Dowd and Shapiro[201] allowed Cbl[I] (from reduction of OH-Cbl with borohydride *in situ*) to react with dimethyl bromomethylmalonate (3-fold excess), obtaining an unstable alkyl-cobalamin (characterized by u.v./vis. spectroscopy) that was assigned structure (32). On standing, in darkness, for 48 h at room temperature, the solution containing (32) gave, after hydrolysis of the esters and chromatography, 3.7% of succinic acid (yield based on OH-Cbl).

$$
\begin{array}{ccc}
\text{COSCoA} & & \text{COSEt} \\
| & & | \\
\text{CHCH}_2\text{Cbl} & & \text{MeCCH}_2\text{Cbl} \\
| & & | \\
\text{CO}_2\text{H} & (\text{MeO}_2\text{C})_2\text{CHCH}_2\text{Cbl} & \text{CO}_2\text{Et} \\
\\
(31) & (32) & (33)
\end{array}
$$

Scott and Kang[202] synthesized a sample of (33) containing an ester grouping and a thioester grouping. On incubation with ~ 3 mol equiv. of Cbl[I] (at pH 8—9, for 1 day, at r.t.), it gave a 50—70% yield (based on OH-Cbl) of rearranged ester. This experiment shows that the thioester grouping has migrated in preference to the ester grouping. (*N.B.* COSCoA rather than CO$_2$H migrates in the enzymatic reaction).

Retey and co-workers[203] synthesized the bridge cobaloxime (34) [structure proved by crystal structure analysis], which was irradiated anaerobically for 12h in methanol. The crude product was saponified and 2-methylsuccinic acid was isolated in 83% yield. The structure of (34) ensures that an organic radical derived from homolysis of the Co—C bond cannot travel far from the Co atom. It was suggested that the Co[II] of an intermediate radical pair assists migration of a thioester group and that this may be a role for Co in all AdoCbl-dependent

198 E.-I. Ochiai, 'Bioinorganic Chemistry', Allyn and Bacon, Boston, 1977.
199 T. M. Vickrey, R. N. Katz, and G. N. Schrauzer, *J. Amer. Chem. Soc.*, 1975, 97, 7248.
200 *cf.* J. N. Rowe and L. L. Ingraham, *J. Amer. Chem. Soc.*, 1971, 93, 3801.
201 P. Dowd and M. Shapiro, *J. Amer. Chem. Soc.*, 1976, 98, 3724.
202 A. I. Scott and K. Kang, *J. Amer. Chem. Soc.*, 1977, 99, 1997.
203 H. Flohr, W. Pannhorst, and J. Retey, *Angew. Chem. Internat. Edn.*, 1976, 15, 561.

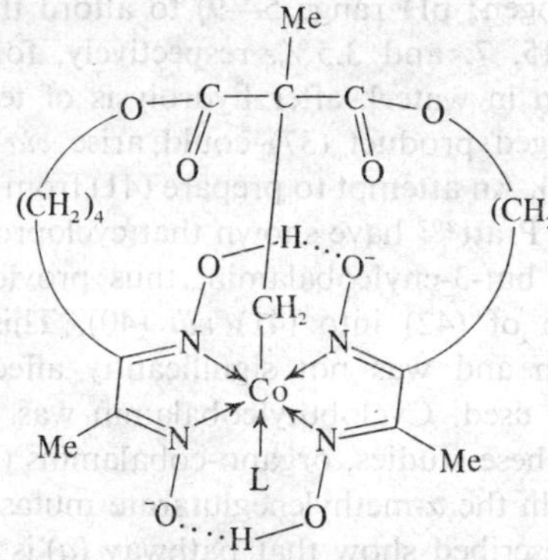

(34) L = MeOH

reactions. Ochiai[204] ascribed a similar role to Co in his review of cobalamin-dependent reactions (see also ref. 205).

Dowd *et al.*[206] allowed Cbl^I to react with the bromide (35) and isolated an organo-cobalamin, assigned structure (36) [characterization by u.v.—visible spectroscopy]. This substance decomposes in ^2H_2O (at 25 °C, during 200—300 h,

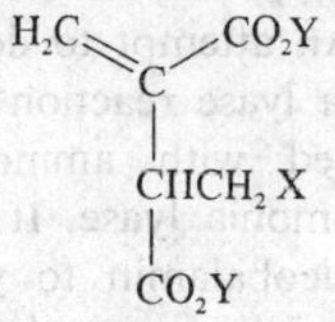

(35) X = Br, Y = THP
(36) X = Cbl, Y = THP
(43) X = Cbl, Y = H

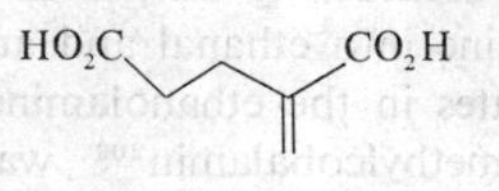

(37)

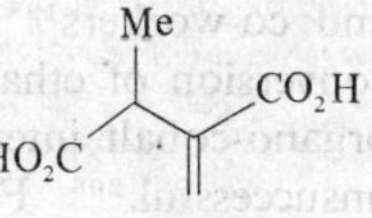

(38)

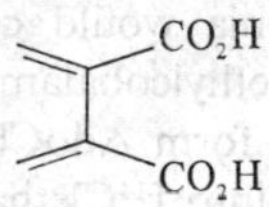

(39)

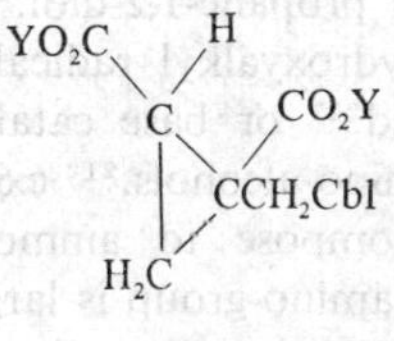

(40) Y = THP
(44) Y = H

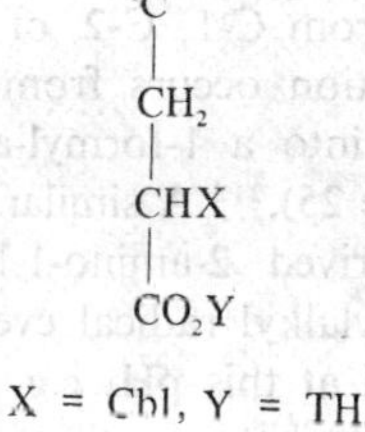

(41) X = Cbl, Y = THP
(42) X = Br, Y = THP
(45) X = Cbl, Y = H

(THP = tetrahydropyranyl)

[204] E.-I. Ochiai, *J. Inorg. Nuclear Chem.*, 1975, **37**, 351.
[205] L. Salem, O. Eisenstein, N. T. Anh, H. B. Burgi, A. Devaquet, G. Segal, and A. Veillard, *Nouveau J. de Chimie*, 1977, **1**, 335.
[206] P. Dowd, B. K. Trivedi, M. Shapiro, and L. K. Marwaka, *J. Amer. Chem. Soc.*, 1976, **98**, 7875.

in darkness, under nitrogen; pH range 5—9) to afford the products (37), (38), and (39) [yields were 15, 7, and 3.5%, respectively, for unlabelled products from an experiment run in water], after hydrolysis of tetrahydropyranyl ester groupings. The rearranged product (37) could arise *via* intermediate organo-cobalamins (40) and (41). An attempt to prepare (41) from (42) was unsuccessful. However, Chemaly and Pratt[207] have shown that cyclopropylcarbinylcobalamin rearranges at 60 °C to but-3-enylcobalamin, thus providing a model for the hypothetical conversion of (42) into (41) *via* (40). This transformation was uninfluenced by oxygen and was not significantly affected by the dielectric constant of the solvent used. Cyclobutylcobalamin was found to be stable at 60 °C. On the basis of these studies, organo-cobalamins (43), (44), and (45) are plausible intermediates in the α-methyleneglutarate mutase reaction.

The model studies described show that pathway (*a*) is chemically reasonable for the reactions catalysed by diol dehydrase, ethanolamine ammonia lyase, methylmalonyl-CoA mutase, and α-methyleneglutarate mutase (no meaningful model studies have been published on glutamate mutases). Proof that these enzymatic reactions proceed *via* organo-cobalt intermediates would obviously be the isolation or detection of such intermediates. The Co^{III} species detected in a rapid kinetic study of ethanolamine ammonia lyase (see above) could be an organo-cobalt intermediate, and this interpretation was favoured by Johnson and co-workers[188] after considering kinetic data and isotope effects for the conversion of ethanolamine into ethanal and ammonia. An attempt to detect organo-cobalt intermediates in the ethanolamine ammonia lyase reaction was unsuccessful.[208] Formylmethylcobalamin[209] was incubated with ammonia, [^{14}C]ethanolamine, deoxyadenosine, and ethanolamine ammonia lyase. It was hoped that ammonia would combine with formylmethylcobalamin to yield 2-amino-2-hydroxyethylcobalamin, which might react with enzyme-bound deoxyadenosine to form AdoCbl. This would then catalyse the conversion of [^{14}C]ethanolamine into [^{14}C]ethanal. However, no ethanal was formed and it could be calculated that $\leqslant 0.0006\%$ of formylmethylcobalamin was converted into AdoCbl.

Model Studies pertaining to Pathway (d). Hydroxyl radicals abstract a hydrogen atom from C-1, C-2, or C-3 of a 1,2-diol such as propane-1,2-diol.[210] When abstraction occurs from C-1, the derived 1,2-dihydroxyalkyl radical is converted into a 1-formyl-alkyl radical by either acid[211] or base catalysis (see Scheme 25).[212] A similar chemistry obtains with amino-alcohols,[212] except that the derived 2-amino-1-hydroxyalkyl radicals decompose to ammonia and 1-formylalkyl radical even at pH 7, because their amino-group is largely protonated at this pH, *e.g.* as shown in reaction (31). Attention was drawn to the similarity between these reactions and those catalysed by diol dehydrase and ethanolamine ammonia lyase.[191, 213] If the conversions of Scheme 25 and

[207] S. Chemaly and J. M. Pratt, *J.C.S. Chem. Comm.*, 1976, 988.
[208] J. S. Krouwer and B. M. Babior, *J. Biol. Chem.*, 1977, **252**, 5004.
[209] Unfortunately, this compound was prepared by an unreliable method (*cf.* ref. 199).
[210] B. C. Gilbert, J. P. Larkin, and R. O. C. Norman, *J.C.S. Perkin II*, 1972, 794.
[211] K. M. Bansal, A. Henglein, and E. Janata, *J. Phys. Chem.*, 1973, **77**, 2425.
[212] T. Foster and P. R. West, *Canad. J. Chem.*, 1974, **52**, 3589.
[213] B. T. Golding and L. Radom, *J.C.S. Chem. Comm.*, 1973. 939.

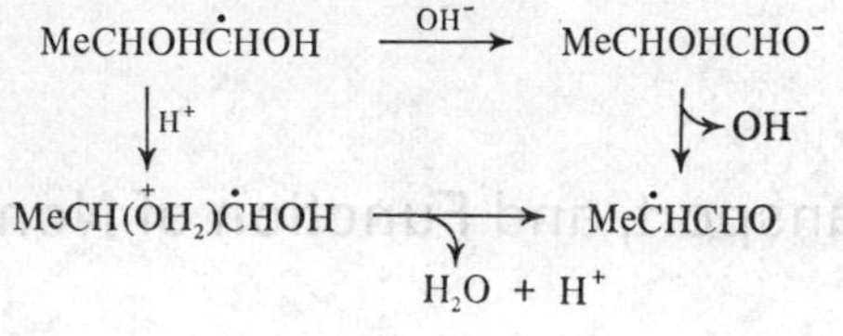

$$\text{MeCHOH}\overset{\bullet}{\text{C}}\text{HOH} \xrightarrow{\text{OH}^-} \text{MeCHOHCHO}^-$$

$$\downarrow \text{H}^+ \qquad\qquad \downarrow\!\!\searrow \text{OH}^-$$

$$\text{MeCH}(\overset{+}{\text{O}}\text{H}_2)\overset{\bullet}{\text{C}}\text{HOH} \longrightarrow \text{Me}\overset{\bullet}{\text{C}}\text{HCHO}$$

$$\searrow$$

$$\text{H}_2\text{O} + \text{H}^+$$

Scheme 25

reaction (31) involve irreversible ejection of H_2O and OH^- from a protonated or deprotonated precursor, such a pathway, if applied to diol dehydrase, cannot explain the ^{18}O-labelling results of Arigoni and co-workers.[178] At the enzyme's active site, H_2O or OH^- that has been eliminated from, *e.g.*, MeCHOH$\overset{\bullet}{\text{C}}$HOH

$$\text{H}_3\overset{+}{\text{N}}\text{CH}.\overset{\bullet}{\text{C}}\text{HOH} \longrightarrow \text{H}^+ + \text{H}_3\overset{+}{\text{N}}\text{CH}_2\text{CHO}^- \longrightarrow \text{NH}_3 + \overset{\bullet}{\text{C}}\text{H}_2\text{CHO} \qquad (31)$$

($>$Me$\overset{\bullet}{\text{C}}$HCHO or MeCH$=\!=$CHOH) might not mix with solvent water, but it may recombine with, say, MeCH$^{+\bullet}$CHOH to yield Me$\overset{\bullet}{\text{C}}$HCH(OH)$_2$. Golding and Radom[192, 213] used *ab initio* calculations to assess the possibility that a radical such as MeCHOH$\overset{\bullet}{\text{C}}$HOH (S·) could be converted by acid catalysis into Me$\overset{\bullet}{\text{C}}$HCH(OH)$_2$ (P·) *via* a bridged intermediate or transition state (46). The calculations indicated that (46) was of comparable stability to protonated S· or P· radicals. This result encouraged model studies designed to show that alkyl radicals could induce the conversion of a 1,2-diol into an aldehyde as in Scheme 25 and Scheme 23. Dihydroxyalkyl-cobaloximes (47a—d) were synthesized and photolysed anaerobically in 0.1M-acetic acid (pH 3). (47b) Gave 10% pentanal, (47c) gave 4% hexanal + 16% hexan-2-one, whereas neither (47a) nor (47d) gave carbonyl products from the dihydroxyalkyl group.[214] These results are explicable

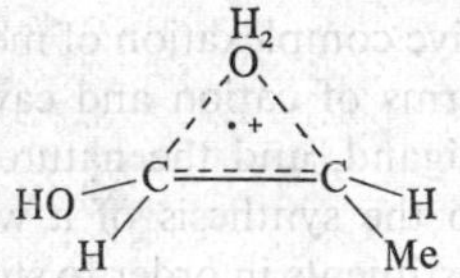

(46) structure with formula:

$$\text{HO}-\overset{\overset{\displaystyle \text{H}_2}{\displaystyle \text{O}}}{\underset{\displaystyle \text{H}}{\text{C}}}=\!=\!=\!=\overset{\displaystyle \text{H}}{\underset{\displaystyle \text{Me}}{\text{C}}}$$

$$(46)$$

[HOCH$_2$CHOH(CH$_2$)$_n$Co(dmgH)$_2$(py)]

(47a) $n = 2$
(47b) $n = 3$
(47c) $n = 4$
(47d) $n = 9$

by a pathway in which initial photoinduced homolysis of the Co—C bond of (47) is followed by rearrangement of the derived dihydroxyalkyl radical (preferred 1,5-*H* shift) in the case of (47b) and (47c). The 1,2-dihydroxypentyl radical from (47b) gives pentanal either by direct elimination of H_2O and then abstraction of a hydrogen atom from dmgH or by acid-catalysed rearrangement to 1-(dihydroxymethyl)butyl radical followed by elimination of H_2O and abstraction of H from dmgH.

Acknowledgement: We thank Dr P. J. Sellars for critical assessment of the manuscript and Prof. J. P. Klinman for comments on section 2.

214 B. T. Golding, T. J. Kemp, C. S. Sell, P. J. Sellars, and W. P. Watson, *J.C.S. Perkin II*, 1978, 839.

3
Storage, Transport, and Function of Non-transition Elements

M. N. HUGHES

1 Introduction

The early part of this chapter is concerned with the interaction of the cations of Groups IA and IIA of the Periodic Table with model ionophoric compounds such as crown ethers, cryptands, and antibiotic ionophores, and with the role of such compounds in mediating cation translocation across artificial and reconstituted membranes. Particular attention is then given to transport systems such as the Na^+–K^+- and Mg^{2+}–Ca^{2+}-activated ATPases from various sources, on which a considerable amount of work has been published. Most of the remaining material is concerned with the role of calcium ions in the control of intracellular processes, but processes such as calcium mineralization and blood clotting are also considered. The use of lanthanide ions as probes for calcium is becoming increasingly important, and this general topic and that of analytical methods are assessed briefly.

2 Interaction of Synthetic Ionophoric Ligands with Group IA and IIA Cations

The Cyclic Polyethers.—While the broad basis of selective complexation of metal ions by these ligands can now be understood[1,2] in terms of cation and cavity sizes, the hydrophilic/hydrophobic properties of the ligand, and the nature of the solvent, increasing attention is now being paid to the synthesis of a wide range of cyclic ethers with varying heteroatoms and substituents in order to study systematically the factors relevant to complex formation. A series of polyether diesters have been synthesized[3] by the reaction of oligo(ethylene glycols) (or their sulphur analogues) with diglycolyl or thiodiglycolyl chlorides. Calorimetric techniques have been used to measure the interaction of Na^+, K^+, Ag^+, and Ba^{2+} with the 2,6-dione of 18-crown-6 (1)[4] and a sulphur analogue (2).[5] While 18-crown-6 shows marked selectivity for Ba^{2+} over K^+ in methanol, ligand (1) shows similiar association constants for Ba^{2+} and K^+. (Log K values are

[1] R. M. Izatt, D. J. Eatough, and J. J. Christensen, *Structure and Bonding*, 1973, **16**, 161.
[2] J. J. Christensen, D. J. Eatough, and R. M. Izatt, *Chem. Rev.*, 1974, **74**, 351.
[3] G. E. Maas, J. S. Bradshaw, R. M. Izatt, and J. J. Christensen, *J. Org. Chem.*, 1977, **42**, 3937.
[4] R. M. Izatt, J. D. Lamb, G. E. Maas, R. E. Asay, J. S. Bradshaw, and J. J. Christensen, *J. Amer. Chem. Soc.*, 1977, **99**, 2365.
[5] R. M. Izatt, J. D. Lamb, R. E. Asay, G. E. Maas, J. S. Bradshaw, J. J. Christensen, and S. S. Moore, *J. Amer. Chem. Soc.*, 1977, **99**, 6134.

(1) X = O
(2) X = S

3.1 ± 0.2 and 2.79 ± 0.02 respectively, at 25 °C.) Thus the presence of the carbonyl groups for complexation has favoured K^+ binding. The cyclic antibiotic valinomycin, which shows selectivity for K^+ over Ba^{2+} in methanol, also contains carbonyl groups suitable for co-ordination.

The normal selectivity of a crown system towards alkali metals can be controlled or even reversed[6] through the electronic effects of substituents. The presence of electron-withdrawing groups on 18-crown-6 ethers results in a marked decrease in their ability to extract K^+ salts from water into methylene dichloride, but has little effect on the extraction of Na^+. Eventually, therefore, such substituents cause a reversal of the normal ion-extraction selectivity of 18-crown-6 ethers. Clearly, the reduced basicity of the oxygen crown system has a much smaller effect on Na^+ binding. This may imply an asymmetric complexation of Na^+ by the ether, so that Na^+ does not interact with the oxygen atoms that are most influenced by the electronic effect. However, no account has been taken of the possible effect of these substituents on solvation.

The effect of metal ions in improving the yields of cyclic ethers has implied a template role for the metal in the synthesis. This effect has been confirmed[7] for the formation of benzo-18-crown-6 in aqueous solution from kinetic studies on the cyclization rate in the presence of cations.

Values of log K, ΔH, and ΔS have been reported for the interaction of Group IA and IIA cations with 15-crown-5, 18-crown-6, and isomers of dicyclohexo-18-crown-6 in aqueous solution,[8] and with benzo-15-crown-5, 18-crown-6, dibenzo-24-crown-8, and dibenzo-27-crown-9 in methanol–water.[9] Methods for measuring formation constants for crown ether complexes have been reviewed,[2] but novel methods continue to be used; for example, the use[10] of c.d. for Na^+ and an optically active benzo-15-crown-5. The complexation of Cs^+ by 18-crown-6 has been studied[11] by ^{133}Cs n.m.r. The chemical shift varied with 18-crown-6:Cs^+ mole ratio in pyridine solution, the resonance shifting downfield as the ligand concentration was increased till a minimum value was reached at 1:1 mole ratio.

[6] K. H. Pannell, W. Yee, G. S. Lewandos, and D. C. Hambrick, *J. Amer. Chem. Soc.*, 1977, 99, 1457.
[7] L. Mandolini and B. Masci, *J. Amer. Chem. Soc.*, 1977, 99, 7709.
[8] R. M. Izatt, R. E. Terry, B. L. Haymore, L. D. Hansen, N. K. Dalley, A. G. Avondet, and J. J. Christensen, *J. Amer. Chem. Soc.*, 1976, 98, 7620.
[9] R. M. Izatt, R. E. Terry, D. P. Nelson, Y. Chan, D. J. Eatough, J. S. Bradshaw, L. D. Hansen, and J. J. Christensen, *J. Amer. Chem. Soc.*, 1976, 98, 7626.
[10] M. P. Mack, R. E. Hendrixson, R. A. Palmer, and R. G. Ghirardelli, *J. Amer. Chem. Soc.*, 1976, 98, 7830.
[11] E. Mei, J. L. Dye, and A. I. Popov, *J. Amer. Chem. Soc.*, 1977, 99, 5308.

Further addition of the ether gave an upfield shift. This was interpreted in terms of the formation of 1:1 and 1:2 Cs⁺–cyclic ether complexes, with $K_1 \approx 10^6$, $K_2 \approx 44$ at 25 °C.

Kinetic Studies.—Previous studies on metal ion–crown ether complexes have been mainly concerned with the decomplexation or 'off' reaction, one that largely determines the specificity of the ether for the cation. It is commonly accepted thst the first stage is a conformational rearrangement of the complex prior to loss of metal. Chock[12] has suggested a complexation mechanism for the reactions of cations with dibenzo-30-crown-10 in methanol, involving a ligand conformational change followed by stepwise substitution of the co-ordinated solvent by the ligand, as shown[13] in equation (1), where CR_2 represents the active conformation of the ether.

$$CR_1 \underset{k_{21}}{\overset{k_{12}}{\rightleftharpoons}} CR_2 \; ; \; CR_2 + M^+ \underset{k_{32}}{\overset{k_{23}}{\rightleftharpoons}} MCR_2^+ \qquad (1)$$

Ultrasonic absorption kinetic studies on the complexation of some singly charged cations by aqueous 18-crown-6[14] and 15-crown-5[15] have allowed an elaboration of this mechanism. For the former ligand, rate constants for both steps are now available $(k_{12} = 6.3 \times 10^8 \text{ s}^{-1}; \; k_{21} = 1 \times 10^7 \text{ s}^{-1}; \; K_{21} = 1.6 \times 10^{-2})$. From the overall formation constant (K_T),[8] rate constant k_{23} could be calculated and shown to be almost two orders of magnitude smaller than the rate constant for a diffusion-controlled reaction. Values of k_{23} are also much smaller than solvent-exchange rates for the M^+ ions, showing that the second stage of equation (1) is probably a multi-step process involving water loss or additional conformational change(s) in the ligand. It is noteworthy that Tl^+ and Ag^+ complex at faster rates, possibly reflecting the need for loss of fewer co-ordinated water molecules. In the case of 15-crown-5, a conformational rearrangement was detected in the absence of cation.[15] The temperature dependence of the ¹³³Cs resonance line in the Cs⁺–18-crown-6 system has allowed[11] the characterization of the decomplexation reaction.

Cyclic Polyethers with Additional Functions: Guest–Host Relationships.—Cram and his co-workers have synthesized 'host' compounds, containing macrocyclic polyethers in their basic structure, which also have structural features designed to complement those of specific 'guest' compounds, *e.g.* with substituents that converge on the functional or binding sites of guest compounds. A series of papers[16] reports results in this area, and includes some information on formation constants with Group IA metals. Another approach has involved the synthesis of crown ethers incorporating physiologically active molecules, in the hope of conferring cation selectivity upon the species and also of controlling the hydro-

12 P. B. Chock, *Proc. Nat. Acad. Sci. U.S.A.*, 1972, **69**, 1939.
13 G. W. Liesegang, M. M. Farrow, N. Purdie, and E. M. Eyring, *J. Amer. Chem. Soc.*, 1976, **98**, 6905.
14 G. W. Liesegang, M. M. Farrow, F. A. Vazquez, N. Purdie, and E. M. Eyring, *J. Amer. Chem. Soc.*, 1977, **99**, 3240.
15 L. J. Rodriguez, G. W. Liesegang, R. D. White, M. M. Farrow, N. Purdie, and E. M. Eyring, *J. Phys. Chem.*, 1977, **81**, 2118.
16 R. C. Helgeson, T. L. Tarnowski, J. M. Timko, and D. J. Cram, *J. Amer. Chem. Soc.*, 1977, **99**, 6411 and preceding papers.

philic/hydrophobic properties of the active molecule. This is illustrated by the synthesis[17] of crowned and double-crowned derivatives of papaverine. A crowned porphyrin has been synthesized[18] that can accommodate a transition-metal ion and a Group IA or IIA ion simultaneously, while a macrocycle (3) with a Hantzsch 1,4-dihydropyridine group is able to serve[19] as a NAD(P)H mimic, in view of its hydride-donating properties to sulphonium salts. This is facilitated by the presence of the ether-bound cation, such an effect of cations on hydride transfer being well known in biochemical systems. Cation complexation with a pyridine-substituted cyclic ether ester (4) has been studied[5] in a similar connection, but the analogy is probably not as close as was claimed.

(3)   (4)

Cryptands.—The synthesis of macrobicyclic ligands (the cryptands) and their selectivity towards cations has been reviewed by Lehn.[20] More recently, he and his co-workers have studied macrotricyclic ligands, which contain two cavities, each of which can bind one cation selectively. These ligands also allow variation of cation–cation interaction by appropriate structural alteration. The interactions of (5), (6), and (7) with cations M^{2+} and M^+ have been studied[21] by n.m.r. techniques, which show the successive formation of a mono- and a di-cationic

(5) X = O   (7)
(6) X = CH₂

[17] F. Vögtle and K. Frensch, *Angew. Chem. Internat. Edn.*, 1976, **15**, 685.
[18] C. K. Chang, *J. Amer. Chem. Soc.*, 1977, **99**, 2819.
[19] T. J. van Bergen and R. M. Kellog, *J. Amer. Chem. Soc.*, 1977, **99**, 3882; *J.C.S. Chem. Comm.*, 1976, 964.
[20] J. M. Lehn, *Structure and Bonding*, 1973, **16**, 1.
[21] J. M. Lehn and J. Simon, *Helv. Chim. Acta*, 1977, **60**, 141.

complex, in some cases allowing the production of a heteronuclear species. The mono-cationic complex undergoes fast cation exchange between the two cavities, but slow intermolecular exchange. Formation constants for mono- and di-cationic complexes show that complexation takes place most readily for 1:1 complexes with (5), probably because the oxygen bridge interacts with the cation, while values for the dinuclear complexes are only slightly lower than K_1 values, showing that the macrotricyclic ligand contains two largely independent sites.

Kinetic Studies with Cryptands.—Relatively little has been done on the kinetics of interaction of cations with cryptands, partly reflecting the difficulties in monitoring the reaction by techniques other than n.m.r. line broadening. Wilkins and co-workers[22] have studied the reactions between Ca^{2+} and the cryptands 2.2.2, 2.2.1, and 2.1.1. [values of l, m, n in (8)]; Sr^{2+} and 2.2.2 and 2.2.1; and Ba^{2+}

(8)

and 2.2.2, by the stopped-flow technique, using the indicators murexide and metalphthalein to monitor free M^{2+}; the reverse reactions were also studied for Ca^{2+}, using alkali-metal ions as scavengers for the dissociated cryptands. A conductance stopped-flow method has been used[23] to measure rates of dissociation of metal cryptate complexes; excess acid was added and the pH changes were monitored. Many of the dissociation reactions are acid-catalysed. Temperature-jump studies have been reported[24] for the crytands 2.2.2 and 2.2.1 (with K^+) and 2.2.1 and 2.1.1 (with Na^+). The ion K^+ recombines about one hundred times faster with cryptand 2.2.1 than does Na^+ with 2.1.1, although the ratios of the ionic to cavity radius are the same. This interesting difference may result from the lower flexibility of the smaller cryptand. The results obtained by different techniques agree well. The formation rate constants are substantially lower than those for the interaction of the same cations with simple ligands. This may reflect a low concentration of the kinetically active form of the cryptand, in which the lone pairs of the bridgehead nitrogen atom point towards the cavity. For a particular cryptate, the rate of the forward reaction is inversely proportional to the radius of the cation, indicating that loss of co-ordinated water is important before or in the rate-determining step of complex formation. The exchange of Na^+ with 2.2.2 cryptate complexes in several solvents has been measured[25] by [23]Na n.m.r. spectroscopy.

The ions Na^+ and K^+ give 1:1 complexes with the pyridinophane cryptands (9) and (10). Two relaxation times are observed[26] for the binding process, and

[22] V. M. Loyola, R. Pizer, and R. G. Wilkins, *J. Amer. Chem. Soc.*, 1977, **99**, 7185.
[23] B. G. Cox and H. Schneider, *J. Amer. Chem. Soc.*, 1977, **99**, 2809.
[24] K. Henco, B. Tümmler, and G. Maass, *Angew. Chem. Internat. Edn.*, 1977, **16**, 358.
[25] J. M. Ceraso, P. B. Smith, J. S. Landers, and J. L. Dye, *J. Phys. Chem.*, 1977, **81**, 760.
[26] B. Tümmler, G. Maass, E. Weber ,W. Wehner, and F. Vögtle, *J. Amer. Chem. Soc.*, 1977, **99**, 4683.

(9)   (10)

these have been attributed to a nearly diffusion-controlled encounter and substitution step and to a subsequent isomerization process, with a frequency of $10^4 s^{-1}$. The formation constants in aqueous solution are similar to those of the 2.2.1 cryptand, despite the effect of the pyridine group on the flexibility of the cryptand. The two pyridinophane cryptands have similar affinity for Na^+, but the more rigid diamide (10) binds K^+ more effectively than (9). Cryptand (10) was successfully incorporated into a biological membrane, and facilitated ion transfer and the depolarization of the nerve axon, suggesting that it may be well suited as a model for biological carriers.

Non-cyclic Ethers and other Ligands.—Ionophoric non-cyclic compounds have been reviewed.[27] In methanol solution, the 1:1 and 1:2 complexes of Na^+ and K^+ with the linear quinoline polyether (11) are of similar stability,[26] a consequence of the flexibility of the tetra(ethylene glycol) chain, which can adapt itself to the different ions. The introduction of a pyridine bridge into the ether chain leads to a much better discrimination, with $K_1(Na)/K_1(K^+)=12$. The effect of chain length and the nature of the terminal groups (*e.g.* polarity and symmetry) on complexing properties have been studied.[28] Thus the quinoline ether with three oxygen functions behaves as a quinquedentate ligand in its complex with RbI, while the septidentate ligand (11) has a helical conformation in the crystalline RbI complex. There has also been considerable interest[29, 30] in the interactions of cations with ligands having amide and ether functions, and evidence has been shown in certain cases for amide bonding. Neutral aromatic and alicyclic 1,2-ethylenedioxydiacetamide ligands selectively extract M^{2+} cations over M^+ cations into organic phases.[31] At low concentrations, the aromatic ligands form 1:1 complexes with the alkaline-earth cations, but other stoicheiometries are found for the cyclohexyl ligands (2:1, 3:2, and 4:3) with Ca^{2+}, Sr^{2+}, and Ba^{2+} respectively. For ligands of this type, the binding constants were in the sequence $Ca^{2+} > Sr^{2+} > Ba^{2+} \gg Mg^{2+}$, *i.e.* following ionic size except for Mg^{2+}, a situation found in aqueous solution for troponin and certain extracellular enzymes. The ligand (12) forms a 2:1 complex with Ca^{2+}, and it is noteworthy that it transports Ca^{2+} across artificial membranes[32] faster than the ligands that only form 1:1 complexes with Ca^{2+}. Possibly the 2:1 complex with Ca^{2+} provides a better

[27] B. C. Pressman, *Ann. Rev. Biochem.*, 1976, **45**, 501.
[28] F. Vögtle and H. Sieger, *Angew. Chem. Internat. Edn.*, 1977, **16**, 396.
[29] R. Buchi and E. Pretsch, *Helv. Chim. Acta*, 1977, **60**, 1141.
[30] N. N. L. Kirsch, R. J. J. Funck, E. Pretsch, and W. Simon, *Helv. Chim. Acta*, 1977, **60**, 2326.
[31] T. C. Wun, R. Bittman, and I. J. Borowitz, *Biochemistry*, 1977, **16**, 2074.
[32] T. C. Wun and R. Bittman, *Biochemistry*, 1977, **16**, 2080.

(11)

$$(12) \quad R = CH_2CNPr^n_2$$

shielding for the cation from the hydrophobic membrane. Crystal structures have been reported[33] for the $CaCl_2$ complex with the non-cyclic ionophore $NNN'N'$-tetrapropyl-3,6-dioxaoctane diamide and for the calcium derivatives of acyclic dicarboxylic acids with repeating ethereal functions.[34]

3 The Ionophore Antibiotics and Related Compounds

The structures of several polyether monocarboxylic acid ionophores have been determined, as their thallium or silver salts. Lonomycin[35] takes up a circular conformation to provide a cavity for the Tl^+ cation, which co-ordinates to six oxygen atoms, with Tl—O bond lengths ranging from 2.56 to 3.01 Å. Most of the oxygen atoms are located in the interior of the molecule. Closely related are the antibiotics A-130A[36] and carriomycin,[37] the former being found to be identical with RO21-6150. In both cases the cyclic conformation is maintained by specific, strong hydrogen bonds, so their structures in solution will depend upon the medium. The antibiotic X-573A (Lasalocid) is well known to be a specific ionophore for Ca^{2+}. This has a linear structure in polar solvents and a cyclic conformation in non-polar solvents in which the outer surfaces are hydrophobic and the inner side is hydrogen-bonded. Metal ions bind to X-573A in polar solvents, but biogenic amines only bind strongly[38] in inert hydrocarbon solvents. A unique property of X-573A is that all the structures determined so far are dimeric, even with univalent cations. The dimer has an almost entirely non-polar exterior, with the two metal ions completely encapsulated by the two ionophore molecules. N.m.r. studies on X-573A in non-polar solvents were consistent with the dimeric structure, but suggested that monomeric forms predominated in

33 K. Neupert-Laves and M. Dobler, *Helv. Chim. Acta*, 1977, **60**, 1861.
34 F. Ancillotti, G. Boschi, G. Perego, and A. Zazzetta, *J.C.S. Dalton*, 1977, 901.
35 N. Otake, M. Koenuma, H. Miyamae, S. Sato, and Y. Saito, *J.C.S. Perkin II*, 1977, 494.
36 N. Otake, H. Nakayama, H. Miyamae, S. Sato, and Y. Saito, *J.C.S. Chem. Comm.*, 1977, 590.
37 H. Koyama and K. Utsumi-Oda, *J.C.S. Perkin II*, 1977, 1531.
38 S. Lindenbaum, L. Sternson, and S. Rippel, *J.C.S. Chem. Comm.*, 1977, 268.

polar solvents.[39] The structures of the sodium salt of X-573A (*N.B.* ionic radii of Na^+ and Ca^{2+} are similar) and the ionophore itself, obtained from methanol solution, now confirm that these are monomeric compounds.[40] The overall conformation is maintained by head-to-tail hydrogen-bonding. The sodium cation is complexed by the same five oxygen atoms that complex the Na^+ in the dimeric form. There is no Na–carboxylate co-ordination, and the remaining co-ordination position is filled by a solvent methanol molecule, which effectively caps the cation. Complexation with Na^+ has little major effect on the structure of the ionophore. The identification of monomeric and dimeric structures for X-573A under different conditions of solvent polarity indicates that metal uptake and release at the polar exterior of a lipid bilayer may involve the monomeric form, and that transport in the lipid bilayer is through the dimer. The ionophorous properties of narasin, a new polyether monocarboxylic acid, have been reported.[41]

The other main group of antibiotic ionophores are the neutral cyclic peptides, typified by valinomycin. An attempt has been made to characterize the different conformational forms of uncomplexed valinomycin in the solid and in solvents of different polarity by the use[42] of Raman spectroscopy. The spectrum of valinomycin that had been recrystallized from n-octane is in accord with the X-ray structure, in which there are two hydrogen-bonded ester $C{=}O$ groups, the four free ester $C{=}O$ groups being highly exposed and so free to initiate co-ordination with K^+. However, the Raman spectrum of valinomycin crystallized from *o*-dichlorobenzene or dioxan indicates that a different structure is present, in which the hydrogen-bonded ester $C{=}O$ groups are absent, a conformation resembling one previously observed in solution. Raman studies in solution confirm that the hydrogen-bonded ester carbonyl group structure is present in non-polar solvents. On increasing the polarity of the solvent, the conformational equilibrium is shifted to forms containing fewer hydrogen bonds

The repeat hexapeptide of elastin (L-Val-L-Ala-L-Pro-Gly-L-Val-Gly) binds cations selectively, with a greater affinity for Ca^{2+}. Related peptides have been synthesized[43] in an endeavour to make the molecule more lipophilic for lipid bilayers (*e.g.* For-MeVal-Ala-Pro-Sar-Pro-SarOMe). These peptides still bound Ca^{2+} preferentially, but did not act as carriers. They did transport K^+. Inhibition by Ca^{2+} may result from irreversible complex formation involving Ca^{2+}, lipid, and peptide, and from the impermeability of the lipid layer to the Ca^{2+}–peptide complex.

The methodology of extraction of ionophoroproteins from mitochondria and the release of the ionophores has been illustrated by the isolation of a new K^+/Ca^{2+} ionophore.[44] This is a neutral peptide of molecular weight 1600, and

[39] D. J. Patel and C. Shen, *Proc. Nat. Acad. Sci. U.S.A.*, 1976, **73**, 1786, 4277.
[40] C. C. Chiang and I. C. Paul, *Science*, 1977, **196**, 1441.
[41] D. T. Wong, D. H. Berg, R. H. Hamill, and J. R. Wilkinson, *Biochem. Pharmacol.*, 1977, **26**, 1373.
[42] I. M. Asher, K. J. Rothschild, E. Anastassakis, and H. E. Stanley, *J. Amer. Chem. Soc.*, 1977, **99**, 2024. K. J. Rothschild, I. M. Asher, H. E. Stanley, and E. Anastassakis, *ibid.*, p. 2032.
[43] R. J. Bradley, W. O. Romine, M. M. Long, T. Ohnishi, M. A. Jacobs, and D. W. Urry, *Arch. Biochem. Biophys.*, 1977, **178**, 468.
[44] G. A. Blondin, R. J. Kessler, and D. E. Green, *Proc. Nat. Acad. Sci. U.S.A.*, 1977, **74**, 3667.

its ionophoric activity (determined in beef heart mitochondria) is close to that of valinomycin under comparable conditions.

4 Cations, Ionophores, and Model Membrane Systems

General accounts of cation transport through membranes[45] and the proteins of membranes[46] have been published. Of rather more specialist interest are kinetic analyses of models for transport.[47] A review of ionophoric activity[48] includes a useful summary of methods for measuring ionophoric activity, encompassing conductivity measurements on bilayer lipid membranes, extraction,[49] transport, and the use of reconstituted vesicles[50] and whole 'organisms', particularly mitochondria. Cation translocation in phospholipid vesicles incorporating cytochrome oxidase has been studied[51] spectrophotometrically, using the dye safranine.

Pore-forming Ionophores.—Unlike the carrier ionophores discussed so far, these induce ion permeability by forming pores or pathways of high dielectric constant through the membrane. Examples are gramicidin A and F-30 alamethicin, which are linear peptides that transport cations M^+ and M^{2+} respectively. Thus cells can be made highly permeable to Na^+ by treatment with gramicidin. Such permeability in pigeon red blood cells can be reversed[52] by treatment with phospholipid, thus opening up the interesting possibility that Na^+ gradients in whole cells can be controlled by successive treatment with gramicidin and phospholipid. Model studies have been reported on cation diffusion and selectivity in hydrophilic pathways.[53]

Synthetic peptides (Leu-Ser-Leu-Gly)$_n$ ($n = 6$, 9, or 12) and the N-formyl derivative for $n = 6$ have been tested[54] conductometrically for the induction of permeability to cations in lipid bilayer membranes, as these compounds would be expected to form $\beta_{6,6}^{12}$ helical ion channels. Positive results were found for the peptide with $n = 12$ and for the N-formyl derivative of the peptide with $n = 6$, acting as a dimer. The smaller peptides were unable to span the membrane interior. The two active peptides are capable of forming helices that are four turns long. These structures have hydrophobic exteriors, resulting from the leucine side-chains, and hydrophilic interiors, owing to the array of serine hydroxy-groups, to which ions traversing the channel may be co-ordinated. This is the first model system having such functional groups in the interior of the helix, a situation postulated[55] for naturally occurring channels.

[45] D. E. Fenton, *Chem. Soc. Rev.*, 1977, **6**, 325.
[46] S. J. Singer, *J. Colloid Interface Sci.*, 1977, **58**, 452.
[47] W. D. Stein and B. Honig, *Mol. and Cell. Biochem.*, 1977, **15**, 27; W. R. Lieb and W.D. Stein, *Biochim. Biophys. Acta*, 1976, **455**, 913.
[48] A. E. Shamoo and D. A. Goldstein, *Biochim. Biophys. Acta*, 1977, **472**, 13.
[49] W. J. Malaisse, G. Devis, and G. Somers, *Experientia*, 1977, **33**, 1035.
[50] A. D. Bangham, M. W. Hill, and N. G. A. Miller, *Methods in Membrane Biology*, 1974, **1**, 1.
[51] H. Gutweniger, S. Massari, M. Beltrame, R. Colonna, P. Veronese, and B. Ziche, *Biochim. Biophys. Acta*, 1977, **459**, 216.
[52] G. A. Vidaver, E. Lee, and W. Lau, *Arch. Biochem. Biophys.*, 1977, **179**, 67.
[53] J. L. Rigoud and C. M. Gary-Bobo, *Biochim. Biophys. Acta*, 1977, **469**, 246.
[54] S. J. Kennedy, R. W. Roeske, A. R. Freeman, A. M. Watanabe, and H. R. Besch, *Science*, 1977, **196**, 1342.
[55] B. Hille, *Fed. Proc.*, 1975, **34**, 1318.

Carrier Ionophores.—Carrier-associated translocation of cations is the result of a number of discrete transport steps. Ultimately, therefore, the use of kinetic techniques becomes essential, and this is reflected in current work.

Group IA Cations. The Ca^{2+}-selective ionophore A23187 forms complexes with univalent cations, but it has not been clear whether such complexes are translocated across membranes. Na^+ binds to A23187 better than K^+, and in the absence of bivalent cations it is transported by it into intact human red cells, although the effect is prevented or reversed by bivalent cations.[56] The ionophoric activities of a wide range of crown ethers to K^+, Rb^+, and Cs^+ have been tested[57] by a range of methods, including uptake by mitochondria, and some useful general conclusions were reached. By far the most effective ionophore was the largest one used, di-t-butyl-dibenzo-30-crown-10, which could completely wrap around the cation. Other ethers giving measurable uptake were those forming sandwich complexes, *i.e.* CKC^+. For K^+, crowns with five oxygens were most effective, association constants showing clearly that all the 15-crown-5 derivatives follow 1:2 association, while solid 1:2 complexes can be isolated. The low uptake rate with 15-crown-5 compounds by mitochondria may reflect kinetic limitations in forming the 1:2 complex. Derivatives of 18-crown-6 were least effective, because two sides of the complex are exposed to the hydrophobic membrane.

Charge-pulse relaxation experiments on valinomycin-mediated Rb^+ transport, using a wide range of glycerolmono-oleate bilayer membranes, have shown[58] the importance of the composition of the membrane and the relative unimportance of the solvent on the translocation rate constants (k_{ms}). An increase in the number of double bonds in the C_{20} fatty acid from one to four increases k_{ms} by a factor of twenty-four. The stability constant of the ion-carrier complex and k_{ms} depends strongly on the nature of the polar head-group of the lipid. The cyclic dodecapeptide PV, *cyclo*(-D-Val-L-Pro-L-Val-D-Pro-)$_3$, a structural analogue of valinomycin, differs[59] from it in its activity to lipid bilayers. PV forms a complex with K^+ in aqueous solution and is then translocated through the membrane (a *cis* mechanism); valinomycin functions by an interfacial *trans* mechanism. Both ionophores have one common kinetic parameter, *i.e.* the translocation rate constant k_{ms}, but the values are 2×10^5 s^{-1} and 5×10^3 s^{-1} in mono-olein membranes for valinomycin and PV respectively. This large difference in k_{ms} values is not a size effect, but may be related to the differences in the nature of the side-chains.

Calcium Ions. Arsenazo III has been used[60] to monitor the ionophore A23187-mediated translocation of Ca^{2+} across phosphatidylcholine membranes. A 2:1 ionophore–Ca^{2+} complex was formed, in accord with the transport of a neutral complex. However, if a neutral ionophore is involved in back transport, there should be a pH gradient, with the low $[Ca^{2+}]$ side at higher pH, which was not observed. The situation may therefore be more complicated, with transfer of species such as $[CaA_2H_2]^{2+}$ 2X$^-$.

[56] P. Flatman and V. L. Lew, *Nature*, 1977, **270**, 444.
[57] E. J. Harris, B. Zaba, M. R. Truter, D. G. Parsons, and J. N. Wingfield, *Arch. Biochem. Biophys.*, 1977, **182**, 311.
[58] R. Benz, O. Fröhlich, and P. Läuger, *Biochim. Biophys. Acta*, 1977, **464**, 465.
[59] R. Benz, B. F. Gisin, H. P. Ting-Beall, D. C. Tosteson, and P. Läuger, *Biochim. Biophys. Acta*, 1976, **455**, 665.
[60] J. Wulf and W. G. Pohl, *Biochim. Biophys. Acta*, 1977, **465**, 471.

On incorporation[61] of the neutral, synthetic ionophore [CH(Me)OCH$_2$CON-(Me)(CH$_2$)$_{11}$CO$_2$Et]$_2$ into phospholipid bilayers, the selectivity of the ionophore towards Ca^{2+} is reduced, and it becomes more selective for Na$^+$. Unlike X-537A and A23187, this ligand carries Ca^{2+} as a charged complex, thus accounting for the enhancement of its activity by uncouplers. A similar situation held for translocation of Ca^{2+} across the membranes of erythrocyte and sarcoplasmic reticulum vesicles, while in the presence of Ruthenium Red (to inhibit the natural carrier) active transport was observed in mitochondria.

Interaction of Metal Ions with Phospholipid Membranes.—Contradictory results for the interaction of cations with phosphatidylcholine membrane vesicles have been clarified by n.m.r. studies.[62] The binding efficiency is controlled by the electrostatic potential produced by the bound cations at the membrane surface, so that the apparent binding constant varies as a function of the metal that is bound. Lanthanides form 2:1 phosphatidylcholine–cation complexes. The ratio of phospholipid groups on outer to inner surfaces of phospholipid vesicles may be measured by n.m.r. spectroscopy in the presence of lanthanide ions bound either to the inside or to the outside of the vesicles. This ratio has been shown conclusively to be $(2.1 \pm 0.1):1$. The ions Ca^{2+} and La^{3+} do not perturb the average headgroup conformation.[63]

Membrane Fusion. Sonicated phospholipid vesicles undergo substantial morphological changes, due to vesicle fusion, in the presence of bivalent ions. This process has been studied by light scattering and freeze-fracture electron microscopy.[64] The calcium-induced interaction of phospholipid vesicles with bilayer membranes[65] and the effect of bivalent cations on lipid phase transitions from fluid to solid[66] have been examined in this context. In a novel approach to the study of the fusion of phosphatidylcholine vesicles, one set was prepared that contained the partially calcium-saturated, calcium-sensitive dye Arsenazo III (AIII.Ca) and another set that contained EGTA. Interaction of EGTA with AIII.Ca gives a colour change, which is therefore an indicator to vesicle fusion, after due allowance for lysis and diffusion. Fusion occurred only slowly, even in the presence of added 'fusogens' lysolecithin and retinol, indicating that such systems may require other membrane constituents before they mimic well the fusion of natural membranes.[66a]

5 Natural Transport Systems

The concentrations of intracellular cations are determined by a number of

[61] P. Vuilleumier, P. Gazzotti, E. Carafoli, and W. Simon, *Biochim. Biophys. Acta*, 1977, **467**, 12; P. Caroni, P. Gazzotti, P. Vuilleumier, W. Simon, and E. Carafoli, *ibid.*, 1977, **470**, 437.
[62] H. Hauser, C. C. Hinckley, J. Krebs, B. A. Levine, M. C. Phillips, and R. J. P. Williams, *Biochim. Biophys. Acta*, 1977, **468**, 364; H. Grasdalen, L. E. Goran Eriksson, J. Westman, and A. Ehrenberg, *ibid.*, 1977, **469**, 151.
[63] W. C. Hutton, P. L. Yeagle, and R. B. Martin, *Chem. Phys. Lipids*, 1977, **19**, 255.
[64] E. P. Day, J. T. Ho, R. K. Kunze, and S. T. Sun, *Biochim. Biophys. Acta*, 1977, **470**, 503; J. G. Stollery and W. J. Vail, *ibid.*, 1977, **471**, 372.
[65] N. Duzgunes and S. Ohki, *Biochim. Biophys. Acta*, 1977, **467**, 301.
[66] D. Papahadjopoulos, *J. Colloid Interface Sci.*, 1977, **58**, 459.
[66a] P. Dunham, P. Babiarz, A. Israel, A. Zerial, and G. Weissmann, *Proc. Nat. Acad. Sci. U.S.A.*, 1977, **74**, 1580

transport systems associated with external membranes and those of intracellular organelles. Such transport systems are reviewed in this section, while the following section deals with the diverse cellular operations controlled by cations, particularly by Ca^{2+}. It is difficult to avoid some overlap between these two sections.

Na^+-K^+-ATPase.—Despite the enormous amount of material published on the sodium pump, it is still not possible to present with confidence any overall hypothesis for its action. It is, however, possible to abandon the attractive circulating carrier mechanism, in which a carrier responsible for moving Na^+ outwards was converted at the outer face of the membrane into a carrier that moved K^+ inwards. This has come about in the light of studies on the availability of internal and external sites for Na^+ and K^+, respectively, before the enzyme is phosphorylated. The review of Glynn and Karlish provides an excellent account of work up to 1975.[67]

Purification and Structure. The enzyme has been purified from pig thyroid glands,[68] cardiac sarcolemma,[69] and human renal tissue,[70] the last example being the first well characterized Na^+-K^+-ATPase from a human source. On SDS gel electrophoresis it gave three protein peaks (117 000, 92 500, and 56 000 dalton). The 92 500 dalton peptide underwent a Na^+-dependent phosphorylation while the 56 000 dalton peptide stained for glycoprotein. These correspond to the catalytic proteins and glycoproteins found in the enzyme from other sources, but there is no explanation at present for the 117 000 dalton protein. The presence and shape of the catalytic proteins and glycoproteins have been tentatively identified by electron microscopy of the enzyme on plasma membrane. A stalked knob protrudes from the catalytic protein, and it has been suggested that it is the catalytic centre.[71]

The controlled hydrolysis of the large, catalytic protein with trypsin and correlation with the associated loss of activity offers a method of associating function of the protein with particular protein sections.[72] The extent of the loss of Na^+-K^+-ATPase activity with trypsin hydrolysis differs in the presence of KCl and NaCl, reflecting the fact that there are two forms of the enzyme in the presence of K^+ or Na^+. In the presence of KCl, for example, loss of activity was proportional to the extent of cleavage of the catalytic chain into fragments of 58 000 and 46 000 daltons. Conformational changes in a limited region of the protein of Na^+-K^+-ATPase have been detected[73] by means of a sulphydryl fluorescence probe. Binding sites for a spin label in ox-brain membranes enriched with Na^+-K^+-ATPase have been characterized; one appears to be in the microsomal membrane.[74] The effect of structural changes in lipids on Na^+-K^+-ATPase has been investigated, using spin labels, in an attempt to account for the non-

[67] I. M. Glynn and S. J. D. Karlish, *Ann. Rev. Physiol.*, 1975, **37**, 13.
[68] Y. Nagai and T. Hosoya, *J. Biochem. (Japan)*, 1977, **81**, 721.
[69] D. C. Pang and W. B. Weglicki, *Biochim. Biophys. Acta*, 1977, **465**, 411.
[70] J. M. Braughler and C. N. Corder, *Biochim. Biophys. Acta*, 1977, **481**, 313.
[71] F. Vogel, H. W. Meyer, R. Grosse, and K. R. H. Repke, *Biochim. Biophys. Acta*, 1977, **470**, 497.
[72] P. L. Jorgensen and J. Petersen, *Biochim. Biophys. Acta*, 1977, **466**, 97; J. R. Lea and C. G. Winter, *Biochem. Biophys. Res. Comm.*, 1977, **76**, 772.
[73] W. E. Harris and W. L. Stahl, *Biochim. Biophys. Acta*, 1977, **485**, 203.
[74] A. F. Almeida and J. S. Charnock, *Biochim. Biophys. Acta*, 1977, **467**, 19.

linearity of Arrhenius plots for the enzyme.[75] Differential scanning calorimetry indicates[76] that the cardiac glycoside inhibitor ouabain induces a structural change in the enzyme, while catechol is suggested to enhance Na^+-K^+-ATPase activity by protecting or favourably modifying the active conformation.[77]

Enzyme Kinetics and Transport Kinetics. Earlier studies with fragmented ATPase were complicated by the fact that Na^+ and K^+ have access to both faces of the membrane, and so compete at the internal Na^+ and external K^+ sites. Of interest then is the reconstruction of the Na^+-K^+-ATPase in phospholipid vesicles, which shows catalysis of ATP-mediated transport of Na^+ and K^+ with a ratio of $3Na^+$:$2K^+$ with ouabain inhibition, similar to red cell and nerve. The full transport system and ATPase activity is clearly contained in the protein-vesicle system.[78] As an alternative approach to the problem, whole cells, such as re-sealed red blood cell ghosts, have been used. In such a system, the sodium pump was reversed by suitable arrangement of concentration gradients, and the value of $K_{0.5}$ at the internal K^+ site was 300 mmol l^{-1}, *i.e.* a site of low affinity for K^+, in accord with the view that its normal function lies in the intracellular discharge of K^+.[79] Inside-out vesicles of human red cells have been used to allow the study of specific interactions of Na^+ and K^+ with particular sides of the transport system.[80] A comparison of inside-out and right-side-out plasma membrane vesicles has shown that proteins (*ca.* 30 000 dalton) are present on the inner face of the membrane which reduce the sensitivity of the enzyme to ouabain.[81] The channel-forming ionophore alamethicin has been used[82] to activate latent Na^+-K^+-ATPase and adenylate cyclase on the inner face of cardiac sarcolemmal vesicles by allowing entry of ATPase into the vesicle.

The study of phosphorylated intermediates continues to contribute to our understanding of Na^+-K^+-ATPase. The enzyme is phosphorylated in the presence of Na^+ and Mg^{2+} (MgATP probably being the substrate) and dephosphorylated by K^+, ATP being produced in the presence of ADP. Synthesis of ATP from ADP and P_i in fragmented membranes (*i.e.* no ion gradients) takes place in two stages,[83] *i.e.* incorporation of phosphate into the protein followed by transfer to ADP, a step that only occurs in the presence of high [Na^+]. It is usually accepted that two phosphorylated forms of the enzyme exist, the first form (E_1P) being converted into a more stable form (E_2P) in a Mg^{2+}-dependent process (2).

$$ E_1 \xrightarrow[Mg^{2+},Na^+]{ATP} E_1P \xrightarrow{Mg^{2+}} E_2P \xrightarrow{K^+} E_2 \longrightarrow E_1 \qquad (2) $$

[75] A. Boldyrev, E. Ruuge, I. Smirnova, and M. Tabak, *F.E.B.S. Letters*, 1977, **80**, 303.
[76] J. F. Halsey, D. B. Mountcastle, C. A. Takeguchi, R. L. Biltonen, and G. E. Lindenmayer, *Biochemistry*, 1977, **16**, 432.
[77] L. C. Cheng, E. M. Rogus, and K. Zierler, *Biochim. Biophys. Acta*, 1977, **464**, 338, 347.
[78] B. M. Anner, L. L. Lane, A. Schwartz, and B. J. R. Petts, *Biochim. Biophys. Acta*, 1977, **467**, 340; S. M. Goldin, *J. Biol. Chem.*, 1977, **252**, 5630.
[79] J. D. Robinson, E. S. Hall, and P. B. Dunham, *Nature*, 1977, **269**, 165.
[80] R. Blostein and L. Chu, *J. Biol. Chem.*, 1977, **252**, 3035.
[81] A. Zachowski, L. Lelievre, J. Aubry, D. Charlemagne, and A. Paraf, *Proc. Nat. Acad. Sci. U.S.A.*, 1977, **74**, 633.
[82] H. R. Besch, L. R. Jones, J. W. Fleming, and A. M. Watanabe, *J. Biol. Chem.*, 1977, **252**, 7905.
[83] C. Hallam and R. Whittam, *Proc. Roy. Soc.*, 1977, **B198**, 109.

Computer simulation of phosphorylation kinetics fitted well a model involving a stepwise sequence of two dephospho- and two phospho-forms of the enzyme,[84] and was consistent with the rate of hydrolysis of ATP. The effectiveness of Mg^{2+}, Na^+, and ATP in preparing Na^+-K^+-ATPase for phosphorylation was checked in experiments in which their order of addition was varied: Na^+ and ATP appear to bind in random order and to exert a mutual potentiation effect.[85] The inhibition of Na^+-K^+-ATPase by higher concentrations of Mg^{2+} has received little attention, but it appears that ATP and K^+ are also required.[86] The Na^+-K^+-ATPase-catalysed ADP–ATP exchange reaction[87] is also inhibited by Mg^{2+} or MgATP, which occupy the low-affinity substrate sites of the enzyme. This inhibition is relieved by K^+. The exchange reaction is postulated to involve two enzyme–ATP complexes and two phosphorylated intermediates.[88]

The cation-binding sites of Na^+-K^+-ATPase have been characterized:[89] *Na^+-binding site;* $K_{0.5}(Na^+) = 1.7$ mmol l^{-1}, $K_i(K^+) = 5.1$ mmol l^{-1}; *β-K^+-binding site;* $K_{0.5}(K^+) = 0.11$ mmol l^{-1}, $K_i(Na^+) = 14$ mmol l^{-1}. From the dependence of $-\log K_{0.5}(Na^+)$ on pH, a group having a $pK \approx 8.1$ is suggested to influence activation at the Na^+-binding site. For the related K^+-dependent phosphatase reaction involving K^+ at α-sites, $K_{0.5}(K^+) = 1.0$ mmol l^{-1}, $K_i(Na^+) = 3.5$ mmol l^{-1}.

Phosphatase activity towards new substrates [2,4-dinitrophenylphosphate and β-(2-furyl)acryloylphosphate] has been demonstrated,[90] and the kinetics have been fitted to a model involving four Na^+ and two K^+ sites per mole of enzyme.

The interaction of Ca^{2+} with Na^+-K^+-ATPase has continued to attract attention. Ca^{2+} reduces[91] the affinity of the enzyme for Na^+, an effect that is enhanced at lower pH values. The effect of Ca^{2+} on the phosphorylation of Na^+-K^+-ATPase by cyclic-AMP-dependent protein kinase has also been studied.[92] Four binding sites for Ca^{2+} on the ATPase were characterized, none of which were affected by phosphorylation.

Inhibitors. Anomalous kinetic behaviour for Na^+-K^+-ATPase has been reported over a number of years. It has now been recognized[93] that this is due to an impurity in 'Sigma Grade' ATP that dramatically increases sensitivity to inhibition by K^+ at high concentration. The impurity has been identified[94] as sodium orthovanadate, which is assumed to substitute for phosphate in phosphate-binding enzymes.

Pb^{2+} in micromolar concentrations inhibits Na^+-K^+-ATPase and K^+-*p*-nitrophenylphosphatase from *Electrophorus* electroplax, but stimulates the phosphorylation of electroplax microsomes in the absence of added Na^+, due,

[84] S. Mardh and S. Lindahl, *J. Biol. Chem.*, 1977, **252**, 8058.
[85] S. Mardh and R. L. Post, *J. Biol. Chem.*, 1977, **252**, 633.
[86] J. B. Fagan and E. Racker, *Biochemistry*, 1977, **16**, 152; L. C Cantley and L. Josephson, *ibid.*, 1976, **15**, 5280.
[87] J. D. Robinson, *Biochim. Biophys. Acta*, 1977, **484**, 161.
[88] M. Yamaguchi and Y. Tonomura, *J. Biochem. (Japan)*, 1977, **81**, 249.
[89] J. D. Robinson, *Biochim. Biophys. Acta*, 1977, **482**, 427.
[90] C. Gache, B. Rossi, and M. Lazdunski, *Biochemistry*, 1977, **16**, 2957.
[91] T. Godfraind, A. de Pover, and N. Verbeke, *Biochim. Biophys. Acta*, 1977, **481**, 202.
[92] K. Kaniike, B. J. R. Pitts, and A. Schwartz, *Biochim. Biophys. Acta*, 1977, **483**, 294.
[93] L. A. Beaugé and I. M. Glynn, *Nature*, 1977, **268**, 355; L. Josephson and L. C. Cantley, *Biochemistry*, 1977, **16**, 4572.
[94] L. C. Cantley, L. Josephson, R. Warner, M. Yanagisawa, C. Lechene, and G. Guidotti, *J. Biol. Chem.*, 1977, **252**, 7241.

it is suggested,[95] to the action of Pb^{2+} at a single independent Pb^{2+}-binding site on the enzyme. The action of the cardiac glycosides has been reviewed.[96]

Calcium-transport Processes.—Calcium concentrations inside and outside the cell are approximately 10^{-6} and 10^{-3} mol l^{-1} respectively. An increase in intracellular $[Ca^{2+}]$ may arise from entry (active or passive) through the cell membrane or by release from internal organelles. Such transport systems must then have the capacity to remove the high Ca^{2+} concentration rapidly. Much attention has been focused on the cellular Ca^{2+} controller. For various types of muscle the sarcoplasmic reticulum is well established as the Ca^{2+} regulator in muscle cells. However, the endoplasmic reticulum of other mammalian cells, while able to take up Ca^{2+}, has a low capacity for it, and so does not appear to be the regulator. Mitochondria, on the other hand, appear to be well qualified for this role, showing considerable capacity for Ca^{2+} uptake, and having an effective calcium pump for the accumulation of Ca^{2+}.

Mitochondria. The mechanism and physiological significance of calcium transport across mitochondrial membranes have been reviewed.[97] The role of mitochondria in regulating intracellular concentrations of Ca^{2+} has been stressed. Mitochondria in squid axons were studied *in situ* by use of the dye Arsenazo III, which showed[98] that the mitochondria in an intact axon can buffer about 99.5% of an imposed Ca^{2+} load, for total loads in the range 50—2500 μmol l^{-1}. Thus mitochondria could maintain the axoplasmic Ca^{2+} at 3—5 μmol l^{-1} despite substantial amounts of Ca^{2+} entering. Low levels of Ca^{2+} (2—3 μmol l^{-1}) were not accumulated, even though these were higher than physiological concentrations, possibly reflecting some co-operative effect in the Ca^{2+}-uptake system. Axons treated with the uncoupler FCCP were much less able to control Ca^{2+} concentrations, while the treatment of fresh axons with FCCP caused a small release of Ca^{2+} from the mitochondria into the axoplasm. A similar role for mitochondria in liver has been demonstrated[99] by the use of Ruthenium Red, a specific inhibitor of mitochondrial electron transport. This was shown to have no effect on transport of Ca^{2+} by microsome and plasma membranes. The Ruthenium-Red-sensitive component of Ca^{2+} transport by liver homogenates is very much greater than the Ruthenium-Red-insensitive component, thus demonstrating the potential of mitochondria to control intercullular $[Ca^{2+}]$ in liver. Removal of Ca^{2+} by mitochondria (and microsomes) in galactosamine-intoxicated rat liver cells has also been demonstrated.[100]

Mitochondria from different cells differ somewhat, and will be considered separately. A warning has been given that secondary lysosomes, which are present in rat kidney mitochondrial preparations, have an inhibiting effect on the uptake of calcium by mitochondria.[101]

[95] G. M. Siegel and S. M. Foot, *J. Biol. Chem.*, 1977, **252**, 5201.
[96] G. T. Okita, *Fed. Proc.*, 1977, **36**, 2225; T. M. Brody and T. Akera, *ibid.*, p. 2219.
[97] L. Mela, in 'Current Topics in Membranes and Transport', ed. F. Bonner and A. Kleinzeller, 1977, **9**, 321.
[98] J. F. Brinley, J. T. Tiffert, L. J. Mullins, and A. Scarpa, *F.E.B.S. Letters*, 1977, **82**, 197.
[99] G. R. Ash and F. L. Bygrave, *F.E.B.S. Letters*, 1977, **78**, 166.
[100] J. L. Farber, S. K. El-Mofty, F. A. X. Schanne, J. J. Aleo, and A. Serroni, *Arch. Biochem. Biophys.*, 1977, **178**, 617.
[101] G. E. Bunce and B. W. Li, *Biochim. Biophys. Acta*, 1977, **460**, 163.

The long-unresolved question[102] of whether the entry of Ca^{2+} into liver mitochondria requires a permeant co-ion appears to have been clarified. The efflux of H^+ from mitochondria occurs with a stoicheiometry of only about $2H^+$ per site, while Ca^{2+} influx involves two Ca^{2+} ions per site, implying, from a charge balance, that each calcium ion is brought in as a singly charged complex. However, the H^+-efflux stoicheiometry has been disputed, and the suggestion made that some H^+ was lost back through the mitochondrial membrane by one of several possible pathways. The stoicheiometry has now been remeasured, with a careful check on possible causes of loss of H^+, and the stoicheiometry has been confirmed,[103] indicating that nett calcium and electric charge translocation is $\leftarrow Ca^+$, the so-called calcium porter involving $[Ca_2]^{4+} [HPO_4]^{2-}$. When other known phosphate pathways are inhibited, phosphate is still transported[104] into the cell with Ca^{2+} (or Sr^{2+}) in correct stoicheiometric ratio, while the transport is lanthanide-sensitive. If the availability of phosphate is limited or inhibited,[105] then mitochondria show a monocarboxylate-dependent import of Ca^{2+}, which is rather specific for β-hydroxybutyrate, an oxidizable substrate. The need for the penetrant anion for Ca^{2+} uptake if endogenous phosphate is unavailable has been confirmed independently[106] for heart and liver mitochondria.

The plot of initial rate of Ca^{2+} transport[107] *versus* concentration for mitochondrial Ca^{2+} transport becomes sigmoidal in the presence of Mg^{2+} and K^+, which bind competitively to low-affinity Ca^{2+} sites on the membrane, and in the presence of Ba^{2+}, which is a competitive inhibitor of Ca^{2+} transport and binding. The value of K_i for inhibition of Ca^{2+} transport by Ba^{2+} increases in the presence of K^+ and Mg^{2+}, indicating that there is competition for the sites. Ruthenium Red, a non-competitive inhibitor, does not affect the shape of the kinetic plot. Thus the presence or absence of surface-binding cations seems to determine whether calcium uptake is sigmoidal or hyperbolic.

The ionophore A23187 increases[108] the permeability of mitochondrial membranes to potassium ions, but this is secondary to the ionophoric action of A23187 on Mg^{2+} permeability, rather than a direct effect. A23187 has also been used[109] to study the steady-state kinetics of energy-dependent Ca^{2+} uptake with β-hydroxybutyrate as the oxidizable substrate. Mg^{2+} is an allosteric inhibitor, increasing $K_{0.5}$ for Ca^{2+} from ~ 5 to 17.8 μmol l^{-1} at 2.0 mmol l^{-1} of added $MgCl_2$. The rate-limiting step is probably associated with the respiratory chain rather than with the calcium carrier, in particular with the substrate dehydrogenases. The normal Ca^{2+} cellular concentrations will be much lower than those present using the ionophore; under such circumstances the Mg^{2+} inhibition may be a regulator of Ca^{2+} transport.

---

[102] G. F. Azzone, T. Pozzan, S. Massari, M. Bragadin, and P. Dell' Antone, *F.E.B.S. Letters*, 1977, **78**, 21.
[103] J. Moyle and P. Mitchell, *F.E.B.S. Letters*, 1977, **73**, 131.
[104] J. Moyle and P. Mitchell, *F.E.B.S. Letters*, 1977, **77**, 2.
[105] J. Moyle and P. Mitchell, *F.E.B.S. Letters*, 1977, **84**, 135.
[106] E. J. Haines and B. Zaba, *F.E.B.S. Letters*, 1977, **79**, 284
[107] K. E. O. Akerman, M. K. F. Wikström, and N. E. Saris, *Biochim. Biophys. Acta*, 1977, **464**, 287.
[108] J. Duszynski and L. Wojtczak, *Biochem. Biophys. Res. Comm.*, 1977, **74**, 417.
[109] S. M. Hutson, *J. Biol. Chem.*, 1977, **252**, 4539.

The Ca^{2+}, ATP-dependent oxidation of succinate[110] by rat liver mitochondria is not lanthanide-sensitive, and is observed in mitochondria lacking the outer membrane. The effect is observed in the absence of Ca^{2+} accumulation, and succinate does not penetrate the membrane. It has been suggested that Ca^{2+} interacts with the low-affinity Ca^{2+}-binding sites on the outer surface of the inner mitochondrial membrane, causing a change in membrane conformation and the activation of the succinate dehydrogenase.

Glucocorticoids are known to alter Ca^{2+} transport in various tissue. The administration of dexamethasone to rats diminishes the initial rate and the extent of substrate-dependent calcium uptake in subsequently isolated liver mitochondria,[111] but has no effect when ATP is employed as the energy source, although ATP translocation from inside to outside is faster. As the ATP in the medium is depleted, the preloaded Ca^{2+} is held in the mitochondria until the ATP reaches 5.7 μmol per mg of protein, when all the Ca^{2+} is released. Dexamethasone thus appears to function by affecting the intramitochondrial ATP content. Physiological control of efflux of Ca^{2+} is an important topic, but while cyclic AMP is reported to stimulate this, effective agents are non-physiological, namely uncouplers and ionophores. Under certain conditions, Ruthenium Red inhibits efflux,[112] but there has been disagreement over this.

Of particular interest is the role of mitochondria in controlling intracellular $[Ca^{2+}]$ in the heart, as this may be associated with its regulation of relaxation–contraction. In the mammalian heart[97] some 60 nmol of Ca^{2+} per gram of tissue would have to be removed during relaxation; only a small fraction of the total heart mitochondrial capacity for Ca^{2+}. Kinetics of uptake and loss of Ca^{2+} from heart mitochondria are therefore of vital importance in resolving this question, and several papers have appeared on this subject. From studies on the energy-independent initial binding of Ca^{2+}, a dissociation constant of 26×10^{-6} mol l^{-1} and a total number of external binding sites of 12.6 nmol of Ca^{2+} per mg of protein have been obtained. A maximum uptake rate was obtained between 10 and 15 μmol l^{-1} of Ca^{2+}. The value of K_m for uptake equals 5.6 μmol l^{-1}, but this is dependent on ATP. Two binding sites for Ca^{2+} were indicated for the trans-membrane binding system.[113]

Palmitoyl-CoA and the adenine translocase inhibitor atractyloside both have an effect[114] in discharging Ca^{2+} from cardiac mitochondria, but the effect is prevented by exogenous ATP. Thus the effect of palmitoyl-CoA may arise from its surfactant property of stimulating an ATPase. The role of external ATP in protecting against Ca^{2+} release has been discussed.[115] The sensitivity of Ca^{2+} uptake to Mg^{2+} seems to rule out a role for mitochondria in beat-to-beat Ca^{2+} cycling.

Efflux of $^{45}Ca^{2+}$ from rat heart mitochondria[116] is induced by Na^+, Li^+, Ca^{2+},

[110] I. Ezawa, E. Ogata, and Y. Sano, *European J. Biochem.*, 1977, **77**, 427.
[111] S. Kimura and H. Rasmussen, *J. Biol. Chem.*, 1977, **252**, 1217
[112] R. Luthra and M. S. Olsen, *F.E.B.S. Letters*, 1977, **81**, 142; T. Pozzan, M. Bragadin, G. F. Azzone, and P. Veronese, *Biochemistry*, 1977, **16**, 5618.
[113] E. A. Noack and E. M. Heinen, *European J. Biochem.*, 1977, **79**, 245.
[114] G. K. Asimakis and L. A. Sordahl, *Arch. Biochem. Biophys.*, 1977, **179**, 200.
[115] E. J. Harris, *Biochem. J.*, 1977, **168**, 447.
[116] M. Crompton, M. Künzi, and E. Carafoli, *European J. Biochem.*, 1977, **79**, 549.

or Sr^{2+}, and is inhibited by La^{3+}, but not by Ruthenium Red. The results are accounted for by an exchange diffusion carrier in the inner membrane that is separate from the energy-linked Ca^{2+}-uptake system.

Other studies on cation transport in mitochondria have included K^+ influx and efflux[117] and the characterization of a Ca^{2+}-Mg^{2+}-ATPase from the outer membrane function of rat spleen mitochondria,[118] while features in the calcium uptake of ascites tumour cells are relevant to their mitochondrial calcium transport.[119]

Ca^{2+}-Mg^{2+}-ATPase from Endoplasmic Reticulum. Rat adipocyte endoplasmic reticulum contains deposits of calcium, and an energy-dependent calcium-uptake mechanism has now been characterized.[120] The transport system was similar to that found in sarcoplasmic reticulum, and is Mg^{2+}, ATP-dependent, with a K_m of *ca.* 3.6 μmol of free Ca^{2+} per litre and a maximum uptake of 5.0 nmol (mg protein)$^{-1}$ min^{-1} at 24 °C. The passive binding of Ca^{2+} to endoplasmic-reticulum-enriched adipocyte microsomes involves three independent sites, which have been characterized over a $[Ca^{2+}]$ range of 1.0 μmol l^{-1} to 10 mmol l^{-1} (Table 1). The site of highest affinity is probably the binding site for the active transport of Ca^{2+}. Mg^{2+} is a non-competitive inhibitor, while calcium binding at a $[Ca^{2+}]$ of 1 μmol l^{-1} was 50% inhibited by concentrations of La^{3+}, Sr^{2+}, or Ba^{2+} of 18 μmol l^{-1}, 600 μmol l^{-1}, and 2.7 mmol l^{-1}, respectively. Although there are substantially fewer Ca^{2+} sites in endoplasmic reticulum compared to sarcoplasmic reticulum, the transport systems appear to operate at similar rates, implying a physiological role for endoplasmic reticulum in $[Ca^{2+}]$ control.

Table 1 *Calcium-binding sites on endoplasmic reticulum*

Site	K/mol l^{-1}	Binding capacity/ mmol (mg protein)$^{-1}$	Dissociation rate constant/s^{-1}	Association rate constant/l mol^{-1} s^{-1}
1	2.1×10^5	0.28	1.6×10^{-3}	8×10^2
2	1.3×10^4	1.1	2.0×10^{-2}	—
3	1.3×10^2	35	—	—

Ca^{2+}-Mg^{2+}-ATPase from Sarcoplasmic Reticulum. The function of sarcoplasmic reticulum (SR) in the uptake or release of calcium ions in muscle for relaxation and contraction respectively is well known. Uptake is *via* a Ca^{2+}-Mg^{2+}-ATPase membrane pump, but the mechanism of efflux of Ca^{2+} is still uncertain, as this process is too rapid for a simple reversal of the pump, and may, for example, involve the formation of ion channels. The Ca^{2+}-Mg^{2+}-ATPase has been purified (102 000 dalton) and has been successfully incorporated into vesicles that show both ATPase and Ca^{2+}-transport properties, confirming that both functions are associated with the protein. Additional calcium-binding proteins have been found in SR membrane: a high-affinity protein (55 000); calsequestrin (44 000—55 000),

[117] D. W. Jung, E. Chávez, and G. P. Brierley, *Arch. Biochem. Biophys.*, 1977, **183**, 452; E. Chávez, D. W. Jung, and G. P. Brierley, *ibid.*, p. 460.

[118] E. K. Vijayakumar and M. J. Weidemann, *Biochem. J.*, 1977, **165**, 355.

[119] R. N. Hines and C. E. Wenner, *Biochim. Biophys. Acta*, 1977, **465**, 391; A. Cittadini, D. Bossi, G. Rosi, F. Wolf, and T. Terranova, *ibid.*, 1977, **469**, 345.

[120] D. E. Bruns, J. M. McDonald, and L. Jarett, *J. Biol. Chem.*, 1976, **251**, 7191; 1977, **252**, 927.

for which a simple method of preparation has been reported;[121] a further protein (30 000); and a lipoprotein (12 000 dalton).

The arrangement of these proteins in the membrane has been studied[122] by cross-linking techniques, which leads to the formation of a series of new proteins having molecular weights in multiples of 100 000 up to 900 000. These products can be accounted for if the Ca^{2+}-Mg^{2+}-ATPase molecules are within 2 Å of each other, while calsequestrin and/or the calcium-binding protein are within 11 Å of each other. No heteropolymers were formed. The aggregation of ATPase has been confirmed independently[123] and suggested to form a Ca^{2+}-efflux channel.[124] The use of enzyme-catalysed iodination has indicated that calsequestrin may be exposed at the outer surface of the membrane.[124,125]

The organization of Ca^{2+}-Mg^{2+}-ATPase has been investigated.[126] Hydrolysis of SR vesicles by trypsin for 1 minute cleaves the enzyme into two fragments of molecular weights 45 000 and 55 000. The latter fragment splits further into 30 000 and 20 000 dalton fragments which contain the site of ATP hydrolysis and the ionophoric activity respectively.[127] The fragments are separated by treatment with sodium dodecylsulphate, but Ca^{2+}-transport activity may be restored to fragments thus treated.[128] Hydrolysis of the enzyme in different states by trypsin has been reported.[129]

Sarcoplasmic reticulum ATPase has been 'probed' by several techniques. Gadolinium, as Gd^{3+}, binds to cardiac and skeletal muscle microsomes at high- and low-affinity sites.[130] For dog heart SR, the constant for the high-affinity site was 10^6 l mol^{-1}, with a capacity of 131 nmol per mg of protein. Low concentrations of ATP increased Gd^{3+} binding, without hydrolysis of the ATP. The Mg^{2+}-Ca^{2+}-ATPase was inhibited by Gd^{3+}, which had no effect on the X-537A-stimulated ATPase activity. The Gd^{3+} was suggested to bind at the ionophoric component of the calcium-transport site rather than the ATPase site. The ionophore X-537A did not cause Gd^{3+} release, unlike the case with Ca^{2+}, indicating that the Gd^{3+} is not free in the intra-cell space. EGTA releases the bulk of the Gd^{3+}, indicating that the site is on the external surface of the membrane.

Electron microscopic examination of Gd^{3+} with SR vesicles shows[131] that little or no Gd^{3+} is taken up within the interior and that the Gd^{3+} is localized at the membrane Ca^{2+}-Mg^{2+}-ATPase sites. A roughly linear relationship exists between the ionic radii of thirteen lanthanide elements and their ability to inhibit the

[121] W. Drabikowski, *Biochem. Biophys. Res. Comm.*, 1977, **75**, 746
[122] C. F. Louis, M. J. Saunders, and J. A. Holroyd, *Biochim. Biophys. Acta*, 1977, **493**, 78.
[123] M. le Maire, K. E. Jørgensen, H. Røigaard-Petersen, and J. V. Møller, *Biochemistry*, 1976, **15**, 5805.
[124] T. Chyn and A. Marlonosi, *Biochim. Biophys. Acta*, 1977, **468**, 114.
[125] C. F. Louis and A. M. Katz, *Biochim. Biophys. Acta*, 1977, **494**, 255.
[126] V. M. C. Madeira, *Biochim. Biophys. Acta*, 1977, **464**, 583.
[127] A. E. Shamoo, T. E. Ryan, F. S. Stewart, and D. H. MacLennan, *J. Biol. Chem.*, 1976, **251**, 4147.
[128] D. H. MacLennan, V. K. Khanna, and P. S. Stewart, *J. Biol. Chem.*, 1976, **251**, 7271.
[129] T. Yamamoto and Y. Tonomura, *J. Biol. Chem.*, 1977, **82**, 653.
[130] N. Krasnow, *Arch. Biochem. Biophys.*, 1977, **181**, 322.
[131] C. dos Remedios, *J. Biochem. (Japan)*, 1977, **81**, 703.

ATPase,[131a] but, in contrast, the sarcolemma appears to be particularly sensitive to those having similar radii to Ca^{2+}.

Addition[132] of ATP to —SH spin-labelled SR vesicular fragments produced a new 'highly constrained' component in the e.s.r. spectrum of the spin-labelled SR. The change is reversible and requires ATP and Ca^{2+} binding, but does not require hydrolysis of the ATP. It was demonstrated that the labelled residue(s) responsible for the effect on the spectrum are not in the immediate vicinity of the ATP-binding site, implying that it results from a conformational change on the ATPase produced by Ca^{2+} and ATP binding. This may define a stage in the translocation of Ca^{2+}. The reaction of 5,5′-dithiobis(2-nitrobenzoate) (DTNB) with all the —SH groups in a Ca^{2+}-Mg^{2+}-ATPase vesicle preparation shows[133] that, in the absence of substrate, four —SH groups reacted with a first-order rate constant some ten times greater than that for the remaining fifteen groups. At 0.1mM-ATP, -ADP, and -MgADP two of the rapidly reacting groups disappeared, and at 1mM-nucleotide the remaining groups reacted more slowly. ATPase activity decreased linearly with the number of modified groups. During phosphorylation, the number of rapidly reacting groups increased from two to three or four. It has been suggested that the binding of nucleotide masks two of the rapidly reacting —SH groups, while the decrease in reactivity of the other groups to DTNB results from conformational changes in the protein.

Binding of Ca^{2+} to SR membranes has been monitored[134] by fluorescence techniques, while [1]H n.m.r. spectroscopy has been used to study[135] the binding of uridine 5′-phosphate to the Ca^{2+}-Mg^{2+}-ATPase of SR vesicles. The rotation movement of the ATPase was characterized from the relaxation times of the bound nucleotide, and found to be rapid compared to that of other membrane-bound proteins.

The uptake and release of Ca^{2+} by SR vesicles under a wide range of conditions (including the use of SR fragments on Millipore filters[136]) has been examined. The presence of ATPase oligomers in vesicles has been demonstrated[137] by fluorescence measurements between ATPase molecules labelled with N-iodoacetyl-N'-(5-sulpho-1-naphthyl)ethylenediamine or iodoacetamidofluorescein (donor and acceptor fluorophores respectively). Dilution of the lipid phase did not affect energy transfer, implying that it derived from complexes of several ATPase molecules rather than from random collisions between them. Such aggregates (in tetrameric association) are postulated[138] to provide hydrophilic channels for the rapid release of Ca^{2+} from SR in muscle activation. The involvement of the calcium pump in the passive efflux of Ca^{2+} from SR vesicles is supported[139] by

[131a] C. G. dos Remedios, *Nature*, 1977, **270**, 750; B. D. Hambley and C. G. dos Remedios, *Experientia*, 1977, **33**, 1042.
[132] C. R. Coan and G. Inesi, *J. Biol. Chem.*, 1977, **252**, 3044.
[133] J. P. Andersen and J. V. Møller, *Biochim. Biophys. Acta*, 1977, **485**, 188.
[134] C. A. M. Carvalho and A. P. Carvalho, *Biochim. Biophys. Acta*, 1977, **468**, 21.
[135] B. A. Manuck and B. D. Sykes, *Canad. J. Biochem.*, 1977, **55**, 587.
[136] G. L. Alonso, D. M. Arrigó, S. E. Terradas, J. M. Mikonov, D. Nespral, and S. E. Palomba, *Biochim. Biophys. Acta*, 1977, **468**, 31.
[137] J. M. Vanderkooi, A. Ierokomas, H. Nakamura, and A. Martonosi, *Biochemistry*, 1977, **16**, 1262.
[138] R. L. Jilka and A. N. Martonosi, *Biochim. Biophys. Acta*, 1977, **466**, 219.
[139] A. M. Katz, D. I. Repke, G. Fudyma, and M. Shigekawa, *J. Biol. Chem.*, 1977, **252**, 4210.

results on the well-known trigger action of increased external Ca^{2+} in releasing stored Ca^{2+}. This efflux cannot be attributed to reversal of the active pump. The reversed pump functions at high internal $[Ca^{2+}]$ (Ca_i) and low external $[Ca^{2+}]$ (Ca_o). As it is coupled to ATP synthesis, it is inhibited by high Ca_o and high ATP. In contrast, the Ca^{2+}-triggered efflux is stimulated by high Ca_o and ATP, and is not coupled to the re-synthesis of ATP. However, several resemblances between Ca^{2+}-efflux and initial Ca^{2+} uptake by the calcium pump suggest that passive calcium efflux may be mediated by the pump.

The calcium permeability of SR vesicles is dependent upon factors in addition to the Ca_i and Ca_o concentration gradient. It appears to undergo time-dependent variations, reflecting the onset of trigger effects. Thus in the absence of oxalate, a transient high accumulation of Ca^{2+} at concentrations up to twice the steady-state accumulation value is observed.[140] Maximum Ca^{2+} accumulation occurs at about pH 6 and is enhanced by Mg^{2+}. Accumulation and release of Ca^{2+} is particularly affected[141] by calcium-precipitating anions such as oxalate and phosphate, which can maintain Ca_i at a low and constant level. The ionophore X-537A and caffeine both cause rapid release of accumulated Ca^{2+} in the absence of these anions, but only a small effect in their presence. Thus in the presence of X-537A and phosphate or oxalate, loss of Ca^{2+} from SR vesicles is greater at higher Ca_o in order to achieve a new steady state; i.e., the effect of X-537A is amplified by high Ca_o, and does not, at constant X-537A concentration, represent a Ca^{2+}-triggered efflux of Ca^{2+}. A second effect of Ca_o (in the absence of iono-phores) is to induce[142] higher calcium-efflux rates with Ca_o at fixed Ca_i. Increasing Ca_i at fixed Ca_o has little effect on calcium-efflux rate.

The role of cyclic AMP in the regulation of Ca^{2+} accumulation in SR from fast skeletal muscle as well as cardiac muscle has been confirmed, and a possible effector site for cyclic AMP characterized.[143]

The Ca^{2+}-Mg^{2+}-ATPase has many similarities with Na^+-K^+-ATPase. During active transport of Ca^{2+}, the ATPase binds Ca^{2+} and is phosphorylated by ATP at the outer surface of the vesicles, giving 2Ca.E.P. This is followed by some change of the enzyme conformation associated with translocation (giving 2Ca.E^1.P) that allows release of Ca^{2+} inside the vesicles prior to hydrolysis of E^1.P and the reverse transition into the original conformation. Calcium accumulation may be uncoupled from ATP hydrolysis by several methods, including treatment with acid, that inhibit[144] the Ca^{2+} translocation and have no effect on ATP hydrolysis under the same conditions. This probably results from partial unfolding of the ATPase protein, which prevents conformational coupling between the catalytic and ionophoric regions. Inactivation with acid distinguishes between the calcium-binding sites that are induced by ATP. When the temperature was reduced to 0 °C, Ca^{2+} uptake and liberation of P_i after the phosphorylation step

[140] M. M. Sorenson and L. D. Meis, *Biochim. Biophys. Acta*, 1977, **466**, 57.
[141] A. M. Katz, D. I. Repke, J. Dunnett, and W. Hasselbach, *J. Biol. Chem.*, 1977, **252**, 1938.
[142] A. M. Katz, D. I. Repke, J. Dunnett, and W. Hasselbach, *J. Biol. Chem.*, 1977, **252**, 1950.
[143] E. P. Bornet, M. L. Entman, W. B. van Winkle, A. Schwartz, D. C. Lehotay, and G. S. Levey, *Biochim. Biophys. Acta*, 1977, **468**, 188; M. L. Entman, E. P. Bornet, A. J. Garber, A. Schwartz, G. S. Levey, D. C. Lehotay, and L. A. Bricker, *ibid.*, 1977, **499**, 228.
[144] M. C. Berman, D. B. McIntosh, and J. E. Kench, *J. Biol. Chem.*, 1977, **252**, 994.

were impaired,[145] possibly because of modification to the lipid environment (the fluidity of which is temperature-sensitive), which slows the conformational change of the enzyme. Another approach to the study of the role of the phospholipid is through the addition of solvents such as DMSO and ethylene glycol, which, however, have quite different effects on several processes.[146] A range of SR preparations[147] show a negative co-operative effect of the ATPase with MgATP, possibly through a single site which shows two affinities through some kinetic mechanism such as a substrate-induced slow transition.

The important question of the effect of univalent ions on calcium uptake has been clarified. Despite earlier reports to the contrary, it now appears[148] that K^+ and other univalent cations (in the sequence $K^+ > Na^+ > NH_4^+ = Rb^+ = Cs^+ > Li^+$) exert a considerable effect on the rate but not on the extent of Ca^{2+} uptake, probably owing to an increased rate of decomposition of the phosphorylated intermediate, a step usually regarded to be rate-determining. Previous results which ascribed an inhibitory role to K^+ were carried out on preparations involving large amounts of K^+. The phosphoprotein formed in the absence of added alkali-metal salts decomposes to give P_i at a very low rate.[149] Two phosphorylated-enzyme intermediates have been identified,[150] one of which is acid-labile and rapidly decomposes to give phosphate. The acid-stable phosphorylated protein is Ca^{2+}-dependent, while the acid-labile species is Ca^{2+}-independent and is formed to a greater extent than that allowed by the number of available enzyme sites, implying that this phosphorylation reaction takes place at sites other than the Ca^{2+}-Mg^{2+}-ATPase sites.

Transport across the Cardiac and Uterine Plasma Membranes. Heart contraction requires extracellular calcium, either to trigger contraction directly or to trigger the release of Ca^{2+} from intracellular stores such as the mitochondrion, as discussed earlier. Rat myocardial sarcolemma binds Ca^{2+}, and the transport is stimulated by ATP. Accumulation of Ca^{2+} was enhanced by phosphorylase kinase-phosphorylation of the membranes, but this was unaffected by cyclic AMP.[151] Binding of Ca^{2+} to the membranes and the characterization of Ca^{2+} ligands have been studied. High- and low-affinity binding sites were detected,[152] with ample binding capacity for the physiological function. Ruthenium Red inhibited both sites, while the high-affinity site was inhibited by La^{3+} and was associated with a 100 000 dalton protein. Another report[153] indicated the presence of one class of binding site, and a 71 400 dalton calcium-binding lipoprotein was purified. Anti-arhythmic agents and anaesthetics have been tested on these preparations. Accumulation of Ca^{2+} by the plasma-membrane fraction of the rat

[145] II. Masuda and L. de Meis, *J. Biol. Chem.*, 1977, **252**, 8567.
[146] R. The and W. Hasselbach, *European J. Biochem.*, 1977, **74**, 611.
[147] K. E. Neet and N. M. Green, *Arch. Biochem. Biophys.*, 1977, **178**, 588.
[148] M. Shigekawa and L. J. Pearl, *J. Biol. Chem.*, 1976, **251**, 6947; P. F. Dunbarr and R. Jacob, *ibid.*, 1977, **252**, 1620.
[149] M. Shigekawa and J. P. Dougherty, *Biochem. Biophys. Res. Comm.*, 1977, **76**, 784.
[150] M. Kurzmack and G. Inesi, *F.E.B.S. Letters*, 1977, **74**, 35.
[151] P. V. Sulakhe and P. J. St. Louis, *Biochem. J.*, 1977, **164**, 457.
[152] C. J. Limas, *Arch. Biochem. Biophys.*, 1977, **179**, 302.
[153] D. A. Feldman and P. A. Weinhold, *Biochem. Pharm.*, 1977, **26**, 2283; *Biochemistry*, 1977, **16**, 3470.

uterus[154] (which also accumulates external Ca^{2+} for regulation of contraction and relaxation) and by mitochondria[155] has been studied.

Erythrocytes. Red blood cells contain a Mg^{2+}, ATP-dependent calcium-extrusion pump. The properties of the red-blood-cell membrane are largely determined by three major proteins, spectrin (220 000 dalton), glycophorin, and Band 3 proteins, which all appear to have some sort of ion-transport properties. ATP-depleted erythrocytes become rigid, leak cations, and accumulate Ca^{2+}. The mechanism of deformability loss and of change of cell shape is of considerable interest, as similar changes are associated with diseased conditions such as Duchenne muscular dystrophy.[156] Deformability loss is reversible at low $[Ca^{2+}]$ but is irreversible at concentrations of 1 mmol l^{-1}. The effect of Ca^{2+} is usually studied by the use of re-sealed cell ghosts (with the appropriate $[Ca^{2+}]$ present) or by the use of[157] ionophore A23187. Calcium exerts specific effects on the erythrocyte membrane.[158] At high $[Ca^{2+}]$, significant changes occur in the peptide components, resulting probably from the cross-linking of spectrin, together with the release of a membrane-bound protease. In addition, a substantial fraction of the spectrin could no longer be extracted by EDTA. The latter effect and the proteolytic cleavage occur at lower calcium concentrations than the aggregation effect. The re-sealing of erythrocyte ghosts in the presence of 4.5mM-Ca^{2+} induces the formation of small membrane vesicles in which the membrane proteins are selectively segregated.[159] The high affinity of the surface of the red blood cell for Ca^{2+} is due to sialic acid (*N*-acetylneuraminic acid), a carbohydrate that binds Ca^{2+} strongly and preferentially. ^1H and ^{13}C n.m.r. confirm a 1:1 complex at pH 7 ($K = 121$ l mol^{-1}), the glycerol side-chain of the sialic acid forming the binding centre.[160]

The erythrocyte membrane Ca^{2+}-Mg^{2+}-ATPase requires[161] an activator (18 000 dalton) which is similar to the calcium regulatory protein of 3′,5′-cyclic nucleotide phosphodiesterase. The activator has now been purified.[162] Parvalbumin and troponin-C were also able to increase the Ca^{2+}-Mg^{2+}-ATPase activity of erythrocyte membranes at high concentration, but phosphodiesterase activator from brain gave[163] very similar results to the erythrocyte cytoplasmic activator, and may be the same protein. The effect of inhibitors on the low-affinity Ca^{2+}-stimulated ATPase and the MgATP-dependent endocytosis of erythrocyte

[154] D. J. Crankshaw, R. A. Janis, and E. E. Daniel, *Canad. J. Physiol. Pharmacol.*, 1977, **55**, 1028.
[155] K. Malmström and E. Carafoli, *Arch. Biochem. Biophys.*, 1977, **182**, 657.
[156] C. A. Dise, D. B. P. Goodman, W. C. Lake, A. Hodson, and H. Rasmussen, *Biochem. Biophys. Res. Comm.*, 1977, **79**, 1286.
[157] D. Taylor, R. Baker, and P. Hochstein, *Biochem. Biophys. Res. Comm.*, 1977, **76**, 205; D. Allen and R. H. Mitchell, *Biochem. J.*, 1977, **166**, 495.
[158] D. R. Anderson, J. L. Davis, and K. L. Carraway, *J. Biol. Chem.*, 1977, **252**, 6617.
[159] E. Weidekamm, D. Brdiczka, G. Di Pauli, and M. Wildermuth, *Arch. Biochem. Biophys.*, 1977, **179**, 486.
[160] L. W. Jaques, E. B. Brown, J. M. Barrett, W. S. Brey, and W. Weltner, *J. Biol. Chem.*, 1977, **252**, 4533.
[161] H. W. Jarrett and J. T. Penniston, *Biochem. Biophys. Res. Comm.*, 1977, **77**, 1210; M. L. Farrance and F. F. Vincenzi, *Biochim. Biophys. Acta*, 1977, **471**, 59.
[162] M. G. Luthra, K. S. Au, and D. J. Hanahan, *Biochem. Biophys. Res. Comm.*, 1977, **77**, 678.
[163] R. M. Gopinath and F. F. Vincenzi, *Biochem. Biophys. Res. Comm.*, 1977, **77**, 1203.

ghosts[164] and the dependence of the kinetics of Ca^{2+} activation on the Ca^{2+} concentration during membrane preparation[165] have also been studied.

Miscellaneous Ca^{2+}-transport Systems. Ca^{2+} is implicated in the visual process, where photons are absorbed by rhodopsin, a protein in the membrane of the photoreceptor cell, and are eventually converted into a neural signal. Ca^{2+}-dependent ATPase activity has been detected in outer segment membranes of receptor cells, and their calcium-binding capacity determined.[166] A simple mechanism for ion translocation involves modification of the ionic conductance of the membrane by rhodospin. This protein has now been incorporated into lipid bilayers; the effect of light is to increase the bilayer permeability, consistent with the formation of a trans-membrane channel of diameter about 10 Å.[167]

Ionophores are among several agents capable of initiating cell proliferation in lymphocytes. Amongst the early changes in the cell is the induction of permeability of the membrane to Ca^{2+} and K^+. Current work centres around the problem of whether the ionophore translocates Ca^{2+} as an initiator for lymphocyte stimulation.[168] A Ca^{2+}-Mg^{2+}-ATPase transport system has been studied in microsomes isolated from the slime-mould *Physarum polycephalum*[169].

Determination of Calcium. The currently used methods[170] for measurement of calcium in cells involve indicators such as Arsenazo III, photoproteins (such as aequorin[171]) which emit light on contact with Ca^{2+}, and Ca^{2+}-sensitive microelectrodes. Recently, electron microprobe techniques have been utilized.[172] The detection limits for Ca^{2+} with aequorin and Arsenazo III are 6×10^{-8} mol l^{-1} and *ca.* 10^{-7} mol l^{-1} respectively. The response time of Arsenazo III is shorter, and so it is of particular value[173] for measuring rapid transient changes in the concentration of free Ca^{2+}. One disadvantage of both methods is the dependence on Mg^{2+} and other cations. Microelectrodes have a detection limit of $\sim 10^{-8}$ mol l^{-1} but a response time of 1—2 s. An ion-selective micropipette has been used to monitor Ca^{2+} modulation in the brain extracellular microenvironment.[174] A striking example[174a] of the use of aequorin involves the measurement of free

[164] H. W. Jarrett, T. B. Reid, and J. T. Penniston, *Arch. Biochem. Biophys.*, 1977, **183**, 498.

[165] O. Scharff and B. Foder, *Biochim. Biophys. Acta*, 1977, **483**, 416.

[166] R. A. Sack and C. M. Harris, *Nature*, 1977, **265**, 465; P. P. M. Schnetkamp, F. J. M. Daemen, and S. L. Bonting, *Biochim. Biophys. Acta*, 1977, **468**, 259; T. Hendriks, P. M. M. Van Haard, F. J. M. Daemen, and S. L. Bonting, *ibid.*, 1977, **467**, 175.

[167] M. Montal, A. Darszon, and H. W. Trissl, *Nature*, 1977, **267**, 221.

[168] T. R. Hesketh, G. A. Smith, M. D. Houslay, G. B. Warren, and J. C. Metcalfe, *Nature*, 1977, **267**, 490; N. Kaiser and I. S. Edelman, *Proc. Nat. Acad. Sci. U.S.A.*, 1977, **74**, 638; P. Jensen and H. Rasmussen, *Biochim. Biophys. Acta*, 1977, **468**, 146; P. Jensen, L. Winger, H. Rasmussen, and P. Nowell, *ibid.*, 1977, **496**, 374; D. Allen and R. H. Mitchell, *Biochem. J.*, 1977, **164**, 389.

[169] T. Kato and Y. Tonomura, *J. Biochem. (Japan)*, 1977, **81**, 207.

[170] T. J. Lea, *Nature*, 1977, **269**, 108.

[171] O. Shimomura and E. H. Johnson, in 'Calcium in Biological Systems', ed. C. J. Duncan, (S.E.B. Symposium XXX), Cambridge University Press, 1976, p. 40.

[172] L. M. Routledge, W. B. Amos, B. L. Gupta, T. A. Hall, and T. Weis-Fogh, *J. Cell. Sci.*, 1975, **19**, 195.

[173] R. Miledi, I. Parker, and G. Schalow, *Proc. Roy. Soc.*, 1977, **B198**, 201.

[174] C. Nicholson, T. Bruggencate, R. Steinberg, and H. Stöckle, *Proc. Nat. Acad. Sci. U.S.A.*, 1977, **74**, 1287.

[174a] E. B. Ridgeway, J. C. Gilkey, and L. F. Jaffe, *Proc. Nat. Acad. Sci. U.S.A.*, 1977, **74**, 623.

Ca^{2+} in the eggs of the medaka, a fresh-water fish, during fertilization. On activation by sperm, the emission of light from the aequorin-injected eggs increases 10 000-fold. Entry of the sperm triggers a calcium wave (released from internal sites) that reaches the opposite side within two minutes.

6 Control of Intracellular Processes by Calcium

The effectiveness of the Ca^{2+}-transport systems in keeping the Ca^{2+} content of the cytoplasm low means that cellular behaviour can be controlled by quite small influxes of Ca^{2+} from intra- and extra-cellular sources. Reports of two symposia on this topic are available.[175] In many cases, calcium-binding or regulatory proteins are important, and their interaction with Ca^{2+} has received much attention. Tb^{3+} fluorescence emission occurs[176] between Tb^{3+} and many Ca^{2+}-binding proteins, due usually to energy transfer from tryptophan groups. Circular polarization of Tb^{3+} emission is observed in some cases, including parvalbumin and troponin C. Emission from Tb^{3+} is thus suggested to be a useful probe for protein structure, although the use of Gd^{3+} e.s.r. to investigate biological sites has been discouraged.[177] Predictions[178] of secondary structure have been made for calcium-binding proteins, based on their amino-acid sequences and the assumption that each Ca^{2+} site involves 'EF hand' architecture (helix–loop–helix).

Calcium-binding Protein Modulator of Cyclic Nucleotide Phosphodiesterase.— This protein undergoes a conformational transition[179] on binding Ca^{2+}, and can then bind to the enzyme. Equilibrium dialysis techniques[180] imply that there are two classes of Ca^{2+}-binding sites, with stoicheiometries of three and one per mole of protein. Chemical modification[181] of the single histidine residue in the protein, the two tyrosine residues, and four out of the six arginine residues has no effect on the activating properties of the modulator, but methionine residues have been implicated. The protein has now been isolated[182] (16 500 dalton). In the presence of 0.1mM-$CaCl_2$, hydrolysis by trypsin gives two large proteins, but the protein is rapidly cleaved into small peptides on removal of the Ca^{2+}. Parvalbumin also activates brain phosphodiesterase to a degree equal to that of the natural regulatory protein.[183] A bovine brain protein has been isolated that effectively inhibits phosphodiesterase activity by binding to the modulator protein.[184] The modulator protein is also an activator for brain adenylate cyclase,

[175] 'Calcium Transport in Contraction and Secretion', ed. E. Carafoli *et al.*, North-Holland, Amsterdam, 1975; 'Calcium in Biological Systems'. (S.E.B. Symposium XXX), Cambridge University Press, 1976.
[176] H. G. Brittain, F. S. Richardson, and R. B. Martin, *J. Amer. Chem. Soc.*, 1976, **98**, 8255.
[177] E. C. N. F. Geraldes and R. J. P. Williams, *J. C. S. Dalton*, 1977, 1721.
[178] P. Argos, *Biochemistry*, 1977, **16**, 665.
[179] C. B. Klee, *Biochemistry*, 1977, **16**, 1017.
[180] D. J. Wolff, P. G. Poirier, C. O. Brostrom, and M. A. Brostrom, *J. Biol. Chem.*, 1977, **252**, 4108.
[181] M. Walsh and F. C. Stevens, *Biochemistry*, 1977, **16**, 2742.
[182] W. Drabikowski, J. Kuznicki, and Z. Grabarek, *Biochim. Biophys. Acta*, 1977, **485**, 124. See also J. T. Russell and N. A. Thorn, *ibid.*, 1977, **491**, 398; J. Weissman, L. E. Schneider, and H. P. Schedl, *ibid.*, 1977, **497**, 358; A. Vandermeers, M.-C. Vandermeers-Piret, J. Rathe, R. Kutzner, A. Delforge, and J. Christophe, *European J. Biochem.*, 1977, **81**, 379.
[183] J. D. Potter, J. R. Dedman, and A. R. Means, *J. Biol. Chem.*, 1977, **252**, 5609.
[184] J. H. Wang and R. Desai, *J. Biol. Chem.*, 1977, **252**, 4175.

which is made up of two components, one of which requires the Ca^{2+}-dependent regulator.[185]

Muscle Contraction.—The best understood Ca^{2+}-regulatory proteins are those involved in contractility, although the mode of action is not common to all muscle systems. The parvalbumins are water-soluble acidic proteins of low molecular weight (12 000) that are found particularly in the white muscle of fish, but also in higher vertebrates. They are characterized by a high content of phenyl-alanine residues relative to tryptophan and tyrosine. Carp parvalbumin contains two non-co-operative Ca^{2+}-binding sites. Irradiation,[186] at 259 nm, of the Tb^{3+}-substituted protein results in dramatic enhancement of the green emission from energy transfer between one Tb^{3+} and the aromatic ring of phenylalanine-57, 4.6 Å away. The half-Ca^{2+}-complexed protein binds Na^+ at the vacant site (five orders of magnitude less). Sodium magnetic resonance competition experiments[187] show the sequence of complexing abilities $Na^+ < K^+ \ll Mg^{2+} < Ca^{2+} \leqslant Mn^{2+}$. The binding of Na^+ is characterized by a surprisingly long residence time on the protein. Calcium-binding proteins from other sources that are similar to parval-bumin have been studied.[188]

Troponin from Skeletal Muscle. This has three components (troponin C, TN-C; troponin I, TN-I; and troponin T, TN-T). With actin and tropomyosin, it makes up the thin filaments of skeletal muscle that interact with myosin, with activation of the Mg^{2+}-ATPase. The calcium-binding unit is troponin C, which shows a structural change on binding Ca^{2+}. TN-C has four sites for binding Ca^{2+}; two high-affinity sites (1,2) also bind Mg^{2+} competitively, while two (3,4) sites of lower affinity, do not bind Mg^{2+}, and are thought to be involved in regulation. Thermodynamic studies[189] indicate that $\Delta H^{\ominus}$ values for Ca^{2+} binding were similar for all sites, but $\Delta S^{\ominus}$ values differed for the two groups. The kinetics of the conformational change induced by Ca^{2+} and H^+ have been followed by temper-ature-jump spectrofluorimetry.[190] E.s.r. spectra of spin-labelled TN-C show that the mobility of the group is decreased on binding of Ca^{2+} or Mg^{2+} to sites (1,2) but that a much greater reduction in mobility was associated with binding of Ca^{2+} at (3,4). Chemical modification studies indicate that the Ca^{2+} site adjacent to Cys-95 is one of the Ca^{2+}-Mg^{2+} sites (1,2).[191] The effect of variation of pH and of pCa on the 250 MHz ^1H n.m.r. spectrum of TN-C reveals[192] three conforma-tional states of the protein (no Ca^{2+} bound; Ca^{2+} bound at sites 1,2; Ca^{2+} bound at 1,2 and a low-affinity site). Binding at the low-affinity site is unaffected by Mg^{2+} and is associated with shifting of upfield phenylalanine and aliphatic resonances. However, tyrosine residues do not appear to be involved in any site, contrary to results obtained from other techniques. Sites of interaction between

[185] C. O. Brostrom, M. A. Brostrom, and D. J. Wolff, *J. Biol. Chem.*, 1977, **252**, 5677.
[186] D. J. Nelson, T. L. Miller, and R. B. Martin, *Bioinorg. Chem.*, 1977, **7**, 325.
[187] J. Grandjean and P. Laszlo, *F.E.B.S. Letters*, 1977, **81**, 376.
[188] J. A. Cox and W. Wnuk, *Experientia*, 1977, **33**, 789; L. Kohler, D. Winge, and J. A. Cox, *ibid.*, p. 795.
[189] J. D. Potter, F. J. Hsu, and H. J. Pownall, *J. Biol. Chem.*, 1977, **252**, 2452.
[190] T. Iio, K. Mihashi, and H. Kondo, *J. Biochem. (Japan)*, 1976, **79**, 689; 1977, **81**, 277.
[191] J. D. Potter, J. C. Seidel, P. Leavis, S. S. Lehrer, and J. Gergely, *J. Biol. Chem.*, 1976, **251**, 7551.
[192] K. B. Seaman, D. J. Hartshorne, and A. A. Bothner-By, *Biochemistry*, 1977, **16**, 4039.

TN-T and TN-C were investigated by monitoring the effects of phosphorylation on their binding properties.[193]

Troponin from Other Sources. An analysis of the amino-acids of TN-C protein from cardiac muscle shows only three regions that are similar to the Ca^{2+}-binding sites of skeletal TN-C. It has now been confirmed[194] that cardiac TN-C binds only three moles of Ca^{2+}, one of which is responsible for changes in the c.d. spectrum. Binding of calcium to the cardiac troponin–tropomyosin complex and the effect of phosphorylation of the TN-I subunit have been examined.[195]

The function of smooth muscle involves myosin-linked regulation, but it appears that troponin is present; thus the physiological action of gizzard muscle depends on a 80 000 dalton component which interacts with Ca^{2+} and which has no kinase activity,[196] while troponin C has been isolated from uterine muscle by affinity chromatography using TN-I, the inhibiting component of the skeletal muscle complex.[197]

Myosin-linked Regulation. In invertebrates, the regulation of muscle by Ca^{2+} binding is linked to the myosin molecule, involving specific regulatory light chains. Sometimes both regulatory systems exist, as in the nematode worm *Caenorhabditis elegans*, on which significant work is being carried out[198] on the assembly of the muscle filaments in mixed systems in association with the isolation of mutants that may be defective in one or other of the regulatory systems. In smooth muscle of higher vertebrates, such as chicken gizzard and pig stomach muscles,[199] the actin–myosin interaction is triggered by the Ca^{2+}-dependent phosphorylation of the 20 000 dalton light chain by a specific kinase, which precedes the rise in actin-activated ATPase. Dephosphorylation occurs by a phosphatase that is insensitive to Ca^{2+}. The light chain does not bind Ca^{2+}, but is phosphorylated in the presence of Ca^{2+}, suggesting that Ca^{2+} acts solely in the phosphorylation step. However, some interaction with the heavy chain may induce Ca^{2+}-binding by the light chain. Tropomyosin amplifies the effect of phosphorylation[200] and so is required for full actin-activated ATPase activity.

Actomyosin is precipitated at physiological ionic strengths,[201] so that much has been done on proteolytic fragments of myosin that do not form aggregates. Mn^{2+}-binding sites on myosin have been characterized by e.s.r. and located on the DTNB light chains. The metal site remains intact through subsequent chymotryptic digestion.[202] Significant differences observed[203] in the e.s.r. spectrum of the Mn^{2+}–heavy meromyosin system above and below ~ 10 °C may be correlated

[193] A. J. G. Moir, H. A. Cole, and S. V. Perry, *Biochem. J.*, 1977, **161**, 371.
[194] L. D. Burtnick and C. M. Kay, *F.E.B.S. Letters*, 1977, **75**, 105.
[195] J. E. Buss and J. T. Stull, *F.E.B.S. Letters*, 1977, **73**, 101.
[196] T. Mikawa, T. Yoyo-Oka, Y. Nonomura, and S. Ebashi, *J. Biochem. (Japan)*, 1977, **81**, 273.
[197] J. F. Head, R. A. Weeks, and S. V. Perry, *Biochem. J.*, 1977, **161**, 465.
[198] H. E. Harris, M. Y. W. Tso, and H. F. Epstein, *Biochemistry*, 1977, **16**, 859.
[199] A. Sobieskek, *European J. Biochem.*, 1977, **73**, 477; J. V. Small and A. Sobieskek, *ibid.*, 1977, **76**, 521.
[200] A. Sobieskek and J. V. Small, *J. Mol. Biol.*, 1977, **112**, 577.
[201] M. Miyahara and H. Noda, *J. Biochem. (Japan)*, 1977, **81**, 285.
[202] C. R. Bagshaw, *Biochemistry*, 1977, **16**, 59.
[203] K. Koga, Y. Kanazawa, and K. Tawada, *J. Biochem. (Japan)*, 1977, **82**, 35.

with a number of observations[204] on the temperature dependence of actin/myosin ATPase activity. Other studies on skeletal muscle relate to the co-operativity of binding of Ca^{2+} to myosins,[205] the use of spin-labelled light chains,[206] conformational changes around a specific thiol group of heavy meromyosin,[207] and the lowered Ca^{2+}-sensitivity of actomyosin from light-chain-deficient myosin.[208] Several aspects of hydrolysis of ATP by myosin have been discussed.[209]

Secretion.—This usually involves the fusion of an intracellular vesicle with the cell membrane, thus releasing its contents into the external medium. This process of exocytosis is dependent on the release of Ca^{2+}, which acts as a link between stimulus and secretion. The topic of calcium and neurosecretion has been reviewed.[210] Cyclic AMP and the guanosine cyclic nucleotide, cyclic GMP, are often associated as second messengers,[211] but can also inhibit certain types of secretion. Calcium is usually thought to facilitate membrane fusion, but may also regulate the movement of the vesicle to the cell surface. In the former context, the granules of the adrenal medullary chromaffin cells have cation-binding sites on their outer membrane which have been probed[212] by the use of lanthanide ions. Tb^{3+} fluorescence measurements implicate a protein as the binding site. It has been suggested that Ca^{2+} binds the granule to its release site on the inner face of the cell membrane. However, Ca^{2+} and Mg^{2+} both cause aggregation of granules, so there may be a second Ca^{2+}-specific process associated with the release of the medullary hormone from the granules.

Polymorphonuclear leukocytes (PMNL) secrete chemicals and enzymes for defence against micro-organisms. Several changes in the cells, including the translocation of granules, are triggered by Ca^{2+}. During the process of phago-cytosis, however, there is an efflux of Ca^{2+} from PMNL.[213] Transport properties of the membrane have also been studied under inactive conditions.[214] The release of histamine from mast cells[215] involves an antigen–antibody reaction on their plasma membrane, which leads to the secretion of insulin, provided that extra-cellular Ca^{2+} is present (0.1—1.0 mmol l^{-1}). Interestingly, Sr^{2+} leads[216] to a secretion twice as great as that induced by Ca^{2+}, while the spontaneous histamine leak that occurs in the absence of the membrane stimulus occurs to a greater extent with Sr^{2+}. Clearly, the resting mast-cell membrane is more permeable to Sr^{2+}. Stimulation of the membrane by antigen induces a channel that is selective

[204] T. Hozumi, *J. Biochem. (Japan)*, 1977, **81**, 329; *European J. Biochem.*, 1976, **63**, 241.
[205] F. Fabian, D. T. Mason, and J. Wikman-Coffelt, *F.E.B.S. Letters*, 1977, **81**, 381.
[206] Y. Okamoto and K. Yogi, *J. Biochem. (Japan)*, 1977, **82**, 835; H. Kuwayama and K. Yogi, *ibid.*, p. 25.
[207] T. Kameyama, T. Katori, and T. Sekine, *J. Biochem. (Japan)*, 1977, **81**, 709.
[208] S. M. Pemrick, *Biochemistry*, 1977, **16**, 4047.
[209] E. W. Taylor, *Biochemistry*, 1977, **16**, 732; P. Dancker and I. Löw, *Biochim. Biophys. Acta*, 1977, **484**, 169; C. R. Bagshaw, *Biochem. Soc. Trans.*, 1977, **5**, 1272; A. G. Weeds, *ibid.*, p. 1274.
[210] P. F. Baker, *Sci. Progr. Oxf.*, 1977, **64**, 94.
[211] M. J. Berridge, *Adv. Cyclic Nucleotide Res.*, 1975, **6**, 1.
[212] S. J. Morris and R. Schober, *European J. Biochem.*, 1977, **75**, 1.
[213] A. Barthelemy, R. Paridaens, and E. Schell-Frederick, *F.E.B.S. Letters*, 1977, **82**, 283.
[214] P. H. Naccache, H. J. Showell, E. L. Becker, and R. I. Sha'afi, *J. Cell. Biol.*, 1977, **73**, 428.
[215] J. C. Foreman, L. G. Garland, and J. L. Mongar, in 'Calcium in Biological Systems', ed. C. J. Duncan, (S.E.B. Symposium XXX), Cambridge University Press, 1976, p. 193; C. M. S. Fewtrell and B. D. Gomberts, *Biochim. Biophys. Acta*, 1977, **469**, 52.
[216] J. C. Foreman, M. B. Hallett, and J. L. Mongar, *Biochem. Soc. Trans.*, 1977, **5**, 879.

116   *Inorganic Biochemistry*

to Ca^{2+} (in the absence of Sr^{2+}), and its function is modulated by cyclic AMP and phosphatidylserine. Ionophore A23187 thus allows external Ca^{2+} to enter in the absence of membrane stimulation.

Ca^{2+} is also necessary for the secretion of insulin, and the fluxes of ^{45}Ca in and out of intact pancreatic islets have been discussed,[217] while the distribution of intracellular Ca^{2+} in pancreatic B-cells has been studied by pyroantimonate and *X*-ray microanalysis techniques.[218] The action of insulin on lipocytes (the activation of oxidation of glucose) involves[219] Ca^{2+}-dependent and -independent paths. In view of the earlier report[220] of a specific, insulin-dependent Ca^{2+}-binding site on lipocyte membranes, it may be that Ca^{2+} also controls the action of insulin by some effect on the lipocyte membrane.

Calcium enters pre-synaptic nerve terminals during depolarization and triggers transmitter release. Studies on pinched-off terminals (which re-seal, giving synaptosomes) indicate that there is an ATP-dependent Ca^{2+}-storage site,[221] in addition to mitochondria,[222] which may accumulate some of this calcium. Phosphorylation of membrane-bound proteins in synaptosomes lowers[223] the rate of Ca^{2+} uptake and efflux, indicative of a possible role for cyclic AMP and protein kinase in the regulation of synaptic transmission.[224]

Calcium–Cyclic AMP Interactions.—The intracellular levels of cyclic AMP and Ca^{2+} oscillate with periods between 0.1 s and 5 min. It has been suggested that these account for oscillations in cardiac pacemaker cells, insulin-secreting cells, and a range of high-frequency biological rhythms.[225]

7 Calcification and Mobilization[226]

Vitamin D_3 promotes the mobilization of calcium from the intestine.[227] The metabolically active form, the 1,25-dihydroxy-vitamin, $1,25-(OH)_2D_3$, is produced in the kidney as necessitated by the need for calcium. Other hydroxy-derivatives have been tested.[228] $1,25-(OH)_2D_3$ stimulates absorption of calcium in the intestine by a receptor-protein-mediated mechanism: the calcium-binding protein[229] has been isolated (10 000 dalton). Intestinal transport of Ca^{2+} has been studied in various membrane vesicles. Those from Golgi membrane preparations showed the highest uptake rate, suggesting that the Golgi apparatus participates in intestinal absorption of Ca^{2+}. Vesicles obtained from vitamin-D-deficient rats

217 W. J. Malaisse, *Biochem. Soc. Trans.*, 1977, **5**, 872.
218 S. L. Howell, *Biochem. Soc. Trans.*, 1977, **5**, 875.
219 D. Bonne, O. Belhadj, and P. Cohen, *European J. Biochem.*, 1977, **75**, 101.
220 J. M. McDonald, D. E. Bruns, and L. Jarett, *Proc. Nat. Acad. Sci. U.S.A.*, 1976, **73**, 1542.
221 N. C. Kendrick, M. P. Blaustein, R. C. Fried, and R. W. Ratzdaff, *Nature*, 1977, **265**, 246.
222 E. K. Silbergeld, *Biochem. Biophys. Res. Comm.*, 1977, **77**, 464.
223 M. Weller and I. G. Morgan, *Biochim. Biophys. Acta*, 1977, **465**, 527.
224 R. J. DeLorenzo and S. D. Freedman, *Biochem. Biophys. Res. Comm.*, 1977, **77**, 1036; L. R. Skirboll, L. Baizer, and K. L. Dietzchen, *Nature*, 1977, **268**, 352.
225 P. E. Rapp and M. J. Berridge, *J. Theor. Biol.*, 1977, **66**, 497.
226 Proceedings, XII European Symposium Calcified Tissue, ed. W. G. Robertson, B. E. C. Nordini, and F. G. E. Pautard *Calcified Tissue Res.*, 1977, 22, (suppl.).
227 L. E. Schneider and H. P. Schedl, *Endocrinology*, 1977, **100**, 928.
228 A. Mahgouband and H. Sheppard, *Endocrinology*, 1977, **100**, 629.
229 M. E. H. Bruns, E. B. Fliesher, and L. V. Avioli, *J. Biol. Chem.*, 1977, **252**, 4145.

showed diminished uptake, confirming the role of the vitamin.[230] Much less is known about the role of 1,25-$(OH)_2D_3$ in bone mobilization, but high-affinity binding proteins for vitamin D_3 metabolites have been found in bone cytosols.[231] A disease similar to vitamin D intoxication (hypercalcaemia and soft-tissue calcification) that is observed in animals grazing on plants such as *Cestrum diurnum* and *Solanum malacoxylon* has been attributed to the presence of a 1,25$(OH)_2D_3$-glycoside.[232]

Calcium salts are readily precipitated (*e.g.* from bile, as gall stones[233]) and are often stored in the tissue as granules, which may be mobilized for shell formation and other processes.[234] Human salivary secretions are usually supersaturated with basic calcium phosphate, so allowing recalcification and protection of the dental enamel. It is extremely difficult to precipitate the phosphate, even by seeding. This necessary physiological situation results from the presence in saliva of precipitation-inhibiting proteins. These include a tyrosine-rich acidic peptide, statherin (5380 dalton), whose amino-acid sequence has been determined and which may be a precursor of dental enamel protein,[235] and a group of proline-rich proteins, whose Ca^{2+}-binding properties have been characterized.[236]

The interaction[237] of Ca^{2+} with dentin phosphoprotein from bovine molars involves specific Ca^{2+}–orthophosphate interaction followed by precipitation. The i.r. spectrum of the product also indicates interaction of Ca^{2+} with carboxylate groups. It has been suggested that such studies may offer a simple model for bone mineralization. Calcified plates (coccoliths) from species of coccolithophoridae contain a Ca^{2+}-binding polysaccharide[238] which would act as a matrix in coccolith formation.

8 Blood-clotting Mechanisms

Aspects of this subject are assessed in the report of a Colloquium on Blood Clotting Mechanisms.[239] Calcium is involved in several stages of the 'cascade' mechanism for coagulation. The *N*-terminal portion of prothrombin and of Factors VII, IX, and X contains γ-carboxyglutamic acid, the biosynthesis of which is dependent on vitamin K. Certain chemicals, *e.g.* the coumarins, can thus prevent[240] coagulation by inducing the production of 'under-carboxylated' prothrombin molecules. These Factors bind to membranes containing acidic phospholipids in the presence of Ca^{2+}. The γ-carboxy-group is required for this, and is also thought to bind Ca^{2+}. The reactions Factor X $\rightarrow$ Factor Xa, and pro-

[230] R. A. Freedman, M. M. Weisner, and K. J. Isselbacher, *Proc. Nat. Acad. Sci. U.S.A.*, 1977, **74**, 3612.
[231] B. E. Kream, M. Jose, S. Yamada, and H. F. Deluca, *Science*, 1977, **197**, 1087.
[232] M. R. Hughes, T. A. McCain, S. Y. Chang, M. R. Haussler, M. Villareale, and R. H. Wasserman, *Nature*, 1977, **268**, 347.
[233] D. J. Sutor and L. I. Wilkie, *Clinica Chim. Acta*, 1977, **79**, 119.
[234] A. S. Tompa and K. M. Wilbur, *Nature*, 1977, **270**, 53.
[235] D. H. Schlesinger and D. I. Hay, *J. Biol. Chem.*, 1977, **252**, 1689.
[236] A. Bennick, *Biochem. J.*, 1976, **155**, 163; 1977, **163**, 229, 241.
[237] S. L. Lee, A. Veis, and T. Glonek, *Biochemistry*, 1977, **16**, 2971.
[238] E. W. DeJong, L. Bosch, and P. Westbrock, *European J. Biochem.*, 1977, **70**, 611.
[239] *Biochem. Soc. Trans.*, 1977, **5**, 1241—1251.
[240] P. A. Friedman, R. D. Rosenberg, P. V. Hauschka, and A. Fitz-James, *Biochim. Biophys. Acta*, 1977, **494**, 271.

thrombin → thrombin, do not involve simple proteases but membrane reactions. These membrane–protein interactions have been assessed[241] by light-scattering intensity measurements, while Ca^{2+}-binding data have been measured.

Prothrombin contains *ca*. 10 binding sites for Ca^{2+}, which are thought to be in two fragments (1,2) that are cleaved off during prothrombin activation and which contain about 6 and 4 sites respectively. Binding of Ca^{2+} to prothrombin 1 induces two separate transitions[242] that lead to phospholipid binding and to dimerization of the fragment 1. Mg^{2+} will partially substitute for Ca^{2+}, and enhances co-operativity. Two sites are non-selective, binding at which leads to a conformational change, while four sites are specific, and lead to dimerization, phospholipid binding, and prothrombin activation.

The production of thrombin allows the proteolytic activation of the fibrinogen molecule, for which Ca^{2+} is required. At pH 7.5 fibrinogen has three Ca^{2+} sites of high affinity, and low-affinity sites that also bind Mg^{2+}. One of the high-affinity sites is eliminated[243] at pH < 7.5. While binding of Ca^{2+} does not involve[244] an overall conformational change, a number of protective effects of Ca^{2+} have been identified which are not shown by Mg^{2+}. The binding of Ca^{2+} by thromboplastin and complex formation between thromboplastin, Factor VII, and Ca^{2+} has been studied.[245]

9 Interaction of Cations with Miscellaneous Biological Molecules

Concanavalin A.—The interaction of this plant lectin with Mn^{2+} and Ca^{2+} has been followed over the pH range 5.3—6.4 by n.m.r. relaxation techniques. The usual sequential binding of Mn^{2+} at site 1 followed by Ca^{2+} at site 2 was observed. The unstable ternary complex first formed undergoes a pH-independent transition to a stable form which has a very low dissociation constant for Ca^{2+}. Evidence was presented that Mn^{2+}, in the absence of Ca^{2+}, can also bind at site 2, with the same unstable–stable conformation change, but is lost much more readily from the stable ternary complex. Early addition of metal ions re-forms the *stable* ternary complex, but the demetallated concanavalin reverts to its initial state over a few days.[246] Of particular interest is the claim[247] that, at pH 6.85, the demetallated concanavalin does bind Ca^{2+} appreciably. This is in contrast to earlier work, and emphasizes that conclusions from results at lower pH may not be relevant to physiological conditions. E.s.r. and equilibrium dialysis studies both show[247] that binding of Mn^{2+} to concanavalin is co-operative in the presence of Ca^{2+} and non-co-operative in the absence of Ca^{2+}.

Control of Polymer Structure.—A sulphated mucopolysaccharide from a marine snail binds potassium as the counter-ion inside mucus cells, but after secretion it releases this and takes up Ca^{2+} and Mg^{2+}. On secretion, the polymer undergoes

241 G. L. Nelsestuen and T. K. Lim, *Biochemistry*, 1977, **16**, 4164; G. L. Nelsestuen and M. Broderius, *ibid.*, p. 4172; T. K. Lim, V. A. Bloomfield, and G. L. Nelsestuen, *ibid.*, p. 4177.
242 F. G. Prendergast and K. G. Mann, *J. Biol. Chem.*, 1977, **252**, 840.
243 G. Marguerie, G. Chagniel, and M. Suscillon, *Biochim. Biophys. Acta*, 1977, **490**, 94.
244 G. Marguerie, *Biochim. Biophys. Acta*, 1977, **494**, 172.
245 G. Wijngaards and J. Immerzeel, *Biochem. Biophys. Res. Comm.*, 1977, **77**, 658.
246 R. D. Brown, F. C. Brewer, and S. H. Koenig, *Biochemistry*, 1977, **16**, 3883.
247 G. M. Alter, E. R. Pandolfino, D. J. Christie, and J. A. Magnuson, *Biochemistry*, 1977, **16**, 4034.

substantial swelling. It has been suggested that binding of K^+ causes the polymer to take up a compact structure inside the cell.[248] The folding[249] and unfolding[250] of transfer RNA is controlled by the binding of bivalent metal,[251] while the conformation of dephospho-Coenzyme A in solution has been studied by n.m.r., with lanthanide probes.[252] The assembly of the protein tubulin into different polymorphic forms[253] is induced by Zn^{2+}. Ca^{2+} is commonly regarded as an inhibitor of assembly of microtubules *in vitro*, and two classes of binding site on tubulin for Ca^{2+} have been characterized.[254] There has been some confusion over the effect of Ca^{2+} on the inhibition of microtubule assembly in recent years,[255] but this probably results from the effect of different ionic strengths.[254]

[248] S. Hunt and K. Oates, *Nature*, 1977, **268**, 371.
[249] J. L. Leroy, M. Guéran, G. Thomas, and A. Favre, *European J. Biochem.*, 1977, **74**, 567.
[250] D. Rhodes, *European J. Biochem.*, 1977, **81**, 91.
[251] D. Labuda, T. Haertlé, and J. Augustyniak, *European J. Biochem.*, 1977, **79**, 293.
[252] G. V. Fazakerley, P. W. Linder, and D. G. Reid, *European J. Biochem.*, 1977, **81**, 507.
[253] F. Gaskin and Y. Kress, *J. Biol. Chem.*, 1977, **252**, 6918.
[254] F. Solomon, *Biochemistry*, 1977, **16**, 358.
[255] E. Nishida and H. Sakai, *J. Biochem. (Japan)*, 1977, **82**, 303.

4
Storage and Transport of Transition-metal Ions

P. M. HARRISON & A. TREFFRY

1 Introduction

In other Reports, the many ways in which transition metals can be exploited to serve the metabolic needs of living organisms are described. To supply these demands, the metals must be acquired from the environment and delivered to the sites where they are needed. In unicellular organisms this may be a relatively simple process, but in higher organisms with many types of specialized cells and a circulatory system, regulatory mechanisms have also been developed to ensure that metals are delivered to cells which most need them, to handle metals within cells, to conserve and reprocess metals at the end of the lifespan of the molecules or cells containing them, and to eliminate un-needed metal either by excretion or by internal storage. Excess free metal may produce toxic symptoms by combining non-specifically with proteins, by distorting the normal metabolism of other metals, and, in the case of iron, by causing the formation of damaging free radicals. Detoxification and reserve or transport functions may be combined within the same molecule.

This Report deals mainly with mammalian metabolism of iron and copper, but zinc also appears, because, although not a transition metal, its intracellular transport and storage may be carried out by the same or similar proteins as those involved in dealing with copper. The emphasis of the Report will be on the molecules involved in transport and storage and their structure–function relationships (respectively, transferrin and ferritin for iron, caeruloplasmin and metallothioneins and related proteins for copper), but overall mechanisms regulating metal homeostasis will also be indicated. A brief summary is given of microbial iron-transport compounds.

For a more complete picture of current experiments and opinions in the field of iron metabolism the reader is referred to three recently published multi-author volumes.[1-3] Many of the individual contributions to these volumes are also cited. Contributions in ref. 3 have been collected from the January and April, 1977, issues of *Seminars in Haematology*. For shorter summaries of the current status

1 'Proteins of Iron Metabolism', ed. E. B. Brown, P. Aisen, J. Fielding, and R. R. Crichton, Grune & Stratton, New York, 1977.
2 'Iron Metabolism', Ciba Foundation Symposium 51 (new series), Elsevier, Excerpta Medica, North Holland, 1977.
3 'Iron Excess: Aberrations of Iron and Porphyrin Metabolism,' ed. U. Muller-Eberhard, P. A. Miescher, and E. R. Jaffe, Grune & Stratton, New York, 1977.

of iron biochemistry and the effects of iron deficiency the reader's attention is called to those of Crosby,[4] Worwood,[5] and Jacobs.[6]

No attempt has been made to cover all the literature on the treatment of toxicity or overload of transition-metal ions. Nickel is not included in this Report. Its metabolism and toxicology have been reviewed.[7]

2 Iron

Iron is an essential trace metal for all living organisms with the exception of lactobacilli. Humans normally contain about 3–4 g of iron, of which 70% is in haemoglobin and myoglobin, and intracellular enzymes account for approximately 0.7%. Most of the remainder is accounted for by storage iron. Fe^{III} tends to form insoluble hydroxides at physiological pH. To circumvent this problem, Fe^{III} is handled by specific transport and storage proteins: transferrin and ferritin.

Transferrin.—The major cycle of iron within the body involves the production, circulation, and degradation of haemoglobin. Since haemoglobin is synthesized in the marrow and broken down in reticuloendothelial cells of spleen and liver, a transport protein is required to return the iron from sites of breakdown to sites of haem synthesis. This transport protein is transferrin. In humans, serum transferrin accounts for only about 4 mg of body iron (about 0.1%), but it delivers nearly ten times this weight of iron daily to the erythropoietic bone marrow. It is not catabolized in so doing, and is thus a true transport protein. In addition, transferrin receives approx. 1 mg of Fe per day, entering the body through the intestine, and it delivers iron to and acquires iron from the parenchymal cells of liver and other tissue (again, about 1 mg of Fe per day in each direction).

Serum transferrin is a glycoprotein of molecular weight about 80 000, a single polypeptide chain, which is synthesized in the liver. It has two Fe^{III}-binding sites, but it is normally maintained at 30% saturation, so that spare sites are available. Each protein molecule contains two identical glycans, ending in sialic acid and conjugated through GlcNAc-Asn linkage, and each Fe^{III}-binding site also binds an anion, normally carbonate or possibly bicarbonate. Fe^{3+} and anion are bound and released synergistically.

The presence of two Fe^{III} on a single polypeptide chain raises the question as to whether the two binding sites are like or unlike. A variety of techniques have been used to resolve this question. It is clear that the two sites are closely similar, but the degree of similarity depends on the techniques used as probe. Of even greater importance is the question of whether the two iron atoms are functionally distinct. About ten years ago, Fletcher and Huehns proposed that the immature erythroid cells of the marrow acquired iron preferentially from one of the two sites, making this site available to iron entering the serum from intestinal mucosal cells, while storage cells mainly acquired iron from the second site. It was also considered that iron could be delivered more readily from sites on diferric trans-

[4] W. H. Crosby, *New England J. Med.*, 1977, **297**, 543.
[5] M. Worwood, *Semin. Haematol.*, 1977, **14**, 3.
[6] A. Jacobs, *Clin. Sci. Mol. Med.*, 1977, **53**, 105.
[7] F. W. Sunderman, jun., *Ann. Clin. Lab. Sci.*, 1977, **7**, 377

ferrin than from molecules with a single Fe atom attached. The Fletcher–Huehns hypothesis has stimulated a great deal of work, and much of the recent research reported below is devoted to studies of structural and functional similarities and of differences in the binding sites. Two related proteins have also been studied. These are lactoferrin, found in milk and other external secretions, and ovo-transferrin or conalbumin, the iron-binding protein of avian egg-white. Lacto-ferrin and ovotransferrin resemble serum transferrin in size, numbers of Fe^{III}-binding sites, and the high affinity of these sites for ferric iron. This last property ensures that little free iron is available in the presence of the unsaturated proteins, and, as a result, all three molecules are bacteriostatic agents.

Transferrin Protein. The amino-acid sequence of human serum transferrin is not yet complete, but large segments have already been sequenced.[8] A comparison of the sequences within the *N*-terminal and *C*-terminal halves of the molecule (residues 1—339 and 340—676 respectively) reveals strong internal homology (about 40%), suggesting that transferrins have evolved after gene duplication from an ancestral molecule with a single metal-binding site. Homology in the *N*-terminal sequences of human serum transferrin and ovotransferrin has also been noted.[8, 9] Sequences around His^{124} and His^{256} are highly conserved in the two proteins including an arginine placed five residues on their *C*-terminal sides, and there are two conserved tyrosine-rich regions about fifty residues towards the *C*-terminus from these. The histidines and tyrosines may be involved in metal-binding and the arginine in anion binding. There are also major differences in the two domains that may have functional importance. In human transferrin, both the carbohydrate chains are attached to the *C*-terminal region, and this is true also of ovotransferrin, but there is no homology in the glycosylation sites of the two proteins. In human lactoferrin, however, carbohydrate is distributed equally between the two halves of the molecule.[10] The structure of asialo-glyco-peptides isolated from human serum transferrin has been studied with the aid of high-resolution ^1H.n.m.r. spectroscopy.[9] Most of the carbohydrate moiety occurs as a branched ('biantennary') structure. Similar structures occur in lactoferrin. Chromatography on DEAE-Sephadex[11] and isoelectric focussing[12] of bovine lactoferrin and human transferrin respectively have revealed a microheterogeneity which has been attributed to variations in sialic acid content. However, four fractions of bovine serum transferrin that were separated by DEAE-Sephadex chromatography did not differ in carbohydrate, and they are thought to have small differences in primary structure.[13]

Tryptic digestion of bovine Fe_2-transferrin gave two fragments (mol. wts. 32 000 and 38 500), the larger of which gave multiple components on electro-

[8] R. T. A. MacGillivray, E. Mendez, and K. Brew, in ref. 1, p. 134.

[9] G. Spik and J. Mazurier, in ref. 1, p. 143.

[10] J. H. Bluard-Deconiuck, J. Williams, R. W. Evans, J. Van Snick, P. A. Osiuski, and P. L. Masson, *Biochem. J.*, 1978, **171**, 321.

[11] A. Cheron, J. Mazurier, and B. Fournet, *Compt. rend.*, 1977, **284**, *D*, 585.

[12] A. Hamann, in 'Electrofocussing & Isotachophoresis', Walter de Gruyter & Co., Berlin and New York (printed in Germany), 1977, p. 229.

[13] M. W. Hatton, *Biochem. Genet.*, 1977, **15** 621.

phoresis on cellulose acetate, as did intact transferrin.[14] Very recent work[15] shows that the tryptic fragments have little sequence homology and that all the carbohydrate in this protein is on one fragment, tentatively identified as the C-terminal.

The Metal-binding Sites. Earlier work on transferrin indicated that the two Fe-binding sites were remarkably similar and that the iron bound randomly. Differences in e.p.r. spectra were observed in the presence of chaotropic agents with Fe^{III}-transferrin, but a clear distinction between the sites could only be made with vanadyl and chromic complexes. The recent finding by Lestas and by Princiotto and Zapolski of differences in pH dependence of Fe^{III} binding at the two sites has opened the way to further structural and functional studies. It has been found recently that the two sites on ovo-transferrin bind metals at different rates, and this has also been further investigated. It is customary to designate the two sites A and B.

Spectroscopic measurements, especially e.p.r. spectroscopy, have continued to be one of the most useful probes of site conformation[16-21] (See Figure 1). At pH 6.0, binding to a single site is found with Fe^{3+}, VO^{2+}, and Cr^{3+} ions,[19] and the second site can be labelled at higher pH. By comparing the spectra of Fe^{III}- and Cr^{III}-transferrins with and without vanadyl at the second site, Harris[19, 22] has deduced that Fe^{3+} binds at the site designated A, while Cr^{3+} binds at site B at pH 6.0. Even at pH 7.5, transferrin is preferentially labelled at site A by iron. Double labelling with ^{59}Fe and ^{55}Fe followed by release of a single Fe at pH 6.0 confirmed the site preferences.[19, 22] The iron was supplied as its nitrilotriacetate (nta) complex in this study. However, iron provided as ferric citrate preferentially occupies site B, while random binding is observed with ferrous ascorbate[23] Recently, Aisen *et al.*[24] have used polyacrylamide–urea gel electrophoresis to show that, while iron from the nta complex preferentially binds to site A, iron provided as ferric citrate, oxalate, or chloride, or as ferrous ammonium sulphate, preferentially occupies site B. Analysis of the electron paramagnetic resonances of vanadyl bicarbonate complexes as a function of pH[17, 18] indicates that the two sites on the protein structure have either conformational state A or B (in which one or more of the ligands and/or their geometry may differ) except that in the pH range 7—8 the conformations cannot be the same. Above pH 9.0 the magnetic environment of site A (the site preferred by VO^{2+} at pH 6.0) alters, and the B resonances gain intensity as the A conformation undergoes a transition to the B conformation. 2.7 ± 0.1 protons are released for each VO^{2+} bound in the pH range 6.0—10.0, so the A and B sites are the same in this respect. Above pH 10.0, new, site C, resonances were observed.[17, 18] Campbell and Chasteen,[21] in a

[14] J. H. Brock, F. R. Arzabe, N. E. Richardson, and E. V. Deverson, in ref. 1, p. 153.
[15] J. H. Brock, F. R. Arzabe, N. E. Richardson, and E. V. Deverson, *Biochem. J.*, 1978, **171**, 73.
[16] J. Mazurier, J. M. Lhoste, G. Spik, and J. Montreuil, *F.E.B.S. Letters*, 1977, **81**, 371.
[17] N. D. Chasteen, L. K. White, and R. F. Campbell, *Biochemistry*, 1977, **16**, 363.
[18] N. D. Chasteen, R. F. Campbell, L. K. White, and J. D. Casey, in ref. 1, p. 187.
[19] D. C. Harris, *Biochemistry*, 1977, **16**, 560.
[20] J. L. Zweier and P. Aisen, *J. Biol. Chem.*, 1977, **252**, 6090.
[21] R. F. Campbell and N. D. Chasteen, *J. Biol. Chem.*, 1977, **252**, 5996.
[22] D. C. Harris, in ref. 1, p. 197.
[23] E. J. Zapolski and J. V. Princiotto, in ref. 1, p. 205.
[24] P. Aisen, A. Leibman, and J. Zweier, *J. Biol. Chem.*, 1978, **253**, 1930.

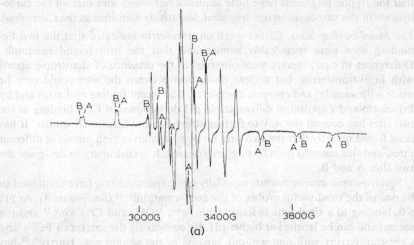

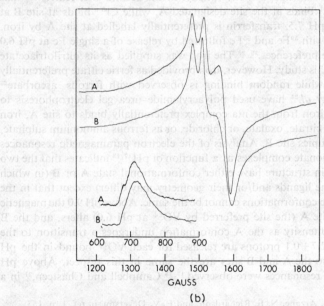

Figure 1 *E.p.r. spectra of transferrin, at 77 K. (a) X-band e.p.r. spectrum of divanadyl transferrin in 90% D_2O, pD = 8.4. Resonances for the A and B metal ion environments are noted. (b) E.p.r. spectra of monoferric transferrin at pH 7.5. A, predominantly Fe$_A$ transferrin, prepared from ferric nitrilotriacetate. B, predominantly Fe$_B$ transferrin, prepared from ferric oxalate. The gain has been increased 6 times in the inset spectra*

(Reproduced by permission from (a) *Biochemistry*, 1977, **16**, 363 (b) *J. Biol. Chem.*, 1978, **253**, 1930)

detailed study of the effects of anions on binding of VO^{2+} in carbonate-free solutions, have concluded, from an analysis of the e.p.r. spectra, that anions could be divided into three groups: (1) non-synergistic, (2) synergistic, promoting binding at site B, (3) synergistic, promoting binding at both A and B sites. Group (3) contains only dinegative anions with two or more carboxy-groups, whereas group (2) anions are mononegative, with one carboxy-group and a second non-ionized electron-withdrawing functional group. The authors explain this anion specificity by the presence of a positively charged functional group (*e.g.* lysine ε-amino) at site A, but not at site B, which must be neutralized before the anion can bind, and this would be consistent with the pH-dependence of the A to B transition which is associated with a non-co-ordinating group, $pK = 10$.[17,18]

Comparison of the transferrin spectra with those of model compounds provides evidence that in the B conformation the anion is directly ligated to the metal through its proximal functional group, and this may also be true of site A, although this is not clear.[21]

The use of a spin-labelled oxalate derivative as anion, which gave no nitroxyl e.p.r. signal on binding to transferrin, confirms that the anion is directly attached to the metal ion.[25] Zweier and Aisen[20] studied the e.p.r. spectra of transferrin with Cu^{2+} in place of Fe^{3+} and suggested that an ionized ε-amino-group of lysine is involved in carbonate binding. As the pH of Cu_2-transferrin is raised above 9.4, the signal due to the carbonate complex is decreased and one of the sites is altered. It is deduced that, in the carbonate complex, there are four planar Cu ligands, one imidazole nitrogen, one carbonate oxygen, and two tyrosyl oxygens, with H_2O as an axial ligand. At high pH, when only one Cu is bound, the e.p.r. spectra showed hyperfine splitting which could be attributed to 3 or 4 equivalent nitrogen ligands. It is possible that some of these bind Fe^{3+}, although they do not participate in the Cu^{2+}-transferrin–carbonate complex. When Cu^{2+} was bound to transferrin in the presence of oxalate, but in the absence of carbonate, in the pH range 8.2—9.3, a single-site ternary complex was obtained, showing that there is a difference in anion-binding requirements at the two sites. A mixed dicupric carbonate–oxalate complex also showed evidence of heterogeneity. Mazurier *et al.*[16] have examined e.p.r. spectra of Cu^{2+} complexes of both human serum transferrin and lactoferrin, before and after modification of histidine by reaction with diethyl pyrocarbonate. In contrast to Zweier and Aisen,[20] they have observed that spectra of unmodified proteins show a 'sharp' component corresponding to one nitrogen ligand at one Cu site and a 'broad' component attributable to 3 or 4 nitrogen atoms bound to the other cupric ion. After carbethoxylation of the apoproteins, only one Cu is bound to one histidine ligand, and the broad component disappears. From pH lability, the 'sharp' and 'broad' copper sites were identified as A and B respectively. Although the A (sharp) site is similar in both proteins, B sites exhibit some differences.

An analysis of resonances due to histidine in the ^1H n.m.r. spectra of transferrin and ovo-transferrin shows the presence of three distinct groups with similar pK values of 6.5 ± 0.01, but giving different chemical shifts on protonation.[26] The four histidines in group A are probably on the molecular surface while the six

[25] C. R. Najarin, D. C. Harris, and P. Aisen, *J. Biol. Chem.*, 1978, **253**, 38.
[26] R. C. Woodworth, R. J. P. Williams, and B. M. Alsaadi, in ref. 1, p. 211.

histidines in each of groups B and C are in the proximity of aromatic residues. Resonances from six histidines in groups B and C are missing from spectra of metal complexes, and it has been proposed that each binding site has one histidine involved in anion binding and two linked (along with tyrosine) to the metal ion. Modification with ethoxyformic anhydride confirms the involvement of histidine in the binding sites of transferrin and ovo-transferrin, since an average of two His per site were protected in the metal complexes,[27] while all of them reacted in the apoproteins. Perturbation of difference spectra[28] and circular dichroism[9] on metal binding has been attributed to the presence of tryptophan as well as tyrosine in the binding sites. From measurements of energy transfer between terbium and iron bound to transferrin, the metal-binding sites are estimated to be 25 ± 2 Å apart.[29] It has been shown that iron bound to transferrin is able to catalyse the oxidation of thiols,[30] which suggests that at least one of the sites is on or close to the surface of the protein.

Aisen, Leibman, and Zweier[24] have obtained overall thermodynamic Fe^{3+}-binding constants from equilibrium dialysis measurements with citrate as complexing agent. At pH 7.4, and at atmospheric $p(CO_2)$, the calculated apparent stability constants K_1' and K_2' are 4.7×10^{20} l mol^{-1} and 2.4×10^{19} l mol^{-1} respectively. By the use of the ability of polyacrylamide – urea gel electrophoresis to separate the two monoferric species and to estimate their proportions, all four intrinsic site constants could be obtained. At pH 7.4, B-site binding is weaker and at pH 6.7 much weaker than at the A-site under the conditions used, so binding must be virtually sequential. At pH 6.7, weak negative co-operativity at each site was found. At pH 7.4 it was less pronounced. The authors distinguish between thermodynamic stability constants and kinetic accessibility. Thus the weaker B-site is more accessible to ferric citrate and a number of other complexes mentioned above but, on standing, B-site iron would be expected to redistribute slowly to A-site.

Donovan has introduced the technique of differential scanning calorimetry to obtain 'thermograms' which show peaks of heat absorption at the denaturation temperatures of the proteins under observation. The number of endotherms indicates the number of non-interacting species present in the system. With human serum transferrin[31] he finds evidence for two domains, with denaturation temperatures of 62 and 72 °C. The effect of binding aluminium on the endotherms shows that the binding of Al to the two different domains is sequential, and this is confirmed in work with Zn, Al, and Ga.[28]

With ovo-transferrin, in contrast, a single endotherm is found, and the effect of Al-binding is the same whether it is bound to one or both sites. When transferrin or ovo-transferrin are fully saturated with iron, added as Fe^{III} (nta), only one endotherm is observed, at a higher temperature than those of the apoproteins. Both monoferric proteins give two main endotherms, but differences in binding and denaturation behaviour suggest that serum transferrin has iron mainly bound

[27] T. B. Rogers, R. A. Gold, and R. E. Feeney, in ref. 1, p. 161.
[28] Y. Tomimatsu and J. W. Donovan, in ref. 1, p. 221.
[29] C. F. Meares and J. E. Ledbetter, *Biochemistry*, 1977, **16**, 5178.
[30] W. L. Anderson and T. B. Tomasi, jun., *Arch. Biochem. Biophys.*, 1977, **182**, 705.
[31] J. W. Donovan, in ref. 1, p. 179.

at a single site, whereas in ovo-transferrin it is apparently distributed between the two binding domains. The thermal denaturation of *N*- and *C*-terminal fragments of ovo-transferrin, produced by tryptic digestion,[32] shows different denaturation temperatures with and without iron, even though they denature simultaneously in the intact molecule, and also shows that the *N*-terminal fragment has the higher affinity for Fe[III] when bound at pH 7.5. The *N*-terminal fragment shows greater thermal stabilization than the *C*-terminal when Al[III] is bound. Evidence for sequential binding in intact ovo-transferrin, which contradicts the random binding given above,[31] has come from double labelling with ^{59}Fe and ^{55}Fe and the analysis of the isotope distribution in the *N*- and *C*-domains after proteolysis.[33] Conditions appear to play a critical role in the mode of binding.

Functional Heterogeneity in Transferrin Iron. According to the Fletcher–Huehns hypothesis, reticulocytes acquire iron more readily from diferric transferrin than from monoferric, and from one of the two sites. Further experiments, to test this hypothesis, continue to give conflicting results. Three types of experiments have been carried out: (1) incubation of reticulocytes with transferrins of various levels of saturation selectively labelled with radio-iron *in vitro* or *in vivo*, (2) uptake of labelled iron by the perfused rat liver, (3) analysis of tissue distribution of iron *in vivo* after selective labelling of plasma.

Zapolski and Princiotto[34] first reported evidence supporting functional heterogeneity; however, the system they used was heterologous, with human transferrin and rabbit reticulocytes. With transferrins of different *average* saturation, but the same concentration of diferric transferrin, they found identical rates of iron uptake during the first 15–20 minutes of the incubation. This indicated preferential uptake from the saturated species. However, the same authors later used two homologous systems, with different results.[35] They found that rabbit reticulocytes acquired iron more rapidly from homologous transferrin than from the human species, but at rates that were independent of the level of saturation. Unlike human transferrin, the two sites in diferric rabbit transferrin were indistinguishable in their ability to release iron at acid pH, and thus they could not be selectively radioactively labelled. This is in agreement with the experiments of Morgan,[36, 37] who found no selective release to ATP plus desferrioxamine with rabbit transferrin. In contrast, selective labelling of human transferrin could be carried out, and the acid-labile site was found also to be the one donating iron most readily to rabbit reticulocytes. However, on examining the uptake of isotopes by human reticulocytes from human transferrin alternately labelled with ^{59}Fe and ^{55}Fe at pH 6.0 and 7.5, Harris could find no preference for uptake from either site in three experiments,[38] confirming earlier work of Harris and Aisen. In rat liver perfusion studies, Zimelman *et al.*[39] found that hepatic iron

[32] R. W. Evans, J. W. Donovan, and J. Williams, *F.E.B.S. Letters*, 1977, **83**, 19.
[33] J. Williams and R. W. Evans, in ref. 1, p. 169.
[34] E. J. Zapolski and J. V. Princiotto, *Biochem. J.*, 1977, **166**, 175.
[35] J. V. Princiotto and E. J. Zapolski, *Biochim. Biophys. Acta*, 1978, **539**, 81.
[36] E. H. Morgan, *Biochim. Biophys. Acta*, 1977, **499**, 169.
[37] E. H. Morgan, in ref. 1, p. 227.
[38] D. C. Harris, *Biochim. Biophys. Acta*, 1977, **496**, 563.
[39] A. P. Zimelman, H. J. Zimmerman, R. McLean, and L. P. Weintraub, *Gastroenterology*, 1977, **72**, 129.

uptake was not influenced by the level of saturation of rat transferrin in the perfusing medium, kept at constant iron concentration.

In previous experiments, E. B. Brown and co-workers have found evidence that rat transferrin selectively labelled *in vitro* preferentially delivers A-site iron to rat recticulocytes, although they, like others, could find no such preference in the human homologous system.[40] They have followed up a further prediction of Fletcher and Huehns in a recent set of experiments. This is that iron absorbed[40, 41] by the intestine is selectively taken up by the site which preferentially delivers iron to erythroblasts. Okada, Chapman, and Brown collected labelled blood through the portal vein during a 30 minute period of intestinal absorption of radio-iron. They assayed the portal plasma with rat reticulocytes and found evidence for site selectivity as compared with uptake of radio-iron from peripheral blood which has been randomly labelled *in vitro* with ^{59}Fe(nta). Marrow erythroblasts, as well as reticulocytes, took up more radio-iron from the labelled portal plasma in incubations *in vitro*. Portal plasma, labelled with ^{55}Fe during absorption, and randomly ^{59}Fe-labelled peripheral blood were also injected into rats, and the tissue distribution of the two labels was measured 30 minutes later. It was found that, in normal rats, the iron from labelled portal plasma was selectively incorporated into red cell precursors and into haem in bone marrow and spleen. The reticulocytes of iron-deficient rats showed greater avidity for iron and less site discrimination than those from normal animals, and the tissue distribution also showed little preference for either isotope. The conclusions of these experiments have been contested by Pootrakul *et al.*,[42] who in similar experiments could find no evidence for functional heterogeneity, and there has been considerable further discussion.[43, 44] The differences are probably methodological in origin. It has now been suggested by Brown[43] that the functional heterogeneity attributed to site differences (A and B) on a single molecule of rat transferrin may in fact be due to the different isotransferrins (S and F) recently rediscovered by Huebers *et al.*,[45] since, in experiments with separated S and F transferrins, S was found to be more efficient than F in delivery of iron to reticulocytes.[46] However, this has been disputed.[44] It is of interest to note that the concentration of S is increased relative to that of F in cases of iron deficiency.[45] Hence, despite the evidence for differences in structure and chemical behaviour accumulated as a result of the Fletcher–Huehns hypothesis, clear evidence for or against the functional heterogeneity predicted by this hypothesis is still elusive.

Release of Transferrin Iron in vitro. The tight binding of iron by transferrin makes it suited to act as a carrier of the element in the circulation, but poses the problem of its delivery. It has been shown that the presence of iron chelators alone does not facilitate iron release. Thus, although desferrioxamine has a higher affinity for iron than transferrin, little transfer from transferrin to desferrioxamine is

[40] E. B. Brown, in ref. 2, p. 125.
[41] S. Okada, B. Chapman, and E. B. Brown, *J. Lab. Clin. Med.*, 1977, **89**, 51.
[42] P. Pootrakul, A. Christensen, B. Josephson, and C. A. Finch, *Blood*, 1977, **49**, 957.
[43] E. B. Brown, *Blood*, 1977, **50**, 1151.
[44] C. A. Finch, *Blood*, 1977, **50**, 1152.
[45] H. Huebers, E. Huebers, S. Linck, and W. Rummell, in ref. 1, p. 251.
[46] S. Okada and E. B. Brown, in ref. 1, p. 261.

observed in the absence of mediators.[36, 37, 47] Pollack *et al.*[47] have investigated the effectiveness of a large number of compounds as mediators of iron release to desferrioxamine at pH 7.4 and 37 °C. Of the compounds tested, by far the most effective was pyrophosphate (PP), with ketomalonate, nitrilotriacetate (nta), ATP, citrate, and others mediating transfer to a lesser extent. No correlation between the ability of anions to replace carbonate in the iron–transferrin–anion ternary complexes and their effectiveness as mediating agents was observed. These results have been confirmed by Morgan,[36, 37] who has also shown that ATP, 2,3-DPG (present in relatively high concentrations in reticulocytes), and GTP were effective in facilitating transfer of iron between human and rabbit transferrin, while pyrophosphate was ineffective. The release of iron from human transferrin in the presence of 4mM-ATP and 5mM-desferrioxamine at pH 7.0 was found to be bimodal, but, interestingly, release of iron from rabbit transferrin followed first-order kinetics, implying that in this species iron was lost at the same rate from both sites. In the absence of iron chelators, ATP, GTP, and 2,3-DPG as well as PP can release iron rapidly from human transferrin, provided mildly acidic conditions (pH 6.1) are used.[48, 49] The release was again bimodal, suggesting preferential loss of iron from one site. It seems that, in the absence of chelator, protonation of the carbonate[17,18] is required for iron removal. The importance of the stability of the iron–transferrin–anion ternary complex has also been emphasized by Graham and Bates,[50] who have also considered other possible factors which might be involved in iron release from transferrin, and they have suggested that reduction of Fe^{3+} to Fe^{2+} occurs only after the release of Fe^{3+} from the transferrin. Pollack *et al.*[47] have suggested that iron release involves a substitution step, displacing carbonate by weaker anion, followed by chelation of the iron released. In experiments in which bicarbonate was added to the ternary iron–ovo-transferrin–EDTA complex labelled with either ^{59}Fe or [^{14}C]EDTA, it was found that both iron and EDTA nearly completely dissociated from the protein within 2 min of addition of bicarbonate,[51] implying that there was not a single displacement of EDTA by carbonate. Graham and Bates[50] have also suggested that attack by bicarbonate on the nta complex proceeds with an intermediate in which the anions' site is not fully occupied. Addition of Ca^{2+} ion seems to stabilize Fe^{3+}–transferrin–CO_3^{2-} in this system, decreasing the rate of iron release, and it has been suggested that Ca^{2+} might act as an allosteric regulator.[50]

Delivery of Iron to Reticulocytes. The half-life of transferrin iron in the serum (1—2 h) is a small fraction of that of the protein (7—8 d). The mechanism by which transferrin releases its iron non-destructively to the cells which require it has been receiving considerable attention. Earlier work by Jandl and associates, and by Morgan, indicated that initial binding to membrane receptors is involved, and that subsequent transfer of iron into the cells is an energy-dependent process.

[47] S. Pollack, G. Vanderhoff, and F. Lasky. *Biochim. Biophys. Acta*, 1977, **497**, 481.
[48] F. J. Carver and E. Frieden, *Biochemistry*, 1978, **17**, 167.
[49] F. J. Carver and D. Frieden, in ref. 1, p. 245.
[50] G. A. Graham and G. W. Bates, in ref. 1, p. 273.
[51] T. B. Rogers, R. E. Feeney, and C. F. Meares, *J. Biol. Chem.*, 1977, **252**, 8108.

Much recent work has been involved with the isolation of the receptors, and with the development of methods for solubilizing membrane components.

Aisen and co-workers[52, 53] have isolated a receptor–transferrin complex from rabbit reticulocytes, incubated with rabbit transferrin, by treatment with Triton X-100 and gel filtration. The estimated molecular weight of the complex was 430 000, giving approximately 370 000 for the membrane receptor after allowing for the molecular weight of transferrin. A component of mol. wt. 400 000 was isolated from reticulocytes that had not been incubated with transferrin. Both yielded fractions of mol. wts. 176 000 and 95 000 in SDS-polyacrylamide gel electrophoresis. Bockxmeer and Morgan[54] obtained a component of mol. wt. 275 000 from rabbit reticulocyte membranes treated with the non-ionic detergent Teric 12A9; it bound plasma transferrin but not other plasma proteins. Binding was reversible and saturable, and was stronger for rabbit than for human transferrin. A Triton X-100 component from rabbit reticulocytes was also able to bind human transferrin,[55] and it seems as if the receptors isolated by those groups of workers are essentially similar. Another approach to the isolation of the receptor has been by Light,[56] who used a rabbit transferrin affinity column to isolate the receptor from rabbit reticulocyte membranes that were solubilized by Triton X-100. Its mol. wt. was only 35 000—40 000, but it had a tendency to aggregate and it also formed complexes of high molecular weight with transferrin. The larger receptor of Aisen[52, 53] and Morgan[54] may be a result of such complex formation. Van der Heul *et al.*[57] estimated the mol. wt. of the rat receptor as 80 000 (isolated with Triton X-100). Possibly it is similar to the 95 000 component of rabbit reticulocyte receptors obtained by Aisen.[52, 53]

Several workers have measured the numbers of receptors present at various stages of red cell maturity. Nunez *et al.*,[58] estimating receptors by binding ^{125}I-labelled transferrin, found a 23-fold loss of receptors during maturation of murine erythroid cells. Uptake of ^{59}Fe from transferrin was proportional to the numbers of receptors present. It has also been shown that the stimulated differentiation of Friend cells which is associated with an increase in the requirement for iron, due to haemoglobin synthesis, is also associated with an increase in the number of transferrin receptors per cell.[59]

The presence of specific transferrin receptors in hepatocytes has also been inferred,[39, 60, 61] and the numbers of these may increase in iron deficiency.[39] The ingestion of lactoferrin by mouse peritoneal macrophages also proceeds *via* a membrane receptor.[62] Although apotransferrin increases the rate of release of

[52] P. Aisen, A. Leibman, H. Y. Y. Hu, and A. I. Skoultchi, *J. Biol. Chem.*, 1977, **252**, 281
[53] P. Aisen and A. Leibman, *Adv. Chem. Series*, 1977, **162**, 104.
[54] F. M. Van Bockxmeer and E. H. Morgan, *Biochim. Biophys. Acta*, 1977, **468**, 437.
[55] B. Ecarot-Charrier, V. Grey, A. Wilczynska, and H. M. Schulman, in ref. 1, p. 291.
[56] N. D. Light, *Biochim. Biophys. Acta*, 1977, **495**, 46.
[57] C. Van der Heul, M. J. Kroos, and H. G. Van Eijk in ref. 1, p. 299.
[58] M. T. Nunez, J. Glass, S. Fischer, L. M. Lavidor, E. M. Lenk, and S. A. Robinson, *Brit. J. Haematol.*, 1977, **36**, 519.
[59] H. Y. Y. Hu, J. Gardner, P. Aisen, and A. I. Skoultchi, *Science*, 1977, **197**, 559.
[60] D. Grohlich, C. G. D. Morley, R. J. Miller, and A. Bezkorovainy, *Biochem. Biophys. Res. Comm.*, 1977, **76**, 682.
[61] D. Grohlich, C. G. D. Morley, R. J. Miller, and A. Bezkorovainy, in ref. 1, p. 335.
[62] J. L. Van Snick, B. Markowetz, and P. L. Masson, *J. Exp. Med.*, 1977, **146**, 817.

iron from isolated hepatocytes,[63] no evidence for the involvement of receptors in release was given.

It is not yet certain what follows the binding of transferrin to its receptors. Iron may be released into the cell at the membrane site[64-67] or transferrin may be taken into the cell by endocytosis and the iron then released internally,[68, 69] possibly at mitochondrial membranes.[70, 71] Attempts to settle this controversy by immobilization of transferrin by covalent attachment to Sepharose[64, 65] or agarose, Enzacryl AA, or latex beads[68] have produced conflicting results. Thus Glass[64] and Loh *et al.*[65] found that reticulocytes could take up iron from immobilized transferrin, while Hemmaplardh and Morgan[68] found that they could not. Morgan and co-workers[72-75] have examined several factors which may affect the uptake and release of transferrin and its iron. Thus Ca^{2+} ion[73] was found to be required for the integrity of receptors or transferrin binding to them, but there was no evidence that Ca^{2+} affected iron uptake directly. The membrane phospholipid composition influenced the interaction of transferrin and receptor, and iron uptake and release were inhibited by phospholipase A.[75] Confirming earlier results of Workman and Bates, reticulocyte lysates were found to mobilize iron transferrin-labelled reticulocyte ghosts, but erythrocyte lysates were also effective, and the removal of a protein of low molecular weight, thought by Workman and Bates to be important in iron release, did not affect iron mobilization.[74] The uptake of iron from transferrin by reticulocytes has been shown to depend on the concentration of cellular ATP,[66, 72] but not of NADH,[72] and, whereas Kailis and Morgan[72] interpreted the results in terms of a need for metabolic energy, Egyed[66] has postulated a specific role for ATP in transferrin iron release, which is also suggested by the studies *in vitro* reported above.

Ferritin.—Ferritin has been well characterized as a protein with a large capacity for iron which acts both as a temporary iron store, making it harmless, and as a long-term mobilizable reserve. The iron is in micellar form as a ferric oxyhydroxide (hydrous ferric oxide)–phosphate complex. A protein shell (apoferritin) encloses the micelles, thus rendering their Fe^{III} soluble in the cell sap. Ferritin, which may contain up to about 4500 Fe atoms, usually maintains a reserve capacity for iron. Apoferritin biosynthesis is stimulated in most cells by the presence of iron, and there is usually an over-production of protein, giving, on average, molecules about half full or less. The inorganic iron component is microcrystalline, one or more small crystals being present per molecule, and the internal structure of these is unrelated to that of the surrounding 24-subunit protein shell. Ferritin

[63] E. Baker, F. R. Vicary, and E. R. Huehns, in ref. 1, p. 327.
[64] J. Glass, *Biochem. Biophys. Res. Comm.*, 1977, **75**, 226.
[65] T. T. Loh, Y. G. Yeung, and D. Yeung, *Biochim. Biophys. Acta*, 1977, **471**, 118.
[66] A. Egyed, in ref. 1, p. 237.
[67] J. Fielding and B. E. Speyer, in ref. 1, p. 311.
[68] D. Hemmaplardh and E. H. Morgan, *Brit. J. Haematol.*, 1977, **36**, 85.
[69] J. Martinez-Medellin, H. M. Schulman, E. de Miguel, and L. Benavides, in ref. 1, p. 305
[70] P. Poňka, J. Neuwirt, J. Borová, and O. Fuchs, in ref. 1, p. 319.
[71] P. Poňka, J. Neuwirt, J. Borová, and O. Fuchs, in ref. 2, p. 167.
[72] S. G. Kailis and E. H. Morgan, *Biochim. Biophys. Acta*, 1977, **469**, 389.
[73] D. Hemmaplardh and E. H. Morgan, *Biochim. Biophys. Acta*, 1977, **468**, 423.
[74] G. W. Blackburn and E. H. Morgan, *Biochim. Biophys. Acta*, 1977, **497**, 728.
[75] D. Hemmaplardh and E. H. Morgan, *J. Membr. Biol.*, 1977, **33**, 195.

formation *in vivo* involves an initial assembly of apoferritin shells, which then accumulate iron. The mechanism of iron uptake has been studied mainly *in vitro*, and until recently this has also been true of iron release.

The structure of apoferritin and its conversion into ferritin is a subject of continuing interest. Considerable attention is being devoted to isoferritins and to their meaning. Questions of concern are: what does the observed electro-phoretic inhomogeneity of ferritin mean in terms of structure and function (what are isoferritins?); how does ferritin arise in serum, and is it altered in moving out of its tissue of origin?

Structure. A number of papers have reviewed ferritin structure in relation to its properties as an iron-storage molecule.[76-79] Ferritins from both animal and plant sources have characteristically similar amino-acid compositions, with about 45% non-polar amino-acids and isoelectric points on the acidic side of neutrality.[76, 80] Not surprisingly, the mammalian ferritins examined resemble each other more closely than they do the phytoferritins. In earlier studies, horse spleen apoferritin was characterized as a molecule of overall mol. wt. 444 000, containing 24 sub-units, each of mol. wt. 18 500. It readily dissolves in dilute salt solutions or buffers. Porcine spleen ferritin and apoferritin form insoluble aggregates in the absence of thiol reagents,[81] suggesting that oxidizable groups are present on its surface. Although it cross-reacts with anti-human ferritin serum and has a similar c.d. spectrum and sedimentation coefficient, $s^0_{20,w} = 17.5$S, its subunit and entire molecular weights (20 000 and 503 000 respectively) are higher.[81] The product isolated from pig spleen is relatively iron-rich. Apoferritin derived from ferritin isolated from *Corbicula sandai*, a bivalve shellfish, has $s^0_{20,w}$ of 18.75 and mol. wt. 503 000, but its c.d. spectrum is distinctive, and it does not react with anti-human-ferritin antisera.[82] Ferritin isolated from pea and lentil[76, 80, 83] may also be larger than the horse spleen molecule (mol. wt. near 500 000, as judged by low-angle neutron-scattering data) and may have a larger capacity for iron, possibly up to 5420 and 6200 Fe atoms per molecule respectively.[76, 80] Subunits from pea and lentil ferritins, of mol. wt. 20 300 and 21 400 respectively, when mixed with those of horse spleen, form separate bands on electrophoresis in SDS-polyacrylamide gels. The molecules contain 24 subunits,[76, 80] like other ferritins. Mammalian heart ferritins have also previously been reported to have a propor-tion of molecules larger than those characteristic of spleen and liver, and these ferritins may also have subunits of different sizes.[84] Subunits of rat liver ferritin resemble those of the horse spleen protein in having a blocked *N*-terminus, the *N*-terminal tripeptide being Ac-Ser-Ser-Gln.[85]

[76] R. R. Crichton, D. Collet-Cassart, Y. Ponce-Ortiz, M. Wauters, F. Roman, and E Paques, in ref. 1, p. 13.
[77] P. M. Harrison, *Semin. Haematol.*, 1977, **14**, 55.
[78] A. Treffry, S. H. Banyard, R. J. Hoare, and P. M. Harrison, in ref. 1, p. 3.
[79] P. M. Harrison, S. H. Banyard, R. J. Hoare, S. M. Russell, and A. Treffry, in ref. 2, p. 19.
[80] R. R. Crichton, Y. Ponce-Ortiz, M. H. J. Koch, R. Parfair, and H. B. Stuhrmann, *Biochem J.*, 1978, **171**, 349.
[81] M. E. May and W. W. Fish, *Arch. Biochem. Biophys.*, 1977, **182**, 396.
[82] A. Baba, M. E. May, and W. W. Fish, *Biochim. Biophys. Acta*, 1977, **491**, 491.
[83] Y. Ponce-Ortiz and R. R. Crichton, *Biochem. Soc. Trans.*, 1977, **5**, 1128.
[84] J. W. Drysdale, in ref. 2, p. 41.
[85] A. Huberman and E. Barahona, *Biochim. Biophys. Acta*, 1978, **533**, 51.

Further advances have been made in the *X*-ray analysis of the three-dimensional structure of horse spleen apoferritin,[77-79, 86] which may be regarded as the prototype structure. Its 24 equivalent subunits, arranged in cubic (432) symmetry, form a compact, fairly smooth shell, as seen at 6 Å resolution, which contains channels of diameter approx. 10 Å.[77-79] These channels could provide the means by which iron enters and leaves the ferritin molecule. An electron-density map at 2.8 Å resolution has now also been published, together with its preliminary interpretation.[86] The new map, part of which is reproduced in Figure 2, shows each of the

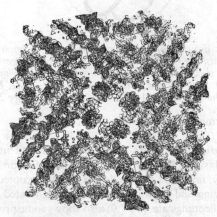

Figure 2 *Electron-density map at* 2.8 Å *resolution of horse spleen apoferritin, The sections of the map shown correspond to a slice near the outside of the molecule, about* 10 Å *thick and* 100 Å *across, perpendicular to a 4-fold axis*
(Reproduced by permission from *Nature*, 1978, **271**, 282)

24 subunits to consist of four long, nearly parallel, α-helical rods (up to 42 Å long), lying perpendicular to the radius vector, and short helices at right angles to these. There are six channels passing through the shell, each delineated by four short helices, from four symmetry-related subunits. The subunits, which approximate to cylinders of diameter about 27 Å and of length 54 Å, pack together as shown in Figure 3, and from this packing it has been deduced that the shell probably assembles *via* dimers and hexamers as intermediates. Isomorphous replacement methods show that the protein shell binds several cations[78, 79, 86] on or near its inside and outside surfaces, but no cations are embedded deeply within the shell, either within, or between, subunits. One or more of these inner sites could represent sites of Fe-binding involved in iron accumulation. Unfortunately, the amino-acid sequence of horse spleen apoferritin is not yet complete, but it is reported[76] that the fragments so far sequenced account for 75% of the primary structure. Hence a more detailed description of the three-dimensional structure and the location of functional side-chains is in sight.

[86] S. H. Banyard, D. K. Stammers, and P. M. Harrison, *Nature*, 1978, **271**, 282.

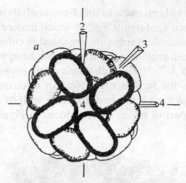

Figure 3 *Schematic drawing of quaternary structure of horse spleen apoferritin, viewed down the 4-fold axis. 24 Subunits pack together in 432 symmetry to form a shell surrounding a central cavity in which iron may be accumulated. Some of the symmetry axes are shown. Iron enters the shell through channels along 4-fold axes. See also Figure 2 (which is approx. 2 times larger)*
(Reproduced by permission from *Nature*, 1978, **271**, 282)

Relatively few structural studies have been published in addition to those reported above. Massover,[87, 88] who has examined ferritin by high-resolution electron microscopy, finds that the shells are partly disrupted by treatment with acetone or by freezing and thawing. Dilute ferritin treated only with negative stain (2% sodium silicotungstate, pH 7.0) also shows some protein fragments, and Massover argues that the belief that ferritin is stable to dilution may have to be revised. Disrupted ferritin shells usually show iron-cores attached to protein fragments containing one or more subunit, implying that there is interaction at this level. Breslow and Silk[89] have made a detailed comparison of ion binding by horse spleen ferritin and apoferritin. Both have the same isoionic pH, *i.e.* 4.58, but when salt is added to suspensions of deionized protein they behave differently. With apoferritin, increasing concentrations of NaCl or KCl produce a progressive increase in pH, indicative of binding of Cl^-. With ferritin, the addition of KCl initially gives a pH decrease, suggesting that K^+ is bound, but as the salt concentration rises, so does the pH, as anions are also bound. The titration behaviour of the two proteins also differs above pH 5.5. Two extra groups (per subunit) now titrate in ferritin, and these are attributed to inorganic phosphate at the core surface. Above pH 8.5, phosphate may be displaced by hydroxyl ions.

In native apoferritin all of the tyrosine and cysteine, two out of nine lysine, and three out of six histidine residues are unavailable to titration. Three of the 26 carboxy-groups are buried and the remainder fall into at least two titration classes. Treffry, Harrison, and co-workers[78, 79, 90] find that [^{32}P]phosphate is taken up by ferritin that has been reconstituted in the absence of this anion. When iron

[87] W. H. Massover, in 35th Annual Proceedings of the Electron Microscopy Society of America, ed. G. W. Bailey, Briton, Mass., 1977, p. 476.
[88] W. H. Massover, in ref. 1, p. 23.
[89] E. Breslow and S. T. Silk, in ref. 1, p. 39.
[90] A. Treffry and P. M. Harrison, *Biochem. J.*, 1978, **171**, 313.

is released from the micelles by reduction with thioglycollic acid at pH 4.25, phosphate is also released. Much of it appears in advance of the iron, suggesting its presence on surface sites, the remainder being occluded in imperfections. Similar behaviour is found with native ferritin. When ferritin is reconstituted in the presence of phosphate, however, the iron and phosphate become more intimately mixed. They are released together. A tentative conclusion has been drawn[79] that, within cells, the addition of iron and phosphate to the native protein is 'compartmentalized'. Both these studies imply that the centre of the ferritin molecule is accessible to inorganic phosphate and other ions, presumably through the channels seen in the high-resolution structure.[86] The channels are of limited selectivity, since they are penetrable by both anions and cations as well as by small neutral molecules. May and Fish[91] have attempted to measure the permeability of the channels by a passive diffusion method. Their method involved incubating apoferritin with radioactive diffusant at 10 °C, removing aliquots at various times, separating protein and free diffusant by gel chromatography, and determining the radioactivity associated with apoferritin. They calculated equilibrium values for the amount of diffusant associated with apoferritin and plotted the 'fraction of equilibrium' *versus* time. Fe^{2+} appeared to diffuse into ferritin at rates several orders of magnitude greater than those for methanol, glucose, methylammonium, and acetate, which, they concluded, experienced considerable hindrance to diffusion. Hence specific means of translocation of iron in and out of the molecule were proposed. Details of the calculation are not provided in the short paper. It would seem from previous low-angle neutron-scattering studies, which indicated much faster penetration rates, that much of the diffusant may have been lost from the interior during chromatography (4 min), and that measurements in this study were indicative more of residual binding to the protein than of penetration rates. Indeed, the authors point out that the Fe^{2+} used in air is rapidly oxidized and bound as Fe^{III} to apoferritin. The fact that Fe^{III} (nta) did not 'penetrate' is consistent with the results of binding studies[78] which showed no binding by apoferritin of Fe from Fe^{III}(nta).

Heavy-atom complexes found inside the apoferritin shell by *X*-ray analysis have been listed.[79] Gallium(III) probably also enters the ferritin molecule[92]

Incorporation of Iron into Ferritin. Since it was first shown by Bielig and Bayer in 1955 that a ferritin-like product could be formed by the addition of $Fe(NH_4)_2$-$(SO_4)_2$ to apoferritin in air, and by Niederer that apoferritin catalyses Fe^{II} oxidation, much work has been done on the mechanism of ferritin formation. Since ferritin contains red-brown polynuclear ferric micelles, a spectrophotometric assay is available which makes the kinetics of iron uptake into ferritin amenable to study, although not easy to interpret. The kinetics are complex (and under certain conditions biphasic), and the reaction involves both oxidation and hydrolysis. It has proved difficult to assign a precise reaction pathway for the oxidation of free Fe^{II} in aqueous solution, even without the catalytic effect of the protein as a further complication. Macara, Hoy, and Harrison have previously

[91] M. E. May and W. W. Fish, in ref. 1, p. 31.
[92] F. N. Hegge, D. J. Mahler, and S. M. Larson, *J. Nucl. Med.*, 1977, **18**, 937.

discussed three alternative models of iron uptake. All of these involve sites for binding and for oxidation of Fe^{II} on the protein.

The mechanism favoured by Macara, Hoy, and Harrison can be summarized as follows. Apoferritin chelates Fe^{II} at sites on its inner surface; this promotes its autoxidation and carboxy-groups and possibly histidine. The bound Fe^{III} starts nucleation of 'FeOOH' during the slow phase of ferritin formation. The reaction speeds up during the second phase because more sites for deposition of iron are being generated as the crystallites grow. When O_2 is used as oxidant, the initial oxidation of protein-bound Fe^{II} could involve the transfer of either a single electron from one Fe^{II} or two electrons from neighbouring Fe^{II} to a molecule of oxygen. Oxidation of Fe^{II} on the crystal surface could involve transfer of electrons from up to four Fe^{II} to one O_2, with reduction of the latter to water. Sequestration of iron in ferritin could thus proceed without the damaging release of free radicals. However, the mechanisms of formation of FeOOH inside or apart from ferritin are essentially similar.

Support for the model of Macara *et al.* has come from studies by Treffry *et al.*[78] They find that Zn^{2+} and Tb^{3+} bind to apoferritin and ferritin and inhibit iron uptake, but inhibition is much less with a ferritin fraction than with apoferritin. The authors suggest that the protein sites are no longer needed, but Zn^{2+} and Tb^{3+} compete with iron for sites on the micelles. The inhibitory effect of Tb^{3+} on binding of Fe^{III} to ferritin from its citrate complex[78] supports this interpretation. Carboxyl modification both inhibits iron uptake and prevents binding of Zn^{II}, but carboxymethylation of two of the histidines gave no inhibition. Treffry *et al.*[78] confirm the conclusion of Crichton that the carboxygroups in question are on the inner surface of the apoferritin shell, and this is consistent with their involvement in micelle nucleation. X-Ray analysis[78, 86] shows Tb^{III} to be close to previously reported UO_2^{2+} sites on the inner surface of apoferritin. The major Tb^{III} sites are near to 2-fold symmetry axes and only 4.3 Å apart. Similarly placed iron atoms might perhaps be able to act concertedly in electron transfer or in oxo- or hydroxy-bridge formation (or both) during the nucleation phase of ferritin formation.[86]

An alternative mechanism of iron accumulation has been proposed by Crichton and co-workers.[76, 93, 94] Like the third model of Macara *et al.*, their model involves both primary catalytic sites, at which all Fe^{II} entering the molecule is oxidized, and secondary heteronucleation sites for FeOOH. The proposed mechanism is shown in Figure 4. An O_2 molecule is bound between two close Fe^{II}, with the formation of a peroxo-complex. Subsequent steps involve hydrolysis and migration of FeOOH to the interior of the molecule, being displaced by incoming Fe^{2+} ions; the driving force is the much greater assumed affinity of Fe^{2+} for the catalytic sites. However, the FeOOH is considered to be capable of migrating back to the oxido-reduction sites to re-form the peroxo-complex.[93] Evidence cited in favour of the mechanism is as follows: (1) the reciprocal of the initial velocity of ferritin formation in borate–cacodylate buffers is approximately proportional to $1/[Fe^{2+}]^2$; (2) pre-incubation of apoferritin with Cr^{3+}, in which 0.5 Cr atoms are bound per subunit, gives rise to 85—90% inhibition; (3) 2,2'-bipyridyl slowly

[93] R. R. Crichton and F. Roman, *Biochem. Soc. Trans.*, 1977, **5**, 1126.
[94] R. R. Crichton and E. P. Paques, *Biochem. Soc. Trans.*, 1977, **5**, 1130.

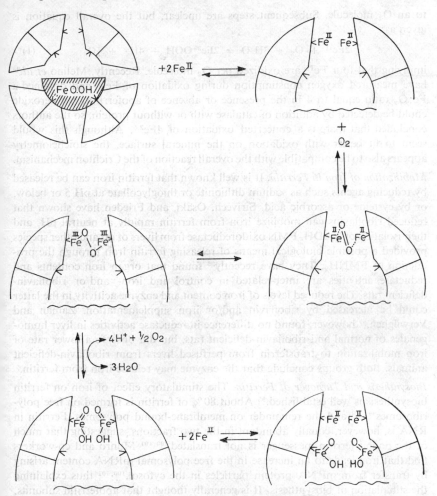

Figure 4 *Model for ferritin iron oxidation and deposition*
(Reproduced by permission from 'Proteins of Iron Metabolism', ed. E. B. Brown,
P. Aisen, J. Fielding, and R. R. Crichton, Grune & Stratton, New York, 1977)

releases Fe^{2+} in two stages, with concomitant formation of 2,2'-bipyridyl *N*-oxide.
The first, more rapid, phase is considered to represent interaction of bipyridyl
with the iron–peroxo-complex of the catalytic sites. The second stage is slower
because FeOOH has to migrate back to the catalytic sites and re-form the peroxo-
complex that will react with bipyridyl. Bipyridyl was observed to react still more
slowly with isolated ferritin micelles. The finding of close pairs of Tb^{III} on
apoferritin[86] might be taken as support for the Crichton mechanism. However,
these pairs lie on the inner surface, approximately 30 Å away from the channels,
and the other evidence outlined above does not clearly distinguish between the
two models. The Crichton mechanism involves initially a two-electron transfer

to an O_2 molecule. Subsequent steps are unclear, but the overall equation is given as:

$$2Fe^{II} + O_2 + 3H_2O = 2Fe^{3+}OOH + 4H^+ + \tfrac{1}{2}O_2 \qquad (1)$$

implying that four Fe^{II} are oxidized per O_2 molecule. Recently, Melino *et al.*[95] have measured oxygen consumption during oxidation of Fe^{II} and obtained a $Fe:O_2$ ratio equal to 4 in the presence or absence of apoferritin. No peroxide could be detected by addition of catalase with or without protein, so the authors concluded that there is a concerted oxidation of $4Fe^{2+}$. Although this would seem to fit better with oxidation on the mineral surface, the stoicheiometry appears also to be compatible with the overall reaction of the Crichton mechanism.

Mobilization of Iron in Ferritin. It is well known that ferritin iron can be released by reducing agents such as sodium dithionite or thioglycollate at pH 5 or below, or by cysteine or ascorbic acid. Sirivech, Osaki, and Frieden have shown that reduced riboflavins can mobilize iron from ferritin rapidly at neutral pH, and their isolation of NADH–FMN oxidoreductase from livers of rat and other species provided a possible biological means of releasing ferritin iron through the production of $FMNH_2$. They have recently[96] found that organ iron contents and reductase activities are inter-related in control and iron- and/or riboflavin-deficient rats. The reduced levels of iron content and enzyme acitivity in the latter could be increased by riboflavin and/or iron supplementation. Zaman and Verwilghen,[97] however, found no difference in reductase activities in liver homogenates of normal and riboflavin-deficient rats, but did observe a lower rate of iron mobilization to transferrin from perfused livers from riboflavin-deficient animals. Both groups conclude that the enzyme may release iron from ferritin.

Biosynthesis and Turnover of Ferritin. The stimulatory effect of iron on ferritin biosynthesis is well established.[98] About 80% of ferritin is formed on free polyribosomes[99–101] and the remainder on membrane-bound polysomes. Ferritin in RNA is, however, equally abundant in the two fractions, suggesting that much of the bound ferritin messenger is not translated.[100,101] Munro and co-workers find that iron leads to an increase in the free polysomal mRNA content arising by transfer from mRNA–protein particles in the cytosol,[100,101] thus explaining the stimulation of biosynthesis. It is generally thought that apoferritin subunits, the first product of biosynthesis, assemble into shells and that the completed shells accumulate iron. Such a pathway is consistent with experiments *in vitro*, but it is difficult to prove. Using sensitive radioimmunoassays to detect ferritin and its subunits, Lee and Richter[99,102,103] have followed the incorporation of

[95] G. Melino, S. Stefanini, E. Chiancone, and E. Antonini, *F.E.B.S. Letters*, 1978, **86**, 136.
[96] S. Sirivech, J. Driskell, and E. Frieden, *J. Nutr.*, 1977, **107**, 739.
[97] Z. Zaman and R. L. Verwilghen, *Biochem. Soc. Trans.*, 1977, **5**, 306.
[98] R. Fargard and R. Saddi, *Biochimie*, 1977, **59**, 765.
[99] S. S. C. Lee and G. W. Richter, *J. Biol. Chem.*, 1977, **252**, 2046.
[100] J. Zahringer, B. S. Baliga, R. L. Drake, and H. N. Munro, *Biochim. Biophys Acta*, 1977, **474**, 234.
[101] M. C. Linder, J. Zahringer, B. S. Baliga, R. L. Drake, R. Barres, and H. N. Munro, in ref. 1, 121.
[102] S. S. Lee and G. W. Richter, *J. Biol. Chem.*, 1977, **252**, 2054.
[103] S. S. C. Lee and G. W. Richter, in ref. 1, p. 91.

[^{14}C]leucine into ferritin in rat liver and in rat hepatoma cells in culture. Incorporation of counts was found first in subunits and subsequently in whole ferritin or apoferritin molecules (about 25 minutes were required for assembly). When ^{59}FeCl$_3$ was also added to suspensions of hepatoma cells, the peak of ^{59}Fe uptake into ferritin molecules occurred well before that of ^{14}C,[102] and little ^{59}Fe was found along with subunits. These experiments support the conclusion that iron is incorporated into intact shells, and this is also suggested by experiments with whole animals[77] which show, in addition, that the specific activity of ^{59}Fe per molecule is dependent on molecular iron content. In animal experiments, ferric ammonium citrate or iron–dextran complexes are often used to stimulate ferritin formation, but ferrocene derivatives can also be an efficient source of iron for this purpose.[104] Although iron stimulates ferritin synthesis in most cells, there are exceptions. Ferritin synthesis in leucocytes, polymorphs, and lymphocytes is not induced by iron, and this is true also of lymphocytes in Hodgkins disease,[105] where ferritin levels are high.

Ferritin molecules are probably degraded and converted into haemosiderin within secondary lysosomes. Iancu and colleagues[106,107] have provided dramatic ultrastructural pictures of such conversion in electron microscographs of liver biopsies from patients with thalassaemia major. Iron-rich ferritin molecules seem to be associated with lysosomal membranes in semi-regular arrays, from which they may be translocated into the lysosomes and there degraded. This raises the interesting possibility that iron-rich molecules can be distinguished from iron-poor by their protein shells. Iron-laden lysosomes, which are also found in primary and secondary haemochromatosis, are more fragile than normal lysosomes, and the disruption of these organelles with release of lytic enzymes may be a cause of the tissue damage associated with these diseases.[108] Lysosomal fragility[108] may result from the production of superoxide and hydroxyl radicals in the presence of iron, and the resulting decompartmentalization of lysosomal iron may be a further cause of tissue damage, again due to the production of free radicals.[109] The interesting suggestion has been made that zinc may protect against the damaging effects of decompartmentalized iron by inhibiting the formation of superoxide.[109]

Isoferritins. Structural work on horse spleen ferritin has led to the picture of a molecule containing twenty-four chemically identical and structurally equivalent subunits. Such a picture may not apply, however, to all tissue ferritins. Ferritins from malignant and normal tissues, or from different organs of the same animal, may migrate in an electric field at different rates. The term 'isoferritin' was coined to denote this different electrophoretic behaviour. Amino-acid analyses of ferritins isolated from different organs or tissues seemed to support the possibility that the isoferritins were products of different genes. A 'second order' of isoferritins has come into the picture recently with reports that ferritin separates

[104] H. Kief, R. R. Crichton, H. Bähr, K. Engelbart, and R. Lattrell, in ref. 1, p. 107.
[105] E. J. Sarcione, J. R. Smalley, M. J. Lema, and L. Stutzman, *Int. J. Cancer*, 1977, **20**, 339.
[106] T. C. Iancu and H. B. Neustein, *Brit. J Haematol.*, 1977, **37**, 527.
[107] T. C. Iancu, H. B. Neustein, and B. H. Landing, in ref. 2, p. 293.
[108] T. J. Peters, C. Selden, and C. A. Seymour, in ref. 2, p. 317.
[109] R. L. Willson, in ref. 2, p. 331.

into multiple bands on isoelectric focussing in Ampholine pH gradients. Some tissue ferritins also give two or more subunits, apparently of different sizes, when the disaggregated molecules are examined by acid–urea or SDS-polyacrylamide gradient pore electrophoresis. An attempt to unify these data has been made by Drysdale and co-workers,[84] who have proposed that ferritins are hybrids of two or more subunits, the relative proportions of these subunits varying from tissue to tissue. This notion has also been hotly contested. Research into the nature and significance of isoferritins continues to be actively pursued in several laboratories. One reason for this is that altered ferritins may be produced in several diseases. Moreover, ferritin, which is normally found in low concentrations in serum, may be present at greatly elevated levels in diseases involving iron overload, liver cirrhosis, and malignant growth. The possibility that serum ferritin could be a tumour marker affords it potential clinical importance. Previously, several sizes have been reported for ferritin subunits, thus presenting a confusing picture. This has now been simplified. It has been agreed by several workers that the two smallest 'subunits' are fragments produced by proteolytic digestion of the 19 000 subunit during preparation,[76,104,110] since they can be eliminated by the addition of protease inhibitors,[76,104,110] to the tissue homogenates from which ferritin is isolated. Drysdale now proposes[84] that the heavy (21 000) subunit, which predominates in heart tissue, should be designated H and the light (19 000) subunit, which predominates in liver, should be designated L. This nomenclature will be assumed in this Report, even though some of the authors quoted use 'HL' for the 19 000 unit. The two classes of subunits have been studied in several tissues, including horse liver,[76, 84] rat heart[84] and liver, [84,111] and human heart,[84] by electrophoresis in gels containing sodium dodecyl sulphate (SDS). In some cases[84] a gradient-pore method was used to increase resolving power. Two subunits have been separated by electrophoresis in acid–urea gels in ferritins from human heart[112,113] and other human tissues.[113] The questions still remain of whether the separated subunits are truly of different sizes, and of whether they come from the same or different molecules and the same or different cell types. The most recent version of the Drysdale hybrid scheme for human ferritins is shown in Figure 5.[84] A similar scheme is implied[84] for other ferritins, although in the case of horse ferritins[79, 84] the pI sequence (but not the subunit sizes) is reversed, with heart being more basic than spleen or liver ferritins. In human ferritins, the basic natural apoferritin from liver or spleen[84,114] and the acidic ferritin isolated from HeLa and other cancer cells lie at the two extremes.

Recent evidence which has been taken to support the hybrid model includes trends in immunological and anion-exchange reactivity. Thus Hazard *et al.*[113] found that antibodies raised in rabbits to crystalline human liver ferritin (reported as being basic, like natural apoferritin) gave reactivites in the order natural apoferritin > liver ferritin > heart ferritin > HeLa ferritin. Liver ferritin fractions of increasing acidity, separated by anion-exchange chromatography,

[110] D. J. Lavoie, D. M. Marcus, K. Ishikawa, and I. Listowsky, in ref 1, p 71.

[111] T. G. Adelman, J. W. Drysdale, and M. Yokota, in ref. 1, p. 49.

[112] A. Bomford, Y. Lis, I. G. Mcfarlane, and R. Williams, *Biochem. J.*, 1977, **167**, 309.

[113] J. T. Hazard, M. Yokota, P. Arosio, and J. W. Drysdale, *Blood*, 1977, **49**, 139.

[114] J. W. Drysdale, T. G. Adelman, P. Arosio, D. Casareale, P. Fitzpatrick, J. T. Hazard, and M. Yokota. *Semin. Haematol.*, 1977, **14**, 71.

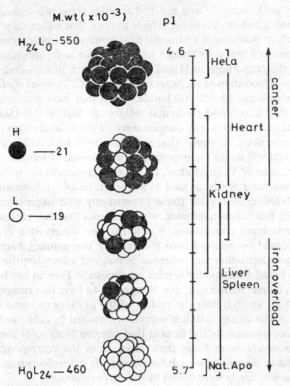

Figure 5 *Model for human isoferritins, showing typical tissue distributions. The size differences in the multimeric structures are exaggerated for emphasis.*
[Reproduced by permission from 'Iron Metabolism', Ciba Foundation Symposium 51 (new series), 1977, Elsevier, Amsterdam]

gave increasing reactivity with an anti-heart-ferritin assay.[115] Horse spleen and liver ferritins have been fractionated, by serial precipitation with increasing concentrations of ammonium sulphate,[79] into components of decreasing pI, suggesting a trend in surface properties, *i.e.* inhomogeneity of molecules within a single tissue. Linking such trends directly with the components obtained by isoelectric focussing, Drysdale proposes that each band represents one of the twenty-five alternative heteropolymers of two subunits.[114] However, it is possible that the discrete bands may be partly artefactual, depending on micro-steps in the pH gradient of the Ampholines.[116] That they arise within a single cell type is shown by the isoelectric focussing patterns obtained from developing red cells.[117]

Formation of hybrid molecules *in vitro* between horse spleen and human liver ferritin subunits and between horse heart and spleen subunits has been reported.[118]

[115] M. Worwood, M. Wagstaff, B. M. Jones, S. Dawkins, and A. Jacobs, in ref. 1, p. 79.
[116] S. M. Russell, P. M. Harrison, and S. Shinjo, *Brit. J. Haematol.*, 1978, **38**, 296.
[117] E. C. Theil and J. E. Brown, in ref. 1, p. 99.
[118] Y. Niitsu, S. Ohtsuka, N. Watanabe, J. Koseki, U. Kohgo, and I. Urushizaki, in ref. 1. p. 65.

On the other hand, human[112] and rat[119,120] ferritins show two separate bands on polyacrylamide gel electrophoresis, which implies that either they are distinct homopolymers or that there is a bimodal distribution of heteropolymers, perhaps among different cell types. In human heart, the two isoferritin bands correlate roughly with the proportions of H and L subunits seen in acid–urea gel electrophoresis.[112] On dissociation of rat heart ferritin by SDS, Vulimiri *et al.*[119] observed the 19 000 subunit and 25—50 000 bands, which they have attributed to aggregates. Since they also found molecular weights as high as 626 000, they have suggested that heart ferritins may contain *more* 19 000 subunits than the usual 24. It does not seem likely, however, that molecules with such different quaternary structures would still retain the properties characteristic of ferritin.

In another paper,[120] Vulimiri *et al.* find sex differences in the relative proportions of the two heart bands, as well as in their rates of synthesis and turnover. Induction of cardiac connective tissue hypertrophy with isoproterenol inverted the ratio of the two bands in females, which suggests that they represent ferritins from independent cell populations.[120] Another new observation is that mouse liver ferritin could be resolved into five narrow iron-staining bands on polyacrylamide gel electrophoresis, whereas horse and other ferritins have only shown single broad bands under similar conditions.[121] Ferritin has been isolated recently from human duodenum, and is reported to have two components, with pI 5.25 and 5.35, which is near the middle of the pI range of human ferritins.[122] The suggestion that cancer cells are necessarily typified by *acidic* isoferritins has been called into question by the finding that ferritins from renal pancreatic and colonic tumours are more *basic* than those from the corresponding normal tissues.[123] Leukaemic cell ferritins have no specific electrofocussing pattern, although very acidic bands were found in the preparations.[124]

The question of whether the H and L subunits represent primary gene products, and whether or not heterogeneity within ferritins is due to H and L hybrid formation or to 24H and 24L molecules in different cell types within a tissue, are difficult to resolve, especially since heart and liver ferritins[76] and their subunits[86] have similar (although not identical) amino-acid compositions and a high proportion of overlapping peptides in their tryptic digests.[111,119] The problem is further complicated by the confirmation of reports that tissue ferritins[110,125] and their isolated subunits[110,125] contain carbohydrate (including amino-sugars, but not sialic acid, and, unusually, glucose), and the amount of carbohydrate seems to be greater in heart ferritin, at least in the horse.[125] The higher proportion of carbohydrate in heart ferritin could be superimposed on genetic differences or could, conceivably, alone account for the electrophoretic differences. Glycosyla-

[119] L. Vulimiri, M. C. Linder, H. N. Munro, and N. Catsimpoolas, *Biochim. Biophys. Acta*, 1977, **491**, 67.

[120] L. Vulimiri, M. C. Linder, and H. N. Munro, *Biochim. Biophys. Acta*, 1977, **497**, 280.

[121] W. H. Massover, *Biochim. Biophys. Acta*, 1978, **532**, 202.

[122] J. W. Halliday, U. Mack, and L. W. Powell, in ref. 1, p. 57.

[123] L. W. Powell, J. W. Halliday, L. B. McKeering, and R. Tweedale, in ref. 1, p. 61.

[124] S. J. Cragg, A. Jacobs, D. H. Parry, M. Wagstaff, and M. Worwood, *Brit. J. Cancer*, 1977, **35**, 635.

[125] M. A. Cynkin and M. Knowlton, in ref. 1, p. 115.

tion of ferritin may represent a means by which ferritin molecules are recognized and processed *in vivo*. Thus iron-rich ferritin molecules may be selectively removed from the cell sap by secondary lysosomes and iron-poor may be selectively released into the serum.[114]

The isolation of 'isoferritins' raises the question as to whether they, like the isoenzymes, are molecules with different functions. The most obvious functional difference to look for is the ability to store iron.[77, 79, 84, 114, 123] Differences in turnover rates of the heart isoferritins have been mentioned.[120] Comparing tissues, there seems to be a rough correlation between the presence of L-type ferritin and high iron content, and it has been suggested that the more basic (L-type) human ferritins have primarily a storage function, whereas the acidic ferritins are secretory.[123] However, there are several observations which do not fit this simple picture. First, Bomford et al.,[112] who examined ferritins from normal and iron-loaded human hearts, found no clear correlation between iron-loading and the proportions of acidic and basic ferritins and their subunits. Secondly, the major component of normal human serum is basic, resembling (and possibly identified with) the most basic component of human liver ferritin, *i.e.* natural apoferritin. Thirdly, there seems to be no evidence that either H or L is synthesized preferentially on bound polysomes for export.[100, 111] Fourthly, although Harrison et al.[79] have found a relationship between surface charge and iron content in horse spleen and liver ferritin fractions and between both of these and the rates of iron uptake *in vitro* by the derived apoferritins, it was the most basic proteins (which in horse should have increasing H content, on the Drysdale pattern) which had the greatest ability to sequester iron. With human ferritins, a similar pattern seems to be emerging:[79] the more basic (L-type) liver ferritin fractions accumulate iron more slowly. Harrison et al.[79] have suggested that the iron-poor molecules could have fewer nucleation centres for iron addition, or, perhaps, channels which are blocked. This could result either from differences in subunit primary structure or from post-transcriptional modifications, which either inhibit or promote iron uptake.

Adelman et al.[111] have found that iron loading in rat liver produces a preferential increase in L subunits. These authors consider that either H and L subunits may represent primary gene products or L is a proteolytic scission product of the larger H subunit. Halliday et al.[122] found that iron-deficient human subjects receiving oral iron produced two *basic* isoferritin components in the duodenal mucosa in addition to the more acidic ones found in normal or iron-loaded subjects.

It is difficult to unify all these observations in a simple scheme. Iron apparently preferentially stimulates L, and yet pure L molecules (natural apoferritin) are least able to accumulate iron. This suggests that L molecules may undergo a conversion into a more active form, which may *parallel* changes in charge or changes in subunit composition (*e.g.* by addition of saccharide chains). Alternatively, L and H may form homopolymers that are segregated in different cell types, each being modified independently. One modification which is found (*in vitro*) to alter charge is reduction/oxidation,[116] but human and horse ferritins are altered in the same direction, and therefore in the opposite sense with respect to iron content.

Serum Ferritin. Serum ferritin has come into prominence with the development of sensitive immunological assays which are able to detect it in low concentration. Serum ferritin levels have been estimated in first-time blood donors, average values being 127 μg l^{-1} in males and 46 μg l^{-1} in females,[126] in general agreement with previously published values. These sex differences probably mainly reflect differences in iron status, high iron-loading in most cases being associated with high levels[123,127-131] and iron deficiency being reflected in low levels of serum ferritin.[130,132-135] A significant correlation of serum ferritin concentration with iron stores has also been found in rats,[136] although the correlation was not good enough to be of predictive value in individual animals. In humans, while a low level of serum ferritin is probably a good indicator of iron deficiency, high levels do not necessarily imply overload. Again, there is too much variation between subjects for serum ferritin levels to be used to predict individual iron-storage levels, but there is a significant correlation within a *single* individual undergoing phlebotomy.[137,138] However, the injection of iron leads to a temporary rise in serum ferritin, even in a subject with low stores.[137]

The level of maternal serum ferritin can be a useful index for deciding if iron therapy is necessary in pregnancy,[133] but it is of interest that levels in cord blood do not reflect the iron status of the mother.[134] High serum ferritin levels are found in many forms of cancer[139] and as a result of fever.[140]

Serum ferritin may be measured by immunoradiometric methods, in which the antibodies are radioactively labelled,[115,129,141] by radioimmunoassay, in which the ferritin used as antigen is tagged,[113,137,142] or by enzyme immunoassays.[143] A new variant of the radioimmunoassays has been reported.[142] One of the problems with either assay is that it involves antibodies raised to a tissue ferritin (usually from liver or spleen). Since tissue ferritins vary in their isoferritin composition, a true estimate may not be obtained if the serum ferritin differs from the original antigen. This has been shown by Hazard and Drysdale[110,113] and by Worwood *et al.*[115] Cancers often, but not always, produce acidic ferritins in serum.[110,115,123]

126 C. A. Finch, J. D. Cook, R. F. Labbe, and M. Culala, *Blood*, 1977, **50**, 441.
127 P. Arosio, M. Yokota, and J. W. Drysdale, *Brit. J. Haematol.*, 1977, **36**, 199.
128 J. W. Halliday, J. L. Cowlishaw, A. M. Russo, and L. W. Powell, *Lancet*, 1977, **ii**, 621.
129 J. W. Halliday, *Brit. J. Haematol.*, 1977, **36**, 395.
130 A. Jacobs, *Fed. Proc.*, 1977, **36**, 2024.
131 C. B. Seelig and D. C. Case, *J. Maine. Med. Assoc.*, 1977, **68**, 364.
132 M. A. Koerper and P. R. Dallman, *J. Pediatr. (St. Louis)*, 1977, **91**, 870.
133 H. G. Van Eijk, M. J. Kroos and G. A. Hoogendoorn, *J. Clin. Chem. Clin. Biochem.*, 1977, **15**, 152.
134 H. G. Van Eijk, M. J. Kroos, G. A. Hoogendoorn, and H. C. S. Wallenburg, *Clin. Chim. Acta*, 1978, **83**, 81.
135 J. R. Walsh and M. Fredrickson, *Amer. J. Med. Sci.*, 1977, **273**, 293.
136 C. Ward, P. Saltman, L. Ripley, R. Ostrup, J. Hegenauer, L. Hatlen, and J. Christopher *Amer. J. Clin. Med.*, 1977, **30**, 1054.
137 G. Birgegard, C. Hogman. A. Killander, H. Lavender, B. Simonsson, and L. Wide. *Scand. J. Haematol.*, 1977, **19**, 327.
138 R. W. Charlton, D. Derman, B. Skikne, S. R. Lynch, M. H. Sayers. J. D. Torrance, and T. H. Bothwell. in ref. 1, p. 387.
139 J. T. Hazard and J. W. Drysdale, *Nature*, 1977, **265**, 755.
140 R. J. Elin, S. M. Wolff, and C. A. Finch, *Blood*, 1977, **49**, 147.
141 D. Lipschitz and J. Cook, in ref. 1, p. 433.
142 D. J. Goldie and M. J. Thomas, *Ann. Clin. Biochem.*, 1978, **15**, 102.
143 L. Theriault and M. Page, *Clin. Chem.*, 1977, **23**, 2142.

Another problem with serum ferritin, relevant to its diagnostic value, is the question of whether it is specifically secreted or whether it reaches the plasma as a result of leakage from damaged cells, or both. Methods for its purification are being developed[115,127,144] which may enable it to be better characterized under different conditions. One of these[144] used immuno-adsorption, so it is likely to be selective for ferritin of the type used to make the anti-ferritin antibodies of the immuno-adsorbant, but it is relatively simple, eliminates the usual heating steps, and makes use of a recently developed electrophoretic desorption method. Serum ferritin has previously been shown to contain carbohydrate, but this seems to be present only in the more acidic components, which bind to columns of concanavalin-A-linked Sepharose.[115] This suggests that ferritin in serum does not result simply from cell damage.[115] The most basic component of serum ferritin, which has a low iron content and which does not bind to the concanavalin A–Sepharose column, may correspond to natural apoferritin.[115,127] It is curious that the major component of serum ferritin in patients with iron overload, whose tissue ferritins are iron-rich, is a molecule of low iron content. On therapy by phlebotomy, patients with haemochromatosis or haemolytic anaemia, with gross secondary iron overload, were found to have greater amounts of acidic components in their serum ferritin.[129] There was no indication, in this study, of the carbohydrate content of these components.

Absorption of Iron. By absorbed iron we mean that fraction of dietary iron which is retained by the body. The process of iron absorption involves its transfer from the gut lumen across the brush borders of the intestinal mucosal cells, through these cells, and across the serosal membrane into the portal blood. An understanding of its regulation is important because iron is unique among the trace metals in that, in the absence of an excretory mechanism, body iron balance must be maintained by keeping out un-needed iron. It is well known that iron-deficient subjects absorb a relatively high proportion of a test dose of iron and that high rates of production of red cells are also associated with high levels of absorption, but the mediators of these homeostatic mechanisms are uncertain. We can state at the outset that the mechanism and control of iron absorption are still largely unsolved problems, and not ones which have attracted most attention in iron metabolism in the past year.

The general feature of iron absorption,[145] its measurement,[146] and its regulation,[147–149] have been the subject of several reviews. The duodenum has been confirmed as the primary site of absorption in rats[150] and several papers describing factors influencing absorption have been published. These factors include nutritional iron bioavailability,[149,151] dietary fat,[152] vitamin C and the common cold,[153]

144 P. J. Brown, M. J. Leyland, and J. P. Keenan, *F.E.B.S. Letters*, 1977, **83**, 256.
145 G. Becker, *Deutsche Apotheker Zeitung*, 1977, **117**, 1651.
146 J. D. Cook and D. A. Lipschitz, *Clin. Haematol.*, 1977, **6**, 567.
147 H. Huebers and W. Rummel, in 'Intestinal Permeation', ed. M. Kramer and F. Lauterback. Elsevier, Excerpta Medica, Amsterdam, 1977, p. 377.
148 M. C. Linder and H. N. Munro, *Fed. Proc.*, 1977, **36**, 2017.
149 H. N. Munro, *Fed. Proc.*, 1977, **36**, 2015.
150 M. S. Ansari, W. J. Miller and M. W. Neathery, *Nutr. Rep. Int.*, 1977, **15**, 37.
151 A. W. Mahoney, *Nutr. Metab.*, 1977, **20**, 222.
152 J. Bowering, G. A. Masch, and A. R. Lewis, *J. Nutr.*, 1977, **107**, 1687
153 J. D. Cook, *Amer. J. Clin. Nutr.*, 1977, **30**, 235.

pregnancy,[154] and subcutaneous infusion of the chelate desferrioxamine in subjects with iron-loading anaemias.[155,156] Here, only papers relating to the mechanism of absorption and its regulation will be described. The work in this area has principally been directed to the search for carrier molecules in membrane or cytosol fractions of mucosal homogenates. Mitochondria play a significant role in iron metabolism within the epithelial cell, but do not participate in iron transport.[157]

Iron-binding Proteins in Cytosol and Membrane Fractions of Mucosal Cells.
The particle-free supernatant fraction of rat and guinea-pig mucosal cell homogenates has previously been shown to contain two proteins, one whose behaviour on gel filtration resembles that of the iron-storage protein ferritin, from liver or spleen, and the other which resembles transferrin. The membrane fraction has received relatively little attention until recently. Furugouri[158] has analysed the iron-binding substances in the intestinal mucose of neonatal piglets after the injection of ^{59}Fe-labelled ferric citrate into the ligated duodenum. About two-thirds of the radioactivity was associated with the membrane fraction and one-third with the supernatant fraction. The supernatant was separated by gel filtration into three radio-iron fractions, probably ferritin, transferrin, and a substance of low molecular weight. The particle fraction also showed evidence for the presence of a ferritin-like component.

The soluble and particulate fractions of intestinal mucosa from rats given ^{59}Fe-labelled FeSO$_4$ by stomach tube have been analysed by Yoshino *et al.*[159] Ferritin and larger components were detected in both fractions. Kaufman *et al.*[160] report studies on the synthesis of microsomal proteins of rat intestinal mucosa. The uptake of [^{3}H]leucine or ^{59}Fe into three microsomal fractions during absorption was examined. Two proteins with molecular weights close to those expected for ferritin and transferrin were found in the ribosomal, but not in the membrane, subfraction. Both could be labelled with ^{59}Fe and both showed increased incorporation of [^{3}H]leucine during absorption. It has been suggested that both proteins are required to handle the absorbed iron. The concentration of a substance of low molecular weight, which did not bind iron, also increased during absorption. Three chromatographic fractions, probably ferritin, transferrin, and compounds of low molecular weight, have also been obtained from particle-free supernatant in mice with X-linked amaemia.[161] The special interest of this report is that the genetic defect linked with transfer of iron is not due to abnormalities in the iron-binding proteins of the supernatant fraction. Edwards *et al.*[162] have suggested that this defect is likely to be due to a disorder of intramucosal iron transport.

[154] R. G. Batey and J. D. Gallagher, *Gastroenterology*, 1977, **72**, 255.
[155] M. J. Pippard, G. T. Warner, S. T. Callender, and D. J. Weatherall, *Lancet*, 1977, **ii**, 737.
[156] R. Batey, *Lancet*, 1977, **ii**, 930.
[157] J. Humphrys, B. Walpole, and M. Worwood, *Brit. J. Haematol*, 1977, **36**, 209.
[158] K. Furugouri, *J. Nutr.*, 1977, **197**, 487.
[159] Y. Yoshino, S. Yamakawa, and Y. Hirai, in ref. 1, p. 397.
[160] N. Kaufman, J. C. Wyllie, and M. Newkirk, *Biochim. Biophys. Acta*, 1977, **497**, 719.
[161] L. S. Valberg, J. Sorbic, and J. Ludwig, *Brit. J. Haematol.*, 1977, **35**, 321.
[162] J. A. Edwards, J. E. Heok, and M. Mattioli, *J. Lab. Clin. Med.*, 1977, **90**, 68.

Previous studies of Rummel and co-workers had implicated mucosal transferrin in the rapid transfer of iron through the mucosa in iron deficiency. They have now followed the uptake of ^{59}Fe from a test injection into transferrin and ferritin in supernatant and SDS-treated membrane fractions in rats during the development of iron deficiency and after rehabilitation.[163] They found that increased ^{59}Fe absorption in progressive iron deficiency was associated with increased uptake into transferrin in both cytosol and membrane fractions, whereas uptake into ferritin decreased. These authors and Furogouri[158] emphasize the importance of both proteins in the regulation of iron absorption. Transferrin seems to have an important role in iron transfer during periods of high absorption. Ferritin probably prevents the transfer of excessive iron into the blood by trapping it in the mucosa.

Pollock and Lasky, who had previously reported an iron-binding protein in guinea-pig mucosa which they judged not to be transferrin, have now[164] reported the purification of a similar protein from several other guinea-pig tissues.

Linder and Munro,[148] who earlier had observed no transferrin in mucosa, have emphasized the importance of low-molecular-weight carriers and drawn attention to the possible regulatory role of serum ferritin in absorption, since its concentration broadly correlates inversely with absorption.[138,165] Cox and Peters[166] have studied iron uptake into duodenal biopsy specimens in patients with iron deficiency anaemia and primary haemochromatosis. Their evidence suggests that iron overload in primary haemochromatosis is due to enhanced intestinal uptake, possibly due to the presence of a carrier with a primary structure abnormality that gives it an increased affinity for inorganic iron. Iron uptake by duodenal mucosa *in vitro* was higher in iron-deficient patients than in controls, but in specimens taken after treatment with iron, uptake was normal, although the total iron concentration in the mucosa remained low.

Other Factors influencing Absorption of Iron. Hegenauer *et al.*[167] have recently proposed that haem synthesis in duodenal cells may participate in the regulation of iron absorption. They have found that the concentration of cytochrome *P*-450 (measured by its e.p.r. signal) in the villous tips of upper duodenal cells of mice increases with increasing dietary iron, whereas absorption decreases. Absorption of cobalt shares a common pathway with that of iron, although it has no storage mechanism and is excreted in the urine. Cobalt, like iron, forms complexes with protoporphyrin-IX. The authors find that turpentine induces inflammation, which reduces plasma iron and plasma iron turnover, leads to an increased absorption of iron and cobalt, and decreases duodenal levels of cytochrome *P*-450. They suggest that diversion of iron towards haem synthesis may deplete the mucosa of transportable iron, but it is not clear whether sufficient quantities could be involved for this to be a significant factor.

[163] F. A. El-Shobaki and W. Rummel, *Res. Exp. Med.*, 1977, **171**, 243.
[164] S. Pollack and F. D. Lasky, in ref. 1, p. 393.
[165] H. C. Heinrich, J. Bruggemann, E. E. Gabbe, M. Gläser, F. Icagic, and E. Pape, *Z. Naturforsch.*, 1977, **32c**, 1023.
[166] J. M. Cox and T. J. Peters, *Lancet*, 1978, **i**, 123.
[167] J. Hegenauer, L. Ripley, and P. Saltman, in ref. 1, p. 403.

Several other metals may be absorbed with iron. Thus, under conditions of increased iron absorption through depletion of iron stores, cobalt,[167] plutonium,[168] lead,[169] and cadmium[169] were found to be absorbed in significantly greater amounts. Conversely, deprivation of nickel impairs intestinal iron absorption, and causes anaemia.[7]

Regulation of the Metabolism of Iron. The major demand for iron is for the synthesis of haemoglobin, and transferrin delivers 30 mg of iron daily to the erythropoietic bone marrow for this purpose. Mature erythrocytes circulate in the blood and at the end of their life-span are engulfed by macrophages and broken down in the reticuloendothelial cells of the spleen, liver, and bone marrow. The iron is then released again into circulation, for binding to transferrin and return to the bone marrow, so that the cycle may begin anew. Neither the erythrocytes nor the reticuloendothelial cells acquire significant amounts of iron from transferrin directly. Within the erythroid and reticuloendothelial cells, and, indeed, within intestinal mucosal cells and hepatocytes, there is growing evidence for the existence of a mobile pool of iron compounds, which may represent iron in transit for its ultimate destination within or outside the cell.[170] This iron pool is probably in equilibrium with storage iron, principally in ferritin and its breakdown product haemosiderin. It probably consists of low-molecular-weight chelate complexes, possibly including small protein molecules, it being exceedingly unlikely that free iron (especially free Fe^{3+}) is present in the cytoplasm.[170,171] This 'chelatable iron pool' probably represents the intracellular form of iron which is most readily released to chelators such as citrate[63] or 2,3-dihydroxybenzoic acid.[172] The latter may rival desferrioxamine as a drug for the treatment of iron overload.[172,173] Fielding and Speyer[174] consider that ferritin itself may play a major role as intermediate in the transport of iron between transferrin and the site of haem synthesis in the mitochondria in developing red cells, and recent work by Nunez *et al.* seems to support this conclusion.[175]

Ponka *et al.*[71,176] have considered ways in which the release of iron from transferrin may be co-ordinated with haem synthesis. Their experiments suggest that haem inhibits release of iron from transferrin, and, conversely, that inhibition of haem synthesis is accompanied by enhanced iron uptake. They consider that transferrin penetrates the reticulocyte membrane (possibly still remaining attached to its receptor) and directly releases its iron to the mitochondria, the rate of release being controlled by the level of 'uncommitted haem'. When haem synthesis is inhibited, some of the unused iron leaves the mitochondria, enlarging the cytoplasmic iron pool and entering ferritin. In contrast, Flatmark[177] has found no evidence for the export of iron from mitochondria, the iron, once accumulated,

[168] H. A. Ragan, E. R. D A. Symposium Series 1977, No. 42, 'Biological Implications of Metals in the Environment', Proc. Annu. Hanford Life Sci. Symp. 15th.
[169] H. A. Ragan, *J. Lab. Clin. Med.*, 1977, **90**, 700.
[170] A. Jacobs, in ref. 2, p. 125.
[171] T. Flatmark and I. Romslo, *Adv. Chem. Ser.*, 1977, **162**, 78.
[172] A. Jacobs, *Biochem. Biophys. Res. Comm.*, 1977, **74**, 1626.
[173] A. Cerami, R. W. Grady, C. M. Peterson, R. L. Jones, and J. H. Graziano, in ref. 1, p. 423.
[174] J. Fielding and B. E. Speyer, in ref. 1, p. 311.
[175] M. T. Nunez, J. Glass, and S. H. Robinson, *Biochim. Biophys. Acta*, 1978, **509**, 170
[176] P. Pŏnka, J. Neuwirt, J. Borovà, and O. Fuchs, in ref. 1, p. 319.
[177] T. Flatmark and A. Tangeras, in ref. 1, p. 349.

being retained within them. Flatmark and colleagues[171,177] have studied the transport of iron into mitochondria. They find that there are two mechanisms, one of which requires energy, and, if iron is supplied as Fe^{III}, reducing equivalents of the respiratory chains are used as well. Thus iron crosses the mitochondrial inner membrane only as Fe^{II}. Mitochondrial iron is divided equally among three types, *viz.* haem, Fe–S centres, and a 'non-haem non Fe–S pool'. The latter is a mitochondrial store which can be drawn on for haem synthesis.[177] Ferritin can act as an iron donor to mitochondria, provided that FMN and succinate are present,[2] and, although transferrin can also act as a source of iron *in vitro*, it is a poor donor, and it is not considered to be the primary source *in vivo*.[171] A study of mitochondrial iron uptake and haem synthesis when there is copper deficiency[178] suggests that cytochrome oxidase may play a direct role in reducing Fe^{III} at the inner mitochondrial membrane.

Reduction to Fe^{II} may also facilitate the passage of iron across the plasma membrane both into[61] and out of[63] hepatocytes and its release from reticuloendothelial and mucosal cells to the serum. There may be another link here between iron and copper metabolism. Thus Frieden has suggested that caeruloplasmin (see below) may catalyse the oxidation of Fe^{II} that has been released into the serum for binding to transferrin. The need for such a ferroxidase activity has, however, been questioned.[179,180]

In pregnancy, another important destination for transferrin iron is the placenta, so that the needs of the developing foetus can be met. The amount of iron crossing to the foetus is determined by a complex interplay of competing factors,[181] in which the foetus does not seem to have prior claim.[182]

The iron compartment which fluctuates most readily with changes in iron status is the stores. Iron is drawn out of the stores in iron deficiency to maintain the iron supply to the bone marrow. This is helped by an enhanced *de novo* synthesis of apotransferrin in the liver,[183] providing transferrin that has a low level of saturation, which thus also has sites available to pick up incoming iron at the mucosa. However, the saturation level of the transferrin may not be directly responsible for controlling release of iron from reticuloendothelial (RE) cells.[184] Factors regulating iron release from hepatocytes and RE cells seem to be different. Thus, although RE cells account for only 5% of the ferritin iron in rat liver, iron is released much more rapidly from these cells than from hepatocytes.[184] Iron release from spleen seems to be dependent on ascorbic acid, and the FMN–NADH-dependent ferrireductase may be absent from this organ. Release of catabolized iron from the RE system occurs in two stages: a fast one, with a half-time of 24 min, and a slow one, representing release from ferritin and haemosiderin, with a half-time of seven days. During inflammation, the fast release is impaired, which may arise from a non-specific stimulation of ferritin synthesis and

[178] D. M. Williams, A. J. Barbuto, C. L. Atkin, and G. R. Lee, in ref. 1, p 341.
[179] P Aisen and E. B. Brown, *Semin. Haematol.*, 1977, **14**, 31.
[180] M. A. Foster, T. Pocklington, and J. R. Mallard, *Phys. Med. Biol.*, 1977, **22**, 531.
[181] Y. Maloth and R. Zaizov, *Biol. Neonate*, 1977, **32**, 43.
[182] V. Fenton, I. Cavill, and J. Fisher, *Brit. J. Haematol.*, 1977, **37**, 145.
[183] A. G. Morton and A. S. Tavill, *Brit. J. Haematol.*, 1977, **36**, 383.
[184] C. Hershko, in 'Progress in Hematology', ed. E. B. Brown, Grune & Stratton, New York, 1977, p. 105.

the consequent diversion of iron from the rapidly exchangeable pool.[185,186] Lactoferrin may play a role in the transfer into the ferritin pool of iron released after erythrocyte catabolism.[62] Copper intoxication interferes with the slow phase of release from ferritin.[187,188] Release of RE iron also seems to show a circadian variation.[189] Like ferritin, the second storage form of iron, haemosiderin, can also provide iron for haemoglobin synthesis.[190] Haemosiderin, which, despite its name, contains no haem,[191] is considered to be mainly a product of ferritin breakdown, occurring only within secondary lysosomes.[192-194] In iron overload, the level of haemosiderin increases relative to that of ferritin, which seems to be maintained at a constant level in the cell sap. Electron micrographs[193,194] of liver biopsies of patients with thalassaemia major show greater numbers of lysosomes containing ferritin and haemosiderin, surrounded by membranes in which arrays of iron-loaded ferritin molecules can be seen, these apparently being in transit into the lysosomes. The sequestration of iron into these bodies thus provides a second defence mechanism against iron overload.

Siderophores.—The insolubility of ferric iron presents a problem to microbial species growing aerobically. To enable them to acquire the iron they need, many of them have been found to synthesize and excrete low-molecular-weight chelate molecules to act as scavengers. These usually have either hydroxamate or pheno-late groups as ligands, the fungal ferrichromes and the enterobactins (enteroche-lins) of enteric bacteria, respectively, being prototypes of these two classes.[195-197]

Like transferrin, the siderophores have a transport role, a high affinity for Fe^{III}, and a much lower affinity for Fe^{II}. The high-spin Fe^{III} is kinetically labile.[198] Biosynthesis of the siderophore is repressed by iron. Membrane receptors have been developed for the assimilation of the ferric siderophores. *E. coli* has receptors not only for enterobactins but also for ferrichrome, which it does not synthesize, and it is curious that these outer membrane receptors are also used by specific phages and colicins.[196,197] The siderophores have recently been fully reviewed by Neilands[195-197] and by Raymond,[198] and they will receive only brief mention here.

Exochelins have recently been isolated from *Mycobacterium smegmatis*.[199] They can obtain iron from both ferric phosphate and ferritin, from the latter with difficulty.[199] Work on enterochelin biosynthesis confirms that it is co-ordinated to the synthesis of three outer-membrane polypeptides and to iron uptake during

[185] A. M. Konijn and C. Hershko, *Brit. J. Haematol.*, 1977, **37**, 7.
[186] C. Hershko and A. M. Konijn, in ref. 1, p. 417.
[187] E. C. Theil and K. T. Calvert, *Biochem. J.*, 1977, **170**, 137.
[188] E. C. Theil, *Fed. Proc.*, 1977, **36**, 2251.
[189] Y. Najean, H. Castro-Malaspina, and P. Colonna, *Eur. J. Nucl. Med.*, 1977, **2**, 189.
[190] F. M. Zuyderhoudt, *Q. J. Exp. Physiol.*, 1977, **62**, 65.
[191] F. X. R. Van Leeuwen, F. J. M. Zuyderhoudt, B. F. Van Gelder, and J. Van Gool, *Biochim. Biophys. Acta*, 1977, **500**, 304.
[192] B. F. Trump and I. K. Berezesky, in ref. 1, p. 359.
[193] T. C. Iancu, H. B. Neustein, and B. H. Landing, in ref. 2, p. 293.
[194] T. C. Iancu and H. B. Neustein, *Brit. J. Haematol.*, 1977, **37**, 527.
[195] J. B. Neilands, *Adv. Chem. Ser.*, 1977, **162**, 3.
[196] J. B. Neilands, in ref. 2, p. 107.
[197] J. B. Neilands and R. R. Wayne, in ref. 1, p. 365.
[198] K. N. Raymond, *Adv. Chem. Ser.*, 1977, **162**, 33.
[199] L. P. Macham, M. C. Stephenson, and C. Ratledge, *J. Gen. Microbiol.*, 1977, **101**, 41.

the growth cycle of *E. coli* k-12.[200] Several further studies relating to membrane receptors have been made on *E. coli*[201-204] and on *S. typhimurium*.[204] A mutant with normal capacity for enterochelin-mediated iron transport, but lacking the colicin Ia receptor, has been reported.[201] The synthesis of ferrichrome sideramines in fungi of the genus *Aspergillus* has been studied with the aid of labelled glycine and serine, which are incorporated into the ferrichrome.[205] Release of iron from sideramines in the fungus *Neurospora crassa* may involve reduction to Fe^{II} with the aid of an NADH-linked reductase.[206] A protein which may be involved in iron transport has been found in cultures of *E. coli*.[207] An iron-binding species involved in transport or storage has been identified in membranes of blue-green algae by Mössbauer spectroscopy.[208]

An ionophore, A23187, previously shown to carry Ca^{II}, has been found to be capable of transferring Fe^{II} but not Fe^{III} across red-cell and synthetic phospholipid membranes.[209]

3 Copper and Zinc

Copper and zinc are both essential for most forms of life. In humans, zinc is quantitatively more important, there being 1—2 g of this element compared with about 0.1 g of copper in the adult human body. Because of similarities in their chemistry, Zn and Cu share some common ligands, and Zn and Cd may antagonize the absorption of Cu by competing for the same binding sites.

Transport of Copper in Serum.—95% of the circulating copper in the blood of normal mammals is present as a blue protein called caeruloplasmin, first isolated by Holberg. It is synthesized in the liver and contains 6 (or 7) Cu atoms per molecule (mol. wt. 132 000). Evidence suggests that caeruloplasmin acts as a transport protein and that its copper is incorporated into cytochrome oxidase and other enzymes. It has also been shown to have oxidase activity. Its ferrioxidase activity may be a step in the release of iron from the tissues for binding to serum transferrin, although this is controversial (see above). Some Cu is also transported by albumin, and this is the principal carrier in the portal blood between the intestine and the liver. Amino-acid–Cu complexes may also play a role in Cu transport.

Caeruloplasmin. The structure of caeruloplasmin and its donation of copper to the tissues have been much less intensively studied than the corresponding features of transferrin. Its copper atoms have previously been classified, in terms of their paramagnetic and spectroscopic properties, into three types. Only about 44% of the Cu is paramagnetic. This Cu probably consists of two type I Cu^{2+} atoms and one type II. Type I Cu^{2+} is characterized by its intense blue colour and the absorption band in the region of 610 nm, Type II Cu^{2+} is colourless and is similar to that

[200] M. A. McIntosh and C. F. Earhart, *J. Bacteriol.*, 1977, **131**, 331.
[201] S. Soucek and J. Konisky, *J. Bacteriol.*, 1977, **130**, 1399.
[202] S. Ichihara and S. Mizushima, *J. Biochem. (Japan)*, 1977, **81**, 749.
[203] A. P. Pugsley and P. Reeves, *Biochem. Biophys. Res. Comm.*, 1977, **74**, 903.
[204] M. Luckey, *Diss. Abs. Internat. (B)*, 1977, **38**, 655.
[205] H. G. Muller and H. Diekmann, *Arch. Microbiol.*, 1977, **113**, 243.
[206] J. F. Ernst and G. Winkelmann, *Biochim. Biophys. Acta*, 1977, **500**, 27.
[207] A. Boyd and I. B. Holland, *F.E.B.S. Letters*, 1977, **76**, 20.
[208] E. Evans, N. G. Carr, and J. D. Rush, *Biochem. J.*, 1977, **166**, 547.
[209] S. P. Young and B. D. Gomperts, *Biochim. Biophys. Acta*, 1977, **469**, 281

present in other blue multi-copper oxidases. Type III Cu is diamagnetic and has an absorption band at 330 nm. Type III Cu is thought to consist of two antiferro-magnetically coupled Cu^{2+} ions, but this presents a problem with caeruloplasmin, since there appear to be three Cu atoms in this class. The third, non-paramagnetic, Cu may be a separate class of Cu^{I}.

Although the ligands of the various types of Cu present in caeruloplasmin are unknown, studies on model compounds[210-212] and other copper proteins have allowed structures to be postulated. Amundsen *et al.*[213] have provided further evidence that the absorption band of type I Cu at 600 nm is due to $RS^{-}(\sigma) \rightarrow Cu^{II}$ charge-transfer, Cu^{II} being in distorted tetrahedral co-ordination. These authors consider that phenolate is a second ligand, with the $Cu^{II} \rightarrow$ phenolate (π^*) transition accounting for the weaker bands observed in stellacyanin and plasto-cyanin, and that the remaining two ligands are probably histidine. Analysis of the electronic spectra of these proteins has led Gray and co-workers[214] to propose that there is distorted tetrahedral co-ordination involving cysteine S, two histidine N, and one deprotonated-peptide N. Similarly, it has been concluded from e.p.r. data of model compounds containing CuS_4 centres that type I has either CuS_2N_2 or $CuSN_3$ co-ordination.[215] Assignments of resonances in the ^1H and ^{13}C n.m.r. spectra of reduced French Bean plastocyanin have indicated that there are two histidine[216,217] and a cysteine and a deprotonated amide[217] as ligands, with methionine and several aromatic and aliphatic side-chains close to Cu.[216] The structure of plastocyanin at 2.7 Å resolution[218] confirms the approximate tetrahedral geometry of the single blue Cu and shows that the ligands are two histidine N, cysteine S, and methionine S. It is not certain that both type I Cu atoms of caeruloplasmin have this arrangement. Possibly only one imidazole ligand is involved.[219] Differences between the two type I Cu^{II} atoms have been observed in their susceptibility to photochemical reduction by laser radiation at wavelengths in the range 360—514.5 nm,[220] in agreement with differences in redox potential, e.p.r. signal, c.d., and other properties found previously. Herve *et al.*[220] have observed changes in the e.p.r. spectra with increasing irradiation, which has enabled them to assign several hyperfine peaks to the two different type I Cu atoms and others to type II. The blue colour is lost during laser irradiation.

There is very little information about the environment of type II Cu in caeruloplasmin. Copper(II) in galactose oxidase may have distorted five-co-ordinate geometry,[213] but this is not necessarily a good model for type II Cu. Pulsed

[210] H. Yokoi and A. W. Addison, *Inorg. Chem.*, 1977, **16**, 1341.
[211] Y. Sugiura, *J. Amer. Chem. Soc.*, 1977, **99**, 1581.
[212] Y. Sugiura, *European J. Biochem.*, 1977, **78**, 431.
[213] A. R. Amundsen, J. Whelan, and B. Bosnich, *J. Amer. Chem. Soc.*, 1977, **99**, 6730.
[214] H. B. Gray, C. L. Coyle, and D. M. Dooley, *Adv. Chem. Ser.*, 1977, **162**, 145.
[215] U. Sakaguchi and A. W. Addison, *J. Amer. Chem. Soc.*, 1977, **99**, 5189.
[216] H. C. Freeman, V. A. Norris, J. A. M. Ramshaw, and P. E. Wright, 'Proceedings of the 4th International Congress on Photosynthesis, 1977', ed. D. O. Hall, J. Coombs, and T. W. Goodwin, The Biochemical Society, London, 1978, p. 805.
[217] E. L. Ulrich, J. L. Markley, and D. W. Krogmann, in ref. 216, p. 815.
[218] P. M. Colman, H. C. Freeman, J. M. Guss, M. Murata, V. A. Norris, J. A. M. Ramshaw, and M. P. Venkatappa, in ref. 216, p. 810.
[219] B. Mondovi, M. T. Graziani, W. B. Mims, and R. Oltzik, *Biochemistry*, 1977, **16**, 4198.
[220] M. Herve, A. Garnier, L. Tosi, and M. Steinbuch, *Biochem. Biophys. Res. Comm.*, 1978, **80**, 797

e.p.r. studies show that, like type I, imidazole is a ligand for type II Cu in caeruloplasmin.[217] Type II Cu in human caeruloplasmin is not susceptible to reduction by laser radiation.[220]

The 330 nm absorption of type III Cu in fungal laccase is probably due to a $R_2S(\sigma) \rightarrow Cu^{II}$ charge-transfer transition of a square-planar or quasi-square-pyramidal Cu^{II} system, which may be similar to the Cu site in oxyhaemocyanin.[213] The structure proposed for the latter contains two Cu atoms 6 Å apart, bridged through tyrosine oxygen.[213] An alternative model for type III Cu has a pair of Cu atoms, each in distorted tetrahedral co-ordination to two S and two N ligands. The two S atoms are joined to both Cu, with S—Cu—S bond angle 97°, so that the whole arrangement is approximately square-planar, the Cu atoms being 5—6 Å apart.[221] The arrangement could account for the antiferromagnetic coupling, and its geometry, intermediate between that preferred by Cu^I and Cu^{II}, may be relevant to the oxidation–reduction behaviour of this type of Cu.

Very few functional studies on caeruloplasmin have been reported. Linder and Moor[222] have found evidence for caeruloplasmin-like oxidase activity in rat heart and other tissues, and the presence of the protein was confirmed immunologically. [³H]Leucine-labelled caeruloplasmin, after intravenous infusion into a Cu-deficient rat, was found in heart and brain within two hours, thus indicating its uptake into these tissues, where it may donate its Cu to cytochrome oxidase. Its transport function may thus be analogous to transferrin, although it is not known whether the protein is degraded or released intact on delivery of its metal.[222] Several workers[223–225] have used e.p.r. measurements to estimate caeruloplasmin and transferrin levels in blood.

In several pathological states, including myocardial infarction,[223] Hodgkin's disease,[225] and some other malignant conditions,[224,226] caeruloplasmin levels rise, with a concomitant decrease in transferrin in some[223,225] (but not all) cases.[224] In stress situations, the opposite may be found, *i.e.* a decrease in caeruloplasmin accompanied by an increase in plasma iron levels.[227] Thus, although a rise in caeruloplasmin might compensate for transferrin catabolism by restoring the level of serum iron for transferrin binding,[223] the overall pattern does not suggest a requirement for this ferrioxidase activity. Foster *et al.*[180] have observed the e.p.r. signals of transferrin and caeruloplasmin in blood after the addition of Fe^{2+} ions. Their evidence shows that both proteins have ferrioxidase activity and that, since there is much more transferrin than caeruloplasmin, it is not likely that the ferrioxidase activity of the latter would be significant for iron mobilization under normal circumstances.

[221] R. H. Lane, N. S. Pantaleo, J. K. Farr, W. M. Coney, and M. G. Newton, *J. Amer. Chem. Soc.*, 1978, **100**, 1610.

[222] M. C. Linder and J. R. Moor, *Biochim. Biophys. Acta*, 1977, **499**, 329.

[223] M. Bomba, S. Camagna, P. L. Cannistraro, P. Indovina, and P. Samoggia, *Physiol. Chem. Phys.*, 1977, **9**, 175.

[224] T. Pocklington and M. A. Foster, *Brit. J. Cancer*, 1977, **36**, 369.

[225] M. A. Foster, L. Fell, T. Pocklington, F. Akinsete, A. Dawson, J. M. S. Hutchison, and J. R. Mallard, *Clin. Radiol.*, 1977, **28**, 15.

[226] A. Scanni, L. Licciardello, M. Trovato, M. Tomirotti, and M. Biraghi, *Tumori*, 1977, **63** 175.

[227] R. Flos and J. Balasch, *Agressologie*, 1977, **18**, 47.

Serum Albumin. The Cu-transport site of human serum albumin involves the α-amino-group of the N-terminal aspartic acid, two intervening peptide nitrogens, and a histidine nitrogen of the third residue.[228] A model tripeptide has been synthesized, L-aspartyl-L-alanyl-L-histidine N-methylamide, and the formation of complexes of this tripeptide with Cu has been studied by equilibrium dialysis and by the analysis of its visible absorption and c.d. spectra.[228] The structure proposed for this complex has four N ligands in square-planar co-ordination to Cu, the ligands being provided by the N-terminal amino-group, two peptide, and one imidazole N atoms.[228] The structure and biosynthesis of serum albumin have been reviewed.[229]

Transport of Zinc in Serum.—Previous work had shown that most of the Zn in venous blood is bound to albumin and a smaller fraction to α₂-macroglobulin. Falchuk *et al.*[230] have found two Zn-binding fractions; one (approx. mol. wt. 800 000) was identified as α₂-macroglobulin and the other was a heterogeneous fraction of mol. wt. about 100 000. Changes in serum Zn levels in pathological states were associated with the smaller fraction. Evans and co-workers have reported that Zn entering the portal blood from the intestine is bound to transferrin. Phillips[231] has shown that human lymphocytes can take up transferrin-bound Zn and that this uptake is stimulated by compounds known to increase cellular levels of cyclic AMP.

Determination of Cu, Zn, and other Metals in Serum.—Complexing agents for Cu and Zn have been tested for their suitability as analytical reagents. Of these, tetraethylthiuram disulphide and its analogue bis-(1-piperidylthiocarbonyl) disulphide are recommended for Cu and 1-(2-pyridylazo)-2-naphthol and 1-[(5-chloro-2-pyridyl)azo]-2-naphthol for Zn.[232] Neutron activation analysis has been used for the determination of chromium and cobalt in human serum.[233]

Metallothioneins and Related Proteins.—A group of low-molecular-weight (6—10 000) proteins have been isolated from liver and kidney and other tissues which have a high affinity for a number of metals, especially zinc, cadmium, and copper. The proteins (thioneins) have an unusual distribution of amino-acids, with a high content of cysteine (but not cystine) and little or no aromatic residues, resulting in the absence of the usual absorption at 280 nm. The properties and functions of these proteins have recently been reviewed.[234,235] The first member of this class to be identified was the Cd-thionein of equine kidney, by Vallee and colleagues some twenty years ago. The finding that Cd and other non-essential metals, including mercury, bismuth, gold, and silver, will not only bind to thioneins but also induce their synthesis in a number of organs (see Table 1) raises the question of their functions, and suggests that for these metals it is primarily one of

228 K. S. Iyer, S. J. Lau, S. H. Laurie, and B. Sarkar, *Biochem. J.*, 1978, **169**, 61.
229 J. Peters, *Clin. Chem.*, 1977, **23**, 5.
230 K. H. Falchuk, J. M. Mathews, and C. Doloff, *New England J. Med.*, 1977, **296**, 1129.
231 J. L. Phillips, *Cell Immunol.*, 1978, **35**, 318.
232 C. S. Feldkamp, R. Watkins, E. S. Baginski, and B. Zak, *Microchem. J.*, 1977, **22**, 347.
233 J. Versieck, J. Hoste, F. Barbier, H. Steyaert, J. DeRudder, and H. Michels, *Clin. Chem.*, 1978, **24**, 303.
234 Y. Kojima and J. H. R. Kägi, *Trends Biochem.*, 1978, **3**, 90.
235 M. G. Cherian and R. A. Goyer, *Ann. Clin. Lab. Sci.*, 1978, **8**, 91.

Table 1 *Increase in thionein-like proteins in organs to response to various metals* (*From ref.* 235)

Metal	Liver	Kidney	Spleen
Cadmium	+	+	+
Zinc	+	−	−
Mercury	−	+	−
Copper	+	+	−
Bismuth	?	+	−
Gold	?	+	−
Silver	+	+	?
Lead	−	−	−

detoxification. Zinc and copper thioneins may also be important in the storage and distribution of these essential metals.[234,235]

Because there has been considerable variation in the reported molecular weights, metal content, and content of cysteine and other amino-acid residues of these proteins, it is probably best to reserve the name of metallothionein to those proteins which have a mol. wt. of approximately 6800, which contain 33 residues per cent cysteine, and which bind 6—7 gram atoms of metal.[234] The amino-acid sequences of human liver,[236] mouse liver,[237] equine liver,[234] equine kidney,[238] and a second equine kidney[234] metallothionein have been completed, to give a total of five primary structures. These are shown in Figure 6. In addition to a marked degree of conservation, these show a striking distribution of cysteine, with seven Cys-X-Cys sequences. These may correspond to the seven, probably separate,[213] metal-binding sites.[234] All of the 20 cysteinyl side-chains participate in metal binding. The proteins all contain 61 amino-acids, with N-terminal N-acetyl-methionine and frequent juxtaposition of basic and hydroxylic amino-acids to cysteine. These studies also show both polymorphism within a single tissue and variation in metal binding to a single polypeptide sequence.[234] Metallothioneins seem to contain a predominance of Zn and Cd over Cu, which may be bound also to Cu-chelatin and other proteins. Cd-thionein and Cu-thionein, however, share common antigenic determinants,[239] and it is possible that some of the higher molecular weights and the lower cysteine and metal contents reported for some of the Cu-binding proteins may result from oxidative cleavage of Cu–cysteine bonds[240,241] present in Cu-thionein and from aggregation during isolation. Gel filtration in 6M-guanidine hydrochloride or in dilute buffers gave respectively 6000 or 9500 for the mol. wt. of human foetal liver Cu-protein, which may be a metallothionein,[240] but a protein isolated from the liver of Cu-injected rats is reported to have a distinct amino-acid composition, including the presence of aromatic residues, as well as a mol. wt. of 11 000.[242] An 8Cu–2Zn-protein, of mol. wt. 11 500, has been obtained from foetal bovine liver and shown to contain

[236] M. M. Kissling and J. H. R. Kägi, *F.E.B.S. Letters*, 1977, **82**, 247.
[237] I. Y. Huang, A. Yoshida, H. Tsunoo, and H. Nakajima, *J. Biol. Chem.*, 1977, **252**, 8217.
[238] Y. Kojima, C. Berger, B. T. Vallee, and J. H. R. Kägi, *Proc. Nat. Acad. Sci. U.S.A.*, 1976, **73**, 3413.
[239] G. Madapallimatam and J. R. Riordan, *Biochem. Biophys. Res. Comm.*, 1977, **77**, 1286.
[240] L. Ryden and H. F. Deutsch, *J. Biol. Chem.*, 1978, **253**, 519.
[241] H. J. Hartmann and U. Weser, *Biochim. Biophys. Acta*, 1977, **491**, 211.
[242] R. D. Irons and J. C. Smith, *Chem. Biol. Interact.*, 1977, **18**, 83.

Human Liver Metallothionein-2
Equine Liver Metallothionein-1A
Equine Kidney Metallothionein-1A
Equine Kidney Metallothionein-1B
Mouse Liver Metallothionein-1

1

N-Ac-MET-ASP-PRO-ASN-CYS-SER-CYS-ALA-ALA-GLY-ASP-SER-CYS-THR-CYS-ALA-
N-Ac-MET-ASP-PRO-ASN-CYS-SER-CYS-PRO-THR-GLY-GLY-SER-CYS-THR-CYS-ALA-
N-Ac-MET-ASP-PRO-ASN-CYS-SER-CYS-PRO-THR-GLY-GLY-SER-CYS-THR-CYS-ALA-
N-Ac-MET-ASP-PRO-ASN-CYS-SER-CYS-VAL-ALA-GLU-SER-CYS-THR-CYS-ALA-
N-Ac-MET-ASP-PRO-ASN-CYS-SER-CYS-SER-THR-GLY-SER-CYS-THR-CYS-THR-

10

40

20

-GLY-SER-CYS-LYS-CYS-LYS-GLU-CYS-LYS-CYS-THR-SER-CYS-LYS-LYS-SER-CYS-CYS-SER-CYS-CYS-PRO-VAL-GLY-
-GLY-SER-CYS-LYS-CYS-LYS-GLU-CYS-ARG-CYS-THR-SER-CYS-LYS-LYS-SER-CYS-CYS-SER-CYS-CYS-PRO-GLY-GLY-
-GLY-SER-CYS-LYS-CYS-LYS-GLU-CYS-ARG-CYS-ALA-SER-CYS-LYS-LYS-SER-CYS-CYS-SER-CYS-CYS-PRO-GLY-GLY-
-GLY-SER-CYS-LYS-CYS-GLN-CYS-ARG-CYS-ALA-SER-CYS-LYS-LYS-SER-CYS-CYS-SER-CYS-CYS-PRO-VAL-GLY-
-SER-SER-CYS-ALA-CYS-LYS-ASP-CYS-LYS-CYS-THR-SER-CYS-LYS-LYS-SER-CYS-CYS-SER-CYS-CYS-PRO-VAL-GLY-

30

50

60   61

-CYS-ALA-LYS-CYS-ALA-GLN-GLY-CYS-ILE-CYS-LYS-GLY-ALA-SER-ASP-LYS-CYS-SER-CYS-SER-CYS-ALA-OH
-CYS-ALA-ARG-CYS-ALA-GLN-GLY-CYS-VAL-CYS-LYS-GLY-ALA-SER-ASP-LYS-CYS-⟨LEU/SER⟩-ASP-LYS-CYS-CYS-SER-CYS-ALA-OH
-CYS-ALA-ARG-CYS-ALA-GLN-GLY-CYS-VAL-CYS-LYS-GLY-ALA-SER-ASP-LYS-CYS-⟨LEU/SER⟩-SER-CYS-ALA-OH
-CYS-ALA-LYS-CYS-ALA-GLN-GLY-CYS-VAL-CYS-LYS-GLY-ALA-SER-ASP-LYS-CYS-CYS-SER-CYS-ALA-OH
-CYS-SER-CYS-LYS-CYS-ALA-GLN-GLY-CYS-VAL-CYS-LYS-GLY-ALA-ALA-ALA-ASP-LYS-CYS-THR-CYS-CYS-ALA-OH

Figure 6 *Amino-acid sequences of metallothioneins, as shown in refs. 234, 236, 237, and 238.*

exclusively Cu[I].[241] Metal-binding proteins of this general class have been described in the liver of foetal sheep[243] and newborn lambs[244] (which also contained components of higher molecular weight) sheep kidney,[245] white and grey matter of brain,[246] rat kidney,[247] and swine liver.[248] A Cd-binding glycoprotein has been isolated from the liver of plaice.[249] Three Cu-proteins isolated from yeast[250] are clearly different from the thioneins of animal origin. They have eight, two, and one Cu per 10 000 mol. wt. and a constant stoicheiometry of Cu:cysteine of 1:2; application of *X*-ray photoelectron spectrometry gave clear evidence for the presence of Cu[I].[250]

The concentration of metallothionein-like proteins has been measured in tissues of normal,[248,251] Cu-poisoned,[252] and Cd-exposed[235,247] animals. In normal tissues, higher concentrations were found in kidney, intestine, and liver, with rat and pig being particularly rich in these proteins.[251] Copper was always found, along with Zn and Cd, as a major metallic constitutent of rat kidney thioneins, even in Cd-injected animals, and the affinity *in vitro* was also found to be higher for Cu than for Zn or Cd ions.[247]

Metallothionein synthesis has been studied in rat liver[253] and in liver-cell culture.[254] The mechanism of Zn stimulation may involve both changes in the amount of metallothionein m-RNA and derepression of the messenger.[253] A half-life of 20 h has been reported for Zn-thionein in rat liver.[253]

Excretion and Absorption of Cu, Zn, and Cd.—Copper is secreted from the liver into the bile, where it forms both low- and high-molecular-weight complexes. The latter, probably in the form of conjugated bilirubin,[255] is available for re-absorption, and so is lost in the faeces. Zn is lost mainly throught exfoliation of intestinal cells, although a portion of this Zn may be re-absorbed.

Both Cu and Zn, and also Cd, may be bound within intestinal mucosal cells by proteins of the metallothionein and related classes, which may regulate their absorption.[256] A defect in these proteins may be a cause of reduced Cu absorption in Menkes kinky hair syndrome.[257] The effect of Cd on Cu absorption has been studied. In previous work it has been found that high levels of Zn and Cd reduced the amount of Cu bound to low-molecular-weight proteins in the mucosal cytosol and also inhibited its absorption into the body. Recently,[258] at lower levels of

[243] I. Bremner, R. B. Williams, and B. W. Young, *Brit. J. Nutr.*, 1977, **38**, 87.
[244] N. E. Soeli, A. Froeslie, and G. Norheim, *Acta Pharmacol. Toxicol.*, 1977, **40**, 570.
[245] I. Bremner and B. W. Young, *Chem. Biol. Interact.*, 1977, **19**, 13.
[246] S. G. Sharoyan, A. A. Shaldzhian, and R. M. Nalbandyan, *Biochim. Biophys. Acta*, 1977, **493**, 478.
[247] K. T. Suzuki, K. Kubota, and S. Takenaka, *Chem. and Pharm. Bull. (Japan)*, 1977, **25**, 2792.
[248] A. Froslie and G. Norheim, *Acta Vet. Scand.*, 1977, **18**, 471.
[249] J. Overnel, I. A. Davidson, and T. L. Coombs, *Biochem. Soc. Trans.*, 1977, **5**, 267.
[250] U. Weser, H. J. Hartmann, A. Fretzdorff, and G. J. Strobel, *Biochim. Biophys. Acta*, 1977, **495**, 465.
[251] A. J. Zelazowski and J. K. Piotrowski, *Experientia*, 1977, **33**, 1624.
[252] G. S. Norheim and N. E. Soli, *Acta Pharmacol. Toxicol.*, 1977, **40**, 178.
[253] K. S. Squibb, R. J. Cousins, and S. L. Feldman, *Biochem. J.*, 1977, **164**, 223.
[254] M. R. Daniel, M. Webb, and M. Cempel, *Chem. Biol. Interact.*, 1977, **16**, 101.
[255] G. M. McCullars, S. O. O'Reilly, and M. Brennan, *Clin. Chim. Acta*, 1977, **74**, 33.
[256] D. S. Norton and F. W. Heaton, *Biochem. Soc. Trans.*, 1977, **5**, 425.
[257] D. M. Williams, *Pediatr. Res.*, 1977, **11**, 823.
[258] N. T. Davies and J. K. Campbell, *Life Sci.*, 1977, **20**, 955.

dietary Cd supplementation, Davies and Campbell found that Cd increased the amount of ^{64}Cu retained in mucosal cells in a low-molecular-weight (10 000) fraction. However, the latter was inversely related to the level of dietary Cd, and the amount of Cd plus Zn found in this fraction was greater in the Cd-supplemented rats than in the unsupplemented controls. Hence the amount of Cu bound may have been affected by both the levels of these proteins and by competition with other metals. The authors consider, however, that it is premature to assume a role for these proteins in the absorption of trace metals.

Recently, it has been shown that human milk, but not cow's milk, contains Zn-binding ligands, of low molecular weight, which may enhance Zn absorption in the neonate.[259] A similar ligand was found in rat's milk and in rat mucosa 16 days after birth, but not earlier, and this may be a means of facilitating absorption before the development of intestinal mechanisms.[260] It is possible that this ligand is a prostaglandin.[261] Prostaglandin E_2 was found to increase Zn-transport into the rat small intestine, whereas prostaglandin F_2 decreased it, and may facilitate its secretion from the small intestine into the lumen.[261] Inability to synthesise prostaglandin E_2, which is required for absorption of dietary Zn, may be responsible for the inherited disorder acrodermatitis enteropathica,[262] the symptoms of which can be alleviated by feeding with human milk, but not with cow's milk.[259]

Mechanisms regulating Zn and Cu homeostasis have been reviewed.[263]

259 C. D. Eckhert, M. V. Sloan, J. R. Duncan, and L. S. Hurley, *Science*, 1977, **195**, 789.
260 L. S. Hurley, J. R. Duncan, M. V. Sloan, and C. D. Eckhert, *Proc. Nat. Acad. Sci., U.S.A.* 1977, **74**, 3547.
261 M. K. Song and N. F. Adham, *Fed. Proc.*, 1977, **36**, 1138.
262 G. W. Evans and P. E. Johnson, *Amer. J. Clin. Nutr.*, 1977, **30**, 611.
263 E. Weigand and M. Kirchgessner, *Nutr. Metab.*, 1977, **22**, 101.

5
Oxygen-transport Proteins

M. BRUNORI, B. GIARDINA, & J. V. BANNISTER

General Introduction

Research on the structural, functional, physiological, and physio-pathological properties of respiratory proteins continues to attract the attention of numerous investigators. During 1977 and early 1978, approximately 400 papers were collected for the purpose of writing this review article, and several important ones may have escaped our attention.

This review has been written as a compromise between a mere list of recent findings related to the subject, which would have been of limited interest to read and very boring to write, and an essay, in which only the fundamental problems and the principal answers are reported. With this scheme in mind, we had to operate a selection among the papers published during 1977, focussing our attention on those which stressed a molecular approach to the problem of structure–function relationships in respiratory proteins. Moreover we have subdivided the subject into two parts, dealing respectively with haemoproteins and with haemocyanins and haemerythrins.

A number of books and review articles have been written on haemoglobin and haemocyanin, and reference to only some of these is given here. The general book on haemoglobin and myoglobin by Antonini and Brunori[1] and the recent one by Bunn et al.[2] focused on the physio-pathological aspects of haemoglobino-pathies; a basic article by Wyman on linked functions[3] and the more recent lucid review by Baldwin[4] are of value. The proceedings of an EMBO Symposium on haemocyanins,[5] edited by J. V. Bannister, include a great deal of the recent work on this subject. The older literature was comprehensively reviewed by van Holde and van Bruggen[6] and by Lontie and Witters.[7]

It is always difficult to condense into a few general sentences a judgment on the state of knowledge in a particular field, and we shall not try to do so. However,

[1] E. Antonini and M. Brunori, 'Hemoglobin and Myoglobin in their Reactions with Ligands', North-Holland, Amsterdam, 1971.
[2] H. F. Bunn, B. G. Forget, and H. M. Ranney, 'Human Hemoglobins,' W. B. Saunders Company, Philadelphia, 1977.
[3] J. Wyman, Adv. Protein Chem., 1964, 19, 223.
[4] J. M. Baldwin, Progr. Biophys. Mol. Biol., 1975 29, 225.
[5] 'Structure and Function of Haemocyanin', ed. J. V. Bannister, Springer-Verlag, Berlin, Heidelberg, and New York, 1977.
[6] K. E. van Holde and E. F. J. van Bruggen, in 'Biological Macromolecules', ed. S. N. Timasheff and G. D. Fasman, Marcel Dekker, New York, 1971, Vol. 5, p. 1.
[7] R. Lontie and R. Witters, in 'Inorganic Biochemistry', ed. G. L. Eichhorn, Elsevier, Amsterdam, 1973, p. 344.

it may be of interest to notice one aspect which transpired during our work. Haemoglobin still represents a challenging biological problem, and the vaste background knowledge already available makes it a system of choice for the application of new and sophisticated biophysical techniques (such as X-ray absorption spectroscopy) to provide an answer to some fundamental question, still unsettled. While the frontier of our knowledge advances, it becomes clearer that available data can be described by-and-large within the conceptual framework of the allosteric model introduced by Monod, Wyman, and Changeux.[8] Thus, although the oversimplified scheme, as originally presented, is not strictly applicable (and for example the tertiary origin of the Bohr effect or the functional chain heterogeneity demand modification), it is surprising how far and how well a great body of equilibrium and kinetic data can be reconciled with the basic ideas of this model. Moreover, extension to giant respiratory proteins (such as haemocyanins and erythrocruorins) of equilibrium and kinetic investigations indicates that, with appropriate minor modifications, the same basic model can be fruitfully applied to describe the functional behaviour of these huge allosteric systems.

PART I: Haemoproteins

1 Simple Haemoproteins

Structural Information.—X-*Ray Analysis*. The stereochemical mechanism of haem–haem interactions proposed by Perutz[9] stressed the displacement of the iron atom from the porphyrin plane. It was suggested that the displacement in deoxyhaemoglobin is greater than in methaemoglobin. Therefore the determination of the relative position of the iron, the proximal histidine, and the porphyrin plane in Sperm whale met- and deoxy-myoglobin was a focal point in the recent work of Takano.[10,11] His data at 2.0 Å resolution clearly indicate that in metmyoglobin the iron atom is displaced from the average porphyrin plane towards the proximal His-F8 by about 0.4 Å, in agreement with previous estimates. As shown by difference Fourier analysis, this displacement increases from 0.40 to 0.55 Å in going from met- to deoxy-myoglobin.

In deoxymyoglobin (deoxy-Mb) the displacement of iron is associated with a lateral movement of His-F8, as well as movements of helix F and of the FG segment towards the EF corner. In addition, the side-chain of the distal histidine moves away from the surface of the molecule and from the site where a water molecule is attached to the haem in metmyoglobin. In deoxymyoglobin a water molecule, bound to the ε-nitrogen atom of His-E7, is in contact with the porphyrin ring but not with the iron, which therefore is five-co-ordinate.[12] Finally, comparing met-Mb to deoxy-Mb, tertiary structural changes, similar to those observed in the case of haemoglobin (but smaller in magnitude), were observed.

[8] J. Monod, J. Wyman, and J. P. Changeux, *J. Mol. Biol.*, 1965, **12**, 88.
[9] M. F. Perutz, *Nature*, 1970, **228**, 726.
[10] T. Takano, *J. Mol. Biol.*, 1977, **110**, 569.
[11] T. Takano, *J. Mol. Biol.*, 1977, **110**, 537.
[12] C. L. Nobbs, H. C. Watson, and J. C. Kendrew, *Nature*, 1966, **209**, 339.

The structural X-ray analysis of oxymyoglobin has recently[13] been carried out under conditions of sufficient stability (low temperature) that the formation of met-Mb is reduced to a minimum during data collection. Some very important results emerge from these new data: (*a*) the ligand is bound with a bent configuration, as originally suggested by Pauling, with an Fe—O—O angle of 121 °; (*b*) in oxy-Mb the metal is still out of plane (by 0.33 Å), this position being intermediate between that of deoxy- and CO-Mb (0.24 Å, preliminary data); (*c*) the position of the O_2 in the haem pocket is fixed, being constrained by steric hindrance between Phe-CD1, Val-E11, and His-E7. Geometrical parameters of the iron for two derivatives of myoglobin are shown in Table 1. Still unpublished, but completed, is the structure of oxy-cobalt-Mb at 1.5 Å resolution obtained by Petsko, Rose, Tsernoglou, Ikeda-Saito, and Yonetani (personal communication). The model shows that the cobalt atom is slightly out of the mean plane of the porphyrin, and that dioxygen is bound with a bent geometry (the angle being ~135°). Moreover, the oxygen bound to the metal is within hydrogen-bonding distance of the ε-nitrogen of the distal histidine.

Table 1 *Geometrical parameters in myoglobin and haemoglobin (distances/Å)*[a]

	Deoxy-Mb	*Oxy-Mb*	*Deoxy-Hb*
Fe–haem plane	0.55	0.33	0.60 (0.63)
F8 N$^\varepsilon$ haem plane	2.60	2.30	2.60 (2.80)

[a] Modified from a paper by S. E. V. Phillips in *Nature*, 1978, **273**, 247.

Spectroscopic Studies. Support for the structural considerations of Perutz comes[14] from Mössbauer and susceptibility experiments on different derivatives of ferric myoglobin. The splitting energies of the 3d orbitals have been correlated to the distances between the iron and its nearest neighbours. Comparisons of the results obtained for $Mb^+(H_2O)$ and Mb^+F^- indicate that in the former the iron is located nearer to the porphyrin ring plane. In Mb^+CN^- the distance from the iron to the porphyrin plane is shorter compared to that in the high-spin compounds.

Apart from the 'hydrophobic effect' of the haem pocket in stabilizing the iron(II)–O_2 complex, it is beyond any doubt that stereochemical factors may be of great importance in modulating binding of a sixth ligand (O_2, CO) to the haem. Co-ordination geometry and electronic structure may play an important role in defining the haem reactivity in different haemoproteins.

Polarized absorption spectroscopy on a single crystal has provided[15] some evidence for differences in the geometry of Fe–O_2 bonding in different haemoproteins. In the case of Sperm whale myoglobin the interpretation is consistent with an O_2–haem structure of the bent type, in agreement with X-ray studies.[14, 16]

[13] S. E. V. Phillips, *Nature*, 1978, **273**, 247.
[14] U. F. Thomanek, F. Parak, S. Formanek, and G. M. Kalvins, *Biophys. Struct. Mechanism*, 1977, **3**, 207.
[15] M. W. Makinen, A. K. Churg, and H. A. Glick, *Nature*, 1979, in the press.
[16] J. P. Collman, R. R. Gagne, C. A. Reed W. T. Robinson, and G. A. Rodley, *Proc. Nat. Acad. Sci. U.S.A.*, 1974, **71**, 1326.

On the other hand, in *Aplysia* myoglobin, in which the distal His is lacking and the haem environment is substantially different,[17] the O—O bond is more nearly coplanar with the haem plane. The X-ray structure of *Aplysia* Mb, which is being carried[18] to a considerably higher resolution, will provide direct information on the differences in the ligand geometry as compared to Sperm whale Mb.

A bent configuration of the iron–O_2 complex in myoglobins and haemoglobins with both proximal and distal histidines present is also consistent with i.r. spectroscopic measurements.[19] The i.r. stretch of O_2 bound to the haem iron, found near 1105 cm^{-1}, is shifted compared to the free-gas value (1555 cm^{-1}), indicating that, as a consequence of binding at the iron, the electron density of the oxygen–oxygen bond increases. The polarity developed on the bound O_2 might influence interactions with the distal histidine: a proton-free nitrogen of the imidazole ring can act as electron donor with respect to the oxygen atom directly bound to the iron.

The influence of the environment experienced by the bound ligand has been monitored also by ^{13}C n.m.r. spectroscopy. Observations of the chemical shift of the bound ^{13}CO were recently performed[20] as a function of the character of the substituents around the porphyrin ring in different haemoproteins. The main conclusions reached with this experimental approach may be summarized as follows: (*a*) The apoprotein seems to influence the degree to which the electronic effects of haem substituents can be transmitted to the iron–ligand bond. Therefore the extent to which the ligand bound to the haem experiences electronic changes occurring in the porphyrin ring is modulated by the apoprotein. (*b*) The nature of the haem pocket imposes differences in ^{13}CO chemical shifts among the various proteins; this finding may reflect changes in the interaction between the ligand and the distal histidine E7. With reference to this aspect of the problem, it should be outlined that haemoproteins lacking His-E7 (*e.g. Glycera*, α-subunits of opossum haemoglobin, β-chains of haemoglobin Zürich) display anomalous shifts of this ^{13}CO resonance. (*c*) In the case of native ^{13}CO-myoglobin the ^{13}C resonance is shifted upfield (from Me_4Si) only below pH 6.5. A similar effect of pH on the ^{13}C resonances has been reported[21,22] in the case of some fish haemoglobins, the shift being more marked for one of the two chains.

Polypeptide Conformation. Natural-abundance ^{13}C n.m.r. spectroscopy has been used[23] to follow the pH dependence of the resonances of Cγ of histidine residues of horse and kangaroo myoglobins. In the case of horse Mb$^+$CN$^-$, eight of the eleven histidines are observable and titrable, while in metmyoglobin only six

17. L. Tentori, G. Vivaldi, S. Carta, M. Marinucci, A. Marra, E. Antonini, and M. Brunori, *Internat. J. Peptide Protein Res.*, 1973, **5**, 187.
18. T. L. Blundell, M. Brunori, B. Curti, M. Bolognesi, A. Coda, M. Fumagalli, and L. Ungaretti, *J. Mol. Biol.*, 1975, **97**, 665.
19. W. S. Caughey, J. C. Maxwell, J. M. Thomas, D. H. O'Keeffe, and W. J. Wallace, 'Metal–Ligand Interactions in Organic Chemistry and Biochemistry', D. Reidel, Dordrecht, Holland, 1977, Part 2, p. 131.
20. R. B. Moon, K. Dill, and J. M. Richards, *Biochemistry*, 1977, **16**, 221.
21. G. M. Giacometti, B. Giardina, M. Brunori, G. Giacometti, and G. Rigatti, *F.E.B.S. Letters*, 1976, **62**, 157.
22. B. Giardina, G. M. Giacometti, M. Coletta, M. Brunori, G. Giacometti, and G. Rigatti, *Biochem. J.*, 1978, **175**, 407.
23. D. J. Wilbur and A. Allerhand, *J. Biol. Chem.*, 1977, **252**, 4968.

histidines give detectable signals; the two lacking resonances are assigned to His-64 and His-97. The sixth histidine ($pK \leq 5$) observed in ^{13}C n.m.r. spectra of ferrimyoglobin does not have a counterpart in the reported ^{1}H n.m.r. data; on the other hand, the ^{13}C n.m.r. data do not show the presence of a histidine residue with higher pK (~ 7.6) that was previously observed by ^{1}H n.m.r. One way of reconciling the inconsistency is represented by the possibility that there is tautomeric equilibrium between the N^{ε_2}—H and N^{δ_1}—H imidazole forms. From these results, it appears that ^{13}C Fourier-transform n.m.r. is a powerful complement to ^{1}H n.m.r. spectroscopy.

Cross-reactivity experiments on myoglobins from a number of different mammals indicate that, in spite of the similarity in the overall three-dimensional structures of the molecules, definite immunological differences are present and may have to be considered, especially for their possible evolutionary significance. Thus,[24] even the presence of identical sequences within the assumed determinant regions does not guarantee cross-reactivity between two native globular proteins. It is possible that identical antigenic sequences in two closely related proteins may differ in conformation, due to amino-acid substitutions that are outside the determinant region but sufficiently close in space for them to take part in the antigenic site.

Sequence Studies. Studies on the primary structure of haemoproteins are of general interest in so far as: (*a*) they may help to elucidate the role of some 'key' residue(s) in relation to specific functional and physiological requirements; (*b*) they may provide a comparison of evolutionary processes occurring in specific regions of the molecule, affording the opportunity of an insight into the structure–function relationships and of prediction of the structure of a 'primordial' respiratory carrier.

Relevant to the second point is a comparison[25, 26] of the sequence of lamprey haemoglobin (which is known to be monomeric in the ligand-bound state and dimeric in the deoxy-form) with those of a 'vertebrate common ancestor' and a 'vertebrate–mollusc common ancestor'. The authors show that in the α_1–β_2 interface (or in the equivalent region of the sequences) there is a high degree of similarity between lamprey haemoglobin and that of the vertebrate ancestor, contrary to what was found between the latter and the vertebrate–mollusc ancestor. This means that lamprey haemoglobin is a rather conservative molecule, presumably because the environmental conditions favoured the maintenance of this structure.

Additional information comes from a comparison of the sequences of myoglobins from various cetaceans, namely Sperm whale, dolphin, Amazon river dolphin, porpoise, and *Orcinus orca*.[27] From the analysis, it appears that the amino-acid residues which differ among the various sequences are restricted to one particular tridimensional area of the molecule. Similar results are also reported in the case of myoglobin from ungulates (horse, zebra, ox, sheep, pig, and red deer). These comparative data suggest that different species display mutations in

[24] J. G. R. Hurrel, J. A. Smith, P. E. Todd, and S. J. Leach, *Immunochemistry*, 1977, **14**, 283.
[25] M. Goodman, G W. Moore, and G. Matsuda, *Nature*, 1975, **253**, 603.
[26] B. Tiplady and M. Goodman, *J. Mol. Evol.*, 1977, **9**, 343.
[27] O. Castillo, H. Lehman, and L. T. Jones, *Biochim. Biophys. Acta*, 1977, **491**, 23.

different and confined structural areas of the polypeptide chain, implying a specific physiological requirement or a specific selection element.

Ligand Reactivity.—*Model Compounds.* The search for porphyrins suitably modified to combine reversibly with gaseous ligands (O_2, CO) and to be, therefore, useful models for the functional properties of haemoproteins has been a field of very active research. Compounds of great interest from this standpoint have been prepared and studied by Traylor and collaborators[28] and by Collman and collaborators.[29] The former group of investigators has been successful in understanding the basic reactivity of porphyrin model systems in terms of penta- and tetra-co-ordinated metal atoms. An application of this idea to the reactivity of myoglobin at very low pH is reported below. Collman and co-workers have studied the structure, spectra, and binding properties of model compounds called 'picket-fence' porphyrins with reference to their chemical structure. Very recently, a series of ferrous (and cobaltous) porphyrins that are capable of reversibly binding O_2 have been characterized[30] in order to obtain information on the effect of steric restraints on the intrinsic affinity of the metal atom for oxygen. The available data reveal two main features. (*a*) The thermodynamic parameters ($\Delta H^\ominus$, $\Delta S^\ominus$) and the $p_{\frac{1}{2}}$ measured in the case of iron porphyrins with unhindered axial bases (*N*-methylimidazole and 'tailed' imidazoles) compare well with those obtained in the case of Mb and the isolated chains (α, β) of human haemoglobin. Thus, these model compounds may be taken as representative of the relaxed state (R) of the haemoglobin molecule. (*b*) The affinity for O_2 of porphyrins with sterically hindered axial bases decreases, becoming more similar to that of the 'T' state of haemoglobin.

The authors conclude that since the affinity of relaxed haemoproteins (Mb, α- and β-chains) for ligands is the same as for the ferrous porphyrin–imidazole system, no additional interactions between the protein and the bound O_2 need be considered in attempts to interpret their properties. With the restraint imposed by the axially hindering imidazoles, the authors have been able to mimic the behaviour of the T state of haemoglobin, thus providing additional evidence for a role played by the proximal histidine in the binding of ligands to deoxy-T-haemoglobin.

Cobalt-substituted Haemoproteins. Substitution of ferrous with cobaltous porphyrins (proto-, meso-, and deutero-) allows a detailed study of the role of the metal atom.

E.p.r. spectroscopy has been used to probe the oxygen interactions in cobalt-substituted Sperm whale Mb,[31] monomeric *Glycera* Hb,[31] and *Aplysia* Mb.[32] The hyperfine structure of the e.p.r. spectrum of oxycobalt myoglobin shows a small reversible pH-dependent change which, in the case of protohaem, has a pK of 5.33 and parallels the pH dependence of the oxygen affinity (26 mmHg at pH

[28] J. Geibel, C. K. Chang, and T J. Traylor, *J. Amer. Chem. Soc.*, 1975, **97**, 5924.
[29] J. P. Collman, *Accounts Chem. Res.*, 1977, **10**, 265.
[30] J. P. Collman, J. I. Brauman, K. M. Doxsee, T. R. Halbert, and K. S. Suslick, *Proc. Nat. Acad. Sci. U.S.A.*, 1978, **75**, 564.
[31] M. Ikeda-Saito, T. Iizuka, H. Yamamoto, F. J. Kayne, and T. Yonetani, *J. Biol. Chem.* 1977, **252**, 4882.
[32] M. Ikeda-Saito, M. Brunori, and T. Yonetani, *Biochim. Biophys. Acta*, 1978, **533**, 173.

7.0 and 42 mmHg at pH 4.8). This pH dependence has been interpreted as being due to a progressive protonation of the δ-nitrogen of the imidazole ring of the distal histidine. Protonation of the δ-nitrogen could, in turn, weaken the hydrogen bond between the ε-nitrogen and the bound ligand. On the contrary, deoxy-cobalt myoglobin does not exhibit any pH dependence of its e.p.r. spectrum.

Of particular significance in this respect are the equilibria and kinetics, as well as the e.p.r. spectra, of cobalt *Glycera* haemoglobin and *Aplysia* Mb, in which the distal histidine is lacking. Thus the e.p.r. spectrum of *Glycera* oxycobalt haemoglobin is pH-insensitive and the substituted protein exhibits a very low oxygen affinity and a large dissociation rate constant. In the same Co-substituted haemoprotein, low-temperature photodissociation experiments have shown that the photolysed species is similar to the deoxy-form, in contrast to Sperm whale myoglobin, which yields a photochemical intermediate with a very specific e.p.r. spectrum. In fact, the e.p.r. features of the photolysed species in the three monomeric haemoproteins investigated are all different, and obviously reflect the nature of the distal site.

In all cases the cobalt-substituted haemoproteins have O_2 affinities and rate constants which parallel those of the natural, iron-containing haemoproteins. This is convincingly illustrated by the data shown in Table 2.

Table 2 *Oxygen affinity and rate constants of iron- and cobalt-substituted single-chain haemoproteins*

		$p_{\frac{1}{2}}$/mmHg	k_{on}/l mol^{-1} s^{-1}	k_{off}/s^{-1}
Sperm whale Mb	Fe (20 °C)	0.51	1.9×10^7	10
	Co (10 °C)	26 (15 °C)	2.1×10^7	0.85×10^3
Aplysia Mb	Fe (20 °C)	2.7	1.5×10^7	70
	Co (4 °C)	330	4.3×10^7	2.8×10^4
Glycera Hb	Fe (5 °C)	1.6^a	—	—
	Co (5 °C)	700	—	$> 8.5 \times 10^4$

^{a} Calculated from $p_{\frac{1}{2}} = 1.6$ mmHg at 5 °C and an assumed $\Delta H = -13.6$ kcal mol^{-1} (identical to *Aplysia* Mb).

Energy Barriers to the Binding of Carbon Monoxide. Frauenfelder and co-workers have investigated the binding of CO to the isolated α- and β-chains of human haemoglobin[33] and compared these findings with those obtained previously on protohaem and myoglobin. Rebinding of the photodissociated complexes has been followed from 5 to 340 K and from 2 μs to 1 ks. The data indicate that in moving from the solvent to the iron, CO encounters three barriers, to which activation parameters have been assigned. Below 30 K, binding proceeds by quantum-mechanical tunnelling. The kinetic information has been correlated with available structural data.

'Tetraco-ordinated' Myoglobin? The reactivity of different haemoproteins for the same ligand may be very variable, covering several orders of magnitude;[1] such

[33] N. Alberding, S. S. Chan, L. Eisenstein, H. Frauenfelder, D. Good, I. C. Gunsalus, T. M. Nordlund, M. F. Perutz, A. H. Reynolds, and L. B. Sorensen, *Biochemistry*, 1978, **17**, 43.

large differences cannot be ascribed solely to steric effects on the distal side. An attempt to get a better insight into the structural basis of intrinsic ligand reactivity is the rationale for a study[34] of the kinetics of the combination of carbon monoxide with ferrous myoglobin at very low pH (2.5—6.0). The experiments represent an extension of investigations[28,35] on model compounds which have shown that tetraco-ordinated ferrous haem (obtained by protonation of the iron-binding imidazole) displays a very high 'on' constant for carbon monoxide. A number of pH-jump experiments (attained by rapid mixing) have shown that, immediately after the pH shift, and before denaturation has time to occur, the absorption spectrum of ferrous Mb is different from that of native Mb at neutral pH, and consistent with an absorption expected for a four-co-ordinated protohaem. The combination of carbon monoxide with this 'acid' form of ferrous Mb was measured by confronting the protein with the ligand immediately after the pH jump.

The intrinsic pK for the protonation step, calculated from the pH dependence of the apparent rate constant for combination, on the basis of a simple scheme which assumes that only one proton is involved in the transition, is ~ 3.45, very close to that previously obtained for a model compound.[28,35] These results support the idea that protonation of the proximal imidazole yields a tetraco-ordinated and very reactive ferrous haem iron. It may be relevant to note that the rate constant for the combination of CO with acid ferrous Mb (1.4×10^7 1 mol^{-1} s^{-1}) is comparable with that found for some haemoproteins, such as *Chironomus* haemoglobin and the β-chains of haemoglobin Zürich.[36]

Different Responses to Different Ligands. It has been generally believed that haemoproteins display a similar behaviour towards ligands of comparable size, such as O_2 and CO. For this reason, in many cases (and especially for haemoglobins susceptible of rapid autoxidation), carbon monoxide has been chosen as a ligand to investigate structure–function relationships. Indeed, in the case of human and horse haemoglobins, quasi-equivalence of these two ligands has been well established. However, in some cases this similarity breaks down dramatically. Thus in the case of monomeric *Chironomus* haemoglobin (component III) considerable differences were found between oxygen and carbon monoxide in respect of the Bohr effect, which is[37] much larger in the case of O_2. In full agreement with the ligand-linked proton-release experiments, the isoelectric points of the various derivatives are different, that for carboxyhaemoglobin being practically coincident with that of deoxyhaemoglobin (I.P. = 5.92), and that for O_2 being ~ 0.05 pH unit lower.

Support for these findings comes[38] from c.d. and i.r. spectroscopy of the same component III. The carbon monoxide stretching band is at ~ 1963 cm^{-1}, considerably shifted to the blue end of the spectrum compared to mammalian

[34] G. M. Giacometti, T. G. Traylor, P. Ascenzi, M. Brunori, and E. Antonini, *J. Biol. Chem.*, 1977, **252**, 7447.
[35] J. Cannon, J. Geibel, M. Whipple, and T. G. Traylor, *J. Amer. Chem. Soc.*, 1976, **98**, 3395.
[36] G. M. Giacometti, E. Di Iorio, E. Antonini, M. Brunori, and K. Winterhalter, *European J. Biochem.*, 1977, **75**, 267.
[37] G. Steffens, G. Buse, and A. Wollmer, *European J. Biochem* , 1977, **72**, 201.
[38] R. Wollmer, G. Steffens, and G. Buse, *European J. Biochem.*, 1977, **72**, 207.

haemoglobin (1951 cm^{-1}). A similarly high value of $\nu(CO)$ has been obtained[39] for Hb Zürich [$\nu(CO)$ = 1958 cm^{-1}] in which the distal histidine of the β-chains is substituted by Arg. Correlating these results, the authors point out[37,38] that, in *Chironomus* Hb, His-E7, though present, does not interfere[40] with the bound CO. A crucial residue on the other hand may be Ile-E11, which, as suggested[41] from the difference Fourier map, may hinder the movement of the iron towards the haem plane.

If this interpretation is correct, the haem iron in *Chironomus* HbCO would have the same geometry as in the deoxy-derivative, and thus the ligand-linked conformational change would be limited to the reaction with oxygen.

Photochemistry. Demma and Salhany have investigated[42] the formation of met-Hb and free superoxide anion (O_2^-) upon flashing a solution of oxyhaemoglobin. Although, for convenience, the experiments were performed with haemoglobin, it is implicit that similar results may be obtained with other oxycomplexes. Even in the absence of an action spectrum, the authors were able to show that only the u.v. component ($\lambda \leq 300$ nm) of the white light of a flash is active in the photochemical oxidation of haemoglobin and the generation of free superoxide. The formation of O_2^- was established by simultaneous reduction of cytochrome c^{3+} and the inhibition of this process by superoxide dismutase. Some correlation between this phenomenon and the photohaemolysis of red blood cells (with a maximum efficiency at ~ 280 nm) is suggested by the authors.

The reversible photodissociation of liganded haemoproteins substituted with different metals (Fe, Mn, or Co) has been investigated in detail by Hoffman and Gibson,[43] who were able to propose a sound explanation for the absolute values of quantum yield. It is well established[1] that the nature of both the ligand and the protein affects the photochemical efficiency, the values of quantum yields ranging from 1 to $\geq 10^{-3}$. These effects have been investigated[44] before by measuring the quantum yield of proteins from different sources, combined with different ligands. Table 3 shows the photodissociation quantum yields for liganded myoglobins and metal-substituted myoglobins in different oxidation states. An overall stereoelectronic classification has been proposed[43] to account for old and new findings on the photochemistry of low-spin haemoproteins, based on the idea that: (*a*) complexes with linear metal–ligand fragments are highly photodissociable, while (*b*) complexes with a bent metal–ligand fragment and higher electron occupancy are relatively photoinert. This view, initially suggested by the finding that Mn-Mb-NO has a quantum yield very close to Fe-Mb-CO (*i.e.* ~ 1), is discussed[43] by Hoffman and Gibson. This interpretation does not, of course, exclude a role of the protein environment in modulating the quantum efficiency within a certain 'class' of metal–ligand fragments.

[39] W. S. Caughey, J. O. Alben, S. McCoy, S. H. Boyer, S. Charache, and P. Hathaway, *Biochemistry*, 1969, **8**, 59.
[40] R. Huber, O. Epp, and H. Formanek, *J. Mol. Biol.*, 1970, **52**, 349.
[41] R. Huber, O. Epp, W. Steigemann, and H. Formanek, *European J. Biochem.*, 1971, **19**, 42.
[42] L. S. Demma and J. M. Salhany, *J. Biol. Chem.*, 1977, **252**, 1226.
[43] B. M. Hoffman and Q. H. Gibson, *Proc. Nat. Acad. Sci. U.S.A.*, 1978, **75**, 21.
[44] M. Brunori, G. M. Giacometti, E. Antonini, and J. Wyman, *Proc. Nat. Acad. Sci. U.S.A.*, 1973, **70**, 3141.

Table 3 *Photochemical quantum yield, ω, of various haemoproteins (from refs. 43 and 44, modified)*

Metal	Ligand	ω	Protein
Fe II	CO	0.25—1	Hb, Mb
Fe II	CN⁻	1	Mb
Fe II	RNC	0.1—0.3	Mb, Hb
Fe III	NO	~1	Mb
Mn II	NO	~1	Mb
Co III	CN⁻	>0.2	Peroxidase
Fe II	NO	<0.001	Hb
Fe II	O₂	0.05—0.01	Hb, Mb
Co II	NO	<0.0001	Mb
Co II	O₂	<0.0001	Mb

A Special Case: Leghaemoglobin and its Significance.—The molecular properties of leghaemoglobin, including its three-dimensional structure, have been actively investigated[45] in the past few years.

Leghaemoglobin is found in the cells of the nitrogen-fixing nodules that are formed in the roots of legumes in response to invasion by bacteria of the genus *Rhizobium*. Within these nodules, leghaemoglobin is required for nitrogen fixation: nitrogenase activity, in fact, is reduced more than ten times in the presence of carbon monoxide, *i.e.* when the oxygen-combining capacity of leghaemoglobin is abolished.

The central part of the nodules is maintained[46] at a very low partial pressure of oxygen (0.006 mmHg) due to the vigorous respiration of the bacteroids. Under these conditions leghaemoglobin is about 20% saturated with oxygen ($p_{\frac{1}{2}}=0.04$ mmHg)[47,48] and hence it is able to facilitate the diffusion of oxygen from the intercellular air-filled spaces to the intracellular bacteria (on the assumption of free diffusion of cytoplasmic leghaemoglobin). At the level of the bacteroids, molecular oxygen delivered by haemoglobin is accepted by a bacterial terminal oxidase system, which supports nitrogenase activity. Therefore an important aspect of leghaemoglobin's function is related to its property of acting as an oxygen buffer which prevents the oxygen pressure (at the bacterial surface) from falling much below 0.04 mmHg, which is the $p_{\frac{1}{2}}$ of the protein. In this respect it is of great significance that the oxidase system operates[46] optimally when leghaemoglobin is about half saturated.

Dimeric Myoglobins.—Two unusual myoglobins, isolated from the radular muscles of the sea-snail *Nassa mutabilis*, have been characterized[49] in some of their physicochemical and functional properties. They show very similar isoelectric points (~10.2) and physical properties. The determination of the molecular weight of both proteins points to a dimeric structure (mol. wt. = 28 000), the minimum molecular weight ranging from 13 000 to 15 000. The

45 N. Ellfolk, *Endeavour*, 1972, **31**, 139.
46 J. B. Wittenberg, in 'Oxygen and Physiological Function', ed. F. F. Jöbsis, Professional Information Library, Dallas, Texas, 1976, p. 228.
47 T. Imamura, A. Riggs, and Q. H. Gibson, *J. Biol. Chem.*, 1972, **247**, 521.
48 J. B. Wittenberg, C. A. Appleby, and B. A. Wittenberg, *J. Biol. Chem.*, 1972, **247**, 527.
49 G. Geraci, A. Sada, and C. Cirotto, *European J. Biochem.*, 1977, **77**, 555.

functional properties are characterized by the absence of a Bohr effect and the presence of considerable homotropic interactions ($n = 1.5$) although the dimer is the stable species in both oxy- and deoxy-forms. The affinity constants for the two steps, calculated on the basis of a two-stage Adair scheme, show that the affinity for the second oxygen molecule is increased by a factor of ~ 2.

Similarly, the characterization[50, 51] of two dimeric intracellular haemoglobins from the blood clam, *Anadara broughtonii* and *A. senilis*, has shown that these stable dimers have no Bohr effect, but significant homotropic interactions

These data make dimeric myoglobins an interesting system from an evolutionary point of view. The determination of the amino-acid sequence should provide information on the specific substitutions occurring in the protein–protein interaction regions that are involved in the stabilization of the dimer and which are also necessary to drive the haem–haem interactions.

2 Haemoglobin

Structural Investigations.—*Magnetic and Spectroscopic Studies.* The electronic structure and the mode of binding of dioxygen (and other ligands) to haemoglobin continue to attract the attention of scientists. Relevant information has been obtained by magnetic,[52] Mössbauer,[53, 54] i.r.,[55] n.m.r.,[56] and e.p.r.[57] studies, as well as theoretical calculations.[58-60]

Magnetic susceptibility measurements of human oxyhaemoglobin carried out between 25 and 250 K have indicated that, contrary to what is widely accepted on the basis of classical measurements,[61] HbO_2 is not diamagnetic. The results of Cerdonio *et al.* show[52] that, above approximately 50 K, HbO_2 has a magnetic moment corresponding to two parallel electron spins per iron, while HbCO under identical conditions is indeed diamagnetic. The temperature-dependent effect involves a transition between a singlet and a triplet state, which lies 1.72 kJ mol⁻¹ [0.41 kcal (mol of Fe)⁻¹] above the singlet and thus becomes thermally populated at higher temperatures. If the two electrons are assigned one to the O—O bond and the other to the Fe—O bond (in a Weiss-type configuration), the spins have to be antiferromagnetically coupled.

This new finding has been commented upon by Pauling,[62] who suggested that freezing may cause segregation of the protein and ice, and who thus questioned

50 H. Furuta, M. Ohe, and A. Kaiita, *Seikagaku*, 1975, **47**, 715.
51 J. S. Djangmah, and E. J. Wood, *Biochem. Soc. Trans.*, 570th meeting, Cardiff, 1977.
52 M. Cerdonio, A. Congiu-Castellano, F. Mogno, B. Pispisa, G. L. Romani, and S. Vitale, *Proc. Nat. Acad. Sci. U.S.A.*, 1977, **74**, 398.
53 T. Kent, K. Spartalian, G. Lang, and T. Yonetani, *Biochim. Biophys. Acta*, 1977, **490**, 331.
54 A. Alfsen, D. Bade, U. van Bürck, H. Eicher, S. Formanek, G. M. Kalvins, F. Lavialle, A. Mayer, F. Parak, J. Tejade, and U. F. Thomanek, *Biophys. Struct. Mechanism*, 1977, **3**, 229.
55 S. Yoshikawa, M. G. Choc, M. C. O'Toole, and W. S. Caughey, *J. Biol. Chem.*, 1977, **252**, 5498.
56 M. E. Johnson, L. W. M. Fung, and Chien Ho, *J. Amer. Chem. Soc.*, 1977, **99**, 1245.
57 M. Chevion, M. M. Traum, W. E. Blumberg, and J. Peisach, *Biochim. Biophys. Acta*, 1977 **490**, 272.
58 B. D. Olafson, and W. A. Goddard, III, *Proc. Nat. Acad. Sci. U.S.A.*, 1977, **74**, 1315.
59 B. H. Huynh, D. A. Case, and M. Karplus, *J. Amer. Chem. Soc.*, 1977, **99**, 6103.
60 A. Warshel, *Proc. Nat. Acad. Sci. U.S.A.*, 1977, **74**, 1789.
61 L. Pauling, and C. E. Coryell, *Proc. Nat. Acad. Sci. U.S.A.*, 1936, **22**, 210.
62 L. Pauling, *Proc. Nat. Acad. Sci. U.S.A.*, 1977, **74**, 2612.

the significance of this result at physiological temperatures. However, magnetic measurements at room temperature by Cerdonio *et al.*[63] have confirmed that, by their technique, HbO_2 is paramagnetic also at 20 °C, as compared to the diamagnetic HbCO. This new paper, which also comments upon and reviews previous experimental data, draws attention to the contradictory facts concerning this crucial point.

Theoretical papers have contributed to this problem. *Ab initio* quantum mechanical calculations on model systems indicate[58] that: (*a*) the Fe—O—O bond angle should be 119°, (*b*) only 0.10 electron is transferred to the O_2 molecule (in contradiction to the Weiss model), (*c*) the triplet state lies 34 kJ mol^{-1} above the singlet (*i.e.* about 20 times more than is indicated by the magnetic studies), (*d*) the movement of the Fe associated with the T to R transition is only 0.04 Å towards the haem plane. Somewhat similar conclusions have also been reached by Karplus and co-workers,[59] whose calculations seem to exclude a net charge transfer to O_2, favouring the Pauling model and emphasizing the role of the porphyrin and imidazole in modulating the iron population.

Various spectroscopic methods have been applied to haemoglobin. Significant results on the interaction of the distal His with bound CO have been obtained[64] by comparing the i.r. spectra of the α-CO and β-CO isolated chains of HbA, HbACO itself, and HbCO from rabbit. In the latter case the α-CO subunits (in the tetramer) exhibit a lower vibrational frequency for the C—O stretch (1928 cm^{-1}) as compared to the β-CO subunit ($\nu = 1951$ cm^{-1}).

A very careful ^{1}H n.m.r. study of both Mb and Hb has confirmed the importance of investigating both the field and temperature dependence of a shifted proton resonance to interpret the linewidth of systems containing paramagnetic centres.[56]

Proton n.m.r. spectroscopy of model compounds (ferrous meso-tetra-*p*-isopropylphenylporphyrin, PFe, with methyl-substituted imidazoles) has made[65] it possible to assign the proton resonances of the proximal imidazole. The corresponding protons of the proximal His of Sperm whale deoxy-Mb, as well as the NH signal of the same residues in deoxy-HbA, have also been assigned. It is found that in HbA the NH proton resonance of the proximal His is shifted to a different extent for the α- and β-chains. Moreover it was reported that, in spite of the large difference in ligand affinity, one of the proximal His NH-shifts is in a very similar position in HbA and Mb, suggesting that there is a very similar environment in the two cases.

Quaternary Interactions and Solution Properties. Assembly of the haemoglobin tetramer from its isolated subunits, and ligand-linked effects, have been re-investigated[66–69] by Ackers and collaborators. The authors have used molecular sieve chromatography and calorimetry, in conjunction with O_2-binding deter-

[63] M. Cerdonio, A. Congiu-Castellano, L. Calabrese, S. Moranti, B. Pispisa, and S. Vitale, *Proc. Nat. Acad. Sci. U.S.A.*, 1978, **75**, 4916.
[64] J. D. Satterlee, M. Teintze, and J. H. Richards, *Biochemistry*, 1978, **17**, 1456.
[65] G. N. La Mar, D. L. Budd, and H. Goff, *Biochem. Biophys. Res. Comm.*, 1977, **77**, 104.
[66] R. Valdes and G. K Ackers, *J. Biol. Chem.*, 1977, **252**, 74.
[67] S. H. C. Ip and G. K. Ackers, *J. Biol. Chem.*, 1977, **252**, 82.
[68] R. Valdes and G. K. Ackers, *J. Biol. Chem.*, 1977, **252**, 88
[69] R. Valdes and G. K. Ackers, *Proc. Nat. Acad. Sci. U.S.A.*, 1978, **75**, 311.

minations and some kinetic experiments, to obtain equilibrium data for the reactions shown in Scheme 1. The study of the self-association of the isolated α^{SH}- and β^{SH}-chains (in the fully oxygenated state) has shown that: (*a*) the α-chains never go above the dimer stage, and (*b*) the β-chains go all the way from monomer to tetramer. Thermodynamic parameters for self-association of the α- and β-chains, given in Table 4, indicate that this process is essentially entropy-driven, and point to a dominant role of the hydrophobic interactions in self-association.

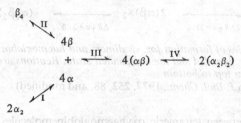

Scheme 1

Table 4 *Thermodynamic data for self-association of oxygenated α- and β-chains*

Reaction	ΔH	$-\Delta G^c$/kcal (mol of haem)$^{-1}$	ΔS^c/e.u.
$2\alpha \rightleftharpoons \alpha_2$	4.34^a	5.28	32.7
$4\beta \rightleftharpoons \beta_4$	23.46^b	21.77	153.5

^{a} Units are kcal (mol of dimer)$^{-1}$. ^{b} Units are kcal (mol of tetramer)$^{-1}$. ^{c} Based on a standard state of 1 mole of haem per litre at 21.5°C, at pH 7.4, in the presence of 0.1 mol l^{-1} of Tris hydrochloride, 0.1 mol l^{-1} of NaCl, and 1 mmol l^{-1} of Na₂EDTA.

While the dimerization of the α^{SH}-chains is not linked with the presence of O_2, the formation of the β^{SH} tetramers is dependent on the presence of O_2, the deoxy-form being[69] more easily dissociable as compared to the oxy-form [$- \Delta G^\ominus = 19.05$ kcal (mol of tetramer)$^{-1}$ for the deoxy- as compared to $- \Delta G^\ominus = 22.43$ kcal (mol of tetramer)$^{-1}$ for the oxy-form]. Given the present accuracy of measuring the isotherm for O_2 binding, the independence of $p_{\frac{1}{2}}$ on protein concentration reported in the past for the β^{SH}-chains is consistent with these new results.

A thermodynamic characterization of the dimer–tetramer association of both unliganded and liganded HbA (reaction IV in Scheme 1) has been reported. These new data have confirmed the large influence of O_2 on the association free energy, and have provided estimates of oxygenation-linked enthalpy and entropy of dimer–dimer association as 32.7 ± 1.7 kcal (mol of dimer)$^{-1}$ for $\Delta H^\ominus$ and 90.2 ± 5.9 e.u. for $\Delta S^\ominus$. These values are consistent with an increased role of hydrophobic bonding in the dimer–dimer contact regions when O_2 is bound.

Equilibrium and calorimetric data have been inserted into a complete linkage scheme (Figure 1) which includes data derived from studies of binding of O_2 as a function of temperature, as well as considerations of energy conservation. It is clear that all the oxygenation-linked enthalpic difference is attributable to the dimer–tetramer step, and no linkage exists in the formation of dimers from chains.

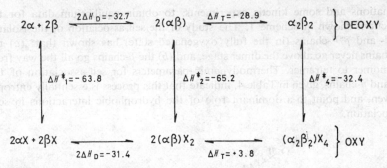

Figure 1 *Enthalpies of formation for αβ dimers and haemoglobin tetramers in both the fully liganded and the unliganded state. Reations are from left to right and from top to bottom*
(Data taken from *J. Biol. Chem.*, 1977, **252**, 88, and modified)

Interactions between tetrameric oxyhaemoglobin molecules in solution have been detected[70] by nuclear magnetic relaxation in H_2O/D_2O. From the dependence of the proton and deuterium longitudinal relaxation rates on protein concentration, the authors imply that self-association of two or more tetramers can be detected by ^1H n.m.r., at concentrations of haem $\geqslant 10$ mmol l^{-1}.

The kinetics of the tetramer-to-dimer dissociation of HbCO have also been investigated[71] in the alkaline range (10.3—11.57), using a laser light-scattering stopped-flow apparatus. The results are consistent with first-order dissociation kinetics, k_{app}, varying from 0.25 s^{-1} to 24 s^{-1} with an apparent pK of 11.07 and a slope $n \simeq 2.56$. The data have been interpreted in terms of a progressive titration of the salt bridge in the α_1-α_2 and β_1-β_2 contacts. At one pH value the first-order rate constant is essentially independent of the derivative (whether CO, O_2, or met-CN).

The physicochemical properties of haemoglobin have been analysed,[72] with special reference to the conditions of high protein and salt (NaCl) concentration. Analysis of non-ideality effects occurring in very concentrated solutions of haemoglobin has been presented[73, 74] by Ross and Minton, providing a theoretical basis for a correct analysis of the equilibrium properties of HbS at high concentrations (see below).

Association with other macromolecular components has been studied[75] for (*a*) the haptoglobin–Hb system; (*b*) various dextrans and dextran derivatives, with special attention to the O_2-linked effects of binding; and (*c*) spectrin, a protein component of the red blood cell membrane which, contrary to previous observations, was clearly shown[76] by Cassoly not to interact at all with native haemoglobin A.

70 J. Brnjas-Kraljevic, J. Maricic, and U. Bracika, *Biophys. Chem.*, 1977, **6**, 191.
71 D. P. Flamig and L. J. Parkhurst, *Proc. Nat. Acad. Sci. U.S.A.*, 1977, **74**, 3814.
72 J. Bernhardt and H. Pauly, *J. Phys. Chem.*, 1977, **81**, 1290.
73 P. D. Ross and A. P. Minton, *J. Mol. Biol.*, 1977, **112**, 437.
74 P. D. Ross and A. P. Minton, *Biochem. Biophys. Res. Comm.*, 1977, **76**, 971.
75 M. Rogard and M. Waks, *European J. Biochem.*, 1977, **77**, 367.
76 R. Cassoly, *F.E.B.S. Letters*, 1978, **85**, 357.

X-*Ray Structural Data.* The structure of horse methaemoglobin at 2.0 Å resolution has been published[77] by Perutz and co-workers. The refined structure shows essentially the features already established[78, 79] from previous work. However, some important and new facts have been clearly established, particularly concerning: (*a*) the set of hydrogen bonds within and between subunits, related to the presence of a considerable number of bound water molecules (41 per asymmetric unit), many of which failed to show up in the previous study. Thus the number of H-bonds between neighbouring subunits increased to 17 (or 19) in the $\alpha_1-\beta_1$ contact, and to 6 (or 7) in the $\alpha_1-\beta_2$ contact. On the basis of these new findings, the altered properties of some abnormal haemoglobins seem to be understandable in terms of perturbations of the hydrogen-bonded set of water molecules at interfaces; (*b*) the iron atom relative to the mean plane of the porphyrin is found to be displaced towards the proximal histidine by 0.07 Å in the α-chains and by 0.21 Å in the β-chains (with a probable error of 0.06 Å). However, the accuracy of the data is not sufficient to allow a decision as to whether the porphyrin ring is flat or ruffled; (*c*) the distances of amino-acid side-chains from the centre of the porphyrin in mct-IIb have been compared with those reported for human deoxy-Hb. The shifts observed are similar for the α- and β-chains, showing that similar reactions occur in both chains upon ligation of the haems.

The structure of human foetal deoxy-Hb (F$_{II}$) has been solved[80] and compared with that of human deoxy-HbA. Very few differences are detectable in the tertiary structures of the β- and γ-chains; amongst these, some perturbations in the central cavity between the two γ-chains, which may account for the lower affinity of HbF for DPG. The data provide some structural understanding of: (*a*) the greater solubility of deoxy-HbF compared to deoxy-HbA [due to the substitution of $\beta22$ (B4) Glu→γ Asp, which perturbs the interaction with $\alpha20$ (B1) His of a neighbouring molecule]; (*b*) the greater stability of HbF in the presence of alkali [related to the substitution in the γ-chains of two buried ionizable groups of the β-chains, *i.e.* $\beta112$ (G14) Cys →γ Thr and $\beta130$ (H8) Tyr →γ Trp]. Other characteristic properties of HbF, such as the smaller acid Bohr effect,[81] cannot be accounted for in conclusive terms.

From crystallographic data it appears that, in human fluoro-met-HbA,[82] which in the presence of IHP is isomorphous with deoxy-HbA without IHP, the iron atoms of both chains are moved towards the distal side, suggesting possible rupture of the bonds to the N$^\varepsilon$ of histidine (similar to what was found previously for the α-chains in HbM Boston and in HbA–nitric oxide derivative in the presence of IHP).

Functional Properties.—*Binding Experiments and Analysis.* In the past few years the quantitative description of haemoglobin kinetics and equilibria has been

[77] R. C. Ladner, E. J. Heidner, and M. F. Perutz, *J. Mol. Biol.*, 1977, **114**, 385.
[78] M. F. Perutz, H. Muirhead, J. M. Cox, L. C. G. Goaman, F. S. Mathews, E. L. McGandy, and L. E. Webb, *Nature*, 1968, **219**, 29.
[79] M. F. Perutz, H. Muirhead, J. M. Cox, and L. C. G. Goaman, *Nature*, 1968, **219**, 131.
[80] J. A. Frier and M. F. Perutz, *J. Mol. Biol.*, 1977, **112**, 97.
[81] E. Antonini, J. Wyman, M. Brunori, E. Bucci, C Fronticelli, M. Reichlin, and A. Rossi Fanelli, *Arch. Biochem. Biophys.*, 1964, **108**, 569.
[82] G. Fermi and M. F. Perutz, *J. Mol. Biol.*, 1977, **114**, 421.

mainly based on the allosteric model[8] of Monod *et al.* Several papers have appeared which are relevant to the problem of the applicability of this model to the functional behaviour of haemoglobin. New sets of O_2 equilibrium data, obtained by the polarographic method[84] of Imai *et al.*, have been published as a function of pH,[85] temperature,[86] and protein concentration.[87] Analysis of binding data has been by the classical Adair equation,[1] or by suitable modifications, such as that given[88] by Imai and Adair. A very clever and useful way of dealing with binding curves has been derived by Gill *et al.*[89]

A problem of major importance in the quantitative evaluation of the O_2 dissociation curves is the linearity of the relationship between combination with O_2 and spectral changes, which are usually employed to determine the ligand-binding isotherm.[1] Imaizumi *et al.*[90] have redetermined: (*a*) the anion-linked spectral changes in the visible and Soret regions and (*b*) the O_2-binding isotherm at different wavelengths. The lack of a dependence of the binding parameters on wavelength, in spite of the clear spectral changes observed on addition of anions to HbO_2, is taken by the authors as evidence for the validity of previous conclusions drawn from analysis of the O_2 dissociation curves.

A model analysis of the available O_2-binding data has been presented[91] by Otsuka and Kunisawa, with the special purpose of proposing a plausible scheme to rationalize the systematic finding that K_2 and K_3 (second and third Adair constants) are not monotonically increasing. A sequence of possible conformational changes during the course of saturation is suggested: the tertiary interactions are ligand-linked in a stepwise manner, and the quaternary structural change occurs after three ligand molecules are bound. The analysis, which would seem consistent with the model advanced by Hopfield (see below), allows a set of data to be described with three parameters.

A set of experiments using laser photolysis has been performed[92] by which both kinetic and equilibrium data have been obtained for the same solution of haemoglobin. This fact, as well as the possibility of controlling contamination by carboxy- and met-haemoglobins, which are very critical, seem to make this method particularly reliable. The experiments, performed in the range from 0 to 20% saturation, are in very good agreement with those previously obtained[93] gasometrically, but show substantial differences when compared with more recent data.[85] The authors believe this difference to be due to a small amount of contamination by carbon monoxide ($\sim 2\%$) which, in the earlier experiments, was not removed. The combination of O_2, following the laser flash, attributed to reaction with deoxy tetramers in a T state, is markedly biphasic, with two

83 R. G. Shulman, J. J. Hopfield, and S. Ogawa, *Quart. Rev. Biophys.*, 1975, **8**, 325.
84 K. Imai, M. Morimoto, M. Kotani, H. Watari, W. Hirata, and M. Kuroda, *Biochim. Biophys. Acta*, 1970, **200**, 189.
85 K. Imai and T. Yonetani, *J. Biol. Chem.*, 1975, **250**, 2227.
86 K. Imai and T. Yonetani, *J. Biol. Chem.*, 1975, **250**, 7093.
87 K. Imai and T. Yonetani, *Biochim. Biophys. Acta*, 1977, **490**, 164.
88 K. Imai and G. S. Adair, *Biochim. Biophys. Acta*, 1977, **490**, 456.
89 S. J. Gill, H. T. Gaud, J. Wyman, and B. G. Barisas, *Biophys. Chem.*, 1978, **8**, 53.
90 K. Imaizumi, K. Imai, and I. Tyuma, *J. Biochem. (Japan)*, 1978, **83**, 1707.
91 J. Otsuka and T. Kunisawa, *Arch. Biochem. Biophys.*, 1977, **179**, 706.
92 C. A. Sawicki and Q. H. Gibson, *J. Biol. Chem.*, 1977, **252**, 7538.
93 F. J. W. Roughton and R. L. J. Lyster, *Hvalradets Skr.*, 1965, **48**, 185.

kinetic components of roughly equal amplitude (the combination rate constants being 11.8 and 2.9 l μmol^{-1} s^{-1} and the dissociation rate constants 2500 and 180 s^{-1}). The authors have interpreted this result in terms of a difference between the two types of chain in a T-state tetrameric haemoglobin. The assignment of a kinetic component to a specific type of chain remains unresolved. Both kinetic and equilibrium data are compatible with a modified two-state model which includes an intramolecular functional heterogeneity, *i.e.* two functionally distinct types of chain. The new results do confirm that at very low saturations in the T state there is a species characterized by a very high dissociation rate constant, similar to what was originally reported by Ilgenfritz and Schuster.[94] Moreover, the new findings are probably not inconsistent with the ^{1}H n.m.r. titrations reported by Huang and Redfield.[95]

It may be important to outline once more the effect of small amounts of contaminants which significantly shift the lower part of the binding curve to the left. This effect, although not very critical for routine affinity measurements, is of enormous relevance in the accurate thermodynamic characterization of ligand-binding behaviour of haemoproteins. In this respect it is necessary to realise that a number of very accurate equilibrium and kinetic experiments have been performed on *unpurified* haemoglobin(s), although attention has been focused on the bottom of the binding curve. The recently established fact that HbA$_{1c}$, which can be present in concentrations between 5 and 10%, has O$_2$-binding properties different from those of HbA constitutes an additional problem for the interpretation of these very accurate data.

The Energy of Homotropic Interactions. The free energy of interaction in haemoglobin, experimentally obtainable from the analysis of the dissociation curve by estimating the asymptotic values of the isotherm in a Hill plot, amounts (at neutral pH) to approximately 3 kcal per site.[3]

Following the nomenclature[8] of Monod *et al.*, this free energy of interaction (ΔF_I) corresponds to the difference in the binding energy between the two extreme states, *i.e.* is related to the ratio of the intrinsic binding constants K_T and K_R.

Whether the two-state allosteric model is strictly applicable to haemoglobin or some modifications need to be introduced, a problem of basic interest in the mechanism of haem–haem interactions in haemoglobin is represented by the localization of the free-energy difference between the two states, R and T. According to Perutz,[96] the low-affinity characteristic of the T state had to be related to the tension which the globin imposes on the haem through the direct bonding between the proximal imidazole and the metal atom. The movement of the metal from the position out-of-plane (a property of the high-spin deoxy-form) to that in-plane (characteristic of the low-spin ligated state) was supposed to be hindered by the interactions with the globin.

An alternative proposal had subsequently been advanced[97] by Hopfield, who challenged the idea that the energy of interaction was stored preferentially in one

[94] G. Ilgenfritz and T. M. Schuster, *J. Biol. Chem.*, 1974, **249**, 2959.
[95] Tai-Huang Huang and A. G. Redfield, *J. Biol. Chem.*, 1976, **251**, 7114.
[96] M. F. Perutz, *Nature*, 1972, **237**, 495
[97] J. J. Hopfield, *J. Mol. Biol.*, 1973, **77**, 207.

critical bond and proposed a model in which this interaction energy is diffused throughout the protein.

This problem has been the subject of recent additional investigations. Eisenberger and collaborators[98] have applied EXAFS measurements to haemoglobin as well as to model compounds. Analysis of the data obtained with this very advanced technique has allowed accurate measurements of the Fe—N (pyrrole) distances in these various compounds. It is found that (a) HbO_2 in the R state and the oxy-picket-fence complex are characterized by an Fe—N bond distance of appoximately 1.99 ± 0.01 Å, and (b) Hb in the T state, deoxy-picket-fence complex, and Hb Kempsey in the R state are characterized by an Fe—N bond distance of approximately 2.02 ± 0.01 Å (personal communication and ref. 97).

The conclusions drawn by the authors are that the Fe—N (pyrrole) bond length is not affected by the interactions of the iron porphyrin with the protein and that in deoxyhaemoglobin the quaternary structure has no effect on the same bond length, which varies by less than 0.02 Å in going from the R to the T state.

An energy calculation reported[60] by Warshel also suggested (within the limitations imposed by the assumptions intrinsic to the method) that the haem–haem interaction energy is not stored significantly at the level of the haem, and therefore changing the spin-state does not represent a major contribution to the energy. Warshel points out that the role of the geometry of the haem as a whole, in its interactions with the globin, should be emphasized, in agreement with a good deal of experimental data.[1]

Perutz and collaborators[99,100] have recently made an attempt to measure the change in free energy associated with the high- to low-spin transition in a six-co-ordinated derivative, in both quaternary states, T and R. This was made possible by using haemoglobins from fish, which are characterized by a pH-dependent R→T transition in the ligated form. The azide derivative of methaemoglobin from carp proved to be a very suitable experimental material because: (a) it can be switched from R to T by adding IHP at a fixed pH, and (b) it shows a thermal spin equilibrium in both these quaternary structures. The experiments involved (a) spectral measurements in the visible, Soret, and u.v. regions, (b) [1]H n.m.r. measurements, (c) paramagnetic susceptibility measurements over the range 90–300 K, (d) studies of the pressure dependence of the spectral properties.

The temperature dependences of the paramagnetic susceptibilities of the R and T states have been fitted to a function which takes into account the thermal spin equilibria. The pressure dependence of the spin state in both quaternary conformational structures has allowed a determination of the volume changes associated with the high- to low-spin transition in either the R or T state. The upshot of all the data is that the high-spin form is favoured in the T structure compared to the R structure by about 1 kcal mol^{-1}. The authors thus conclude that the tension at the haem builds up when, in a T quaternary state, the ligand tries to

[98] P. Eisenberger, R. G. Shulman, G. S. Brown, and S. Ogawa, *Proc. Nat. Acad. Sci. U.S.A.*, 1976, **73**, 491.
[99] M. F. Perutz et al., *Biochemistry*, 1978, **17**, 3640.
[100] C. Messana et al., *Biochemistry*, 1978, **17**, 3652.

force the haem into its low-spin conformation, and this energy term amounts to approximately 1 kcal mol^{-1}. This value is only $\sim\frac{1}{3}$ of the energy of interaction, and although the authors point out that this is a minimum estimate, it may be possible that the remaining part of this energy term is localized at the protein rather than at the haem.

In this respect, it may be relevant to point out that studies by resonance Raman scattering have revealed that differences in the Fe—N$^\epsilon$, Fe—O$_2$, and Fe—N(porphyrin) stretching frequencies between the T and R states are very small in both deoxy- and oxy-Hb.[101] The differences are smaller (by approximately 10 times) than those expected if the interaction energy was stored in one of these bonds as strain energy.

The conclusion that very little strain is present at the unliganded haem has also been reached[102] by Gelin and Karplus on the basis of conformational energy calculations on a single-chain molecule. The authors have stressed that the strain is present in a ligand-bound haem in a T quaternary structure, and that this restraint is a reason for the low affinity of deoxy-(T) structure.

Kinetics of Conformational Changes. Many of the experimental results obtained in the past have clearly shown that ligand-linked conformational transitions in haemoglobin are indeed very fast, generally faster than the binding of ligands.[1]

In the past two years the application of photochemical methods to the study of the reaction of human haemoglobin with O$_2$ or CO has made it possible to determine the rate constants for the conformational changes, by making use of the differences in chemical reactivity and/or spectral properties between the R and T allosteric states.

Laser pulse experiments.[92,103] These are conventional single-perturbation experiments which make use of a powerful, short-lived, laser pulse which allows partial or complete photodissociation of oxygen from oxyhaemoglobin. In the alkaline range (8.4—9.6), in borate buffer, the data collected with both O$_2$ and CO could be made to fit a Monod–Wyman–Changeux model. The rate constant for the R→T transition (*i.e.* from the quickly reacting deoxy-Hb in the R state to the slowly reacting deoxy-Hb in the T state) was found to be 6400 ± 600 s^{-1} at 20 °C, with an apparent activation energy of 17 kcal mol^{-1}. The rate constant for the conformational R→T transition decreases as the number of ligand molecules that are bound to a species immediately after the pulse increases, according to the expression $k_{R\to T}/d^x$, where $k_{R\to T} = 6400$ s^{-1}, $d = 2.3 \pm 0.2$ (as obtained from a fitting of the data), and $x = 1$—4, being the average number of ligands bound to the molecule.

At lower pH values, unexpected and hitherto unexplained complications do not allow a good fit of the data to the simple two-states model.

Modulated excitation experiments. These have been[104] carried out with HbCO at pH 7 and 22 °C. With this method a very small level of photodissociation (no more than 1 %) is achieved, and thus the data refer to the triply liganded molecule

[101] K. Nagai, T. Kitagawa, and H. Morimoto, *Sixth International Biophysics Congress, Kyoto (Japan)*, 1978, 1V-1-(553).

[102] B. R. Gelin and M. Karplus, *Proc. Nat. Acad. Sci. U.S.A.*, 1977, **74**, 801.

[103] C. A. Sawicki and Q. H. Gibson, *J. Biol. Chem.*, 1977, **252**, 5783.

[104] F. A. Ferrone and J. J. Hopfield, *Proc. Nat. Acad. Sci. U.S.A.*, 1976, **73**, 4497.

7

[$Hb_4(CO)_3$]. Phase-shift data, obtained at different frequencies of the photo-dissociating beam (from 100 to 1000 Hz), indicate that a single relaxation process dominates. Analysis of the data, which involved discrimination of interfering effects (*e.g.* chain heterogeneity or the presence of dimers), allowed the rate constants to be obtained for the allosteric transition at the level of three ligands bound ($R_3 \rightleftharpoons T_3$). The relevant rate constants are $k_{RT} = 780 \pm 40$ s^{-1} and $k_{TR} = 2500 \pm 200$ s^{-1}, yielding $k_{RT}/k_{TR} = L_3 = 0.31 \pm 0.04$. This value is in good agreement with L_3 obtained from O_2 equilibrium data.

A photochemical method has again been employed[105] to measure the kinetics of reaction with CO, using changes in magnetic susceptibility to measure the extent of reaction. This new technique uses a superconducting magnetometer to detect rapid changes of magnetic susceptibility. The rate constant for combination of the last ligand with $Hb_4(CO)_3$ was in very good agreement with the results obtained by photometric methods; the magnetic moment of this species was determined to be 4.9 ± 0.1 μ_B. The determination of this parameter is relevant if compared with the magnetic moment (per haem) of deoxyhaemoglobin Hb ($\mu_B = 5.3$) and with that of the isolated α- or β-chains of haemoglobin ($\mu_B = 4.9$). It is evident that the μ of the triply ligated species is essentially identical to that of the chains, thus showing that co-operative interactions have a corresponding effect on the electronic state of the iron.

Modified Haemoglobins. Careful oxygen equilibrium experiments have been[106] performed recently on valence hybrid haemoglobins in which one chain is in the reduced and the other in the oxidized form. These intermediates were examined[1] several years ago, using more conventional O_2-binding methods. The new data indicate that all the hybrids display an asymmetric oxygen equilibrium curve, which is not consistent with the presence of only two O_2-binding sites per tetramer. This inconsistency has been rationalized on the basis of a small amount of oxidation of the ferrous subunits, which indicates, once again, how critical it is to have complete control of the purity of the materials under investigation. An important observation which originates from these new experiments is that in the $\alpha_2\beta_2^+$ hybrid the affinity of one ferrous chain for O_2 depends on the type of ferric ligand, decreasing in the order $CN^- > N_3^- > H_2O > F^-$. Thus the affinity for O_2 is lower in high-spin derivatives, implying that the T state of the molecule is more stable for high-spin hybrids, in agreement with Perutz's stereochemical interpretation of co-operative binding of ligands. However, this correlation is not at all clear in the $\alpha_2^+\beta_2$ hybrid, indicating that some other factors may be involved in the control of ligand affinity. This possibility is also suggested by the effect of organic phosphates, which seems to abolish the correlation outlined above between affinity for O_2 and the spin state.

A naturally occurring valence hybrid, *i.e.* one of the haemoglobins M [Milwaukee: $\beta 67$ (E11) Val$\rightarrow$Glu] has been re-investigated[107] by ^{1}H n.m.r. spectroscopy. The haem environment of the normal ferrous α-chains and the subunit interactions in HbM Milwaukee are similar to those of HbA. Stepwise addition of

[105] J. S. Philo, *Proc. Nat. Acad. Sci. U.S.A.*, 1977, **74**, 2620.
[106] K. Nagai, *J. Mol. Biol.*, 1977, **111**, 41.
[107] L. W. N. Fung, A. P. Minton, T. R. Lindstrom, A. V. Pisciotta, and Chien Ho, *Biochemistry*, 1977. **16**, 1452.

O_2 to the system is associated with the presence of n.m.r.-detectable intermediates, which are stable, and which indicate that a two-state model cannot account for the functional and structural properties of this system. The authors have proposed a sequential model to fit the data quantitatively, and stress their belief that these results are relevant to the behaviour of normal HbA.

Cobalt-substituted haemoglobin. Valuable information on the allosteric mechanism underlying the regulation of the haemoglobin function has come from studies[108] on functionally active cobalt-substituted haemoglobin. Oxygen equilibrium curves determined over a wide range of pH, both in the presence and the absence of allosteric effectors, show that the functional behaviour of Co-Hb is analogous to (but clearly different from) that of Fe-haemoglobin. Thus, although the overall O_2 affinity for Co-Hb is 10 to 20 times lower than that for ferrohaemoglobin, co-operativity is maintained to appreciable levels ($n_{\frac{1}{2}} =$ 1.6—2.0), and heterotropic interactions are also substantially preserved.

Analysis of the data in terms of the two-state model points to (a) a drastic change of the allosteric equilibrium constant in favour of the R state ($L_0 = 2.3 \times 10^3$, *cf.* 3.1×10^9), (b) an increase in the intrinsic affinity of the T structure for O_2, as indicated by the value of c ($= K_T/K_R$), which is 0.048 for Co-Hb and 0.00052 for natural haemoglobin, (c) a small influence of the metal substitution on the relative stabilities of the fully oxygenated T and R structures (as indicated by the product $L_0 c^4$, which is similar for both proteins).

It should be pointed out that the free energy of haem–haem interactions is only a third of that characteristic of ferrohaemoglobin. This fact may be in agreement with the smaller displacement of the proximal histidine with respect to the haem plane, as estimated[109] for Co-Hb.

Different porphyrins. The ligand-binding behaviour of haemoglobins reconstituted with different haems has been studied for a long time, in attempts to discriminate between the inductive electronic effects of substituents on haem and the stereochemical effects originating from their interaction with the globin.[1]

Recently, horse haemoglobins reconstituted with 2,4-dimethyl-deuterohaem and 2,4-dibromo-deuterohaem were characterized[110] in their functional properties. The reason for choosing these modified haems resides in the identical van der Waals radii (2 Å) of the two types of substituents and in the marked difference in their inductive effects. Thus, the stereochemical interactions of the haems with the globin should be practically identical. The data obtained indicate a substantially different functional behaviour of the two reconstituted proteins, pointing out the role of the inductive electronic effects in controlling the functional properties of the haemoglobin molecule. The functional alterations seem to reside almost entirely in the ligand-dissociation reactions, the association rate constants being only slightly different in the two cases. However, since the dissociation and the association processes were studied using, respectively, O_2 and CO as ligands, the conclusions drawn should be taken with some caution.

Heterotropic interactions. Bohr effect. A mass-spectrometric method, which

[108] K. Imai, T. Yonetani, and M. Ikeda-Saito, *J. Mol. Biol.*, 1977, **109**, 83.
[109] R. G. Little and J. A. Ibers, *J. Amer. Chem. Soc.*, 1974, **96**, 4452.
[110] D. W. Seybert, K. Moffat, and Q. H. Gibson, *J. Biol. Chem.*, 1977, **252**, 4225.

allows an estimate of the pK's of individual histidine residues, has been applied[111] to haemoglobin to measure the pK value of β146 (H24) His, which is known[112] to be responsible for approximately half of the alkaline Bohr effect. The pH-dependent hydrogen–deuterium exchange at the C-2 position of the imidazole ring was determined under several conditions, after converting the amino-acid residue into the methylthiohydantoin derivative. In agreement with the results previously reported,[113] the pK_a values for the imidazole of β146 (H24) His were found to be 7.0 in oxyhaemoglobin and 8.2 in deoxyhaemoglobin (at 36.5 °C). These values were practically unaffected by IHP, while blocking the reactive sulphydryl groups at β93 causes[114] a reduction in pK in both liganded and unliganded states. The same type of approach was used to determine the pK_a of β2 (NA2) His, which is one of the residues involved in the binding of organic phosphates. In the absence of IHP, the pK_a of this His was found to be 6.9 in both oxy- and deoxy-haemoglobins. Therefore, as already proposed,[112] this residue is not involved in the alkaline Bohr effect. However, in the presence of IHP this pK_a is shifted from 6.9 in oxyhaemoglobin to $\sim$7.7 in deoxyhaemoglobin. The result provides direct evidence for a role of this residue in the increased Bohr effect that is observed in the presence of organic phosphates. The Bohr effect upon binding of CO and NO to human haemoglobin, both in the presence and absence of IHP, has been re-examined[115] by de Bruin *et al.* By using the differential proton titration technique, the authors found that, in the absence of IHP, the maximum number of protons released upon binding of NO is even greater than that induced by CO (by approximately 25%); the situation is reversed in the presence of IHP (0.6 mM for a 0.4 mM solution of haemoglobin tetramer). This result acquires significance because it was previously shown that HbNO in the presence of a saturating concentration of IHP is in a T quaternary structure. Thus the Bohr effect observed with NO under these conditions reflects the ligand-linked proton release within the T state, which should be considered to be of tertiary origin.

The difference in the IHP-induced Bohr effect, as between the ligands CO and NO, is attributed by the authors to a restoration of the salt bridge between 146-(H24) His and β94 (FG1) Asp. It may be of interest, however, to recall that addition of NO breaks the bond between the proximal His and the haem iron in the α-chains, and thereby causes a large perturbation of the tertiary structure of the molecule, which may be the cause of these complexities.

Binding of carbon dioxide. Binding of CO_2 to the N-terminal residues of the α- and β-chains of human haemoglobin yields the corresponding carbamino-compounds, which have been carefully investigated.[116,117]

111 M. Ohe and A. Kajita, *J. Biochem.* (*Japan*), 1977, **81**, 431.
112 M. F. Perutz, M. Muirhead, L. Mazzarella, R. A. Crowther, J. Greer, and J. V. Kilmartin, *Nature*, 1969, **222**, 1240.
113 J. V. Kilmartin, J. J. Breen, G. C. K. Roberts, and Chien Ho, *Proc. Nat. Acad. Sci. U.S.A.*, 1973, **70**, 1246.
114 M. Ohe and A. Kajita, *J. Biochem.* (*Japan*), 1977, **82**, 839.
115 S. H. de Bruin, F. J. C. Boen, H. S. Rollema, and G. G. M. van Beek, *Biophys. Chem.*, 1977, **7**, 169.
116 J. B. Matthew, J. S. Morrow, R. J. Wittebart, and F. R. N. Gurd, *J. Biol. Chem.*, 1977, **252**, 2234.
117 M. Perrella, G. Guglielmo, and A. Mosca, *F.E.B.S. Letters*, 1977, **78**, 287.

The overall binding constant (λ) is related to proton concentration by the following expression:

$$\lambda = K_C \, K_Z/(K_Z[H^+] + [H^+]^2)$$

where K_C is the equilibrium constant of the carbamate reaction and K_Z is the ionization constant of the N-terminal α-amino-groups.

A re-investigation of CO_2 binding[117] has yielded K_C and K_Z for the α- and β-chains of deoxy- and carboxy-haemoglobins. The newly calculated binding constants (λ_α and λ_β) agree well with those previously obtained. The approach used was essentially based on blocking the α-amino-groups of the β-chains with IHP and measuring the binding of CO_2 to the α-chains, the amount of CO_2 bound to the β-chains being calculated by difference.

Binding of anions. The effects of anions on the functional properties of haemoglobin, known for a long time, have acquired more interest since the discovery[118] of the physiological role played by organic phosphates. Inorganic anions and organic phosphates (*e.g.* DPG, ATP, IHP) have comparable effects on the affinity[1] of HbA for oxygen, the main differences being in the concentration of the effector required to achieve a given change in the value of $p_{\frac{1}{2}}$. In addition, experiments performed on monomeric haemoprotein (Mb) indicate that the quaternary structure of Hb is a necessary requirement for specific effects of these anions. The effects of different polyanions, such as DPG (2,3 diphosphoglycerate), IHP (inositol hexaphosphate), IPP (inositol pentaphosphate), and IHS (inositol hexasulphate), on haemoglobin function have been re-investigated[119] by Benesch *et al.* Two main facts appear from the data. (*a*) The allosteric effect on the affinity of Hb for O_2 is directly related to the number of negative charges carried by the specific effector (this number being 8, 6, 6, 4 respectively for IHP, IPP, ISH, DPG). (*b*) The binding of IHP, IPP, and IHS to oxyhaemoglobin is strong enough to permit the measurement of the binding constant even with this derivative.

In view of the small binding constant of DPG to HbO_2, the identity between the 'Bohr' coefficient ($\Delta \log p_{\frac{1}{2}}/\Delta$ pH) and the 'Haldane' coefficient ($\Delta[H^+]$), though verified for stripped haemoglobin, is no longer valid in the presence of DPG.

In the case of IHP, IPP and IHS, in view of their higher binding constants, it is feasible to achieve complete saturation of the binding site (or sites) in both oxy- and deoxy-forms and thus to restore the identity between the two coefficients (as defined above). The additional Haldane effect observed in the case of DPG has implications for transport of carbon dioxide, which occurs both through carbamate formation and through the neutralization of carbonic acid. Thus the increased Haldane effect will compensate for the displacement of CO_2 by DPG from the β-chains.

The dependence of $p_{\frac{1}{2}}$ obtained from experiments in which the concentration of Cl^- was varied may be analysed according to linkage principles.[3] A fit of these[120] new data yields an apparent association constant for chloride of 11 ± 2 l mol^{-1}. The difference between the free energy of O_2 binding in the absence and

[118] R. Benesh and R. E. Benesh, *Nature*, 1969, **221**, 618.
[119] R. E. Benesh, R. Ledalji, and R. Benesh, *Biochemistry*, 1977, **16**, 2594.
[120] R. N. Haire and B. E. Hedlund, *Proc. Nat. Acad. Sci. U.S.A.*, 1977, **74**, 4135.

in the presence of chloride is about 4.9 ± 0.2 kcal (mole of haemoglobin tetramer)$^{-1}$. This value should be taken as a limiting one, expressing a potential maximal change in free energy that chloride ion could induce. The results are compatible with the existence of one or several (but independent) sites that bind Cl^-.

The effect of organic macromolecular anions other than polyphosphates has been studied[121] with the aim of discriminating between the effects of the negative charge density and that of their spatial distribution.

Similarly to polyphosphates (DPG, ATP, IHP), macromolecular sulpho-anions, such as dextran sulphate and heparin, interact preferentially with the low-affinity quaternary structure of the protein. Polyphosphates and poly-sulphates affect the functional properties of HbA to a similar extent, in spite of the differences in their overall three-dimensional structures. The main difference between these two classes of effectors lies in the apparent stoicheiometry for the polyanion–Hb complex, which, in the case of polysulphate, is well below 1. This fact may be ascribed to the dimensions and flexibility of polysulphates, which may be large enough to provide a 'binding site' for two or more haemoglobin molecules, as supported[121] by sedimentation data.

Binding of zinc. Human haemoglobin binds zinc with an apparent association constant of 1.3×10^7 l mol^{-1} and a stoicheiometry of two atoms of zinc per tetramer.[122] Binding of Zn^{2+} produces an increase in the affinity for O_2, implying that HbO_2 binds the cations more easily than Hb.

Studies under various conditions and with various modified haemoglobins have suggested that $\beta_1 139$ (H17) Asp and $\beta_2 143$ (H21) His are involved in the site that binds zinc. Since this reaction is only possible in the liganded quaternary state, this structural interpretation is consistent with the increase in affinity for O_2 diplayed when Zn^{2+} is bound.

Ligand Binding in the Red Blood Cell.—O_2 *Binding of Human Blood.* The O_2-binding curve of normal human blood under conditions of full control of pH (7.4), $p(CO_2)$ (40 mmHg), and DPG concentration [0.78—0.95 μmol (μmol of tetramer)$^{-1}$] in the sample has been redetermined.[123] The methodology employed makes use of the technique described by Rossi-Bernardi *et al.*[124] for the middle range of saturation, and of a new polarographic method which allows a very accurate determination of the initial (up to 10% saturation) and the final (above 90%) sections of the binding curve. The dissociation curve differs significantly from that reported[125] by Roughton *et al.*, especially in the lower region of saturation. The authors provide possible explanations for this discrepancy with older data, such as the absence of control of pH and total [HCO_3^-], and the unknown concentration of DPG in the material studied by Roughton *et al.*[125] Moreover,

121 G. Amiconi, L. Zolla, P. Vecchini, M. Brunori, and E. Antonini, *European J. Biochem.*, 1977, **76**, 339.
122 J. M. Rifkind and J. M. Heim, *Biochemistry*, 1977, **16**, 4438.
123 R. M. Winslow, M. Swenberg, R. L. Berger, R. I. Shrager, M. Luzzana, M. Samaja, and L. Rossi-Bernardi, *J. Biol. Chem.*, 1977, **252**, 2331.
124 L. Rossi-Bernardi, M. Luzzana, M. Samaja, M. Davi, D. Da Riva-Ricci, J. Minoli, B. Seaton, and R. L. Berger, *Clin. Chem.*, 1975, **21**, 1747.
125 F. J. W. Roughton, E. C. Deland, J. C. Kernohan, and J. W. Severinghaus, in 'Oxygen Affinity of Hemoglobin and Red Cell Acid Base Status', ed. P. Astrup and M. Rorth, Academic Press, New York, 1972, p. 73.

as outlined above, the bottom asymptote is the one which is more critically affected by small amounts of contaminants, such as carboxyhaemoglobin or other haemoglobin components present in the blood.

A very careful analysis of the data, carried out using an Adair oxygenation scheme, allows the authors to state that the results can be described with only three parameters rather than four. Their startling conclusion is that, with the currently available precision of data collection, an unequivocal determination of all four Adair constants is very unlikely. In so far as the number and precision of the data points obtained[123] by Winslow *et al.* is comparable to that obtained by other authors on purified haemoglobin, this result is of widespread relevance.

Kinetics of the Reaction in Human Erythrocytes. It is known that the rate of uptake of oxygen by human erythrocytes is about 20—40 times slower than that measured in the case of haemoglobin in solution. As a possible interpretation of this phenomenon, earlier analysis attributed[126] the observed slow-down to a high resistance of the erythrocytic membrane to O_2 diffusion. However, in the past few years this conclusion has been questioned in a number of papers.[127,128] In view of its obvious physiological interest, the problem of the relationships between diffusion and chemical reaction in erythrocytes has been re-investigated recently, using different techniques, *i.e.* rapid mixing and microspectrophotometry.

Stopped-flow data. Extensive rapid-mixing experiments with red blood cell suspensions have been carried out,[129] results being obtained as a function of both oxygen concentration and intra-erythrocytic haemoglobin concentration. The results have been analysed in terms of four possibilities,[126] *i.e.* (i) The existence of an unstirred layer of fluid around the cell; (ii) the high resistance of the erythrocytic membrane to O_2 diffusion; (iii) the possibility that diffusion of O_2 within the cell may be very slow, owing to the high protein concentration; (iv) the possibility that ligand-binding properties of haemoglobin may be markedly different inside the red blood cell.

Computational analysis of the data has shown that the rate-limiting step in O_2 uptake is a diffusion barrier due to the presence of an unstirred layer of fluid surrounding the cell membrane. The hypothesis of a particularly high membrane resistance has been ruled out. In fact, as oxygen diffuses into the cell and combines the haemoglobin, the solvent layer near the membrane becomes depleted of the ligand, giving rise to a drop in the rate of diffusion into the cell. The thickness of this layer of unstirred fluid is estimated to be $\sim 3\mu$m at infinite time (after the flow has stopped).

Microspectrophotometric observations. A different type of approach has been introduced[130] by Antonini *et al.*, based on a photochemical technique coupled to microspectrophotometric observations. By this method the kinetics of reaction of intra-erythrocytic haemoglobin with carbon monoxide may be followed in

[126] F. J. W. Roughton, *Proc. Roy. Soc.*, 1932, **B111**, 1.
[127] W. Moll, *Respir. Physiol.*, 1969, **6**, 1.
[128] H. Kutchai, *Respir. Physiol.*, 1975, **23**, 121.
[129] J. T. Coin and J. S. Olson, in Proceedings of the Symposium on the Clinical and Biochemical Aspects of Hemoglobin Abnormalities, ed. W. Caughey, Academic Press, New York, 1977, p. 559.
[130] E. Antonini, M. Brunori, B. Giardina, P. A. Benedetti, G. Bianchini, and S. Grassi, *F.E.B.S. Letters*, 1978, **86**, 209.

single red blood cells, making use of the reversible photodissociation of the carbon monoxide–haemoglobin complex. When a cell, immersed in an 'infinite' volume of fluid containing carbon monoxide at a fixed concentration, is continuously irradiated with light at 546 nm, photodissociation of HbCO and relaxation to the steady state can be followed. An analysis of the data at different carbon monoxide concentrations has shown that the apparent rate of combination of CO with haemoglobin is about thirty times smaller than that measured with haemoglobin in solution and ten times smaller than that obtained by parallel stopped-flow experiments. Thus the combination of CO with haemoglobin in a single immobile red blood cell is rate-limited by a diffusive process, as also indicated by the zero-order time course of the observed absorbance change. Therefore it appears that the solvent layer around the cell is a major factor in determining the rate of equilibration with the ligand. The maximal thickness of the unstirred layer of fluid around the erythrocyte, calculated on the basis of simple diffusion laws and of the diffusion coefficient of CO in water, is approximately 10 μm.

It may be relevant to point out that if the situation '*in vivo*' is intermediate between the violent stirring present in the mixing chamber of the stopped flow and the stagnancy of the experiments described above, then the unstirred layer may play an important role in controlling the relationships between diffusion and chemical reaction in erythrocytes.

3 Mutant and Abnormal Haemoglobins

General Considerations.—The list of new variants of the α- and β-chains of haemoglobin A increases very rapidly, owing to the large number of people screened in haematological departments. Several of these new variants do not have substantially different O_2-binding properties, while others show significant alterations. In some cases abnormal haemoglobins already known (and well characterized), such as the haemoglobins M, are being 'rediscovered'.

In view of the large number of variants nowadays identified, it does not seem pertinent to this review to provide a partial list of those recently discovered. A complete, up-to-date list (with references) is available at the 'International Hemoglobin Information Center (IHIC)' (R. N. Wrightstone, Medical College of Georgia, Augusta, Georgia 30901, U.S.A.).

In what follows, we shall deal in detail with a few selected abnormal proteins which have proved to be of particular interest because of their physio-pathological and/or molecular properties.

Some Special Cases.—*Sickle-cell Haemoglobin.* (HbS, $\alpha_2^A \beta_2^{6\ Glu \rightarrow Val}$) The ever growing interest in the structural properties and the assembly of HbS is documented by the numerous publications dealing with the molecular basis of this deasease. Some of the latest views are published in the proceedings of a symposium on 'The Clinical and Biochemical Aspects of Hemoglobin Abnormalities' (ed. W. Caughey), held in Colorado in October 1977.

The functional properties of HbS at concentrations below the M.G.C. (Minimum Gelling Concentration) have been carefully re-examined.[131] No differences

131 R. R. Pennelly and R. W. Noble, in ref. 129, p. 401.

can be detected in the behaviour of HbS as compared to HbA, both in the presence and absence of organic phosphates, thus substantiating earlier equilibrium results.[132]

Research on HbS has been focused on four aspects of the problem:

(i) Structural information on the assembly and molecular contacts of HbS tetramers in the fibres. Earlier crystallographic information by Love and co-workers[133] will not be reviewed; however, it is relevant to point out that poly(ethylene glycol), used to obtain crystals suitable for *X*-ray analysis, was found[134] to have no significant effect on the functional properties of either HbS or HbA.

Bundles of HbS fibres, obtained in the presence of IHP, were examined[135] by electron microscopy. They are composed of ordered arrays of fibres in an eight-stranded form, with adjacent strands staggered by 37 Å. Since similar reflections are observed for the isolated fibres and for the fibres in the bundles, it has been concluded that the fibre structure is unperturbed when forming the bundles. Some difficulties of reconciling this model with the pairing of strands observed in the HbS crystals have been pointed out.[135]

With a different approach, Nagel and co-workers have obtained[136] evidence for the participation of a number of external amino-acid residues in the contacts between HbS molecules. The experimental approach is based on measuring the changes in M.G.C. that are due to alteration in the chemical nature of some surface residues, largely using mixtures of HbS and other mutant haemoglobins. The T quaternary conformation is an absolute requirement for the formation of the polymer; moreover, the β 93 thiol groups are not directly involved in the contacts. Only one β6 (A3) Val per tetramer participates in the contacts, and thus experimentation with asymmetric hybrid molecules ($\alpha_2^A \beta^S \beta^A$) acquires significance. The following residues seem to be contributed by the β-chain, which is in a *trans* position relative to β6 (A3) Val: 19, 66, 73, 80, 83, 87, 95, and 121. By a similar approach,[136] the α-chain adjacent to β6 (A3) Val participates in contacts with two residues, *i.e.* 23 and 78.

The identification of a number of contacts besides β6 (A3) Val was also studied[137] by *X*-ray diffraction of HbS crystals obtained from poly(ethylene glycol), extending previous crystallographic observations[133] by the same authors.

(ii) The study of the kinetics of HbS gelation has revealed peculiar features, characterized by a lag phase (or delay time, t_d) followed by a concerted polymerization into fibres and alignment of the fibres to form a semi-crystalline phase. Eaton and collaborators[138,139] carried out an investigation of the time course of gelation, using several techniques (viscosity, heat absorption, turbidity . . .) and initiating the reaction by a rapid temperature increase, which causes a solution of

[132] D. W. Allen and J. Wyman, *Rev. Hematol.*, 1954, **9**, 155.
[133] B. C. Wishner, K. B. Ward, E. E. Lattman, and W. E. Love, *J. Mol. Biol.*, 1975, **98**, 179.
[134] G. Amiconi, C. Bonaventura, J. Bonaventura, and E. Antonini, *Biochim. Biophys. Acta*, 1977, **495**, 279.
[135] R. H. Crepeau, G. Dykes, and S. J. Edelstein, *Biochem. Biophys. Res. Comm.*, 1977, **75**, 496.
[136] R. L. Nagel *et al.*, in ref. 129, p. 195.
[137] W. G. Love *et al.*, in ref. 129, p. 165.
[138] P. D. Ross, J. Hofrichter, and W. A. Eaton, *J. Mol. Biol.*, 1977, **115**, 111.
[139] H. R. Sunshine, J. Hofrichter, and W. A. Eaton, *Nature*, 1978, **275**, 238.

deoxy-HbS to gel. Similar results may be obtained by rapidly inducing the release of a ligand from liganded HbS, either by deoxygenation or by photochemically displacing CO.[130] The delay time was found to be dramatically dependent on experimental variables, such as temperature and haemoglobin concentration; in particular, it is[140] inversely proportional to the 30th power of the protein concentration.

The data have been treated within the framework of a mechanism involving: (*a*) a nucleation step, *i.e.* the formation of thermodynamically unstable nuclei by a series of bimolecular combinations, and (*b*) a propagation step, whereby monomers (of mol. wt. 64 000) are added to the nucleus in a series of thermodynamically favourable discrete processes, leading to polymer formation. The results can be described by a simple equation:

$$1/t_d = \gamma S^n$$

which relates the delay time (t_d) to the so-called supersaturation ratio S, which is the ratio of the total concentration (C_t) of haemoglobin S to the solubility (C_s), measured after gelation is complete; γ and n are experimental constants ($\gamma = 10^{-7}$ s^{-1}, $n = 30$). The 30th-power dependence of the delay time implies, within this mechanism, that a 'stable' nucleus is contributed by approximately 30 monomers. Measurements performed following viscosity changes led to an estimate of the activation energy for the formation of the critical nucleus, the value being[141] approximately 100 kcal mol^{-1} (and somewhat dependent on the concentration of IHP).

These kinetic data have obvious clinical implications if one considers the length of the delay time in relation to the transit time of an erythrocyte in the capillaries (which is approximately 1 second). The delay time of intracellular gelation may be increased in three different ways: (1) by direct inhibition by molecules that competitively bind to sites of intermolecular contact, (2) by reduction of the intracellular haemoglobin concentration, (3) by indirect inhibition *via* an increase of the affinity of the Hb molecule for O_2. Keeping these points in mind, Sunshine *et al.* have computed [139] the delay time necessary to obtain a satisfactory therapeutic effect and the requirements to produce such a delay time. Table 5 shows

Table 5 *Correlation between the increase in the delay time, the requirements for achieving it, and the corresponding clinical course of sickle-cell disease (modified from ref.* 139)

$t_d/t_{d(SS)}$	$\overline{X}$	Δc/g dl^{-1}	Δp_{50}/mmHg	Clinical course of SS disease
10^3—10^4	0.50—0.60	4—8	> 10—13	much less severe
10^6	0.75	7.5—11	>17	no disease

Δt_d is the factor by which the delay time for gelation (t_d) should be increased to achieve the clinical course indicated in the last column. $\overline{X}$, Δc, and Δp_{50} are, respectively, the fractional saturation of the inhibitory site, the decrease in intracellular Hb concentration, and the decrease in p_{50} required to produce the increase in delay time indicated in the first column, if considered one at a time.

[140] J. Hofrichter, P. D. Ross, and W. A. Eaton, *Proc. Nat. Acad. Sci. U.S.A.*, 1974, **71**, 4864.
[141] S. Kowalczykowski and J. Steinhardt, *J. Mol. Biol.*, 1977, **115**, 201.

some of the results of this analysis. Case-by-case, the conditions reported in the table will be difficult to achieve, but it is pointed out that one may act simultaneously along all those lines in an attempt to improve the chances of overcoming the severity of the disease.

The thermodynamic aspects of gelation have also been re-examined by calorimetry and solubility measurements. Considering the non-ideality effects that are due to the excluded volume,[138] the standard-state thermodynamic quantities for gelation at 37 °C are: $\Delta G^{\ominus} \simeq -3$ kcal mol^{-1}, $\Delta H^{\ominus} \simeq 0$ kcal mol^{-1}, $\Delta S^{\ominus} \simeq 10$ cal K^{-1} mol^{-1}. The important observation that the heat of polymerization is strongly dependent on the temperature ($\Delta C_{p} = -197$ cal K^{-1} mol^{-1}) has been confirmed.

The kinetics of polymerization have also been monitored[142] by following the transverse relaxation times of water protons after deoxygenation. When measured by this technique, the kinetics of polymerization follow pseudo-first-order behaviour at a concentration of HbS of 300 mg ml^{-1}. The half-time of polymerization, which by this method is about 1.5 s at pH 7, in 0.25M-phosphate buffer and at 37 °C, becomes as long as 37 s when a 1:1 mixture of HbS and HbA is studied.

(iii) Investigations have been directed towards the design and testing of anti-sickling agents. Chemical modifications of amino-groups (*e.g.* by nitrogen mustards or by nitrotropone[143]) have a definite effect; treatment[144] with mono- and di-saccharides and also modification[145] of the β-93 thiol groups by bulky reagents (cystamine) seem to increase the minimum gelling concentration. A very promising development is represented by attempts towards an intracellular modification of HbS, for instance[146] by DL-glyceraldehyde. The modification involves up to two Lys residues per tetramer, and at concentrations of ~ 10 mmol l^{-1} causes a definite reduction in sickling at zero O_2 tension, without drastic modification of either the affinity for O_2 or the co-operativity. New synthetic peptides have been proposed[147] and explored as a class of anti-sickling agents, and the role of inorganic (*e.g.* quaternary ammonium ions) and organic (*e.g.* aromatic) solutes has been investigated.[148]

(iv) At concentrations above the M.G.C., the O_2-binding curve of HbS is very difficult to determine. Winslow[149] has clearly shown that the binding isotherm of SS blood displays hysteresis, *i.e.* that the binding curve measured starting from deoxy-HbS lies definitely to the right compared to that measured starting from *oxy*-HbS. This observation implies that kinetic effects come into action and shows that most of the curves for binding of O_2 *in vivo* are, at best, approximate.

142 G. L. Cottam, M. R. Waterman, and B. C. Thompson, *Arch. Biochem. Biophys.*, 1977, **181**, 61.

143 D. L. Currell, E. Benitez, C. Ioppolo, B. Giardina, S. G. Condo, F. Martini and, E. Antonini, *European J. Biochem.*, 1978, **91**, 285.

144 P. M. Abdella, J. M. Ritchey, J. W. O. Tam, and I. M. Klotz, *Biochim. Biophys. Acta*, 1977, **490**, 462.

145 W. Hassan, Y. Beuzard, and J. Rosa, *Proc. Nat. Acad. Sci. U.S.A.*, 1976, **73**, 3288.

146 M. Nigen and J. M. Manning, *Proc. Nat. Acad. Sci. U.S.A.*, 1977, **71**, 367.

147 A. N. Schechter, C. T. Noguchi, and W. A. Schwartz, in Proceedings of the Symposium on the Clinical and Biochemical Aspects of Hemoglobin Abnormalities, ed. W. Caughey, Academic Press, New York, 1977, p. 129

148 P. D. Ross and S. Subramanian, in ref. 147, p. 629.

149 R. M. Winslow, in ref. 147, p. 369.

Gill and co-workers[150, 151] have recently measured O_2 equilibria of very concentrated HbS solutions by their original thin-film technique. The binding isotherm (see Figure 2) shows that at a given $p(O_2)$ the binding curve becomes very steep, owing to the onset of polymerization. The relationship between ligand binding and polymerization, the so-called polysteric effect,[152] is clearly evident.

Finally, the gelation of HbS that is partially liganded with CO has been studied[153] by measuring the solubility and linear dichroism of oriented polymer.

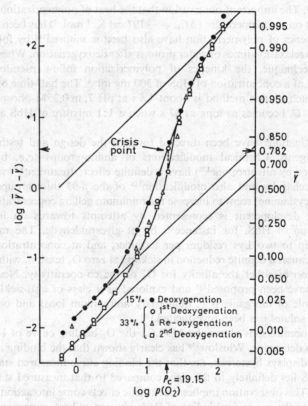

Figure 2 *Hill plots of binding of oxygen to HbS at two different protein concentrations, 15% and 33%. In the latter case the linkage between ligand binding and polymerization (polysteric effect) is clearly evident. The so-called 'crisis point' may be taken as representative of the initiation of aggregates of low affinity for oxygen. At low concentration, the 'crisis point' does not appear, while at high concentrations it moves to higher degrees of oxygenation.*
(Reproduced by permission from *Science*, 1978, **201**, 362)

[150] J. D. Mahoney, S. C. Ross, and S. J. Gill, *Analyt. Biochem.*, 1977, **78**, 535.
[151] S. J. Gill, R. Sköld, L. Fall and T. Shaeffer, *Science*, 1978, **201**, 362.
[152] A. Colosimo, M. Brunori, and J. Wyman, *J. Mol. Biol.*, 1976, **100**, 47.
[153] F. A. Ferrone, J. Hofrichter, and W. A. Eaton, in 'Frontiers of Biological Energetics', ed. P. L. Dutton, Academic Press, New York, 1978, p. 1085.

It has been shown that the polymer consists largely of deoxy-HbS monomers, even when the total saturation with CO is above 60%. Thus the monomer–polymer equilibrium can be looked upon as involving largely (but not exclusively) the deoxygenated derivative.

Haemoglobin Zürich. (HbZh, $\alpha_2{}^A \beta_2{}^{63\ His \to Arg}$). The substitution of the distal His of the β-chains is associated[154,155] with alterations in the haem environment, in the absorption spectra, and in the stability of the molecule. In addition, oxygen binding to Hb Zürich has been shown[156] by Winterhalter *et al.* to be characterized by both homo- and hetero-tropic interactions, although the Hill constant (n) at neutral pH is smaller than that characteristic of HbA ($n=2.3$, *cf.* $n=2.8$).

Recently, accurate oxygen equilibrium curves were measured[157] in the presence and in the absence of DPG; some of the results obtained are shown in Table 6

Table 6 *Equilibrium parameters for the reaction of HbZh with* O_2 *(in* 0.1M-*bis-Tris, at* 25 °C).

	Stripped	*With* 2mM-DPG
p_m/mmHg	1.20	4.80
n	1.83	2.30
K_1/(mmHg)$^{-1}$	0.462	0.165
K_2/(mmHg)$^{-1}$	0.377	0.04
K_3/(mmHg)$^{-1}$	0.718	0.113
K_4/(mmHg)$^{-1}$	3.85	2.65

in terms of the four Adair constants. Comparison with human haemoglobin A shows that K_1 (the first Adair constant) increases by about one order of magnitude in HbZh while K_4 decreases by a factor of about 2. This fact suggests that the oxygen affinity of HbZh in the deoxy quaternary state is higher than that of HbA, possibly due to the lack of the distal histidine. This interpretation would be in agreement with the kinetic model for the reaction with carbon monoxide proposed[158] by Giacometti *et al.* on the basis of extensive rapid-mixing and flash-photolysis experiments. The reaction scheme, illustrated in Figure 3, is based on the following experimental evidence: (*a*) the time course of the reaction depends on the observation wavelength; (*b*) the second-order rate constant and the spectral properties of the initial binding phase are similar to those of the isolated β^{Zh}-chains; (*c*) at very low CO saturations ($\bar{Y} < 0.1$), only a fast process is detected (12×10^6 l mol^{-1} s^{-1}); (*d*) the switch from T to R was shown to occur before the α-chains in the T state have a chance to combine with CO; moreover, the T→R transition is not rate-limiting; (*e*) at a wavelength that is isosbestic for the β^{Zh}-chains, the rate constant determined by mixing experiments (rate

[154] K. H. Winterhalter and K. H. Wütrich, *J. Mol. Biol.*, 1972, **63**, 477.

[155] K. H. Winterhalter, E. Di Iorio, J. G. Beetlestone, J. B. Kushimo, H. Velhack, H. Eicher, and A. Mayer, *J. Mol. Biol.*, 1972, **70**, 665.

[156] K. H. Winterhalter, N. M. Anderson, G. Amiconi, E. Antonini, and M. Brunori, *European J. Biochem.*, 1969, **11**, 435.

[157] M. Ikeda-Saito, T. Yonetani, M. Brunori, and K. H. Winterhalter, *Biochim. Biophys. Acta*, 1979, in the press

[158] G. M. Giacometti, E. Di Iorio, E. Antonini, M. Brunori, and K. H. Winterhalter, *European J. Biochem.*, 1977, **75**, 267.

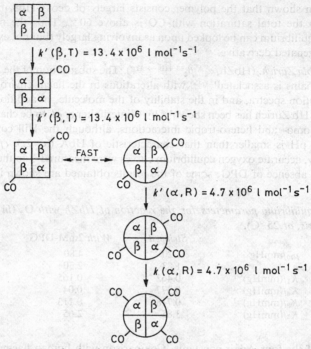

Figure 3 *Scheme for the binding of CO to tetrameric HbZh. Squares and circles represent respectively the low-affinity state (T) and the high-affinity state (R) of the molecule*
(Drawn using data given in *European J. Biochem.*, 1977, **75**, 267)

constant for α-sites) is, after a visible lag, very similar to that of the normal R state of HbA (4.4×10^6 l mol^{-1} s^{-1}).

Therefore binding is strictly sequential, the abnormal β^{Zh}-chains reacting very rapidly even when the tetramer is the T state, and the α-chains behaving normally in the R state. From the data it was concluded[158] that the role of the distal histidine in modulating the reactivity of the haems is considerably more important in the low-affinity T state of the molecule than in the R state.

Further support for these conclusions is provided[159] by recent X-ray work. The bottom half of the haem is normally in contact with two atoms of the distal histidine (*i.e.* N$^\epsilon$ and C$^\epsilon$) that are important in maintaining the haem in its correct orientation. Substitution of the distal histidine by arginine, which is H-bonded to the propionates of the haem, removes these two atomic contacts, allowing the haem to tilt in the direction normally associated, on ligand binding, with shortening of the Fe–N bonds. If this tilt were allowed in deoxy-HbZh, the reactivity of the haems associated with the abnormal chains would be intrinsically high. Therefore in the β-subunits the distal histidine helps to maintain the tension at the haem, and hence the intrinsic reactivity characteristic of the T structure.

159 P. W. Tucker, S. E. V. Phillips, M. F. Perutz, R. Houtchens, and W. S. Caughey, *Proc. Nat. Acad. Sci. U.S.A.*, 1978, **75**, 1076.

Substitutions at the α_1-β_2 *Interface.* This region of haemoglobin is of crucial importance, since a number of critical molecular contacts among dissimilar chains alter in going from oxy- to deoxy-haemoglobin. Several amino-acid substitutions are known to occur at position 99(G1) of the β-chain, which is normally occupied by an aspartic acid residue. This residue forms an H-bond with α42 (C7) Tyr in the deoxy-state which is broken[9] upon oxygenation. The substitutions are: Hb Yakima (replacing residue is His),[160] Hb Kempsey (Asn),[161] Hb Ypsi (Tyr),[162] and recently Hb Radcliffe (Ala).[163] All these haemoglobins have similar functional properties in so far as all display a very high affinity for oxygen and almost completely lack homotropic interactions ($n_1 \approx 1$). In the case of Hb Radcliffe, the Bohr effect and the DPG effect are somewhat reduced, the former being about two-thirds of that of HbA.[163] These data have been interpreted on the basis of the X-ray structures of HbA; replacement of aspartate is very probably associated with a perturbation of chain–chain interactions, which could affect both the quaternary and tertiary structures of the molecule, leading to a stabilization of the R state even in the unliganded form. In agreement with this, ^{1}H n.m.r. data have shown[163] that in deoxy-Hb Radcliffe the resonance at -9.4 p.p.m. from H_2O is missing, and the one normally observed at -12.4 p.p.m. is shifted to -13.0 p.p.m. The lack of the former resonance confirms its previous assignment[164] to the hydrogen bond between α42 (C7) Tyr and the aspartic acid β99(G1), while the shift of the latter, which was assigned to a methyl group on the haem of α-chain,[165] strongly implies that structural changes at the level of α_1-β_2 contact may propagate to haem groups of both chains. In the same interface (α_1-β_2), substitution of β37 (C3) Trp also leads to dramatic changes in the molecular properties of haemoglobin, as exemplified by Hb Hirose, where this residue is replaced by Ser. The properties of this haemoglobin differ[166] from those of haemoglobin A as follows: low homotropic interactions ($n \approx 1.3$) and affinity for O_2 about three times higher than that of HbA; the alkaline Bohr effect and the 2,3-DPG effect clearly reduced; oxygen dissociation kinetics biphasic and dependent on wavelength. The sedimentation data show that the tetramer-to-dimer dissociation is greatly enhanced; $K_{4,2}$ in deoxy-Hb Hirose is ~ 1.2 μmol l^{-1}, *i.e.* 1 000 000 times greater than that of deoxyhaemoglobin A. Therefore the tryptophan β37 at the α_1-β_2 interface seems to play a crucial role in holding the dimers together in the tetramer.

Glycosylated Haemoglobins. Within the category of the so-called minor haemoglobin components, a number of species have been identified which correspond to

[160] R. T. Jones, E. E. Osgood, B. Brimhall, and R. D. Koler, *J. Clin. Invest.*, 1967, **46**, 1840.
[161] C. S. Reed, R. Hampson, S. Gordon, R. T. Jones, M. J. Novy, B. Brimhall, M. J. Edwards, and R. D. Koler, *Blood*, 1968, **31**, 623.
[162] D. L. Rucknagel, *Ann. Rev. Med.*, 1971, **22**, 221.
[163] D. J. Weatherall, J. B. Clegg, S. T. Callender, R. M. G. Wells, R. E. Gale, E. R. Huehns, M. F. Perutz, G. Viggiano, and Chien Ho, *Brit. J. Haematol.*, 1977, **35**, 177.
[164] L. W. M. Fung and Chien Ho, *Biochemistry*, 1975, **14**, 2526.
[165] D. G. Davis, T. R. Lindstrom, N. H. Hock, J. J. Baldassare, S. Charache, R. T. Jones, and Chien Ho, *J. Mol. Biol.*, 1971, **60**, 101.
[166] J. Sasaki, T. Imamura, T. Yanase, D. H. Atha, A. Riggs, J. Bonaventura, and C. Bonaventura, *J. Biol. Chem.*, 1978, **253**, 87.

chemically modified haemoglobin A.[167-169] Amongst these, HbA_{1c} corresponds to normal HbA which contains one residue of glucose or mannose bound per $\alpha\beta$ dimer, the chemical bond being formed at the level of the *N*-terminal valine of the β-chains. This modified protein has been synthesized *in vitro* by treating HbA with an excess of D-glucose; it was also isolated in particularly large proportions (up to $\sim 10\%$) in diabetics.

This modified haemoglobin is of interest from at least two viewpoints. In the first instance it may become a very useful diagnostic tool for judging the average- to long-term levels of blood glucose in diabetics, since its fluctuations are obviously very small compared to those of blood sugar levels. Moreover, the possibility that, in diabetics, similar modifications may occur for proteins which have a long turn-over (such as the crystalline proteins) may acquire significance in considering the physio-pathology of complications of diabetes mellitus.

In the second place, it has been found that glycosylated Hb is characterized by O_2-binding properties significantly different from those of HbA. Thus, although both homotropic and heterotropic interactions are still present, the affinity of stripped HbA_{1c} for O_2 is higher, and the effect of DPG on the affinity for O_2 is lower than those characteristic of HbA. The presence of this modified protein in the haemolysate of normal individuals may cast some doubts on the significance of the O_2-binding parameters determined on unpurified haemoglobin.

4 Comparative Aspects

Haemoglobins.—Comparative studies on haemoglobin function have acquired special interest, since they may lead to an understanding of those changes which, in the course of evolution, have developed in different organisms to meet specific physiological requirements. Besides yielding information on the features of the molecule mainly involved in the evolutionary processes, a comparative approach may substantially contribute to verify the generality of the molecular mechanism proposed for mammalian haemoglobin. Along these lines, species having multiple haemoglobins in their blood are of interest in so far as multiplicity of components is an evolutionary response to the special needs of the individual.

Fish Haemoglobins. Root-effect haemoglobins. Haemoglobins from teleost fishes are generally characterized by a special type of Bohr effect, known as the Root effect. It consists of a pronounced and parallel reduction of both ligand affinity and co-operativity as the pH is brought towards acid values (generally below pH 7.5). In addition, at acid pH's, these haemoglobins are only partially saturated with oxygen, even in air. This special behaviour is related to the function of the swim bladder, *i.e.* to the control of the buoyancy of the fish at various depths.[170,171]

[167] H. F. Bunn, and D. N. Haney, *Biochem. Biophys. Res. Comm.*, 1975, **67**, 103.
[168] R. Flückiger and K. H. Winterhalter, *F.E.B.S. Letters*, 1976, **71**, 356.
[169] R. J. Koenig, S. H. Blobstein, and A. Cerami, *J. Biol. Chem.*, 1977, **252**, 2992.
[170] M. Brunori, *Current Topics Cell. Reg.*, 1975, **9**, 35.
[171] A. Riggs, in 'Fish Physiology', ed. W. S. Hoar and D. J. Randall, Academic Press, New York and London, 1970, Vol. 4, p. 209.

For Hb IV, the major haemoglobin component from trout blood (*Salmo irideus*), the data for binding of oxygen and of carbon monoxide[172,173] display the following characteristics: (*a*) a dramatic dependence of the shape of the binding curve on pH (Figure 4), (*b*) a clear difference in behaviour between O_2 and CO, since the latter does not show antico-operative binding while the former does.

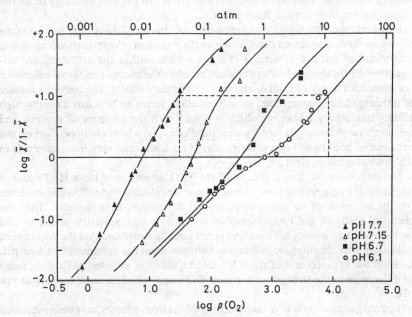

Figure 4 *Oxygen-binding curves of purified stripped trout Hb IV at different pH values in 0.05M-bis-Tris buffer at 14°C. The partial pressure of O_2 (in Torr) has been used to calculate the lower scale and is shown in atm (top scale)*
(Modified from *Proc. Nat. Acad. Sci. U.S.A.*, 1978, **75**, 4310)

As to the mechanistic interpretation of the Root effect, the stabilization of a low-affinity state of the molecule brought about by protons has been proposed and well documented not only for trout Hb IV, but also for other fish haemoglobins, such as those of carp[174] and menhaden.[175]

This conclusion has been substantiated by a thermodynamic study[173] of the reaction of CO with trout Hb IV. Three parallel sets of experiments, comprising microcalorimetric measurements, CO-binding equilibria, and differential titrations, were performed and analysed in terms of a modified two-state allosteric model. From the analysis it appeared that the essential feature of the func-

[172] M. Brunori, M. Coletta, B. Giardina, and J. Wyman, *Proc. Nat. Acad. Sci. U.S.A.*, 1978, **75**, 4310.
[173] J. Wyman, S. S. Gill, L. Noll, A. Colosimo, B. Giardina, H. A. Kuiper, and M. Brunori, *J. Mol. Biol.*, 1978, **124**, 161
[174] A. M. L. Tan and R. W. Noble, *J. Biol. Chem.*, 1973, **218**, 2880.
[175] W. A. Saffran and Q. H. Gibson, *J. Biol. Chem.*, 1978, **253**, 3171.

tional behaviour of trout Hb IV lies in a proton-induced shift of the allosteric equilibrium constant.

The apparent enthalpy change corresponding to the allosteric equilibrium is dependent on pH, becoming more positive as the pH decreases. This provides additional insights for a proton-linked quaternary conformational change as the basis of the special functional behaviour of trout Hb IV, and probably of all the haemoglobins that show a Root effect.

However, the functional properties of these haemoglobins may not be completely understood without considering another class of allosteric effects involving proton-linked tertiary structural changes which, within the tetramer, are different[21,22,175] for the two kinds of chain. Therefore the overall functional behaviour is a complex average of both quaternary and tertiary effects. The common feature of haemoglobins characterized by a Root effect seems to be a low affinity $\rightleftharpoons$ high affinity quaternary transition which is linked with the presence of protons and organic phosphate, while tertiary structural changes, which are dependent on the presence of both ligand and protein, play[172] an additional very important role in the physiological context.

In this respect, ^{13}C n.m.r. spectra of the ^{13}CO derivative of trout Hb IV and of the single Hb component from *Osteoglossum bicirrhosum* yielded[21] clear evidence for the existence of an intramolecular, pH-dependent, heterogeneity. Thus the chemical shifts of the two observable resonances representative of the α- and β-chains change with pH. This effect is not equally distributed on the two chains, indicating that the intrinsic differences between them are enhanced[22] at low pH. In addition, i.r. spectra[176] of trout Hb IV CO, taken at different pH values, point to a decrease in the strength of the iron–CO bond, which is more marked at low pH.

The oxygen equilibria of several haemoglobins from teleosts have been measured as a function of pH in the presence and absence of ATP, which, together with GTP, is the major intracellular polyphosphate.[177-179] The results have shown that the effects of H^+ and ATP are essentially equivalent and may be interpreted, on the basis of the model outlined above, as a preferential binding to the low-affinity state. In addition, similarly to what is observed for mammalian haemoglobins, the Bohr effect is substantially enhanced by organic phosphates ($\Delta \log p_{\frac{1}{2}}/\Delta pH \simeq 0.3$) in all fish haemoglobins so far examined.

As regards the nature of the phosphate-binding site, it is presumably in the same location as in mammalian haemoglobins, although significant differences in some of the residues involved in the binding have been pointed out.[179,180] The effect of IHP on the spectra and spin equilibria of carp Hb[99,100] has been mentioned above.

Temperature dependence of haem–haem interactions. In some fish haemoglobins (*e.g.* Hb I from *Salmo irideus*[164] and Hb from *Thunnus thynnus*[181]) it has recently

[176] F. Ascoli, E. Gratton, F. Riva, P. Fasella, and M. Brunori, *Biochim. Biophys. Acta*, 1978, **533**, 534.
[177] R. G. Gillen and A. Riggs, *Comp. Biochem. Physiol.*, 1971, **38B**, 585.
[178] R. E. Weber, G. Lykkeboe, and K. Johansen, *J. Exp. Biol.*, 1976, **64**, 75.
[179] R. G. Gillen and A. Riggs, *Arch. Biochem. Biophys.*, 1977, **183**, 678.
[180] D. Barra, F. Bossa, J. Bonaventura, and M. Brunori, *F.E.B.S. Letters*, 1973, **35**, 151.
[181] F. G. Carey and Q. H. Gibson, *Biochem. Biophys. Res. Comm.*, 1977, **78**, 1076.

been reported that co-operative effects in ligand binding are strongly influenced by temperature. In the Hb I from trout, CO binding has a positive ΔH at fractional saturations below 10% and a negative ΔH at higher saturations; the ligand-binding curves cross at $\sim 10\%$ saturation, and at lower temperatures the free energy of interaction is[182] significantly higher. In the case of O_2, similar overall behaviour is also observed,[183] although the apparent ΔH is approximately zero at very low saturations and increases as a function of fractional saturation.

A key element which emerges from the analysis of the binding data for Hb I from trout is that the allosteric quaternary transition is associated[182] with a large positive enthalpy change $[\Delta H_{L_0} = +28$ kcal (mol of tetramer)$^{-1}]^2$ and a substantial entropy contribution $[T\Delta S_{L_0} = +23$ kcal (mol of tetramer)$^{-1}$ at 20 °C].

An opposite behaviour is displayed by tuna haemoglobin, whose co-operativity diminishes as temperature decreases, a fact which is considered advantageous for a warm-bodied fish, as indicated[181] by the authors.

These findings are interesting examples of the flexibility of the haemoglobin molecule and of the danger of regarding O_2 and CO as absolutely interchangeable ligands. In addition, they demonstrate that, in order to obtain enthalpy data for binding, an accurate analysis of the binding curve and parallel microcalorimetric data are highly desirable.

Other Species. An interesting example[184] of the physiological importance of organic phosphates in regulating oxygen transport appears from a study of haemoglobin from a primitive amphibian (the Congo eel, *Amphiuma means*), which in some environmental conditions may burrow and hibernate in the mud.

In the presence of organic phosphates, the single haemoglobin from *Amphiuma* displays a small positive Bohr effect; on the other hand the stripped molecule shows a very marked negative Bohr effect, *i.e.* the affinity for O_2 increases when the pH is lowered. This finding, besides suggesting the presence of additional Bohr groups or a drastic modification of the residues involved in human HbA, may be the reflection of a molecular adaptation to meet respiratory requirements during hibernation, when organic phosphates are practically absent and the animal has to extract oxygen from an environment of low pH.

In addition, experiments with the same haemoglobin after it has been enzymatically modified (by digestion with carboxypeptidase A and/or B) demonstrate that homo- and hetero-tropic interactions can vary independently of one another. Thus, the Bohr effect is still present in the absence of co-operative oxygen binding, and may be either positive or negative without appreciable alterations of haem–haem interactions.

Haemoglobins from other species have been investigated, but very often the functional data refer to whole blood rather than to the purified proteins. It may be of interest to draw attention to the paper[185] by Pough, who examined the O_2-binding properties of blood of 15 species of snakes and attempted a correlation between the affinity of blood for O_2 and the body size.

[182] J. Wyman, S. J. Gill, L. Noll, B. Giardina, A. Colosimo, and M. Brunori, *J. Mol. Biol.*, 1977, **109**, 195.
[183] B. Giardina, M. Coletta, M. Brunori, S. J. Gill, and J. Wyman, 1979, to be published.
[184] C. Bonaventura, B. Sullivan, J. Bonaventura, and S. Bourne, *Nature*, 1977, **265**, 474
[185] F. M. Pough, *Physiol. Zool.*, 1977, **50**, 77.

The Alpha Helix *Expedition in the Amazon.* A systematic study of the functional and structural properties of the haemoglobins from the fishes living in the Amazon basin has been undertaken during a scientific expedition supported by the National Science Foundation of U.S.A. The results obtained 'on the field' by making use of the facilities of the research vessel *Alpha Helix* (of the Scripps Oceanographic Institute) are presently in the press in a single volume of *Comparative Biochemistry and Physiology*. The very large number of species which inhabit the waters of the Amazon River system include, side by side with water breathers, several species which are obligatory or facultative air breathers. The classification of the haemoglobins of individuals of different families, the characterization of their basic properties (spectra, chain composition, ligand-binding equilibria and kinetics, haematological parameters, *etc.*), and the correlation of these properties with the physiological needs and the preferred environment constitute the matter dealt with in approximately 30 papers, which represent the contribution of the scientists who took part in the expedition (see *Comp. Biochem. Physiol.*, January 1979).

Giant Haemoproteins.—Extracellular haemoglobins are widely distributed among annelids, molluscs, and arthropods. They are large multi-subunit proteins with sedimentation constants of approximately 60S and molecular weights of the order of 10^6. Their isoelectric points are generally lower than those measured for intracellular haemoglobins of annelids.

Assembly and Subunit Structure. A detailed description of the quaternary structure demands accurate determination of the molecular weight of the whole moleule and of the various dissociation products, as well as analytical data leading to the minimum molecular weight. In general, the assembled molecules have[186,187] molecular weights ranging from ~2 to 4×10^6. In several cases their overall structure is organized as two superimposed hexagonal discs, composed of six globular subunits called subunit A. Each subunit A, in turn, is composed of smaller units (subunit B). The diameter and the height of this structure are, respectively, 220—270 Å and 105—174 Å. The central region of the whole molecule appears to be empty, except in one case.[188]

The minimum molecular weight, based on the haem content, shows[189] some variability, ranging from 20 000 to 29 000 for erythrocruorins and from 25 000 to 35 000 for chlorocruorins (extracellular haemoproteins carrying chlorohaem). These differences obviously imply different molecular arrangements and variations in the total number of haems carried by the whole molecule. In addition, and in contrast to vertebrate haemoglobins, there are indications, based on the haem–protein stoicheiometry, that the polypeptide chains are not always associated with a haem group.

Looking at the mode of assembly of the subunits, several models have been proposed (see ref. 189 for a review). In *Lumbricus* erythrocruorin, within the frame of the three-fold symmetry model, each subunit A appears to be formed by three subunits B, each containing four haem groups. A schematic representation

[186] R. L. Garlick, and R. C. Terwilliger, *Comp. Biochem. Physiol.*, 1977, **57B**, 177.
[187] R. C. Terwilliger, *Comp. Biochem. Physiol.*, 1974, **48A**, 745.
[188] E. P. J. van Bruggen and R. E. Weber, *Biochim. Biophys. Acta*, 1974, **359**, 210.
[189] E. Antonini and E. Chiancone, *Ann. Rev. Biophys. Bioeng.*, 1977, **6**, 239.

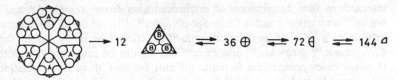

Sed. coeff.　60 S　　　　　11 S　　　5.5 S　　　3.5 S　　　　2.3 S

Mol. wt.　3.3×10^6　　　2.7×10^5　　9.2×10^4　4.6×10^4　　2.3×10^4

Figure 5 *Scheme for the dissociation processes of earthworm erythrocruorin. The dissociation of the whole molecule into A subunits is represented as irreversible*
(Reproduced by permission from *Ann. Rev. Biophys. Bioeng.*, 1977, **6**, 239)

of the molecule and its dissociation products is shown in Figure 5. A different structural model has been proposed[190] by Vinogradov *et al.* as the basis of the dissociation of *Lumbricus* erythrocruorin, studied in the presence of denaturants and at alkaline pH, by gel filtration and polyacrylamide gel electrophoresis. Obviously, only a detailed knowledge of the number, type, and amino-acid sequences of polypeptide chains, still not available, will tell us whether or not the subunit A is built in a similar way in different species.

Dissociation of annelid erythrocruorin into subunits occurs mainly at alkaline pH values ($\geqslant 8$); it is generally complex, involving several steps, some of which appear to be irreversible. In the case of *Lumbricus* erythrocruorin the distribution of dissociation products and the pH dependence of the interconversion rate constants are such that separation of the dissociated and undissociated molecules is easily achieved. This finding has allowed[191] a detailed study of the functional properties of both the assembled molecules and its products of dissociation. Even more complex, although qualitatively similar, is the behaviour[192] of *Spirographis* chlorocruorin.

A special case is represented[193] by the erythrocruorin from *Artemia salina*, whose molecular weight (240 000—260 000) is intermediate between those of giant and intracellular haemoglobins. Its molecular architecture is made up by two globin subunits of mol. wt. 122 000—130 000, and the minimum molecular weight is ~18 000. These findings suggest that the whole molecule carries 14 haems, 7 for each chain. Another peculiarity of this erythrocruorin appears from its amino-acid composition, which lacks (or has a very small content of) cystine. Generally, cystine is present in large proportions in extracellular erythrocruorins, where it is known[194-196] to be involved in the molecular arrangement. In this

[190] S. N. Vinogradov, J. M. Shlom, B. C. Hall, O. H. Kapp, and M. Mizukami, *Biochim. Biophys. Acta*, 1977, **492**, 136.
[191] B. Giardina, E. Chiancone, and E. Antonini, *J. Mol. Biol.*, 1975, **93**, 1.
[192] E. Chiancone, F. Ascoli, P. Vecchini, B. Giardina, M. L. Massa, and E. Antonini, manuscript in preparation, 1979.
[193] L. Moens and M. Kondo, *Biochem. J.*, 1977, **165**, 111.
[194] L. Waxman, *J. Biol. Chem.*, 1971, **246**, 7318.
[195] L. Waxman, *J. Biol. Chem.*, 1975, **250**, 3790.
[196] J. M. Shlom and S. N. Vinogradov, *J. Biol. Chem.*, 1973, **248**, 7904.

respect, *Artemia salina* erythrocruorin resembles the intracellular haemoglobins from annelids and haemoglobins of higher vertebrates.

Functional Properties. Annelid extracellular haemoglobins vary in their oxygen-binding properties, being in some cases characterized by strong haem–haem interactions (see descriptions of erythrocruorins from *Arenicola*[194] and *Lumbricus*[191] and chlorocruorin from *Spirographis*[192,197]), while in other cases they exhibit very little (if any) homotropic interaction (see references to those of *Cirraformia*,[198] *Neanthes*,[199] and *Pista*[200]). It should be remarked that, although in some cases comparison is made difficult because there is dissociation into subunits of lower molecular weight, the differences mentioned above are intrinsic, and not related to differences in the state of aggregation.

In the case of highly co-operative proteins ($n \geq 4$) (such as *Lumbricus* erythrocruorin or *Spirographis* chlorocruorin), the native molecules can be converted into a stable species with different ($n \approx 2$) and reversible O_2-binding properties by changing the pH either towards alkaline or acid values. Once achieved, this conversion is completely irreversible, and, as suggested,[191,192,201] might reflect a metastable state of the protein as isolated.

Oxygen binding and structural domains. Large respiratory proteins, containing ~ 100 or more ligand-binding sites, generally display[202] a very large value of $n_{\frac{1}{2}}$ and a relatively small value of the total interaction free energy (≤ 2 kcal per site).

An interpretation of the functional data, particularly of the kinetics of reaction, is based on the following model:[203] (*a*) the binding sites of the whole molecule are segregated into constellations that are essentially independent of one another, (*b*) within each constellation, the sites may strongly interact *via* ligand-linked conformational changes, (*c*) the overall co-operativity is the resultant of the very strong interactions within each constellation of sites and the opposing effect of the heterogenicity of the functional constellations, if present.

A general problem that is open to investigation is concerned with the relationships between functional constellations and structural domains. Generally, functional interactions occur in the assembled molecule, and are absent in its dissociation products. It may thus be reasonable to postulate that the regions of the molecule which are more tightly held together are the ones in which co-operative functional effects are operative. However, since a group of subunits acquires allosteric properties only in the whole molecule, it cannot be excluded that functional constellations extend over the boundaries that are identifiable by the structural domains. An interesting case is represented[204] by the extracellular haemoglobin of the planorbid snail *Helisoma trivolvis*. The molecule

[197] E. Antonini, A. Rossi-Fanelli, and A. Caputo, *Arch. Biochem. Biophys.*, 1962, **97**, 336.
[198] J. B. Swaney and J. M. Klotz, *Arch. Biochem. Biophys.*, 1971, **147**, 475.
[199] A. P. Econonides and R. M. G. Wells, *Comp. Biochem. Physiol.*, 1975, **A51**, 219
[200] R. C. Terwilliger, N. B. Terwilliger, and R. Roxby, *Comp. Biochem. Physiol.*, 1975, **50B**, 225.
[201] F. Ascoli, E. Chiancone, and E. Antonini, *J. Mol. Biol.*, 1976, **105**, 343.
[202] A. Colosimo, M. Brunori, and J. Wyman, *Biophys. Chem.*, 1974, **2**, 338.
[203] G. M. Giacometti, A. Focesi, B. Giardina, M. Brunori, and J. Wyman, *Proc. Nat. Acad. Sci. U.S.A.*, 1975, **72**, 4313.
[204] R. C. Terwilliger, N. B. Terwilliger, C. Bonaventura, and J. Bonaventura, *Biochim. Biophys. Acta*, 1977, **494**, 416.

has a molecular weight of 1.75×10^6 and a quaternary structure represented by a decameric ring with ten-fold symmetry. Its subunits are very large (mol. wt. = 175 000), about ten times those of most haemoglobins (mol. wt. = 15 000— 17 000). Gentle proteolysis with subtilisin cleaves the protein into haem-containing fragments, of molecular weight 15 000—17 000 and integral multiples of this value. Isolated fragments consisting of one haem group bind oxygen reversibly, with high oxygen affinity. Furthermore, heterotropic and homotropic interactions, present in the intact protein, are lost in the subunits, indicating that the molecule needs to be intact if the full functional behaviour is to be expressed. Although more detailed studies are needed to elucidate the structural similarity of the various fragments, it may be significant that the minimum units have molecular weights and amino-acid compositions which are comparable with those of the myoglobin of the radular muscle of the same animal.

General Considerations. It is obvious that a better understanding of the structure–function relationships in these giant respiratory proteins must await the elucidation of the fine structure, and hence deeper chemical and physico-chemical knowledge.

From the data now available, it appears that highly co-operative extracellular haemoglobins are molecules that are very sensitive to changes in environmental conditions, which may sometimes induce irreversible changes in their functional properties. Again, the presence of metastable states revealed by transient functional and structural properties has been stressed[189] to be a common characteristic of the highly co-operative proteins; however, the molecular basis of this effect and its physiological significance are questions open to further investigation. Some of the general problems relating the structural features and the functional properties of these giant haemoglobins are obviously similar to those encountered and extensively studied in the case of haemocyanins, which are dealt with later in this review.

The differences observed in the molecular control of the physiological function among the various giant respiratory proteins may certainly be taken as an example of the very different molecular adaptations by which closely related species are allowed to survive in their own environment. Finally, it is suggestive that, in terms of packaging haem units (number of haems per cm³), extracellular haemoglobins (2.0×10^{19} haems per cm³) compare favourably with vertebrate erythrocytes (1.3×10^{19} haems per cm³).[190]

PART II: Haemocyanin and Haemerythrin

5 Haemocyanin

Introductory Remarks.—Haemocyanin, a copper protein of high molecular weight, occurs freely dissolved in the haemolymph of invertebrates belonging to the phyla Mollusca and Arthropoda. Next to haemoglobin, it is the most widely distributed pigment in the animal kingdom. It occurs mainly in the marine members of these two phyla, but it is found in terrestrial and freshwater species as well. In the phylum Mollusca, the pigment is present in the classes Amphineura, Gastropoda, and Cephalopoda, but has not so far been demonstrated in the three remaining classes, the Monoplacophora, Scaphopoda, and Bivalvia. In the phylum Arthropoda, the presence of haemocyanin has been clearly demonstrated for the orders Decapoda, Stomatopoda, Isopoda, and Amphipoda of the class Crustacea, in the class Chilopoda, in the orders Araneae and Scorpiones (both belonging to the class Arachnida), and in the horseshoe crab (*Limulus poly-phemus*), which is one of the four remaining species of the order Xiphosura, which belongs to the nearly extinct subclass Merostomata.

Our knowledge of the structure and function of haemocyanin is still very incomplete. The presence of this pigment, particularly in the phylum Mollusca, raises an important question. Why should these animals mainly synthesize haemocyanin when they possess[205,206] a complete electron-transport system made up of flavoproteins and iron porphyrin compounds, and thus should be able to synthesize haem-containing respiratory proteins? It is also of interest that in most molluscs in which haemocyanin has been found, the presence of myoglobin, particularly in the radular muscles, has also been demonstrated.

Relatively little is known about the biosynthesis of haemocyanin. In *Limulus polyphemus*, haemocyanin is formed in a special type of blood cell known[207] as the cyanoblast, whilst in the cephalopod *Octopus* the site of synthesis has been shown[208] to be the paired branchial glands. Other investigations suggest that, in opisthobranchs, haemocyanin is synthesized in a blood gland, whilst in pulmonates it is synthesized[209] in pore cells. The fine structure of these pore cells is reported[210] by Sminia.

Excellent reviews on haemocyanin[6,7,211] have been produced in recent years. In 1977 the proceedings of an EMBO Workshop on the structure and function of haemocyanin were edited by Bannister.[5] A report on the meeting has also been published.[212]

General Properties.—Haemocyanin accounts for almost all the protein present in the haemolymph of both molluscs and crustaceans. Apo-haemocyanin is[213] best prepared by dialysis against cyanide. Removal of the copper does not appear[214] to disrupt the physical properties of the protein, and it is[215] easier to remove all the copper from arthropod than from molluscan haemocyanin. Reconstitution of apo-haemocyanin to functional protein has been achieved,[216] using a copper(I)–acetonitrile complex. The kinetics of the reaction between haemocyanin and

205 A. Ghiretti-Magaldi, A. Giuditta, and F. Ghiretti, *J. Cell. Comp. Physiol.*, 1958, **52**, 389.
206 F Ghiretti, A Ghiretti-Magaldi, and L. Tosi, *J. Gen. Physiol.*, 1959, **42**, 1185.
207 W. H. Fahrenbach, *J. Cell. Biol.*, 1970, **44**, 445.
208 P. N. Dilly and J. B. Messenger, *Z. Zellforsch. Mikrosk. Anat.*, 1972, **132**, 193.
209 T. Sminia, and H. H. Boer, *Z. Zellforsch. Mikrosk. Anat.*, 1973, **145**, 443.
210 T. Sminia, in 'Structure and Function of Haemocyanin', ed. J. V. Bannister, Springer-Verlag, Berlin, Heidelberg, and New York, 1977, p. 279.
211 F. Ghiretti 'Physiology and Biochemistry of Hemocyanins', Academic Press, New York and London, 1968.
212 J. V. Bannister and H. A. O. Hill, *Nature*, 1976, **263**, 280.
213 H. Fernandez-Moran, E. F. J. van Bruggen, and M. Ohtsuki, *J. Mol. Biol.*, 1966, **16**, 190.
214 A. Ghiretti-Magaldi, C. Nuzzolo, and F. Ghiretti, *Biochemistry*, 1966, **5**, 1943.
215 W. N. Konings, E. J. Siezen, and M. Gruber, *Biochim. Biophys. Acta*, 1969, **194**, 376.
216 R. Lontie, V. Blaton, M. Albert, and B. Peeters, *Arch. Internat. Physiol. Biochim.* 1965, **73**, 150.

cyanide and of the reconstitution reaction have been investigated.[217] In the reaction between haemocyanin and cyanide, the presence of two types of kinetically different copper at the active site is proposed. The reconstitution reaction appears to be complex, with a slow start and a subsequent rapid increase in rate, possibly in agreement with the presence of two kinetically different types of copper.

The molecular weights and copper contents of arthropod and molluscan haemocyanins show considerable variation. A copper content of 0.166—0.180% and a molecular weight of up to 1.0×10^6 have been reported for arthropod haemocyanin, whilst a copper content of 0.245—0.266% and a molecular weight of up to 9.0×10^6 have been reported for molluscan haemocyanin. The copper content corresponds[6,7] to an average minimum molecular weight of 36 700 and 25 000 for arthropod and molluscan haemocyanins respectively. It has, however, been established[216] that, during the binding of oxygen, two copper atoms per molecule of oxygen are involved in oxygenation. This means that the smallest functional subunits of arthropod and molluscan haemocyanin have molecular weights of 73 400 and 50 000 respectively.

Association–Dissociation Behaviour.—Haemocyanins form a remarkable variety of subunit aggregates. These aggregates are observed to be present under certain conditions of pH, ionic strength, and temperature. The properties of these aggregates were first investigated[218] by Ericksson-Quensel and Svedberg. Arthropod haemocyanins occur[6] as structures with sedimentation coefficients of around 5S, 16S, and 25S. In a few cases, particularly in *Limulus*, sedimentation coefficients of about 34S and 60S are also observed.[6,218] Molluscan haemocyanins have different subunit aggregation patterns. Subunits of 11S, 20S, 60S, and 100S have been reported.[6] Larger aggregates of 130S, 155S, and 175S have been detected[218] in *Busycon canaliculatum*. The larger aggregates of both arthropod and molluscan haemocyanins are the stable forms at nearly neutral pH. Dissociation occurs at low or high pH. The changes taking place with change in pH are not, as yet, clearly understood, and in general they are decreased when bivalent cations are present. Some of the observed dissociations have been found to be reversible. However, the processes involved do not appear to follow the mass-action law, since changes in the concentration of what is supposed to be an equilibrating mixture do not give the expected shift in composition. This behaviour has been observed by various workers, and the explanation supplied is based[219] on there being microheterogeneity. The nature of microheterogeneity is unknown, although it should involve charge differences between the chains. Siezen and van Driel have suggested[219] that differences in the carbohydrate content may be related to the microheterogeneity. The first direct evidence for this phenomenon has been reported[220] by Sullivan *et al.*, who separated the 5S subunits of *Limulus* haemocyanin into 5 major chromatographic zones. Recent

[217] B. Salvato and P. Zatta, in 'Structure and Function of Haemocyanin', ed. J. V. Bannister, Springer-Verlag, Berlin, Heidelberg, and New York, 1977, p. 245.
[218] I. B. Ericksson-Quensel and T. Svedberg, *Biol. Bull.*, 1936, **71**, 498.
[219] R. Siezen and F. van Driel, *Biochim. Biophys. Acta*, 1973, **295**, 131.
[220] B. Sullivan, J. Bonaventura, and C. Bonaventura, *Proc. Nat. Acad. Sci. U.S.A.*, 1974, **71**, 2558.

studies have confirmed the presence of heterogeneous monomers, which may be correlated with the effects attributed to microheterogeneity. In *Panulirus inter-ruptus* haemocyanin, two chromatographic zones are reported,[221] whilst three monomeric species are reported for the haemocyanin of the spiny lobster (*Jasus edwardsii*) and the swimming crab (*Ovalipes catharus*).[222] Dissociation of tarantula (*Dugesiella californica*) haemocyanin leads[223] to the formation of five monomers, which can be separated by preparative polyacrylamide gel electrophoresis, while six 5S monomers, which can be separated by ion-exchange chromato-graphy, are reported[224] from scorpion haemocyanin. A detailed analysis of the association–dissociation behaviour of this haemocyanin has been reported.[224]

The haemocyanin of the thalassinid shrimps *Callianassa californiensis*, *C. gigas*, and *Upogebia pugettensis* also exhibits heterogeneity among the monomers. The electrophoretic pattern of *U. pugettensis* is markedly different from those of *C. californiensis* and *C. gigas*, which show[225] similarities. The association–dis-sociation patterns of the haemocyanins of *C. californiensis* and *U. pugettensis* also show[225,226] differences. Dissociation of *C. californiensis* haemocyanin leads to the 17S subunit, which can associate either directly to the 39S tetramer or to the 25S dimer. The association to tetramer, which is the favoured reaction at room temperature, appears to be blocked at low temperature; instead, 25S dimers are stabilized by bivalent cations. Dimers are also formed in the absence of cations at pH 4.5 and pH 5.5. at room temperature. The stability conditions reported[226] for this haemocyanin are:

		[Magnesium]	
		High	Low
Temperature	High	Tetramer	Monomer
	Low	Dimer	Monomer

Dissociation of *U. pugettensis* haemocyanin does not lead directly to the 17S subunit, as in *C. californiensis*, but there is[225] a two-step transition, first from 39S tetramer to 25S dimer and then from 25S dimer to 17S monomer. Therefore, whilst *U. pugettensis* can form stable 25S dimers during dissociation at room temperature, this intermediate can only be formed in *C. californiensis* from 17S monomers at low temperature.

The presence of microheterogeneity no doubt complicates any quantitative analysis of the association–dissociation reactions. A model to interpret the microheterogeneity present during the dissociation of *Helix pomatia* haemocyanin has been proposed by Kegeles.[227] Using this model, Tai *et al.*[228] reported very satisfactory agreement between the equilibrium constant obtained from light scattering and that derived from relaxation kinetics.

[221] A. A. van den Berg, W. Gaastra, and H. A. Kuiper, in ref. 217, p. 6.
[222] H. A. Robinson and H. D. Ellerton, in ref. 217, p. 55.
[223] B. Linzen, D. Angersbach, R. Loewe, J. Markl, and R. Schmid, in ref. 217, p. 31.
[224] J. Lamy, J. Lamy, M. C. Baglin, and J. Weill, in ref. 217, p. 37.
[225] K. I. Miller, N. W. Eldred, F. Arisaka, and K. E. van Holde, *J. Comp. Physiol.*, 1977, **115**, 171.
[226] K. E. van Holde, D. Blair, N. Eldred, and F. Arisaka, in ref. 217, p. 22.
[227] G. Kegeles, *Arch. Biochem. Biophys.*, 1977, **180**, 530.
[228] M. J. Tai, G. Kegeles, and C.-H. Ke Huang, *Arch. Biochem. Biophys.*, 1977, **180**, 537.

Chemical and Enzymatic Investigations.—The haemocyanin in gastropods has been shown to be a glycoprotein. Carbohydrate side-chains have been suggested[229] as the links between the polypeptide chains, having[6] a functional role at the copper-binding site or being involved in determining the microheterogeneity (see above). However, little is known about the carbohydrate moieties. Glycopeptides solubilized with citraconic anhydride have been obtained[230] following cleavage of *Buccinum undatum* haemocyanin with cyanogen bromide. The carbohydrates found were generally glucosamine, fucose, and mannose, together with traces of galactosamine.

Various attempts at characterizing the size of the haemocyanin polypeptide chain by chemical and enzymatic means have been reported. There is, with one exception, universal agreement that the size of the arthropod chain is around 75 000 dalton. This has been confirmed for the haemocyanins of the tarantula,[223, 231] scorpion,[224] and crayfish.[232] Salvato and Ricchelli, however, claim,[233] from reconstitution and 1-anilino-8-naphthalenesulphonate (ANS) binding experiments, that the size of the minimal subunit of arthropod haemocyanin is 50 000 dalton. This is the size that is expected for the molluscan subunit from its copper content and oxygen binding. Several attempts at dissociating molluscan haemocyanin, using a variety of chemical dissociating reagents, have been reported.[229, 234–237] There is, however, a great discrepancy between the sizes obtained for the polypeptide chain. Brouwer and Kuiper reported[236] that the size of the polypeptide chain of *Helix pomatia* haemocyanin is 265 000 dalton, whilst Waxman reports[237] a value of 290 000 dalton for various molluscan haemocyanins. In the past year two reports were published on the size of the molluscan polypeptide chain. A chain of 50 000 dalton has been reported[238] for *Murex trunculus* haemocyanin exposed to denaturing conditions. This subunit appears to re-associate very easily. The other chain size reported[239] is for *Octopus vulgaris* haemocyanin. In the presence of 3M-urea, this haemocyanin dissociates into a 250 000 dalton chain. Further experimental evidence is required to elucidate whether the chains of high molecular weight described are the result of incomplete dissociation or whether the chains of low molecular weight are the result of breakage of peptide, or protein–sugar, or sugar bonds. Clearly, the chain size of molluscan haemocyanins remains to be elucidated.

The study of the structure of the polypeptide chain of molluscan haemocyanin has also been carried out, using a variety of proteolytic enzymes. A remarkable

[229] J. Dijk, M. Brouwer, A. Coert, and M Gruber, *Biochim. Biophys. Acta*, 1970, **221**, 467.
[230] R. L. Hall and E. J. Wood, in 'Structure and Function of Haemocyanin', ed. J. V. Bannister, Springer-Verlag, Berlin, Heidelberg, and New York, 1977, p. 95.
[231] H. J. Schneider, J. Markl, W. Schartau, and B. Linzen, *Z. Physiol. Chem.*, 1977, **358**, 1133.
[232] H. D. Ellerton, L. B. Collins, J. S. Gale, and A. Y. P. Yung, *Biophys. Chem.*, 1977, **6**, 47.
[233] B. Salvato and F. Ricchelli, in ref. 230, p. 113.
[234] W. N. Konings, J. Dijk, T. Wichertjes, E. C. Beuvery, and M. Gruber, *Biochim. Biophys. Acta*, 1969, **188**, 43.
[235] J. Cox, R. Witters, and R. Lontie, *Internat. J. Biochem.*, 1972, **3**, 283.
[236] M. Brouwer and H. A. Kuiper, *European J. Biochem.*, 1973, **35**, 428.
[237] L. Waxman, *J. Biol. Chem.*, 1975, **250**, 3796.
[238] J. V. Bannister, J. Mallia, A. Anastasi, and W. H. Bannister, in 'Structure and Function of Haemocyanin', ed. J. V. Bannister, Springer-Verlag, Berlin, Heidelberg, and New York, 1977, p. 103.
[239] B. Salvato, and L. Tallandini, in ref. 238, p. 217.

degree of success has been obtained with subtilisin. With this enzyme, functional subunits of molecular weight 50 000 were obtained from *H. pomatia* and *M. trunculus* haemocyanin.[240,241] These subunits were, however, heterogeneous in polyacrylamide gel electrophoresis. Bonaventura *et al.*[242] digested the haemocyanin from several molluscan species with subtilisin, and the subunits obtained were separated by ion-exchange chromatography. Some of the chromatography peaks obtained were shown to be homogeneous by electrophoresis. In all the digests, molecular weights of about 50 000 were present.

Other enzymes have been used to obtain 50 000 dalton subunits. Some degree of success has been obtained with papain, but this enzyme does not appear[238] to be as effective as subtilisin. Gielens *et al.*, by digesting 1/20 molecular weight subunits from *H. pomatia* (using a combination of plasmin, staphylococcal protease, and trypsin), found[243] that eight oxygen-binding sites are present in this polypeptide chain. The digestion of haemocyanin with trypsin has also aroused a great deal of interest. When haemocyanin from *H. pomatia* is digested with trypsin in the presence of calcium, the formation of aggregates of very high molecular weight, called tubes, was observed.[244] Similar material was obtained using *Lymnaea stagnalis* haemocyanin. In this haemocyanin a shift in the spectral bands due to the copper–oxygen complex was observed.[245] It is interesting to note that other haemocyanins do not form tubular structures when digested with trypsin.[238,246]

Arthropod haemocyanins behave totally differently when subjected to proteolytic digestion. When *Limulus* haemocyanin subunits are digested with subtilisin, there is little evidence of proteolytic cleavage in regular electrophoresis, but extensive cleavage is reported to occur under denaturing conditions.[242] Clearly, proteolytic digestion of arthropod haemocyanins requires further investigation.

Extensive amino-acid analyses of haemocyanins from very different sources have been obtained[247] and analysed by Ghiretti-Magaldi *et al.* The authors have drawn interesting conclusions on the evolutionary aspects of haemocyanin from this approach.

Oxygen-binding Equilibria and Kinetics.—The ability of a variety of haemocyanins to bind oxygen has been studied.[248–250] Besides the usual protein band at 280 nm, oxyhaemocyanin exhibits copper bands around 350 and 580 nm. The absorption at 350 nm is generally used as a measure of oxygen binding, since it is more intense than the 580 nm band. It is assumed that the intensity of this band is proportional to the extent of oxygen binding. A detailed study of oxygen binding by *Murex trunculus* haemocyanin has been published by Bannister *et al.*[248]

[240] R. Lontie, M. DeLey, H. Robberecht, and R. Witters, *Nature*, 1973, **242**, 180.
[241] J. V. Bannister, A. Galdes, and W. H. Bannister, *Comp. Biochem. Physiol.*, 1975, **51B**, 1.
[242] J. Bonaventura, C. Bonaventura, and B. Sullivan, in ref. 238, p. 206.
[243] C. Gielens, G. Preaux, and R. Lontie, in ref. 238, p. 85.
[244] J. F. L. van Breemen, T. Wichertjes, M. F. J. Muller, R. van Driel, and E. F. J. van Bruggen, *European J. Biochem.*, 1975, **60**, 129.
[245] E. J. Wood, in 'Structure and Function of Haemocyanin', ed. J. V. Bannister, Springer-Verlag, Berlin, Heidelberg, and New York, 1977, p. 77.
[246] E. J. Wood and L. J. Mosby, *Trans. Biochem. Soc.*, 1977, **5**, 694.
[247] A. Ghiretti-Magaldi, C. Milanesi, and G. Tognon, *Cell Differ.*, 1977, **6**, 167.
[248] J. V. Bannister, A. Galdes, and W. H. Bannister, in ref. 245, p. 193.
[249] E. J. Wood, G. R. Cayley, and J. S. Pearson, *J. Mol. Biol.*, 1977, **109**, 1.
[250] R. van Driel and E. F. J. van Bruggen, *Biochemistry*, 1974, **13**, 4079.

Oxygen binding was investigated in the absence and in the presence of calcium ions and as a function of ionic strength. In the absence of bivalent ions, a small but statistically significant reverse Bohr effect was reported, and the value of n (Hill's constant) was not significantly different from unity. In a similar study, however, Linzen *et al.*, measuring[231] the oxygen-binding properties of *Dugesiella californica* haemocyanin, observed a strong positive Bohr effect and a four-fold change in the value of n within one pH unit.

For haemocyanin of the whelk (*Buccinum undatum*) oxygen is bound[249] non-co-operatively at pH 7.2 and 8.2 in the absence of bivalent ions, whilst the addition of calcium or magnesium at pH values near 8.0 resulted in the appearance of co-operativity. The results obtained with *M. trunculus* haemocyanin for the effect of ionic strength in the absence of calcium ions indicate[249] that the presence of bivalent ions is not essential for co-operative behaviour at high ionic strength. Co-operativity in the absence of calcium ions has also been shown to occur with *Helix pomatia* haemocyanin,[250] with the 250 000 dalton subunit obtained[239] from *Octopus* haemocyanin in 3M-urea, and with three arthropod haemocyanins.[251] According to Klarman and Daniel, the haemocyanins of three arthropods, *i.e.* a scorpion (*Leinus quinquestriatus*) and the crabs *Telphusa fluviatilis* and *Ocypoda cursor*, behave[251] co-operatively in the absence of calcium and magnesium. The effect of bivalent ions on arthropod haemocyanins seems to be restricted[251] to modifying the affinity and degree of co-operativity in a way reminiscent of the role played by diphosphoglycerate in the haemoglobin–oxygen system. It also appears that, in molluscan haemocyanins, co-operativity can also be mediated[248,250] by other effects besides calcium concentration.

Data on oxygen binding by subunits have been presented.[242,252] In *Dugesiella* haemocyanin, association of 6S to 16S subunit does not alter the oxygen affinity, and there is very little change in the value of n, whilst in *Cupiennius* the same effect results in a strong decrease of affinity, with a significant increase in co-operativity. In both species, dissociation of 27S and 24S particles to 6S results in a significant decrease in p_{50} and in the value of n. Oxygenation studies[244] on the various chromatographic zones obtained after separation of the subtilisin digest of *Limulus* haemocyanin showed variation in p_{50} values, and the value of n was frequently less than 1, suggesting that there is heterogeneity in oxygen-binding material within some of the chromatographic zones.

The two-state allosteric model[8] of Monod, Wyman, and Changeux has been applied, from time to time, to describe the O_2-binding properties of haemocyanins. The canonical model cannot generally describe the ligand-binding curve under different conditions, and modifications have been proposed (see, for example, ref. 253). The oxygen binding of *Murex trunculus* haemocyanins has been qualitatively interpreted[248] according to a two-state model. It was found that, in the absence of bivalent ions, haemocyanin exists mainly in the R state and hence binds oxygen non-co-operatively at all pH values; on addition of bivalent ions, dissociation is prevented and a state that has low oxygen affinity is stabilized. The T and R states bind oxygen non-co-operatively, with p_{50} values of approxi-

[251] A. Klarman and E. Daniel, *J. Mol. Biol.*, 1977, **115**, 257.
[252] R. Loewe, R. Schmid, and B. Linzen, in ref. 245, p. 50.
[253] K. I. Miller and K. E. van Holde, *Biochemistry*, 1974, **13**, 1668.

mately 42 and 2, respectively, the observed co-operativity being related[248] to the transition between the T and R states. Two papers dealing quantitatively with the binding of ligands by haemocyanins have recently appeared. Brouwer *et al.* have reported[254] on the effect of Cl^- on the O_2 dissociation curve of whole *Limulus* haemocyanin, and described their data on the basis of an effect of the anion concentration of the T state itself, as well as general stabilization of this state by chloride ions. The pH dependence of O_2 binding of the β-haemocyanin from *Helix pomatia* has been reported[255] by Zolla *et al.* Over a small pH interval (from ~ 6.5 to ~ 8) there is a complete transition from almost pure R state (at lower pH) to a pure T state (at higher pH), with an additional pH effect on T itself, as demonstrated by the dependence on pH of the bottom asymptote.

The kinetics of oxygen binding have been investigated following the initial work using the temperature-jump technique.[256] Relaxation measurements of dissociated haemocyanin from *Panulirus interruptus* show[257] one relaxation process, with rate constants of $k_{on} = 4.6 \times 10^7$ $1\,mol^{-1}$ s^{-1} and $k_{off} = 1500$ s^{-1}. Taking the properties of the dissociated protein as characteristic of the T state, as indicated by the equilibrium behaviour, this result showed that the kinetic basis of the low affinity in haemocyanins is in the very high value of the dissociation rate constant. This becomes clear by comparing the O_2 combination and dissociation rate constants for the T state (given above) with those determined[257] for the R state of the undissociated molecule, *i.e.* at very high saturations ($k_{on} = 3.1 \times 10^7$ $1\,mol^{-1}$ s^{-1} and $k_{off} = 60$ s^{-1}).

A similar conclusion has been reached by comparing the kinetics of the R and T states of *Helix pomatia* α-haemocyanin at pH 8.4 and in the presence of 10mM-$CaCl_2$. Under these conditions the molecule is highly co-operative, and its ligand-binding kinetics have been carefully investigated[258, 259] by temperature-jump and stopped-flow techniques. The two major conclusions reached by these workers are: (*a*) that the low-affinity characteristic of the T state is due to the very high value of the O_2 dissociation rate constant, the combination rate constant being similar for both states, (*b*) that the switch from T to R is rate-limiting under some conditions, the half time for the allosteric transition being approximately 10—100 ms. The latter finding, which has been verified[260] by a different approach for the same system, has led the authors to propose that, when the size of the allosteric unit involves many binding sites (~ 15 for *Helix* haemocyanin), the conformational switch may become rate-limiting.

The Active Site.—The active site of haemocyanin has been the subject of a number of investigations which involved studies either of the copper or of the ligands interacting with the metals. The general view is that in deoxyhaemocyanin the copper is available in the copper(I) state and that in oxyhaemocyanin the copper forms a peroxo-complex in the copper(II) state: $Cu^{II}–O_2{}^{2-}–Cu^{II}$. The cupric state

[254] M. Brouwer, C. Bonaventura, and J. Bonaventura, *Biochemistry*, 1977, **16**, 3897.
[255] L. Zolla, H. A. Kuiper, P. Vecchini, E. Antonini, and M. Brunori, *European J. Biochem.*, 1978, **87**, 467.
[256] M. Brunori, *J. Mol. Biol.*, 1971, **55**, 39.
[257] H. A. Kuiper, E. Antonini, and M. Brunori, *J. Mol. Biol.*, 1977, **116**, 569.
[258] R. van Driel, M. Brunori, and E. Antonini, *J. Mol. Biol.*, 1974, **89**, 103.
[259] R. van Driel, H. A. Kuiper, E. Antonini, and M. Brunori, *J. Mol. Biol.*, 1978, **121**, 431.
[260] H. A. Kuiper, M. Brunori, and E. Antonini, *Biochem. Biophys. Res. Comm.*, 1978, **82**, 1062.

of copper in oxyhaemocyanin has been confirmed[261] by X-ray photoelectron spectroscopy, whilst an interpretation[262, 263] of the optical spectra, magnetic circular dichroism,[264] and resonance Raman spectra[265] demonstrate the presence of a peroxo-bridge.

Haemocyanins show[266] no e.p.r. signal in the native state, either oxy or deoxy. This does not necessarily imply that the copper is in the diamagnetic state. Two paramagnetic ions which would give e.p.r. signals in mononuclear centres may become 'silent' if they are very close to each other, whether or not they are bridged by a common ligand. Evidence has been obtained for dipolar interaction in e.p.r. spectra of haemocyanin treated with nitric oxide.[267] Seven hyperfine lines are observed on the half-field signal, showing that the triplet spectrum arises from two copper(II) centres interacting with each other. It appears that the nitric oxide oxidizes the copper(I) ions of the deoxy-protein and binds to each of the resulting copper(II) ions, thus preventing an exchange interaction between them. Recent evidence, however, indicates[268] that the nitric oxide is not a ligand of the copper. The distance between the two copper centres was calculated[268] as approximately 6 Å. This distance was considered to fit an exchange interaction of the two coppers through a peroxo bridge in the native oxy-protein.

Witters *et al.* investigated[269] the e.p.r. spectrum of fluoride- and azide-generated methaemocyanin. The spectra obtained by either treatment indicate the presence of copper pairs. The e.p.r. spectrum of fluoride-generated methaemocyanin vanishes on further addition of fluoride, indicating that the solution behaves as fresh haemocyanin. As a working hypothesis, these authors suggest that copper(I), a bridging ligand, low-spin copper, and peroxide are present, as follows:

$$Cu^{I}—L—Cu^{III}—O \diagdown O \quad ^{2-}$$

This hypothesis is further extended[270] by Lontie and is not inconsistent with spectral data.[271]

Studies on the ligands involved in metal binding indicate the presence of histidine and tryptophan. Photo-oxidation studies and fluorescence spectra indicate[272-274] that these two amino-acids are involved.

[261] H. van der Deen, R. van Driel, A. H. Jonkman-Beuker, G. A. Sawatzky, and R. Wever, in 'Structure and Function of Haemocyanin', ed. J. V. Bannister, Springer-Verlag, Berlin, Heidelberg, and New York, 1977, p. 172.
[262] K. E. van Holde, *Biochemistry*, 1967, 6, 93.
[263] W. H. Bannister, and E. J. Wood, *Nature*, 1969, 223, 53.
[264] W. Mori, O. Yamaguchi, Y. Nakas, and A. Nakahara, *Biochem. Biophys. Res. Comm.*, 1975, 66, 725.
[265] E. I. Solomon, D. M. Dooley, R. H. Wang, H. B. Gray, M. Cerdonio, F. Mogno, and G. L. Romani, *J. Amer. Chem. Soc.*, 1976, 98, 1029.
[266] T. Nakamura and H. S. Mason, *Biochem. Biophys. Res. Comm.*, 1960, 3, 297.
[267] A. J. M. Schoot-Uiterkamp, *F.E.B.S. Letters*, 1972, 20, 93.
[268] H. van der Deen and H. Hoving, *Biochemistry*, 1977, 16, 3519.
[269] R. Witters, M. DeLey, and R. Lontie, in 'Structure and Function of Haemocyanin', ed. J. V. Bannister, Springer-Verlag, Berlin, Heidelberg, and New York, 1977, p. 239.
[270] R. Lontie, in ref. 269, p. 150.
[271] L. Calabrese, and J. Rotilio, in ref. 269, p. 180.
[272] E. J. Wood and W. H. Bannister, *Biochim. Biophys. Acta*, 1968, 154, 10.
[273] G. Jori, B. Salvato, and L. Tallandini, in 'Structure and Function of Haemocyanin', ed. J. V. Bannister, Springer-Verlag, Berlin, Heidelberg, and New York, 1977, p. 156.
[274] M. DeLey and R. Lontie, in ref. 273, p. 164.

Models.—There are as yet no available studies on the *X*-ray structure of haemocyanin. Models have to be constructed from electron micrographs. A preliminary *X*-ray study on *Limulus* haemocyanin subunits has been reported[275] by Magnus and Love, but, as stated by these workers, the crystals obtained at the time were unsuitable for high-resolution studies. Conditions should therefore be sought for obtaining more suitable crystals, and these seem to have been found very recently (Love; personal communication). Hopefully, the first results may be available in a short while.

Electron microscope studies have provided useful information on the structure of haemocyanin. The first model, based on a three-dimensional reconstruction from electron micrographs, was published[276] by Mellema and Klug. The undissociated molecule from *Helix* was shown to be cylindrical and composed of sixty morphological units; six layers of ten units each. Besides the wall, two collars are present, each containing five blobs of material. The diameter of the cylinder is about 260—300 Å and the height is 340—380 Å. Dissociation into half molecules occurs in a plane perpendicular to the axis of the cylinder. Therefore, in the whole molecule, two of these asymmetrical half-size molecules join their open ends to form a hollow cylindrical whole, more or less limited at both ends by a collar cap.

Siezen and van Bruggen have presented[277] a refinement to the model of Mellema and Klug by putting forward a model for the dissociation of the half cylinder into compact one-tenth molecules. The dissociation occurs along one set of the helical grooves, releasing a piece of the cylinder wall with one blob of the collar attached to it. These compact one-tenth molecules can change their conformation, to become a loose one-tenth molecule. Van Breemen *et al.* proposed[278] an arrangement of two one-twentieth molecules within the compact one-tenth molecule. The assumption is that each one-twentieth molecule is a polypeptide chain of eight domains (oxygen-binding), six belonging to the wall of the cylinder and two to the collar.

6 Haemerythrin

Haemerythrin has been found in a few species belonging to four different phyla – the Sipunculids, the Polychaetes, the Priapulids, and the Branchiopods. The pigment has been found in all the Sipunculids examined, and the species belonging to this phylum have been studied the most. Haemerythrin is located in erythrocytes circulating in the vascular system and in the coelom. A muscle form, termed myohaemerythrin, is located in the retractory muscles.

Haemerythrin does not have an iron–porphyrin system but has instead iron directly bound to the protein. The molecular weight of haemerythrin is 108 000 and the iron content is[279] 0.82%, which gives a value of 6800 dalton for the minimum subunit. However, the smallest functional subunit has a molecular weight of 13 600, since two iron atoms have been found[279] to bind one oxygen

[275] K. A. Magnus and W. E. Love in ref. 273, p. 70.
[276] J. E. Mellema and A. Klug, *Nature*, 1972, **239**, 146.
[277] R. J. Siezen and E. F. J. van Bruggen, *J. Mol. Biol.*, 1974, **90**, 77.
[278] J. F. L. van Breemen, G. J. Schuurhuis, and E. F. J. van Bruggen, in ref. 273, p. 122.
[279] I. M. Klotz, T. A. Klotz, and H. A. Fiess, *Arch. Biochem. Biophys.*, 1957, **68**, 284.

molecule. The protein, therefore, has eight functional subunits. Whilst the coelomic and vascular haemerythrins occur[280] as octameric molecules, myo-haemerythrin occurs as a monomer of 13 600 dalton.

The spectral properties of oxyhaemerythrin, which is violet-red, show bands at 330 and 380 nm and at 550 nm. In deoxyhaemerythrin, the band at 550 nm vanishes. Spectroscopic studies on haemerythrin have been reported.[281, 282]

The most extensively characterized haemerythrin is that of the Sipunculid *Goldfingia gouldii*. The primary structure of this haemerythrin is known.[283-285] The primary structure of myohaemerythrin from *Dendrostomum pyroides* is also known.[286]

The mechanism of reversible oxygen binding of haemerythrin is of great interest *vis-à-vis* the binding of oxygen by haem proteins. Various studies have shown no Bohr effect, with only a slight site–site interaction, the value of *n* (Hill's constant) being[287,288] between 1.0 and 1.4. In all the haemerythrins investigated, a strong temperature dependence for the oxygenation reaction has been found. The nature of oxygen binding is[287, 289] more than a simple bimolecular liganding process, although there is some disagreement about this.[290] At any rate, it has been shown that in the oxy-form two iron environments are present.[291, 292]

Since there is no haem or other prosthetic group in haemerythrin, the binding and electronic state of the iron have aroused considerable interest. Modification of haemerythrin by a number of group-specific reagents has yielded information about the amino-groups involved in iron and subunit binding. Histidine, tyrosine, and tryptophan have been implicated.[293-295]

Haemerythrins show a tendency for autoxidation whereby the oxy-form changes into the iron(III) methaemerythrin form. This process is of fundamental importance since it leads to the loss of oxygen-carrying capacity. Oxyhaemerythrin changes slowly into the methaemerythrin form in solution. The transformation is accelerated[296-298] by various anions.

[280] G. L. Klippenstein, D. A van Riper, and E. A. Oosterom, *J. Biol. Chem.*, 1972, **247**, 5959.
[281] K. A. Garbett, D. W. Darnall, I. M. Klotz, and R. J. P. Williams, *Arch. Biochem. Biophys.*, 1969, **103**, 419.
[282] A. W. Addison and R. E. Bruce, *Arch. Biochem. Biophys.*, 1977, **183**, 328.
[283] A. R. Subramaniam, J. W. Holleman, and I. M. Klotz, *Biochemistry*, 1968, **7**, 3859.
[284] G. L. Klippenstein, J. W. Holleman, and I. M. Klotz, *Biochemistry*, 1968, **7**, 3868.
[285] G. L. Klippenstein, *Biochemistry*, 1972, **11**, 372.
[286] G. L. Klippenstein, J. L. Cote, and S. E. Ludlam, *Biochemistry*, 1976, **15**, 1128.
[287] G. Bates, M. Brunori, G. Amiconi, E. Antonini, and J. Wyman, *Biochemistry*, 1968, **7**, 3016.
[288] R. M. G. Wells and R. P. Dales, *Comp. Biochem. Physiol.*, 1974, **49A**, 57.
[289] F. Bossa, M. Brunori, G. W. Bates, E. Antonini, and P. Fasella, *Biochim. Biophys. Acta*, 1970, **207**, 41.
[290] J. A. De Waal and R. G. Wilkins, *J. Biol. Chem.*, 1976, **251**, 2339.
[291] J. L. York and A. J. Bearden, *Biochemistry*, 1970, **9**, 4549.
[292] M. Y. Okamura, J. M. Klotz, C. E. Johnson, M. R. C. Winter, and R. J. P. Williams, *Biochemistry*, 1969, **8**, 1961.
[293] R. L. Rill and I. M. Klotz, *Arch. Biochem. Biophys.*, 1971, **147**, 226.
[294] J. L. York and C. C. Fan, *Biochemistry*, 1971, **10**, 1659.
[295] J. L. York and M. P. Roberts, *Biochim. Biophys. Acta*, 1976, **420**, 265.
[296] M. Y. Okamura, and I. M. Klotz, in 'Inorganic Biochemistry', ed. G. L. Eichhorn. Elsevier, Amsterdam. 1973, vol. 1, p. 320.
[297] S. Keresztes-Nagy and I. M. Klotz, *Biochemistry*, 1965, **4**, 919.
[298] Z. Bradic, R. Conrad, and R. G. Wilkins, *J. Biol. Chem.*, 1977, **252**, 6069.

6

Electron-transport Proteins

C. GREENWOOD & D. BARBER

Oxidation–reduction reactions are fundamental to all life as we know it, and are intimately involved in the energy-transducing process of anaerobes as well as aerobes. Although not exclusively so, copper- and iron-containing proteins dominate in the electron-transfer role, with the polypeptide component appearing to tune the metal centre for the redox role, as well as providing a vehicle which will allow for specific interaction between chosen partners and, in some cases, a means of location in a membrane or in the cytosol. There are also good reasons for supposing that the protein enables electrons to move over considerable distances from one redox centre to another, although the mechanism for this is still somewhat speculative.

In attempting to cover the enormous range of electron-transporting proteins the arbitrary decision has been made to consider in this chapter only those metalloproteins involved in electron transfer where this is not directly coupled to an enzymatic reaction. This will be recognized as the prejudice which does not consider the electron as a substrate. In order to get complete coverage it will be necessary for the reader to study Chapter 7. Despite this restriction, the diverse properties of the various classes of metalloprotein have necessitated sub-division of material within the chapter, so that cytochromes, iron–sulphur proteins, copper proteins, and photosynthetic aspects are dealt with separately.

1 Cytochromes

Mammalian Cytochromes c.—In 1930 Keilin[1] first obtained a crude preparation of cytochrome c from Delft yeast and, in so doing, provided protein chemists, biochemists, and inorganic biochemists with a research vehicle second only to haemoglobin in popularity. In the case of cytochrome c, and in spite of the enormous amount of structural and functional information that has been accumulated in the past fifty years, it is a sobering thought that there is as yet no consensus as to how the protein carries out its deceptively simple role of accepting an electron from cytochrome c_1 and transferring it to cytochrome c oxidase.

The capacity to function in a redox role is an inherent property of ferric and ferrous ions in solution and also of iron porphyrins; thus the role of the protein is to 'tune' these basic properties, and no haem protein, including cytochrome c, could be effective without this. A major objective in cytochrome c research, so

---

[1] D. Keilin, *Proc. Roy. Soc.*, 1930, **B106**, 418.

210

far largely unattained, is an understanding at the mechanistic level of these quite remarkable phenomena. Undeterred (and perhaps even spurred) by this relative lack of success, research on cytochrome *c* continues apace, but, before going on to review the current year's activity, it is perhaps appropriate to mention some germinal and current reviews. The most comprehensive review of cytochrome *c* to appear to date is that by Margoliash and Schejter[2] in 1966 and is particularly valuable for its coverage of the early literature. Since then Dickerson and Timkovich[3] have produced an admirable contribution which is especially useful for its presentation of structural information on the molecule. Two valuable reviews have recently appeared which give particular emphasis to the electron-transfer properties of cytochrome *c*[4] on the one hand and the molecular variations of cytochrome *c* as a function of the evolution of species[5] on the other.

Sequence and Structure. Because of the relative ease of purification of cytochrome *c* it has become the example, *par excellence*, of the newly emerging subject of protein taxonomy, in which species are compared on the basis of protein sequence. It seems reasonable to expect that a thorough understanding of structure–function relationships for cytochrome *c* on the basis of amino-acid sequences from a wide range of species could lead to a view of evolution that is more precise than has been possible so far. It is not surprising therefore that the year has seen its new crop of sequences to add to the growing list. Niece *et al.*[6] have determined the complete amino-acid sequence of the South American tylopod, the guanaco (*Llama guanicoe*) and, interestingly, this turns out to be identical to those of the whale and the camel. In another study Carlson and co-workers[7] report on the primary structure of mouse, rat, and guinea-pig cytochromes *c*. They conclude that rodent cytochromes *c* are evolutionarily conservative and that their findings permit them to test the generation-time hypothesis, according to which molecular evolution is faster in organisms with short generation times (*i.e.* rodents) than in organisms with long generations (*i.e.* primates). The results of Carlson *et al.*[7] would seem to support the converse view, that cytochrome *c* appears to have experienced faster evolution in primates than in rodents. The questions of phylogenetic trees based on protein sequence data, of the rates of evolutionary change, and of the role of evolutionary selection, including the spatial structure of cytochrome *c* and constraints on evolutionary variation, are all well covered by Margoliash,[5] in Volume 29 of *Advances in Chemical Physics*, which is devoted to membranes, dissipative structures, and evolution.

With respect to evolutionary considerations, it is important not to lose sight of the fact that the amino-acid sequence of cytochrome *c* is, operationally, near the genetic end of the scale of life and, by itself, has no phenotypic meaning. Only when the polypeptide chain folds into its own particular tertiary structure

[2] E. Margoliash and A. Schejter, *Adv. Protein Chem.*, 1966, **21**, 113.
[3] R. E. Dickerson and R. Timkovich, in 'The Enzymes', ed. P. D. Boyer, 1975, Vol. XI, p. 397.
[4] F. R. Salemme, *Ann. Rev. Biochem.*, 1977, **46**, 299.
[5] E. Margoliash, *Adv. Chem. Phys.*, 1977, **29**, 191.
[6] R. L. Niece, E. Margoliash, and W. M. Fitch, *Biochemistry*, 1977, **16**, 68.
[7] S. S. Carlson, G. A. Mross, A. C. Wilson, R. T. Mead, D. W. Lawrence, S. F. Bowers, N. T. Foley, A. O. Muijsers, and E. Margoliash, *Biochemistry*, 1977, **16**, 1437.

and incorporates the prosthetic group (haem *c*) does it become a functioning molecular machine. Since evolutionary selection operates on the basis of fitness to fulfil a particular task, our understanding of this and of the actual molecular mechanism of electron transfer depends on knowledge of the tertiary structure of the protein. In terms of three-dimensional structure, *X*-ray crystallography remains our most successful experimental approach, and in the case of cytochrome *c*, Dickerson's group in California are leaders in the field. In a series of papers appearing in the *Journal of Biological Chemistry*[8-10] this group report the structures of oxidized and reduced tuna cytochrome *c* at 2.0 Å resolution through the use of four isomorphous heavy-atom derivatives. They conclude that there are no significant differences to be observed between the two oxidation states and that the difference-map studies using re-oxidized crystals of ferrocytochrome *c* confirm the absence of a conformation change. A detailed analysis of H-bonding (see Figure 1) shows the presence of six β or 3_{10} bends of type II, with obligatory glycines in the third residue position which account for six of the ten nearly invariant glycine residues in the molecule.

With the increased knowledge of cytochrome *c* structure it has become possible to design and interpret experiments concerned with the antigenicity of cytochrome *c*. Urbanski and Margoliash[11,12] have begun an immunological investigation in part concerned with rabbit, mouse, and guanaco cytochromes *c*, which differ from each other by two amino-acid residues. Because of this small difference between these molecules, no shift in the polypeptide backbone conformation can have occurred, and thus the immune system must be responding to local changes in the surface topography of the molecule, and all of the sites except one centre around differences of single amino-acid residues between antigen and host cytochrome *c*.

Chemical Modification. The chemical modification of amino-acid residues in cytochrome *c* continues to exert its attraction for workers interested in probing the possible specific role of a particular residue as well as the overall effects of modification. It is important to recognize the limitations of this approach and the possible dangers in over-interpretation of data, particularly that concerned with reactivity, which can arise from non-specific effects of the reagents employed.

Several papers have appeared in the literature concerned with the modification of lysine residues in the molecule and the consequent effect on reactivity with cytochrome *c* oxidase,[13] cytochrome *c* peroxidase,[14] and cytochrome b_5.[15] Trifluoroacetylation of lysine residues 13, 55, and 99 has been reported by

[8] R. Swanson, B. L. Trus, N. Mandel, G. Mandel, O. B. Kallai, and R. E. Dickerson, *J. Biol Chem.*, 1977, **252**, 759.

[9] T. Takano, B. L. Trus, N. Mandel, G. Mandel, O. B. Kallai, R. Swanson, and R. E. Dickerson, *J. Biol. Chem.*, 1977, **252**, 776.

[10] N. Mandel, G. Mandel, B. L. Trus, J. Rosenberg, G. Carlson, and R. E. Dickerson, *J. Biol. Chem.*, 1977, **252**, 4619.

[11] G. J. Urbanski and E. Margoliash, *J. Immunol.*, 1977, **118**, 1170.

[12] G. J. Urbanski and E. Margoliash, in 'Immunochemistry of Enzymes and their Antibodies', 1977, p. 203.

[13] N. Staudenmayer and N. G. Siong, *Biochemistry*, 1977, **16**, 600.

[14] E. Stellwagen, L. M. Smith, R. Cass, R. Ledger, and H. Wilgus, *Biochemistry*, 1977, **16**, 4975.

[15] S. Ng, M. B. Smith, H. T. Smith, and F. Millett, *Biochemistry*, 1977, **16**, 4975.

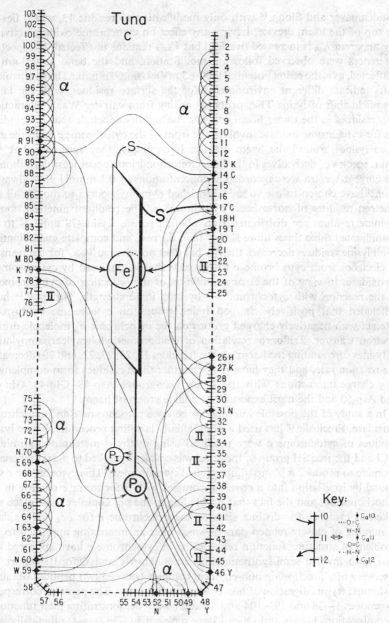

Figure 1 *Hydrogen-bonding diagram for tuna ferri- and ferro-cytochrome* c.
*Arrows run from hydrogen donor to hydrogen acceptor. Fifth and sixth
iron ligands are shown as darker arrows. S = thioether haem links
via cysteines 14 and 17. P_I and P_0 are inner and outer propionic groups
on haem. α and II indicate regions of α-helix and type II 3_{10} bend*

(Reproduced by permission from *J. Biol. Chem.*, 1977, **252**, 776)

Staudenmayer and Siong,[13] with only modification to residue 13, which lies at the top of the haem crevice, having any effect on cytochrome oxidase activity. The apparent K_m is increased five-fold but V_{max} remains unaffected. No spectral differences were observed following modification, and the band at 695 nm is unaffected, as is the redox potential (260 ± 5 mV). Using [19]F n.m.r., large chemical shifts indicate different environments for the surface residues, with the large downfield shift of lysine TFA probably arising from van der Waals interaction with residues in the upper haem crevice. The authors conclude that the binding site for cytochrome oxidase involves the front of the cytochrome *c* molecule and those lysines around the haem crevice. Interestingly, they speculate that [19]F n.m.r. spectra of derivatives in the intact mitochondrion appear feasible, although the current spectra were recorded at concentrations of 0.1 mmol l^{-1}. Stellwagen *et al.*[14] have chosen to use 90% ^{13}C-enriched *O*-methylisourea to modify all the nineteen residues of horse heart cytochrome *c*. The resultant nineteen homo-arginine residues are distributed over about 1 p.p.m. Lysine-79 appears to be guanidinated three times more slowly than the rest, and complete guanidination of all lysine residues decreases the K_m but does not change V_{max} for the transfer of electrons from cytochrome *c* to peroxide that is catalysed by cytochrome *c* peroxidase. In view of the strong inhibition of the reduction of cytochrome *c*, in the reaction with cytochrome b_5, by high ionic strength, Ng *et al.*[15] have concluded that positively charged lysine groups on cytochrome *c* normally interact with negatively charged groups on the cytochrome b_5 molecule during electron transfer. Trifluoroacetylation or trifluoromethylphenylcarbamoylation of lysines surrounding the haem crevice (residues 13, 25, 27, 72, and 79) decreased the reaction rate, and they have proposed that these residues form complementary charge interactions with the b_5 carboxy-groups Asp-48, Glu-43, Glu-44, and Asp-60 and the most exposed propionate group of haem.

In a study of the possible interactions between cytochrome *c* and cytochrome *c* oxidase, Erecinska[16] has used a photoaffinity labelling procedure. Three lysine residues of cytochrome *c* were modified using methyl-4-mercaptobutyrimidate HCl and the free SH group of the latter was covalently linked to *p*-azidophenacyl bromide to produce a photoaffinity-linked cytochrome *c*. This cytochrome *c* was bound by irradiation into a covalent complex with cytochrome *c* oxidase in the mitochondrion, and the fact that no binding to the bc_1 complex took place was taken as evidence for distinct sites on the cytochrome *c* for bc_1 and oxidase. A number of closely related papers which offer an interesting approach to the problem of structure–function relationship in cytochrome *c* have appeared this year, and involve semi-synthetic procedures. Harris and Offord[17] report the discovery of a functioning non-covalent complex between two peptides obtained by limited tryptic digestion of horse heart cytochrome *c*. The peptides correspond to residues 1—38 and 39—104 and, in addition to demonstrating their functional recombination, Harris and Offord have replaced Lys-39 semi-synthetically with ornithine and with *p*-fluorophenylalanine, and have removed it entirely in a third case. The absorption spectrum of the T1 peptide (haem peptide 1—38) is closely similar to that of ferricytochrome *c* with respect to position and intensity of the

[16] M. Erecinska, *Biochem. Biophys. Res. Comm.*, 1977, **76**, 495.
[17] D. E. Harris and R. E. Offord, *Biochem. J.*, 1977, **161**, 21.

main bands but differs from ferricytochrome c at 280 nm, where the bands due to tyrosine and tryptophan are missing; the 695 nm band is also lost. Various modifications of the structure have an effect on the ability to restore activity to a cytochrome-c-depleted mitochondrial preparation. It is clear that the formation of such non-covalent complexes, coupled to semi-synthetic modification of the component peptides, offers an exciting approach to the problem of relating structural features to molecular function.

Hantgen and Taniuchi[18] have used cyanogen bromide to prepare the haem fragment (1—65) from horse heart cytochrome c which has been complexed to the apoprotein (1—104), and the redundant portions of the ferrous complex have been removed by limited tryptic digestion. Figure 2 shows, schematically, the nature of this approach. The complementary peptide fragments which constitute the derived complex have been isolated, and they correspond to four apoprotein fragments (39—104, 40—104, 54—104, and 56—104) in addition to a single haem fragment (1—53H). The apparent K_d for this latter fragment and for each of the apoprotein fragments is 3×10^{-7} mol l^{-1}, and it is apparent from this and from activity assays that the region between residues 38 and 57 of the amino-acid sequence is not vital to the formation of an ordered structure. This is an interesting finding in view of the fact that in the cytochrome c-551 isolated from *Ps. aeruginosa* the sequence 38—57 is missing.

The cyanogen bromide fragments (1—65H) and 66—104 of horse heart cytochrome c have been complexed and the peptide link has been restored by Barstow *et al.*[19] The product is indistinguishable from native cytochrome c, and these authors have gone on to create, synthetically, the sequence 66—104 (using the Merrifield solid-phase procedure) and to use this to produce apparently native cytochrome c by complexation and bond re-formation. Clearly this is a potentially useful approach to delineating the role of aromatic amino-acids and, moreover, allows the introduction of isotopically labelled or structurally altered amino-acids into the molecule. It is interesting to note in connection with this type of work that the haem peptide fragment (residues 1—18) has recently been prepared,[20] and the pH-dependent properties of the ferric undecapeptide of cytochrome c have been described.[21] In this latter study Wilson *et al.*,[21] using a variety of spectrophotometric approaches, have assigned a number of the pH-induced transitions, the first of these being due to the binding of a deprotonated imidazole group to the iron, whilst the second and third originate from the binding of two groups, the $^{\alpha}NH_2$ of valine and the εNH_2 group of lysine, which, because of steric restraints on the peptide, can only co-ordinate to the iron *via* intermolecular bonds, which leads to polymerization of the peptide. Their results are summarized diagrammatically in Figure 3.

Metal-substituted Cytochromes c. In an attempt to try to evaluate the role of the central metal ion and the role of the protein in electron-transferring proteins, a

[18] R. R. Hantgen and H. Taniuchi, *J. Biol. Chem.*, 1977, **252**, 1367.
[19] L. E. Barstow, R. S. Young, E. Yakale, J. J. Sharp, J. C. O'Brien, P. W. Berman, and H. A. Harbury, *Proc. Nat. Acad. Sci. U.S.A.* 1977, **74**, 4248.
[20] N. L. Alarkon, G. A. Vasileva, G. E. Kirlova, A. F. Mironov, and R. P. Evstigneeva, *Zhur. obshchei Khim.*, 1977, **46**, 2635.
[21] M. T. Wilson, R. J. Ranson, P., Masiakowski, E., Czarnecka, and M. Brunori, *European J. Biochem.*, 1977, **77**, 193.

number of metal-substituted cytochrome *c* molecules have been prepared. So far, their application has not been widespread, but, from the range already available, it is apparent that the protein moiety, acting as a macrocyclic ligand, exerts a strong influence on the co-ordination chemistry of the metal. One immediate and potentially powerful use of metal-substituted cytochromes *c* is in their application to the problem of protein structure determination in solution through the use of n.m.r. Williams *et al.*[22] report the use of cobalt(III) cytochrome *c* in which the isomorphous replacement is very precise and where Co^{III} can serve as a diamagnetic reference state for the Fe^{III} low-spin state of the protein. Using a difference procedure, a direct determination of the paramagnetic shifts of the n.m.r. lines in the Fe^{III} state (essential data for structure determination) is possible. Cobalt(II) porphyrin is an excellent shift reagent, having strictly axial symmetry, whilst relaxation studies will become possible through the use of Mn^{II} high-spin or Cr^{III} high-spin cytochromes *c*.

A dramatic demonstration of the power of n.m.r. structure determination in relation to cytochrome *c* has come from a comparison of diamagnetic Co^{III} cytochrome and diamagnetic Fe^{II}. Within the limits of n.m.r. studies, the two states indicate the absence of any conformational changes in the haem pocket or elsewhere in the molecule. This result is in direct contradiction with the deductions based on general physical properties (thermal stability, gel binding, *etc.*) and on the early *X*-ray crystallographic studies of the two oxidation states. The recent refined structures from Dickerson's laboratory[8-10] for tuna cytochrome *c* amply confirm this original and important observation, which carries the mechanistic implication that electron transfer changes the metal–ligand distances by only a small amount.

Nickel cytochrome *c* has been prepared from metal-free porphyrin plus Ni^{II} in 0.6M-glycylglycine and 4M-KSCN by Findlay and Chien.[23] Electronic spectra and susceptibility measurements showed the central metal ion to be high-spin octahedral, clearly indicative of the strong influence of the protein moiety on the co-ordination chemistry of the Ni. The substituted protein has the same electrophoretic mobility, helicity, and pK_a as the normal iron-containing protein but reacts readily with NO to give a nitrosyl derivative, the e.p.r. parameters of which are $g_{||} = 2.187$ and $g_\perp = 2.140$. It is interesting to note that, under similar conditions, nickel protoporphyrin dimethyl ester in pyridine does not react with NO; the protein environment thus enables the formation and stabilization of the nitrosyl product. Nickel cytochrome *c* exhibits no e.p.r. spectrum down to 77 K, and, because of this property, it is possible to study the high-resolution contact-shifted n.m.r. spectra.

Copper cytochrome *c*[24] provides further evidence of the power of the protein to act as a macrocyclic ligand, since e.p.r. and electronic spectra indicate that, over the pH range 4—11, the central metal ion is six-co-ordinate; most copper porphyrins do not usually complex with two axial Lewis-base ligands. At

[22] R. J. P. Williams, G. R. Moore, and P. E. Wright, in 'Biological Aspects of Inorganic Chemistry', ed. A. W. Addison, W. R. Cullen, D. Dolphin, and J. James, Wiley Interscience, New York, 1977, p. 369.
[23] M. C. Findlay and J. C. W. Chien, *European J. Biochem.*, 1977, **76**, 79.
[24] M. C. Findlay, L. C. Dickinson, and J. C. W. Chien, *J. Amer. Chem. Soc.*, 1977, **99**, 5618.

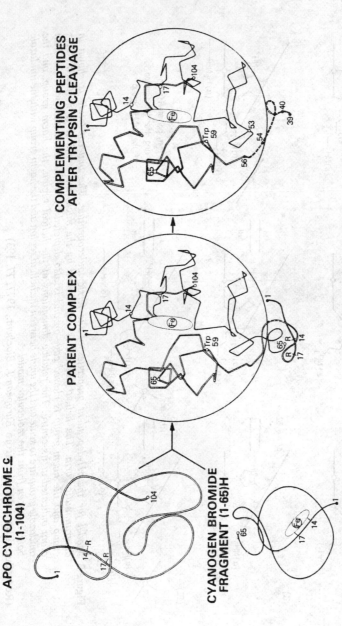

Figure 2 *Schematic diagram of the results of experiments in which a parent complex, formed from disordered apoprotein and (1—65)H haem fragment, is converted by limited trypsin digestion into the complementing system depicted at the right.*
(Reproduced by permission from *J. Biol. Chem.*, 1977, **252**, 1367)

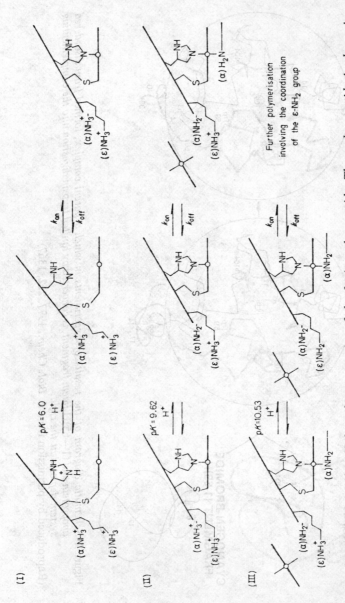

Figure 3 *Model for the pH-dependent properties of the ferric undecapeptide. The undecapeptide is shown in diagrammatic form. The helical peptide is represented by the thick line terminating in the α-NH_3^+ group and the haem group is viewed edge-on and is also depicted by a thickened line, with the iron atom indicated at its centre. The relative dimensions of the helical region, the haem group, and the imidazole group are to scale and the angle between the helical region and the haem group with the imidazole bond is taken from the molecular model*

(Reproduced by permission from *European J. Biochem.*, 1977, **77**, 193)

extremes of pH the molecule dimerizes, but it probably retains one axial ligand.

Manganese cytochrome c is of interest for a variety of reasons, not least of which is the fact that, along with iron and cobalt among the first-transition-series metalloporphyrins, it has two stable oxidation states. As a consequence, manganese cytochrome c could function as an electron-transfer protein. Oxidized Mn Cyt $^+$ and reduced Mn Cyt c have been prepared[25] by insertion of Mn^{2+} into porphyrin c. The reduced form of Mn Cyt c required the addition of a two-fold excess of dithionite to the oxidized material and was stable only in the presence of excess reductant. Both forms of the protein were found to have identical ion-exchange, and electrophoretic properties. The oxidized form does not react with F^-, CN^-, or N_3^- and the reduced form will not bind CO or O_2. The fact that Mn Cyt c^+ does not bind these strong ligands suggests, in part, that the metal ion is too far out of plane to feel the π-donation effect of the anion, and that this is combined with steric hindrance by Met-80 or Lys-79. The reduced protein is rapidly autoxidized by O_2 even at $-50\ ^\circ C$, and it has been suggested that oxygen molecules approaching the top of the porphyrin ring are capable of effecting electron transfer, but the orbital overlap is not sufficiently large to cause spin pairing of the oxidized metal, leading to the assumption of an in-plane position and a stabilizing interaction with the superoxide ion. The half-reduction potentials for Mn Cyt c and Mn Hb are $+60$ and $+40$ mV. The Mn Cyt c and Mn Cyt c^+ are high-spin complexes, with $3d^5$, $S = \frac{5}{2}$ and $3d^4$, $S = 2$ configurations respectively, and each reacts with NO (reversibly in the case of the oxidized protein) and abolishes the $g = 6$ signal. Since no superhyperfine interaction is observable with the NO ligand, it is assumed that the unpaired electron density is mostly in the d_{xy} orbital of the metal ion. The authors further suggest that the NO is bound to the manganese on the side of His-18, the central metal ion being out of plane towards this particular residue. Because of the rapid autoxidation of Mn Cyt c, the redox potential was not stable, the considerable uncertainty being reflected in the measured value of $+60 \pm 40$ mV. Thermodynamically, Mn Cyt c should reduce mammalian oxidase (aa_3) although, because of the aforementioned autoxidizability, it is not possible to use a conventional oxygen-electrode assay. Titrating the oxidase anaerobically with Mn Cyt c revealed no evidence of electron transfer, as judged by the absence of the spectrum characteristic of reduced cytochrome oxidase, and in experiments in which Mn Cyt c^+ was titrated with NADH and NADPH it was found to be non-reducible. The absence of enzymatic activity is disappointing, but is taken as evidence that manganese cytochrome c is in fact five-co-ordinate, with the central metal ion significantly out of plane towards His-18. This points to the Met-80 link as an important contributor to oxidoreductase activity.

Vanderkooi *et al.*[26] have used tin(IV) and zinc(II) cytochromes c to investigate the interaction of cytochrome c with mitochondria and cytochrome oxidase, using the fact that the fluorescence yields of tin and zinc cytochrome c fall on interaction. Based on the known spectral overlap and quantum yield, and using

[25] L. C. Dickinson and J. C. W. Chien, *J. Biol. Chem.*, 1977, **252**, 6156.
[26] J. M. Vanderkooi, R. Landerberg, G. W. Hayden, and C. S. Owen, *European J. Biochem.*, 1977, **81**, 339.

the Förster theory, they estimate the distance between the porphyrin rings of cytochrome c and cytochrome a to be of the order of 3.5 nm.

Reactivity of Cytochromes.—Moore and Williams[27] have considered at some length those factors which are most likely to be important in determining the redox potential of cytochromes, including cytochrome c, and conclude that the nature of the axial ligand is clearly a major determinant, since electrostatic charge and donor and acceptor power all contribute to the redox potential in the following ways:

(1) The higher the negative charge on the ligand, the lower the redox potential.
(2) The higher the pK_a (as a guide to donor power), the lower the redox potential.
(3) The greater the acceptor power (using unsaturation as a guide), the higher the redox potential.

Changes in spin state of the central metal ion will differentially alter the importance of (2) and (3) above, whilst steric factors can be chosen so that one or other spin or oxidation state is favoured, thus overriding other considerations.

Table 1 summarizes their classification of cytochromes into groups A, B, and C according to the axial ligands and the redox potentials whilst Table 2 gives the redox potentials of various model systems. Williams, Moore, and Wright[22] have developed their ideas concerning the oxidation–reduction properties of cytochrome c, and they conclude that the constraints imposed by the protein fold in cytochrome c have placed the ligands at such distances and bond angles as to hold this protein close to the cross-over between the spin states (an entatic state). This relatively rigid structure does not limit the mobilities of the side chains, and these fluctuations are probably vital in the actual mechanism of electron transfer, which they believe to be by a tunnelling process through a preferred solvent, of which the hydrophobic channels seem most probable.

An interesting observation regarding the temperature dependence of the redox potential of horse heart cytochrome c in 0.1 M-NaCl has been reported by

Table 1 *Classification of cytochromes*

Group	Porphyrin	Axial ligands	E_0'/mV	Examples
A	Protohaem	Histidine, Histidine	+80 to −200	b_5, b-562
B	Haem c	Histidine, Histidine	−200 to −500	c_3, c_7
C	Haem c	Histidine, Methionine	+400 to ∼ −60	c, c_2

Table 2 *Redox potentials of model systems*

System	E_0'/mV
Mesohaem	−158
Protohaem	−115
Bishistidine mesohaem	−220
Histidine–methionine mesohaem	−110
Bismethionine mesohaem	+20

[27] G. R. Moore and R. J. P Williams, *F.E.B.S. Letters*, 1977, **79**, 229.

Anderson *et al.*[28] The value of E_0' decreases linearly to 42 °C in H_2O, at which point there is a break, after which the plot is again linear but with an increased negative slope. No such biphasic behaviour is seen in D_2O, and this has been interpreted in terms of the ability of NaCl to cause destructuring of bulk H_2O above 42 °C but not of D_2O. In a study of the effect of pH on the mid-point potentials of cytochrome c_2 *in vivo*, Prince and Dutton[29] found that no net change in protonation occurs during oxidation–reduction.

The influence of salts and salt concentration on the reactivity of cytochrome c is a recurring theme, and this year has seen the usual crop of studies in this field. Goldkorn and Schejter[30] have studied the influence of ionic strength on the reduction of cytochrome c by ascorbate and find that, as $I \to 0$, two protein conformations seem to be present in solution. Eighty to ninety per cent of the total cytochrome in solution is rapidly reduced and the rest reacts slowly or not at all. These experiments were conducted at pH 7.2 in Tris cacodylate buffer (0.002 M), with NaCl added to adjust the ionic strength. Goldkorn and Scheijter[31] have also studied the reduction of cytochrome c at low ionic strength, using sodium dithionite as the reductant. Using this more powerful reductant, the slow, or ascorbate-irreducible, conformer previously observed is reduced to give an intermediate species, which decays at a rate of $10\ s^{-1}$ to give the normal reduced protein. Figure 4 summarizes their conclusions for the reduction of cytochrome c with ascorbate and with dithionite, and they have further suggested that the limiting rate of opening of the haem crevice is a salt-dependent phenomenon. In the absence of salt the crevice opening is facilitated, and they report rates of $150\ s^{-1}$ for reduction by dithionite, *i.e.* far in excess of the previously supposed limit of $\sim 60\ s^{-1}$. On the other hand, 1 M-NaCl imposes a more rigid structure on the molecule, and the rate of crevice opening falls to $45\ s^{-1}$.

The effect of ions, specifically K^+ and Cl^-, on the apparent equilibrium constant of the reaction between horse heart cytochrome c and ferricyanide has been the subject of an investigation carried out by Morton and Breskvar.[32] These

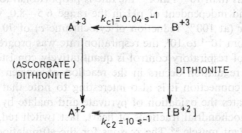

Figure 4 *Suggested scheme for the reduction of cytochrome* c *at pH 7, $I \to 0$, with dithionite and with ascorbate*
(Reproduced by permission from *F.E.B.S. Letters*, 1977, **82**, 293)

[28] C. W. Anderson, H. B. Halsall, W. R. Heineman, and G. P. Kreishman, *Biochem. Biophys. Res. Comm.*, 1977, **76**, 339.

[29] R. C. Prince and P. L. Dutton, *Biochim. Biophys. Acta*, 1977, **459**, 573.

[30] T. Goldkorn and A. Schejter, *F.E.B.S. Letters*, 1977, **75**, 44.

[31] T. Goldkorn and A. Schejter, *F.E.B.S. Letters*, 1977, **82**, 293.

[32] R. A. Morton and K. Breskvar, *Canad. J. Biochem.*, 1977, **55**, 146.

authors have made an attempt to identify possible ion-binding sites in the molecule through the use of guanidinated and trinitrophenylated forms of the protein. Both chemically modified derivatives have somewhat larger equilibrium constants than native cytochrome *c*, although the redox properties are not dramatically different. The redox-linked binding of K^+ was unaffected by modification of the lysines, whereas the equivalent binding of Cl^- (more Cl^- bound to ferricytochrome *c*) was reduced by modification, a finding which was interpreted as being due to the loss of a single anion site which appears to depend on one or a few lysine residues.

Lebon and Cassatt[33] report some kinetic and thermodynamic parameters for the oxidation of trifluoroacetylated ferrocytochrome *c* by ferricyanide; the value is $3 \times 10^5 \, l \, mol^{-1} \, s^{-1}$ for the rate constant, with an enthalpy of activation of $0.4 \, kcal \, mol^{-1}$ and an entropy of -33 e.u. The low enthalpy of activation is thought to occur because there is a pre-equilibration of the ferrocytochrome *c* derivative with a more easily oxidized conformer.

The influence of cytochrome *c* on the reactivity of mammalian cytochrome *c* oxidase has been the subject of two interesting papers. Kornblatt's[34] has shown that a four-fold molar excess of ferrocytochrome *c* increases the rate at which the 428 nm species of oxidized cytochrome oxidase converts into its conformational isomer with a Soret absorption maximum between 418—423 nm, whereas ferricytochrome *c* has no effect. The position of the oxidized Soret peak has long been known to depend on the past history of the enzyme, and Kornblatt's work has added further pieces to the jigsaw. If oxidase is reduced by ferrocytochrome *c* or other reductants, it forms the 418—423 species after oxidation if its last contact was with ferricytochrome *c*, but if the last contact before oxidation was with ferrocytochrome *c* then the 428 nm species is formed.

The dependence of mitochondrial respiratory rate on the level of reduction of cytochrome *c* has been measured as a function of the [ATP]/[ADP] [Pᵢ] ratio and of pH by Wilson, Owen, and Hobian.[35]

At ratios of less than $10^{-1} \, l \, mol^{-1}$ the rate is proportional to the reduction of cytochrome *c* and independent of pH in the range 6.5—8.0, with a maximum turnover number (at 100% reduction of cytochrome *c*) of $70 \, s^{-1}$. As the ratio was increased from 10^{-1} to 10^4, the respiration rate was progressively inhibited. This behaviour of respiratory control is quantitatively consistent with a mechanism in which regulation occurs in the reaction of oxygen with cytochrome oxidase. In this connection it is also interesting to note that exogenous cytochrome *c* stimulates the oxidation of pyruvate and malate by 120, 66, 65, and 35% for the mitochondrial fractions of heart, fast twitch red, mixed, and fast twitch white rodent muscle.[36] The reasons for the stimulation may reflect the fact that high salt concentrations in buffers could have removed cytochrome *c* during the preparation of mitochondria, and thus previous reports may have under-estimated the respiratory capacity of these tissues. Several papers have

[33] T. R. Lebon and J. C. Cassatt, *Biochem. Biophys. Res. Comm.*, 1977, **76**, 746.
[34] J. A. Kornblatt, *Canad. J. Biochem.*, 1977, **55**, 458.
[35] D. F. Wilson, C. S. Owen, and A. Hobian, *Arch. Biochem. Biophys.*, 1977, **182**, 749.
[36] P. J. Van Handel, W. R. Sandel, and P. Mole, *Biochem. Biophys. Res. Comm.*, 1977, **74**, 1213.

appeared this year in which the response of cytochrome c, in terms of concentration and turnover, has been measured in relationship to various external factors. Exercise has been shown to increase the concentration of cytochrome c in rat skeletal muscle[37] and, from the time it took for the concentration to increase the new higher steady state, the half life was estimated to be six days. The half life for the decrease after cessation of training was seven to eight days. Sidell[38] has measured cytochrome c turnover in the skeletal muscle of the green sunfish during thermal acclimatization. The concentration of cytochrome c changes from 0.98 nmol g^{-1} at 25 °C to 1.51 nmol g^{-1} at 5 °C, and the synthesis and degradation during this change were studied using [^{14}C]δ-aminolaevulinic acid, which is a non-reutilizable haem precursor. At 25 °C the T_{50} for cytochrome c was 7.1 days, whilst at 5 °C this increased to 13.7 days. Although the transfer from 25 to 5 °C results in a rapid 40% decrease in the rate of synthesis of cyto-chrome, the rate of degradation falls by 60%; the disproportionality of the two processes leading to an apparent 50% increase in cytochrome c concentration.

Porphyrin c.—De Kok et al.[39] have made a thorough study of the reduction of porphyrin cytochrome c by hydrated electrons from a pulse radiolysis source. At 21 °C and pH 7.4, a bimolecular rate constant of 3×10^{10} l mol^{-1} s^{-1} was observed, and this was followed by an absorbance change in the range 430—470 nm which had a T_{50} of 5 μs and which was followed by a more stable intermediate, decaying with a T_{50} of 15 s. The spectrum of the intermediate observed 50 μs after generation of the hydrated electrons shows a broad absorption between 600 and 700 nm and a Soret peak at 408, and has been attributed to a protonated form of the initially produced anion radical. Reduced porphyrin c reacts with ferricytochrome c with a bimolecular rate of 2×10^5 l mol^{-1} s^{-1} at pH 7.4 and 21 °C, and this increases to 2×10^6 l mol^{-1} s^{-1} in 1 M-NaCl. It has been proposed that electron transfer occurs *via* the exposed part of the haem.

Other Cytochromes, and Cytochrome–Membrane Interactions.—Magnetic circular dichroism continues to make major contributions in the field of haemoprotein biochemistry, and a recent publication extends the technique into the near-i.r. region. Rawlings et al.[40] have measured the near-i.r. spectra for the cytochromes c_1 from *R. rubrum*, *C. vinosum*, and *C. palustris*, which, in the case of the reduced protein, turn out to be very similar to that of deoxyhaemoglobin. Spectra in the pD range 1—13 correspond to four species (designated A, B, C, and D), and not the nine previously reported as a result of e.p.r. studies. B is high-spin, with C and D low-spin, A being close to high-spin proteins not supportive of previously postulated mixed spin state. The experiments are technically difficult since the weak transitions in the wavelength range 0.8—2.0 μm necessitate the use of very concentrated solutions. However, the energies of the near-i.r. bands are sensitive to, and (in principle) provide a means of elucidating, axial ligation of the haem.

[37] F. W. Booth and J. O. Holloszy, *J. Biol. Chem.*, 1977, **252**, 416.
[38] B. D. Sidell, *J. Exp. Zool.*, 1977, **199**, 233.
[39] J. De Kok, J. Butler, R. Braams, and B. F. Van Gelder, *Biochim. Biophys. Acta*, 1977, **460**, 290.
[40] J. Rawlings, P. J. Stephens, and L. A. Nafic, *Biochemistry*, 1977, **16**, 1725.

The observed similarity between ferrocytochrome c_1 and deoxyhaemoglobin suggests an identical ligation, *i.e.*

His–Fe–oxygen co-ordinated ligands

Another relatively new technique, the linear electric field effect, in e.p.r. has been applied to two bis-imidazole haem complexes which are considered to be models for B and H hemichromes of haemoglobin and cytochrome b_5.[41] Bis-imidazole ferric haem is considered to be the structure at the haem site of cytochrome b_5 and of two different low-spin ferric hemichromes that are formed from methaemoglobin. Strong base alters the absorption spectrum of bisimidazole ferric haems in organic solvents, but the linear electric field effect in e.p.r. shows that the ligated imidazole is not exchanged for the OH anion. The two hemichromes that are formed differ only in the state of protonation of the N-1 in the bound imidazoles.

The cytochrome-b_5-like domain from chicken liver sulphite oxidase has been isolated by Guiard and Lederer,[42] using limited chymotryptic digestion. A haem-binding fragment of molecular weight 11 000 can be isolated from the digest and it has a spectrum identical to that of the native enzyme, although the ability to oxidize sulphite is lost. The N-terminal sequence of this core protein (34 residues) shows a strong similarity to those of liver microsomal b_5 and bakers' yeast cytochrome b_2 core. These observations are suggestive of a common origin for the three b-type cytochromes. In a related study, Ozoles and Gerard[43] have determined the complete covalent structure of the membranous fragment of horse liver cytochrome b_5, which spans residues 91—133.

The relationship between cytochromes and membranes is of fundamental importance since all eukaryotic cytochromes are associated with the mitochondrial inner membrane in some way and many prokaryotic cytochromes seem to function in association with the limiting membrane or some membrane-like structure. Organization of the redox components within the lipoprotein matrix clearly imposes constraints on their mobility and in turn affects their reactivity; thus our understanding of electron-transfer processes must take account of these factors. Brown and Wuthrich[44] have studied the interactions of cytochrome c with artificial membranes in the form of mixed cardiolipin–phosphatidylcholine vesicles. Vesicles were prepared using a 1:4 ratio of cardiolipin to phosphatidylcholine, incorporating bound ferricytochrome c, and were characterized for size and homogeneity by chromatography, ultrafiltration, electron microscopy, and ^{1}H n.m.r. spectroscopy. The diameter of the vesicles both with and without bound cytochrome c was found to be ~ 300 Å, and the experiments based on ^{1}H n.m.r. and the use of e.p.r. spin labels covalently bound to methionine-65 of cytochrome c suggest that cytochrome c binds to the lipid bilayer mainly by ionic interactions. Figure 5 shows a conceptual model of the interaction between cytochrome c and the vesicular membrane, based on the experiments of Brown and Wuthrich, which seem to indicate that binding of cytochrome c results in the

[41] J. Peisach and W. W. Mimms, *Biochemistry*, 1977, **16**, 2795.
[42] B. Guiard and F. Lederer, *European J. Biochem.*, 1977, **74**, 181.
[43] J. Ozols and C. Gerard, *J. Biol. Chem.*, 1977, **252**, 8549.
[44] L. R. Brown and K. Wuthrich. *Biochim. Biophys. Acta*, 1977, **468**, 389.

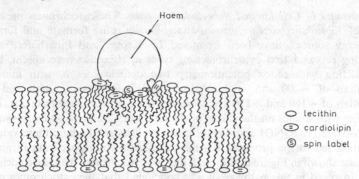

Figure 5 *A conceptual model of the interaction of cytochrome c with a cardiolipin–phosphatidylcholine membrane. The spin label is attached to Met-65 of cytochrome c. A suggested orientation of the cytochrome c haem group relative to the lipid membrane is also shown*
(Reproduced by permission from *Biochim. Biophys. Acta*, 1977, **468**, 389)

two lipids becoming segregated. Cardiolipin apparently becomes preferentially localized at the surface immobilized boundary layer, with the cytochrome c separating it from the fluid bulk of the lipid.

The haem group appears to point away from the surface of the bilayer. Resonance Raman spectroscopy has been used by Adar and Erecinska[45] to look for evidence of interactions between membrane-bound cytochromes b and c_1 isolated from mitochondria of pigeon breast muscle. Spectra were run at liquid-nitrogen temperature on samples which had been adjusted to different redox potentials prior to freezing. At a redox potential of -105 mV, at which all the components are reduced, the bands due to b are at 1306 and 1342 cm^{-1} and 1315 cm^{-1} for cytochrome c_1, and these bands decrease as the redox potential is raised and the sample becomes oxidized. The resonance Raman spectra as a function of redox potential indicate that the contribution from cytochrome c_1 is constant over spectra where it is fully reduced, whilst at potentials which indicate that only c_1 is reduced the b bands show structure which might be consistent with two inequivalent forms (b-561, b-565). The existence of b bands in the state where only c_1 is reduced is interpreted as being due to coupling between b and c_1, possibly through quantum-mechanical mixing of the vibrational levels. Similar interactions have been suggested from c.d. studies of the bc_1 complex, although the physical mechanism of the interactions is not clear. Rigorous theoretical analysis may yield information relating to the functioning of the haems in the electron-transport chain within the mitochondrial membrane.

Bacterial Cytochromes.—The rich diversity of the cytochromes found in bacteria continues to provide a fertile research area for those engaged in this field. Studies have continued on the cytochromes present in both cellular and subcellular systems, as well as cytochromes which have been isolated from their cellular environments. These two aspects provide an appropriate division for the work reported in this section.

---

[45] F. Adar and M. Erecinska, *F.E.B.S. Letters*, 1977, **80**, 195.

Cytochromes in Cellular and Subcellular Systems. The cytochromes present in cells of *Vibrio succinogenes*, grown anaerobically using formate and fumarate as energy sources, have been examined by Kröger and Innerhofer.[46] Spectroscopy revealed that cytochromes of types *a*, *d*, and *o* were absent, but in conjunction with redox potentiometry two cytochromes *b*, with mid-point potentials of −200 and −20 mV, and two cytochromes *c*, with mid-point potentials of −160 and +170 mV, were identified in membrane fractions. Taken together with studies on the effects of the inhibitors 2-n-heptyl-4-hydroxyquinoline *N*-oxide (HOQNO) and 4-chloromercuriphenylsulphonate, these values of potentials led to the proposed scheme for electron transport from formate to fumarate shown in Figure 6. Although the two cytochromes *b* appeared definitely to be involved in this pathway, it was concluded that the cytochromes *c* were probably not involved. De Vries *et al.*[47] have found that a cytochrome *b* is also associated with electron transport to fumarate in *Propionibacterium freudenreichi* and *P. pentosaceum.*

The *b*-type cytochromes of *Rhodopseudomonas capsulata* were the principal objects of a study by Zannoni *et al.*[48] This species is a facultative photosynthetic

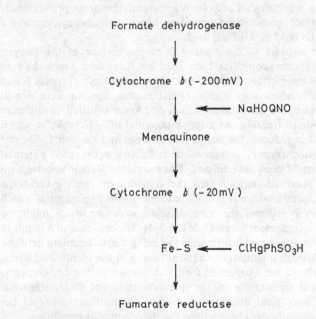

Figure 6 *Sequence of the components of the formate–fumarate reduction of* V. succinogenes. *The* b *cytochromes are differentiated by their mid-point potentials (given in brackets)*
(Reproduced by permission from *European J. Biochem.*, 1976, **69**, 497)

[46] A. Kröger and A. Innerhofer, *European J. Biochem.*, 1976, **69**, 497.
[47] W. De Vries, M. I. H. Aleem, A. Hemrika-Wager, and A. H. Stouthamer, *Arch. Microbiol.*, 1977, **112**, 271.
[48] D. Zannoni, B. A. Melandri, and A. Bacarini-Melandri, *Biochim. Biophys. Acta*, 1976, **449**, 386.

bacterium, and in this case the study was of those cytochromes present in sphero-plasts prepared from cells grown aerobically in the dark. Two mutant strains of the organism are known, one of which, M7, where NADH and succinate respiration are inhibited only by high KCN concentrations and CO, lacks a cytochrome c oxidase. The other mutant, M6, has an active cytochrome c oxidase and is highly sensitive to KCN, but insensitive to CO. By use of redox potentiometry at different pH values, both on wild and mutant preparations, and in the presence and absence of CO, no less than five b-type cytochromes were identified. In the wild strain these cytochromes had potentials of $+413 \pm 5$, $+270 \pm 5$, $+148 \pm 5$, $+56 \pm 5$, and -32 ± 5 mV at pH 7.0, and three of them (b_{148}, b_{56}, and b_{-32}, where subscripts refer to potential) possessed pH-dependent mid-potentials. Similar behaviour was also observed with the mutants, though one of them, M7, lacks b_{413}. In the presence of CO, the apparent potential of b_{270} shifted to higher values ($+355$ mV); this was most clearly seen in the M7 mutant (see Figure 7), as in the wild strain there is some overlap with b_{413}. The M6 mutant, however, did not show any difference of behaviour in the presence of CO. Control experiments on spheroplasts treated with lysozyme argued against any relationship between b_{270} and the CO-binding cytochrome cc' of this organism. Such treatment released virtually all the cytochrome cc', as well as a large amount of cytochrome c_2, and was found to impair the respiratory chain that is sensitive to low KCN concentration, thus confirming the previously proposed role of cytochrome c_2 in this pathway. Conversely, the lack of effect on the CO-sensitive pathway indicated that neither cytochrome c_2 nor cc' was involved in that route. In total, the results obtained supported the authors' previously advanced concept of branched electron transport to two oxidases, one of which (b_{417}) is dependent on cytochrome c_2 and one of which (b_{270}) is not. It was suggested that b_{270} should be classified as a cytochrome o.

A number of papers have appeared this year concerned with the involvement of cytochromes in the photosynthetic systems possessed by some bacteria. Shioi *et al.*,[49] as part of a continuing investigation of the green sulphur bacterium *Prostheochloris aestuarii* have studied light-induced reactions in both intact cells and vesicle fractions. With intact cells, at least three cytochromes were observed, *i.e.* cytochrome *c*-555 (rapid and slow components) and cytochrome *c*-552 (intermediate in rate). Comparisons of their light-minus-dark difference spectra with the reduced-minus-oxidized difference spectra of the previously[50] purified cytochromes *c*-555(550) and *c*-551.5 suggested a correspondence of *c*-555 (rapid) to *c*-555(550) and of *c*-552 to *c*-551.5. It was proposed that *c*-555 (slow) might be a membrane-bound form of *c*-555(550). Studies of the reactions of 'starved' cells [where only the *c*-555 (rapid) component remained in the reduced state in the dark, and became oxidized on actinic illumination] together with studies of the effects of inhibitors, and previous estimates of the reduction potentials of cytochrome *c*-555(550) and cytochrome *c*-551.5, led to the proposal of a tentative scheme for electron transfer in the bacterium. In this scheme, shown in Figure 8, *c*-555 (rapid) is positioned in a cyclic electron-transfer pathway. Interestingly,

[49] Y. Shioi, K. Takamiya, and M. Nishimura, *J. Biochem. (Japan)*, 1976, **80**, 811.
[50] Y. Shioi, K. Takamiya, and M. Nishimura, *J. Biochem. (Japan)*, 1972, **71**, 285.

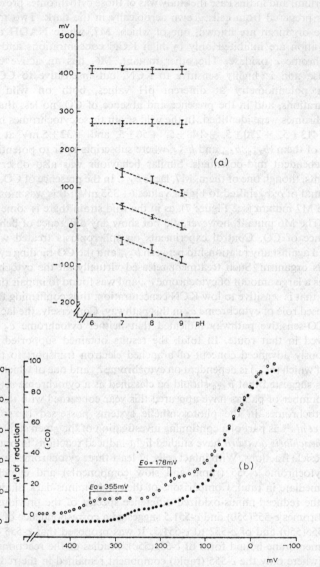

Figure 7 (*a*) *Dependency of the mid-point potential of cytochromes* b *on pH, measured under nitrogen in a buffer whose pH varied in the range 6.0— 9.0. The protein concentration in the assay was* 2.5 mg ml⁻¹. (*b*) *Potentiometric titration of cytochromes* b (561—570 nm) *in membranes from M7, in the presence or absence of carbon monoxide. The measurements were performed at pH 7.0 under nitrogen (●) or under CO/nitrogen (○) atmosphere. The protein concentration in the assay was* 2.1 mg ml⁻¹

(Reproduced by permission from *Biochim. Biophys. Acta*, 1976, **449**, 386)

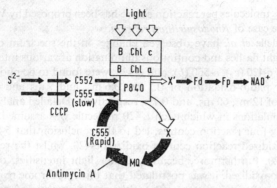

Figure 8 *A tentative scheme for the electron-transfer systems in* P. aestuarii. *Open arrows indicate processes driven directly by light. Closed arrows indicate dark electron-transfer processes.*
[Reproduced by permission from *J. Biochem. (Japan)*, 1976, **80**, 811]

no protohaem has been detected in pyridine haemochrome preparations of *Prostheochloris aestuarii*.

The light-induced reactions of cytochromes have also been investigated in examples of purple non-sulphur bacteria. Hochman and Carmeli[51] have examined some properties of the cytochrome c_2 associated with 'heavy chromatophores' prepared from *Rhodopseudomonas capsulata* cells grown anaerobically under illumination. It was found that the cytochrome c_2 in these preparations could be easily removed by washing, suggesting that, in contrast to normal chromatophores (from which cytochrome c_2 cannot be removed by washing), the membrane was oriented with the cytochrome attached to the outside surface. The washed heavy chromatophores possessed a decreased ability for phosphorylation which could be restored by the addition of cytochrome c_2, but not by other cytochromes c. Examination of the capacity of these preparations to accomplish the light-induced electron transfer from ascorbate to oxygen showed that this activity was also affected adversely, but that in this case other cytochromes c as well as c_2 could restore the activity. The fact that only cytochrome c_2 could be both oxidized and reduced by this membrane system led to the suggestion that it was involved in a cyclic electron-transport system with bacteriochlorophyll.

The cytochromes c_2 from *Rhodopseudomonas capsulata*, and from *Rhs. sphaeroides*, were the objects of a redox-potentiometric study by Prince and Dutton.[52] However, in this case normal chromatophores, again prepared from cells grown anaerobically under illumination, were used to assess the effects of pH. Their results indicated that, unlike the isolated protein, the reduction potential of the cytochrome does not vary between pH 5 and 11, and hence that, when bound to the membrane, it is unlikely that it acts as a redox-coupled hydrogen carrier. It was also found that there were two thermodynamically and kinetically equivalent cytochrome c_2 molecules per bacteriochlorophyll reaction centre. This is interesting in view of the fact that a rather different number of

51 A. Hochman and C. Carmeli, *Arch. Biochem. Biophys.*, 1977, **179**, 349.
52 R. C. Prince and P. L. Dutton, *Biochim. Biophys. Acta*, 1977, **459**, 574.

cytochrome c_2 molecules per reaction centre has been proposed by Van Grondelle *et al.*[53] for the case of *Rhodospirillum rubrum*.

Van Grondelle *et al.* have observed changes in the spectrum of whole cells induced by light flashes and continuous illumination at various intensities. Three cytochromes, c-420 (c_2), c-560 (b), and c-428, were found to be photoactive, with half lives for photo-oxidation of 0.3 ms, 0.6 ms, and 7 ms and half lives for re-reduction of 12 ms, 60 ms, and 0.7 s, respectively. Detailed analysis, including computer simulations in which each c-420 molecule was assumed to be able to diffuse among four reaction centres, led to the conclusion that 5% of the total number of oxidized reaction centres oxidized c-428, whilst the remaining 95% oxidized c-420. Furthermore, because, at low light intensities, only the c-428 became photo-oxidized, it was postulated that this cytochrome reacted preferentially at a type of reaction centre (P_2) with greater numbers of antenna chlorophyll molecules than those reaction centres (P_1) which reacted with c-420, *i.e.* there were two types of reaction centre. It was estimated that the number of molecules of c-420 was half the number of P_1 reaction centres. Once photo-oxidized, c-420 was reduced in a competitive way by c-560 and by an electron-donor pool. HOQNO was observed to inhibit both pathways, and the fact that (in the presence of this inhibitor) c-560 was in the oxidized state in the dark and became reduced after a light flash was taken as evidence that this cytochrome was involved in a cyclic electron-transport chain. Figure 9 shows part of the scheme put forward to explain these observations, the P_2 system having been omitted from this diagram.

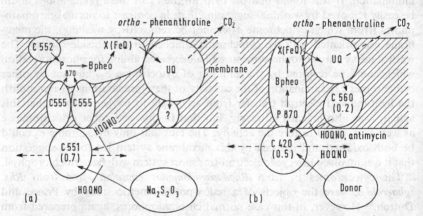

Figure 9 *Comparison of the electron-transport schemes for whole cells of (a) C. vinosum and (b) R. rubrum. The second photosystem present in a small amount in R. rubrum has been omitted. The function of c-552 is hitherto unknown. It may function at very low light intensities, as suggested for c-428 in the second photosystem of R. rubrum. The number written below c-551, c-560, and c-420 represents the number of cytochrome molecules present per reaction centre*

(Reproduced by permission from *Biochim. Biophys. Acta*, 1977, **461**, 188)

53 R. Van Grondelle, L. N. M. Duysens, and H. N. van der Wal, *Biochim. Biophys. Acta*, 1976, **449**, 169.

Van Grondelle and co-workers[54] have also conducted an analogous series of experiments to those described above, on whole cells of the purple sulphur bacterium *Chromatium vinosum*. This organism differs from *Rhodospirillum rubrum* in a number of ways. Two types of membrane-bound cytochrome *c* are present, a low-potential *c*-552 and a high-potential *c*-555. Both of these proteins are able to donate electrons efficiently to bacteriochlorophyll and seem to occur in a stoicheiometry of (at least) two per reaction centre. However, the re-reduction of *c*-552 after a light flash is so slow that this cytochrome can hardly contribute at all to electron transport at normal light intensities, and in fact under the conditions used it was permanently oxidized. It was found that *c*-555 was reduced by another, soluble, cytochrome, *c*-551 and it was at this level that the system showed analogies to *R. rubrum*. There were 0.6—0.7 molecules of *c*-551 per reaction centre and the observed kinetics were consistent with a certain mobility of the cytochrome among reaction centres, although less so than with the corresponding *c*-420 in *R. rubrum*. Cytochrome *c*-551 did not react with *c*-552, and its behaviour in the presence and absence of external electron donors indicated that it was reduced partly by a cyclic pathway and partly by a non-cyclic pathway. The rate of the latter was increased between 5- and 10-fold upon addition of thiosulphate, suggesting a role for *c*-551 between the final donor pool and the oxidized membrane-bound cytochromes *c*. The scheme in Figure 9 was proposed to account for the results.

Isolated Cytochromes. This year has seen the publication of a number of papers describing the isolation of bacterial cytochromes and aspects of the characterization of the isolated proteins. Bruschi *et al.*[55] have reported the amino-acid composition of the cytochrome c_3 from *Desulphovibrio desulphuricans* (Norway strain) and found that the spectral properties of the protein (mol. wt. 16000) were very similar to those of other cytochromes c_3. The molecule possessed four haems and had a very low potential, acting as an electron carrier that couples the hydrogenase and the thiosulphate reductase present in extracts of both *D. gigas* and *D. desulphuricans*. Interestingly, the c_3 of *D. gigas* was not effective in extracts of *D. desulphuricans*. A comparison of the activities of these two homologous cytochromes c_3 suggested that they had different specificities in the electron-transfer chains of these bacteria.

It is appropriate to mention two other papers concerned with cytochrome c_3 at this point. Sperry and Wilkins[56] have reported the possible identification of a cytochrome c_3 in *Desulphomonas pigra* and Niki *et al.*[57] have studied the electrode reaction of the cytochrome c_3 from *Desulphovibrio vulgaris*. The latter workers concluded, on the basis of polarographic and cyclic voltammetric results, that the electrochemical kinetics of the redox reactions were diffusion-controlled. They suggested that a detailed analysis of their data supports the idea that the four haems in this cytochrome c_3 are not independent, but constitute an interlocked haem cluster, *i.e.* there is haem–haem interaction.

[54] R. Van Grondelle, L. N. M. Duysens, J. A. van der Wel, and H. N. van der Wal, *Biochim. Biophys. Acta*, 1977, **461**, 188.
[55] M. Bruschi, C. E. Hatchikian, L. A. Golovleva, and J. Le Gall, *J. Bacteriol.*, 1977, **129**, 30.
[56] J. F. Sperry and T. D. Wilkins, *J. Bacteriol.*, 1977, **129**, 554.
[57] K. Niki, T. Nagi, H. Inokuchi, and K. Kimura, *J. Electrochem. Soc.*, 1977, **124**, 1888.

Van Beeuman *et al.*[58] have sequenced the cytochromes *c*-555 from the green sulphur bacteria *Chlorobium thiosulphatophilum* and *C. limicola* (a component of the syntrophic mixed culture '*Chloropseudomonas ethylica* 2K'). Both proteins were found to be single polypeptide chains, the *C. thiosulphatophilum* cytochrome having 86 residues and the *C. limicola* cytochrome having 99 residues, with 47 residues being common to both sequences. Although the two proteins were found to be unreactive with mitochondrial cytochrome *c* reductase, the *c*-555 of *C. thiosulphatophilum* exhibited a relatively high degree of activity with bovine cytochrome oxidase, a property which is not shared by any other prokaryotic cytochrome (including *c*-555 of *C. limicola*) except the cytochrome *c*-550 from *Thiobacillus novellus*. It was suggested that the cytochrome oxidase reactive site of the *c*-555 of *C. thiosulphatophilum* had arisen independently by parallel evolution in this organism. The determination of the X-ray structure of this protein was reported to be in progress.

A cytochrome c_7, namely *c*-551.5, from the anaerobic sulphur-reducing bacterium *Desulphuromonas acetoxidans* has been purified and characterized by Probst *et al.*[59] The molecular weight was estimated as 9800 by gel filtration, and a determination of the amino-acid composition and N-terminal sequence showed the protein to be identical with the three-haem low-potential cytochrome c_7 isolated from '*Chloropseudomonas ethylica* 2K'.

Fischer and Trueper[60] have reported the isolation of a highly thermoresistant cytochrome *c*, cytochrome *c*-550, from the soluble fraction of cells of *Thiocapsa roseopersicina*. This protein had a molecular weight of approximately 34000 and was capable of reducing sulphide even after heating to 80—100 °C for five minutes. Another highly thermoresistant cytochrome *c*, cytochrome *c*-552, has been the subject of a study by Hon-nami and Oshima.[61] In this case the protein was isolated from the soluble fraction of the extreme thermophile *Thermus thermophilus* HB 8 and was reported to have a molecular weight of approximately 15000, an isoelectric point of 10.8, and a reduction potential of +0.23 V. Cytochrome *c*-552 was found to react rapidly with the nitrite reductase (cytochrome oxidase) of *Pseudomonas aeruginosa*, but only slowly with bovine cytochrome oxidase and with the hydroxylamine cytochrome *c* reductase of *Nitrosomonas europea*. Interestingly, the spectrum of the reduced form of this single-haem protein showed a split α band at 77 K. A band at 690 nm was also observed in the oxidized form.

The reaction of cytochrome *c*-552 from *Thermus thermophilus* HB8 with ferrocyanide has been the subject of a temperature-jump kinetic investigation.[62] A parallel study of horse heart cytochrome *c* was also undertaken for comparison. The results showed that the activation enthalpy for oxidation of cytochrome *c*-552 was considerably lower than that of the eukaryotic protein, whilst the activation entropies of oxidation and reduction were both more negative

[58] J. Van Beeuman, R. P. Ambler, T. E. Meyer, and M. D. Kamen, *Biochem. J.*, 1976, **159**, 757.
[59] J. Probst, M. Bruschi, N. Pfennig, and J. Le Gall, *Biochim. Biophys. Acta*, 1977, **460**, 58.
[60] U. Fischer and H. G. Trueper, *FEMS Microbiol. Letters*, 1977, **1**, 87.
[61] K. Hon-nami and T. Oshima, *J. Biochem. (Japan)*, 1977, **82**, 769.
[62] H. Kihara, H. Nakatani, K. Hiromi, K. Hon-ami, and T. Oshima, *Biochim. Biophys. Acta*, 1977, **460**, 480.

than for horse heart cytochrome c. The rate of reduction of c-552 by ferrocyanide was found to be lower than that of horse heart cytochrome c, as was the redox potential, which, at $+0.19$ V, would appear to differ somewhat from the value given earlier.

Two publications have appeared in the past year connected with the nature of the binding of the flavin component to two flavocytochromes c. Kennedy and McIntire[63] have shown that the cytochrome c-553 from *Chlorobium thiosulphatophilum* contains an 8α-S-cysteinyl FAD thioether, and Kennedy and Singer[64] have reported that a similar group is present in *Chromatium* cytochrome c-552.

As with all the other classes of electron-transfer protein, considerable effort has been expended in the study of the spectroscopic aspects of isolated bacterial cytochromes. Resonance Raman studies have been reported for a number of cytochromes c. Adar[65] has investigated the ferrocytochromes c-557 from *Crithidia oncopelti* and c-558 from *Euglena gracilis*. Resonance Raman spectra of haem proteins are sensitive to unsaturated side-group substituents on the deuterohaem core because the porphyrin vibrational states that are enhanced are the in-plane deformations of the conjugated haem core, and only unsaturated side groups have the possibility of introducing extra fundamental bands. Thus in two regions of resonance Raman spectra single bands of deuterohaem (1324 and 1545 cm^{-1}) are replaced by two bands of the same respective symmetry in protohaem (1305 and 1342 cm^{-1}, 1538 and 1563 cm^{-1}). In mesohaem and cytochrome c these regions maintain single bands, but chemical evidence suggests that cytochromes c-557 and c-558 have only one thioether linkage to the protein, with the other one characteristic of normal cytochromes c being replaced by a vinyl group. These two proteins should therefore exhibit the side-group effect and, as shown in Table 3, the observed spectra were consistent with the chemical evidence.

Table 3 *Raman bands (wavenumbers/cm^{-1}) that respond to side-group substitutions at positions 2 and 4*

	Cyto-chrome c	Meso-haem in cyto-chrome b_5	Deutero-haem in cyto-chrome b_5	Proto-haem in cyto-chrome b_5	Cyto-chrome c-557 (br)	Cyto-chrome c-556
Anomalously polarized bands	1315	1313	1324	1306 1342	1312 1339	1312 1339
Depolarized bands	1548	1548	1545	1538 1563	1543 1560	1543 1558

The pH dependencies of the resonance Raman spectra of a number of cytochromes, namely c, c_2, c_3, c-551, and c-555, have been investigated by Kitagawa *et al.*[66] The particular feature of their approach was the use of a ligand-sensitive

[63] W. C. Kennedy and W. McIntire, *Biochim. Biophys. Acta*, 1977, **483**, 467.
[64] W. C. Kennedy and T. P. Singer, *J. Biol. Chem.*, 1977, **252**, 4767.
[65] F. Adar, *Arch. Biochem. Biophys.*, 1977, **181**, 5.
[66] T. Kitagawa, Y. Ozaki, J. Teraoka, Y. Kyogoku, and T. Yamanaka, *Biochim. Biophys. Acta*, 1977, **494**, 100.

marker band, which may be detected in both proteins and haem complex models, and which appears at 1539 cm^{-1} in oxidized and 1565 cm^{-1} in reduced horse heart cytochrome c. The Raman spectra of the ferric forms of cytochromes c, c_2, c-551, and c-555 all possessed lines which were sensitive to pH, but, despite the fact that all these proteins have histidine and methionine as the two axial ligands, the observed frquency changes all occurred at different pH values. Notably, horse heart cytochrome c yielded a pK_a of 9.4. Oxidized cytochrome c_3 showed no change over the pH range 3.1—11; n.m.r. evidence has revealed that the four haems in the c_3 molecule have, as axial ligands, two histidines each. Conversely, the resonance Raman spectra of the ferrous cytochromes c (pH 3— 12.6), c_2 (pH 5.3—12.1), c-551 (pH 4—10.5), and c-555 (pH 2.5—12.5) did not change over the ranges indicated, but the ligand marker line of ferrocytochrome c_3 at 1539 cm^{-1} did. The results in this case were interpreted as a stepwise change in the ligation of the four haems in the molecule, whereby one of the histidines at each iron atom is replaced by another group (lysine ?) such that between pH 7 and 8 two haems are altered, but at pH 9 all the haems are converted

Kitagawa and co-workers[67] have also studied the resonance Raman spectrum of the cytochrome c' from *Rhodospirillum rubrum*. Five states of this protein have been characterized by absorption spectra, two in the reduced form (types a and n) and three in the oxidized form (types I, II, and III). Type a is high-spin and is the state of the reduced protein around neutral pH, whilst type n predomi- nates above pH 11. Type III is low-spin and is the only form present above pH 12, whilst type II, which exists principally over the pH range 9.3—12.1, is high-spin. The type I form, predominating over the pH range 2—8.4, has been proposed[68] to be a mixture of spin $\frac{5}{2}$ and $\frac{3}{2}$ states. Kitagawa *et al.*[67] have examined the Raman spectra of all the above species and suggest that the frequency of the ligand-sensitive Raman line is consistent with the co-ordination of a lysine nitrogen at the sixth position in type n and that the sixth ligand in type III is either lysine or histidine, but would not be methionine. Although the features of the Raman spectrum of type I cytochrome c' were not typical of other high-spin ferric haemoproteins, they nevertheless concluded, on the basis of results from model systems, that it was not of intermediate spin. They suggest that it is an unusual high-spin species. In this regard Kitagawa *et al.* agree with the work of Rawlings *et al.*[40] mentioned earlier, but the latter workers identify four oxidized species and conclude that oxygen appears to be the most likely candidate for the sixth ligand in all of them.

From the above discussion, some discrepancy may be discerned in the number of species of cytochrome c' observed by different techniques. The problem is further complicated by a Mössbauer study carried out by Emptage *et al.*[69] Using the *R. rubrum* protein, they have found that three oxidized species are distinguishable by Mössbauer spectroscopy over the pH range 6—9.5, with two pK_a values of 6 and 8.5. All these three species were found to exhibit Mössbauer parameters in the range found for high-spin ferric proteins, but the data were

[67] T. Kitagawa, Y. Ozaki, Y. Kyogoku, and T. Horio, *Biochim. Biophys. Acta*, 1977, **495**, 1.
[68] M. M. Maltempo and T. H. Moss, *Quart. Rev. Biophys.*, 1976, **9**, 181.
[69] M. H. Emptage, R. Zimmermann, L. Que, E. Munck, W. D. Hamilton, and W. H. Orme- Johnson, *Biochim. Biophys. Acta*, 1977, **495**, 12.

fitted to a spin Hamiltonian which took into account a weak mixing of excited $s = \frac{3}{2}$ states with the $s = \frac{5}{2}$ ground state. In contrast to previous reports, the quadrupole interactions were typical of high-spin ferric haems, the value of the quadrupole coupling constant being positive.

Some reports of the application of n.m.r. spectroscopy to bacterial cytochromes have appeared in the past year. Moura *et al.*[70] have used the technique to study the cytochrome c_3 from *Desulphovibrio gigas*, both in the absence and presence of its ferredoxin electron acceptor Fd II (see Section 4). Spectra of oxidized cytochrome c_3 were collected over the pH range 5.8—11 and a large number of contact-shifted resonances were found to titrate. In addition, shifts of haem resonances were induced when Fd II was present, but only in unbuffered solutions; these shifts were not found in 0.2M-phosphate buffer, thus implying an electrostatic interaction between the two proteins. Although e.p.r. has shown that the four haems of cytochrome c_3 have different redox potentials, the n.m.r. spectra produced on mixing oxidized and reduced cytochrome c_3 in various proportions were too complex to allow conclusions as to the existence of a particular order of oxidation of haem. The major effects observed were the immediate effect of oxidation on the resonances of some individual haem groups, followed by co-operative shift effects on neighbouring haem groups, with inter- and intra-molecular electron exchange affecting the appearance of the haem resonances in different ways. However, some resonances of a distinct intermediate were observed. When reduced cytochrome c_3 was mixed not with the oxidized protein but with oxidized Fd II, the re-oxidation spectra were different, suggesting that the electron-exchange rates between the haems are altered in the presence of ferredoxin, and thus providing further evidence of an interaction between the two proteins.

Two papers concerned with n.m.r. spectroscopy of *Pseudomonas* cytochrome c-551 have been published recently. Keller *et al.*[71] reported that the protein had both structural similarities and differences with other cytochromes c, although methionine and histidine were clearly implicated as the axial ligands of the haem iron. The reduced protein did not have the typical distribution of haem ring methyl resonances found for other cytochromes c, but the absence of methyl resonances, other than from methionine-61, at higher field than 0.6 p.p.m. indicated that no aliphatic methyl groups were near the haem plane. For ferri-cytochrome c-551 the principal features of the local magnetic fields were similar to other low-spin ferric haemoproteins, with pseudocontact and ring-current shifts (as indicated by thioether resonances) having opposite sign. The presence of 10 well resolved methyl lines between 0.5 and −0.5 p.p.m. also showed that numerous aliphatic side-chains were positioned near the haem edges. One important difference to horse heart cytochrome c was observed; the inter-molecular electron-exchange rate between oxidized and reduced c-551 (1.2×10^7 l mol^{-1} s^{-1}) was found to be 2—3 orders of magnitude faster than that of the mammalian protein.

[70] J. J. G. Moura, A. V. Xavier, D. J. Cookson, G. R. Moore, R. J. P. Williams, M. Bruschi, and J. Le Gall, *F.E.B.S. Letters*, 1977, **81**, 275.
[71] R. M. Keller, K. Wuthrich, and I. Pecht, *F.E.B.S. Letters*, 1976, **70**, 180.

Moore *et al.*[72] have extended the assignments of resonances in the n.m.r. spectra of cytochrome *c*-551 by use of convolution difference spectra, spin decoupling, and spin echo double-resonance techniques as well as examination of pH, temperature, and redox mixture effects. Interestingly, they found that the n.m.r. spectrum of tryptophan-56 was very similar to that of tryptophan-59 in mammalian cytochrome *c*, implying a similar orientation in both proteins.

2 Copper Proteins

Copper and iron seem to occupy the same biochemical niche in that, in combination with a protein moiety, their properties can be 'tuned' to fulfil the same range of roles from electron transfer to oxygen binding and final reduction of oxygen to water. Like the cytochromes they offer an attractive vehicle for the experimentalists from a number of disciplines, and the nature of the blue or 'type I' copper-containing proteins is one of the fascinating problems in biological inorganic chemistry. These type I copper centres possess a combination of properties, such as an intense absorption band near 600 nm that is approximately two orders of magnitude greater in extinction than most complexes and an e.p.r. spectrum with a very small hyperfine splitting constant $A_{||}$ coupled to a high redox potential, none of which have been satisfactorily mimicked in model complexes. Type I centres, which occur in such proteins as azurin and plastocyanin, and which appear to be associated with an electron-transfer function, also occur in other proteins in association with other types of copper centre namely type II and type III, which possess different but distinctive properties.

Sequence and Structure.—The sequence of the thirteen-residue glycopeptide containing the sole cysteine residue, and being considered to be a putative copper-site peptide, in the blue copper protein stellacyanin has been sequenced by Wang and Young[73] along with a pentapeptide containing a possible copper-site histidine, using the dansyl Edman method. Little sequence homology exists between stellacyanin, plastocyanin, or azurin in the cysteine region (see Figure 10), and the adjacent histidine proposed as the ligand for Cu in plastocyanin and azurin is absent in stellacyanin. Interestingly, there are homologies between the copper-containing subunit of cytochrome oxidase and these blue copper proteins in this region. The histidine pentapeptide of stellacyanin shows good homology with the sequence around the invariant histidine, although this is somewhat less good in the case of azurin. Haslett *et al.*[74] have begun a survey of plant plastocyanins, using sequence studies, in the hope of assembling evolutionary data along the same lines as has been done for cytochrome *c*. They note that the rate of evolution of plastocyanin is greater than of cytochrome *c* and feel that differences between the plastocyanin sequences will probably be adequately reflected in the first third (the *N*-terminal sequence) of the total sequence. Sequences from 1 to 40 have been obtained by a rapid automatic sequencing procedure, and those obtained from a wide range of members of the Compositae are compared.

[72] G. R. Moore, R. C. Pitt, and R. J. P. Williams, *European J. Biochem.*, 1977, **77**, 53.
[73] T. T. Wang and N. M. Young, *Biochem. Biophys. Res. Comm.*, 1977, **74**, 119.
[74] B. G. Haslett, T. Gleaves, and D. Boulter, *Phytochemistry*, 1977, **16**, 363.

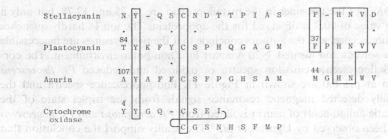

Figure 10 *Alignment of stellacyanin peptide sequences with portions of plastocyanin from the elder* (Sambucus nigra), *azurin from* Pseudomonas denitri- ficans, *and a 36-residue peptide from the copper subunit of bovine cardiac cytochrome oxidase. Invariant residues are in boxes and dots indicate some residues that can be related by single base changes* (Reproduced by permission from *Biochem. Biophys. Res. Comm.*, 1977, **74**, 119)

Natural-abundance [13]C n.m.r. has been used by a number of workers to probe the structural features of blue copper proteins, in particular plastocyanins and azurin. Differences between the Cu^I and Cu^{II} forms of spinach plastocyanin were investigated by this technique[75] at 67.9 MHz, using proton noise decoupling. At this level of resolution sixteen single carbon peaks are diluted in the spectrum of Cu^I plastocyanin which vanish when the protein is oxidized. These correspond to carbons which are very close to the metal and whose resonances become broadened beyond detection by the paramagnetic Cu^{II}. Markley *et al.*[75] claim that their spectra confirm that His-38 and His-91 are copper ligands, and further demonstrate that the co-ordination is by the $N^{\delta 1}$ of both imidazole rings. On the basis of their evidence and the work of others, these authors have proposed a model of the copper-binding site of plastocyanins. The n.m.r. data are consistent with cysteine sulphur and a peptide nitrogen as the other two ligands for copper. Ugurbil *et al.*[76] have also used natural-abundance [13]C Fourier-transform n.m.r. at 15.18 MHz, using spinning 20 mm sample tubes, to study the individual carbon sites of azurin isolated from *Pseudomonas aeruginosa*. In particular, the environ- ments of the aromatic residues, along with the single arginine residue, were investigated. The pH dependence of chemical shifts is used to identify the resonances of C^γ of titrating histidines and of C^γ and C^ζ of the two tyrosines. Two of the four histidine residues do not titrate in the pH range 4—11, and these are probably copper ligands with $N^{\delta 1}$ co-ordinated to the metal. A single carbon amide carbonyl resonance with an unusual chemical shift is observed and is thought to arise from an amide group that probably is co-ordinated to the copper. Ugurbil and Bersohn[77] have used the tryptophanless azurin obtained from *Pseudomonas fluorescens* to study tyrosine emission, using fluorometric titration, fluorescence quantum yield, fluorescence polarization, and I^- quenching. The presence of copper evidently reduces the quantum yield to $\sim 60\%$ of that of the apoprotein, although this latter has a quantum yield greater than free tyrosine.

[75] J. L. Markley, E. L. Ulrich, and D. W. Krogmann, *Biochem. Biophys. Res. Comm.*, 1977, **78**, 106.
[76] K. Ugurbil, R. S. Norton, A. Allerhand, and R. Bersohn, *Biochemistry*, 1977, **16**, 886.
[77] K. Ugurbil and R. Bersohn, *Biochemistry*, 1977, **16**, 895.

The pK_a's of the tyrosines in the metalloprotein are 10.75 and 12.78, but only a single value of 10.9 is observed for the apoprotein. The ion I$^-$ hardly quenches the fluorescence, and the authors conclude that the two tyrosines are inaccessible to the solvent and located $\gtrsim 20$ Å apart in a non-polar environment. The corrected fluorescence emission spectra of oxidized and reduced *Ps. fluorescens* azurin at pH 6.0 are shown in Figure 11. Phosphorescence spectra and the optically detected magnetic resonance signals from the triplet state of the aromatic amino-acids of azurin isolated from *Ps. aeruginosa* and *Ps. fluorescens* have been observed by Ugurbil *et al.*[78] The results support the conclusion that tryptophan cannot effectively quench the singlet energy of both tyrosines. O.D.M.R. spectra for oxidized and reduced azurin from *Ps. fluorescens* (no tryptophan) aud from the reduced azurin of *Ps. aeruginosa* (containing tryptophan) are presented in Figure 12; they are all very narrow, indicating that the chromophores are buried in the interior of the protein.

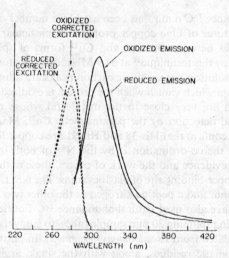

Figure 11 *Fluorescence emission and corrected excitation spectrum of oxidized and reduced* P. fluorescens *(ATCC 13430)* azurin, *pH 6.0, 25 °C. Emission was monitored at 310 nm for the excitation spectrum. Excitation for the emission spectrum was by light of wavelength 278 nm*
(Reproduced by permission from *Biochemistry*, 1977, **16**, 895)

Interesting as these results are, the most exciting developments in the structural studies have come from the application of X-ray crystallography to a blue copper protein. Chapman *et al.*[79] surveyed the plastocyanins isolated from many different plants with a view to finding those which would crystallize and give crystals suitable for X-ray crystallographic analysis. Crystals of French bean and cucumber plastocyanins were found to be long thin needles, unsuitable for

[78] K. Ugurbil, A. H. Maki, and R. Bersohn, *Biochemistry*, 1977, **16**, 901.
[79] G. V. Chapman, P. M. Colman, H. C. Freeman, J. M. Guss, M. Murata, V. A. Norris, J. A. M. Ramshaw, and M. P. Venkatappa, *J. Mol. Biol.*, 1977, **110**, 187.

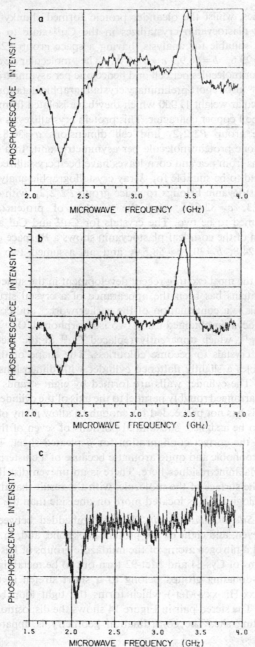

Figure 12 *D − E and 2E slow-passage ODMR signals observed for the tyrosine residues of (a) oxidized azurin A; (b) reduced azurin A; (c) reduced azurin B*
(Reproduced by permission from *Biochemistry*, 1977, **16**, 901)

diffraction studies, whilst the oleander protein formed chunky but disordered crystals. Poplar plastocyanin crystallizes in the Cu^{II} state to give deep blue rhombic prisms suitable for analysis, having a space group $P2_12_12_1$, with cell dimensions $a=29.6$, $b=46.9$, $c=57.6$ Å. The molecular weight is 10500 and there are four molecules per cell and hence one per asymmetric unit. Colman and co-workers[80] also report preliminary crystallographic data for a basic copper protein of molecular weight 11000 which they have isolated from cucumber and which shows type I copper character. This protein crystallizes as thin olive-green plates of space group $P2_12_12_1$ and cell dimensions $a=30.8$, $b=45.6$, and $c=66.6$ Å, with one protein molecule per asymmetric unit. In a separate study, the plastocyanins from pea and corn leaves have been crystallized by Chergadze *et al.*[81] and found to be suitable for X-ray crystallographic analysis. The crystal form of pea plastocyanin belongs to space group $P2_12_12_1$, with cell dimensions $a=49.0$, $b=53.3$, $c=82.6$ Å, and the number of protein molecules per unit cell is assumed to be two. The crystals for Cu^{II} and Cu^I are isomorphic. The crystal form of the corn leaf plastocyanin shows a P_1 space group, with cell parameters $a=24.8$, $b=30$, $c=58.5$ Å and an assumed two molecules per unit cell.

Undoubtedly the most exciting recent development in the structural studies on blue copper proteins has been the appearance of a crystal structure at 2.7 Å resolution.[82] This is a continuation of the earlier work[79] and concerns data on poplar leaf plastocyanin obtained using two isomorphous UO_2^{2+} derivatives and a third with Hg^{II}, which apparently replaced Cu^{II} at the metal-binding site and caused the crystals to become colourless. The shape of the plastocyanin molecule resembles a slightly flattened cylinder with dimensions $40 \times 32 \times 28$ Å (see Figure 13). The cylinder walls are formed by eight strands of polypeptide chain which are arranged roughly parallel to the axis of the cylinder, and although structure analysis has not proceeded far enough to allow many of the hydrogen-bond contacts to be assigned, substantial portions of seven of the eight strands appear to be in the correct configuration for β pleated sheet. The core of the molecule is hydrophobic and quite aromatic because of a clustering of six out of the seven phenylalanine residues here. There is an uneven distribution of polar side-chains on the surface of the molecule, with the result that the nett negative charge at physiological pH is located more on one side than the other.

Copper-binding Site of Plastocyanin. This is embedded between the ends of strands three, seven, and eight of the protein backbone, and, in addition to the widely predicted δ-nitrogen atoms of the imidazole groups of His-37 and His-87, the sulphur atoms of Cys-84 and Met-92 turn out to be metal ligands. Three of the four ligand-donating groups belong to a short stretch in the amino-acid sequence (-Cys-xx-His-xxx-Met-) which forms the tight loop between strands seven and eight. The stereo-pair in Figure 14 shows the disposition of residues in the copper-binding site. The co-ordination geometry is apparently irregular,

[80] P. M. Colman, H. C. Freeman, J. M. Guss, M. Murata, V. A. Norris, J. A. M. Ramshaw, M. P. Venkatappa, and L. E. Vickery, *J. Mol. Biol.*, 1977, **112**, 649.
[81] Y. N. Chergadze, M. B. Garber, and S. V. Nikonov, *J. Mol. Biol.*, 1977, **113**, 443.
[82] P. M. Colman, H. C. Freeman, J. M. Guss, M. Murata, V. A. Norris, J. A. M. Ramshaw, and M. P. Venkatappa, *Nature*, 1978, **272**, 319.

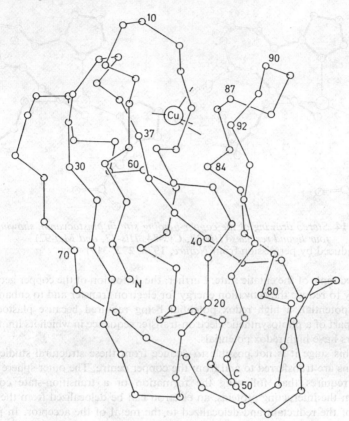

Figure 13 *The polypeptide chain in poplar plastocyanin. The circles represent α-carbon positions derived by applying the model-fitting procedure of Diamond to atomic co-ordinates measured on the Watson–Kendrew model. Every tenth residue and the four copper-binding residues are numbered following the scheme of Boulter et al. for higher plant plastocyanin sequences. The letters N and C denote the N-terminal and C-terminal residues, respectively. The approximate directions of the Cu–ligand bonds are indicated*

(Reproduced by permission from *Nature*, 1978, **272**, 319)

with bond angles deviating by as much as 50° from the values for a tetrahedron, although this situation may change with structure refinement. Accessibility of solvent to the side seems limited, but from one direction, only the imidazole group of His-87 restricts this to about 6 Å.

The ligand groups represent a compromise between the requirements of copper in its two oxidation states, and this presumably means that reduction of Cu^{II} can proceed without changes in co-ordination, although this remains to be proved by means of structure analysis of the Cu^{I} protein. In respect of the nature of the ligands and the distorted co-ordination geometry, electron transfer *via* an outer-sphere mechanism would seem to be favoured and plastocyanin seen to provide a

9

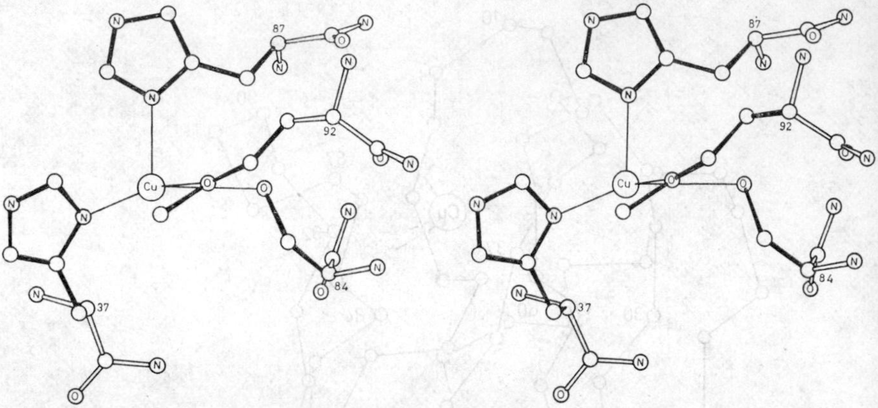

Figure 14 *Stereo drawing of the copper-binding site in plastocyanin, showing the four ligand residues (His-37, Cys-84, His-87, and Met-92)*
(Reproduced by permission from *Nature*, 1978, **272**, 319)

good example of the entatic state. Further, the distortion of the copper geometry is likely to reduce the activation energy for electron transfer and to enhance the redox potential, a high redox potential being required because plastocyanin forms part of a photosynthetic electron-transfer sequence in which its immediate partners have high redox potentials.

At this stage it is not possible to deduce from these structural studies how electrons are transferred to and from the copper centre. The outer-sphere hypothesis requires that, following the formation of a transition-state complex between the interacting proteins, an electron can be delocalized from the metal atom of the reductant and delocalized to the metal of the acceptor. In plastocyanin this type of electron transfer could occur by way of the co-ordinated imidazole ring of His-87, which is all that separates the copper centre from the surrounding medium. The absence of negatively charged side-chains in this area of the molecule also makes it an attractive site for interaction with negatively charged inorganic oxido-reductants such as $[Fe(CN)_6]^{3-}$ and $[Fe(CN)_6]^{4-}$, which are considered to transfer electrons *via* an outer-sphere mechanism.

The competing mechanism for electron transfer that involves quantum-mechanical tunnelling along hydrophobic channels is a viable alternative in plastocyanin. Several such channels can be identified in plastocyanin, one of which is particularly interesting since it starts at the periphery of the molecule in a region which carries predominantly negative charges. Originating near Tyr-83, this channel is lined by the side-chains of Phe-82, Val-93, Gly-94, and Phe-14, which in plant plastocyanins are all invariant residues, there being minor variations only in the algal proteins.

The observation that there are at least two potential pathways for electron transfer in plastocyanin gibes well with the needs for directionality in electron transfer, since the protein is apparently firmly bound in the membranes of photo-synthetic apparatus. Electrons coming from cytochrome *f* could use one pathway, with a second path going to the oxidant, most probably *P*-700. The distinctly hydrophilic and hydrophobic patches on the exterior of the molecule may thus provide specific sites of interaction between plastocyanin and its two redox partners.

In spite of the fact that the availability of suitable crystals dictated that poplar plastocyanin was chosen for study, many of the present findings are likely to prove general throughout the group. Of the 99 amino-acid residues in poplar plastocyanin, 28 are invariant in all plastocyanins, and the high-field n.m.r. spectra of a number of plant plastocyanins prove that the environment of the copper atom is highly conserved. The implications for other type I copper proteins are less clear, although the copper site of stellacyanin must differ from that of plastocyanin because stellacyanin contains no methionine residues. The type I copper protein which is apparently most closely related to plastocyanin is azurin, a bacterial electron-transfer protein of approximately 130 amino-acid residues. If allowance is made for the difference between the polypeptide chain lengths of azurin and plastocyanin, it is possible to align the sequences in such a way as to bring a significant number of identical or similar residues into correspondence. The availability of the tertiary level structure for plastocyanin, in combination with sequence homologies between the proteins, permits an investigation of possible relationships at the tertiary level. By slightly rearranging the structural features of plastocyanin into a two-dimensional topological framework, Colman *et al.* have been able to make a direct comparison with azurin. All those amino-acids which are conserved in all the plastocyanins and azurins studied so far turn out to be either copper ligands or to be in parts of the sequence which are close to the copper site. The positions where additional residues occur in the azurins are nearly all remote from the copper site in the plastocyanin structure and are located where they can be accommodated without causing serious disturbance to the entire structure. The implication of this study seems clear, namely that azurin and plastocyanin have very similar copper sites.

Synthetic Copper Complexes: Analogues of the Active Site.—Prior to the X-ray crystallographic solution of the nature of the copper site, there was considerable speculation regarding the amino-acid ligands for copper in the type I, blue copper proteins. Copper–N(His) and/or copper–S bonds were inferred from a number of studies, and some very ingenious synthetic chemistry was undertaken in an attempt to create analogues of the active site. Thompson, Marks, and Ibers[83] report the synthesis of $[Cu^{I}N_3(SR)]$ and $[Cu^{II}N_3(SR)]$ complexes by the reaction of $[Cu(SR)]$ or $[Cu(SR)]$ ClO_4 derivatives, in which $SR = p$-nitrobenzene-thiolate or O-ethylcysteinate, with potassium hydrotris-(3,5-dimethyl-1-pyra-zolyl)borate. Figure 15 shows a view of the co-ordination geometry about the copper atom for such a complex and, as Figure 16 shows, the optical spectra of the $[Cu^{II}N_3(SR)]$ species are not dissimilar from that of the native system (*Pseudomonas* azurin). However, in marked contrast to the narrow hyperfine couplings seen in the e.p.r. spectrum of the native system, the couplings for $[Cu^{II}N_3(SR)]$ are not unusually small, with $A_{\parallel} = 17.1$ mK when SR is p-$NO_2C_6H_4S$ and 17.0 mK when SR is O-ethylcysteinate ($A_{\parallel} = 3.3$—9.0 for the

[83] J. S. Thompson, T. J. Marks, and J. A. Ibers, *Proc. Nat. Acad. Sci., U.S.A.*, 1977, **74**, 3114.

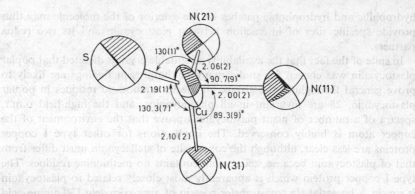

Figure 15 *View of the co-ordination geometry about the copper atom in complexes*
[CuN₃(SR)]
(Reproduced by permission from *Proc. Natl. Acad. Sci., U.S.A.*, 1977, **74**, 3114)

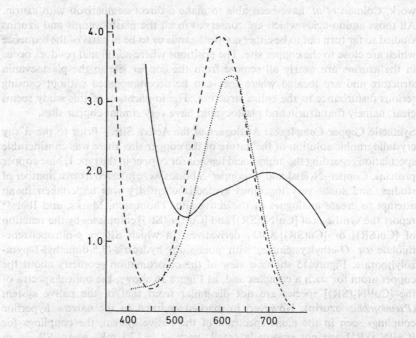

Figure 16 *Optical spectra of:* (– – –) [Cu-HB(3,5-Me₂pz)₃{(p-NO₂C₆H₄S}] *at*
78 °C *in tetrahydrofuran;* (———) [Cu{HB(3,5-Me₂pz)₃}(O-ethyl-
cysteinate)] *at* −78 °C *in tetrahydrofuran;* (· · · ·) Pseudomonas aeruginosa *azurin*
(Reproduced by permission from *Proc. Natl. Acad. Sci., U.S.A.*, 1977, **74**, 3114)

native system). Sakaguchi and Addison[84] have studied the e.p.r. of $[Cu^{II}S_4]$ chromophores and suggest that the e.p.r. parameters of type I copper centres are compatible with CuS_2N_2 or $CuSN_3$ co-ordination, although the low value of $A_{\parallel}$ found in type I centres is not accounted for by a charge, tetrahedral distortion, or environmental effect alone. Yokoi and Addison,[85] in a related study, have looked at the visible absorption and e.p.r. spectra, along with the redox potential, for a series of complexes (1) between pyrrole-2-carboxaldehyde Schiff-base and Cu^{II}, and they have found that there are smooth correlations among d–d band energies, $A_{\parallel}$, $g_{\parallel}$, A_0, and g_0 values. As the dihedral angle between the chelate rings increases from 0 to 90°, $g_{\parallel}$ increases and $A_{\parallel}$ decreases in an antiparallel fashion, whilst the redox potential shifts to more positive values.

(1)

These observations are consistent with tetrahedral distortion at the metal-binding sites of blue copper proteins. Peeling and co-workers[86] report on the S_{2p} binding energies of a variety of compounds, using ESCA spectroscopy, and conclude that the component of a number of plastocyanins that has the higher binding energy arises from sulphur present in a high oxidation state rather than from the co-ordination of sulphur-containing amino-acids to the metal, as had been claimed. In a recent study, using pulsed e.p.r., Mondovi *et al.*[87] have shown that the electron spin-echo decay envelopes for type I and II coppers of *Rhus vernicifera* laccase and for type II copper porcine caeruloplasmin are consistent with imidazole being a ligand in all cases. Additionally, the linear electric field effect in e.p.r. for the type I copper in laccase from which the type II copper had been removed indicated that the symmetry of the site is nearly tetrahedral and that the magnitude of the L.E.F.E. is correlated with the intensity of the blue colour.

Reactivity of Copper Proteins.—As part of a large-scale investigation into the kinetics of oxido-reduction of metalloproteins with small redox-active molecules, McArdle *et al.*[88] have looked at the oxidation of *Ps. aeruginosa* azurin, bean plastocyanin, and *Rhus vernicifera* stellacyanin by tris complexes of 1,10-phenan-

[84] U. Sakaguchi and A. W. Addison, *J. Amer. Chem. Soc.*, 1977, **99**, 5189.
[85] H. Yokoi and A. W. Addison, *Inorg. Chem.*, 1977, **16**, 1341.
[86] J. Peeling, B. G. Haslett, I. M. Evans, D. T. Clark, and D. Boulter, *J. Amer. Chem. Soc.*, 1977, **99**, 1025.
[87] B. Mondovi, M. T. Graziani, W. W. Mimms, R. Oltzik, R., and J. Peisach. *Biochemistry*, 1977 **16**, 4198.
[88] J. V. McArdle, C. L. Coyle, H. B. Gray, G. S. Yoneda, and R. A. Holwerda, *J. Amer. Chem. Soc.*, 1977, **99**, 2483.

throline and its 5-chloro-, 5,6-dimethyl-, 4,7-dimethyl-, and 4,7-diphenyl-4-sulphonate-derivatives with cobalt(III). The electron-transfer pathways from azurin Cu^I and plastocyanin Cu^I to $[Co(phen)_3]^{3+}$ are characterized by large enthalpic activation requirements of 14.3 and 14.0 kcal mol^{-1} coupled to favourable activation entropies of 5 cal mol^{-1} deg^{-1}. Reduced stellacyanin 'prefers' an oxidation mechanism for which the $\Delta S^{\ddagger}$ is negative (-13 cal mol^{-1} deg^{-1}). A summary of the kinetic parameters for the three proteins with a number of oxidizing agents is contained in Table 4. As a result of their investigations the authors have concluded that, in azurin and plastocyanin, oxidant-induced conformational changes expose the active sites which are, by comparison with stellacyanin, inaccessible to attack by reagents. Although the relevance of such studies on small molecules can be questioned in terms of their relation to the interactions between macromolecular redox partners, some comfort can perhaps be drawn from the fact that, using $[Co(5,6-Me_2phen)_3]^{3+}$ and $[Co(4,7-Me_2-phen)_3]^{3+}$ as oxidants for azurin Cu^I, it has been possible to confirm the existence of a difficult-to-oxidize azurin isomer previously discovered by Wilson *et al.*[89] in the interaction of azurin Cu^I with ferricytochrome c_{551}.

Goldberg and Pecht[90] meanwhile have studied the electron-transfer reaction between azurin and the hexacyanoiron(II/III) couple in some detail. Equilibrium constants for the reduction of azurin were measured spectrophotometrically in the temperature range 5—33 °C, and the constant K was found to be 1.1×10^2 at 25 °C, with $\Delta H^{\ominus} = 10.9$ kcal mol^{-1}. Temperature-jump perturbation of the equilibrium revealed only a single relaxation, the reciprocal of the relaxation time increasing linearly as oxidized azurin reacted with increasing amounts of ferrocyanide but reaching a saturation level when reduced azurin was titrated with ferrocyanide. This behaviour, coupled with an analysis of the relaxation amplitudes, has allowed these authors to draw up a scheme for the system:

$$Az^{II} + Fe^{II} \underset{}{\overset{K_1}{\rightleftharpoons}} Az^{II} \cdot Fe^{II} \underset{k_{-3}}{\overset{k_3}{\rightleftharpoons}} Az^I \cdot Fe^{III} \overset{K_2}{\rightleftharpoons} Az^I \cdot Fe^{III}$$

At 25 °C the rate constants were $k_3 = 6.4$, $k_{-3} = 45$ s^{-1}, with association constants $K_1 = 54$ l mol^{-1} and $K_2 = 610$ l mol^{-1}.

Organezova and Nalbandyan[91] have looked at the reduction of plastocyanin by solvated electrons in a non-aqueous medium and find that the environment of the copper is changed in organic solvents, although it is not clear whether it is a change in ligand or just a change in symmetry of the ligand environment of the metal.

3 Photosynthetic Electron Transfer

Electron transport requires that several redox components, usually at least three, must be gathered together, most (if not all) of these components being bound together in a more or less solid matrix, such as a membrane. This arrangement ensures that the electron-transfer processes which occur are efficient and moreover can be made directional and, in certain cases, coupled to energy-yielding

[89] M. T. Wilson, C. Greenwood, M. Brunori, and E. Antonini, *Biochem. J.*, 1975, **145**, 449.
[90] M. Goldberg and I. Pecht, *Bioinorg. Chem. Symp.*, 1977, 179.
[91] E. P. Organezova and R. M. Nalbandyan, *F.E.B.S. Letters*, 1977, **82**, 147.

Table 4 *Kinetic parameters for the oxidation of blue copper proteins at 25 °C*

Oxidizing agent	Stellacyanin[a]			Plastocyanin[e]			Azurin[f]		
	k/l mol^{-1} s^{-1}	$\Delta H^{\ddagger}$/ kcal mol^{-1}	$\Delta S^{\ddagger}$/ cal mol^{-1} deg^{-1}	k/l mol^{-1} s^{-1}	$\Delta H^{\ddagger}$/ kcal mol^{-1}	$\Delta S^{\ddagger}$/ cal mol^{-1} deg^{-1}	k/l mol^{-1} s^{-1}	$\Delta H^{\ddagger}$/ kcal mol^{-1}	$\Delta S^{\ddagger}$/ cal mol^{-1} deg^{-1}
[Co(phen)$_3$]$^{3+}$	1.80(5)×10^5	6.0(2)	−13(1)	4.87(5)×10^3	14.0(5)	5(1)	3.20(5)×10^3	14.3(5)	5(1)
	1.8(1)×10^5 [b]	6.1(2)	−14(1)						
	1.3(1)×10^5 [c]	5.9(2)	−15(1)						
[Co(5,6-Me$_2$phen)$_3$]$^{3+}$	1.85(5)×10^4	9.5(2)	−7(1)	7.97(5)×10^3	13.6(5)	1(1)	1.54(5)×10^3	11.6(5)	−5(1)
[Co(5-Cl-phen)$_3$]$^{3+}$	d	—	−10(1)	6.96(5)×10^2	10.9(5)	−8(1)	4.21(5)×10^2	8.9(5)	−17(1)
[Co(4,7-(PhSO$_3$)$_2$phen)$_3$]$^{3-}$	2.31(5)×10^6	5.9(2)	—	2.59(5)×10^1	7.8(5)	−26(1)	d	—	—
[Co(4,7-Me$_2$phen)$_3$]$^{3+}$	d	—	—	d			8.41(5)×10^1	9.9(5)	−17(1)

(a) For pH 7.0 (phosphate), $\mu=0.1$ M (NaCl) unless otherwise specified. (b) For pH 7.0 (phosphate), $\mu=0.5$ M (NaCl). (c) pH 5.1 (acetate), $\mu=0.5$ M (acetate). (d) Experimental determinations were not made. (e) For pH 7.0 (phosphate), $\mu=0.1$ M [(NH$_4$)$_2$SO$_4$]. (f) For pH 7.0 (phosphate), $\mu=0.2$ M (NaCl).

reactions. Photosynthetic electron-transfer systems provide a good example of this type of arrangement and are particularly appealing to the physically minded experimentalist, since the driving force is not an oxidizable substrate which needs to be 'fed' to the system but the clean, clinical, and eminently quantifiable quantum of light energy.

Structural Features.—Most of the components of the photosynthetic electron-transfer pathway are associated with the thylakoid membrane of the chloroplast, and the question of sidedness arises in regard to the arrangement of the carriers. Smith *et al.*[92] have used the hydrophobic covalent chemical modifiers diazobenzene[^{35}S]sulphonic acid and [^{14}C]glycine ethyl ester plus 1-cyclohexyl-3-(2-morpholinoethyl)carbodi-imide metho-toluene-*p*-sulphonate to probe the location of plastocyanin. Their results suggest that plastocyanin is partially exposed at the external surface of the thylakoid membrane rather than completely buried in or behind the lipoprotein membrane.

Reactivity.—Plastocyanin has recently been invoked by Hardt and Kok[93] as the possible site of inhibition of photosynthetic electron transport by glutaraldehyde. Spinach chloroplasts treated with this bifunctional reagent exhibit marked inhibition of electron transport between the two photosystems (PS I and PS II). Measurements of O_2 flash yield, pH exchange, and fluorescence induction show that the O_2-evolving apparatus and its electron-acceptor pool are unaffected. The behaviour of *P*-700 indicates that its reduction but not its oxidation is severely inhibited whereas cytochrome *f* is still reducible by PS II but more slowly oxidized by PS I. The sensitivity of isolated plastocyanin to glutaraldehyde supports the conclusion that glutaraldehyde inhibits at the level of plastocyanin, thereby inducing a break between *P*-700 and cytochrome *f*.

The question of the relationship between *P*-700, plastocyanin, and cytochrome *f* has been probed, using the kinetics of electron transfer following a brief actinic flash that are listed by Bouges-Bocquet in companion papers.[94, 95] In the first paper[94] the oxidation kinetics in *Chlorella pyrenoidosa* which had been dark-adapted in the presence of hydroxylamine were measured after a single flash. The results are consistent with a linear scheme:

$$P\text{-}700^+ + PC \xrightarrow{k_1} P\text{-}700 + PC^+ + \text{Cyt } f \underset{k_{-2}}{\overset{k_2}{\rightleftharpoons}} PC + \text{Cyt } f^+$$

where $k_1 = 4.6 \times 10^3$, $k_2 = 1.4 \times 10^4$, $k_{-2} = 7 \times 10^3 \text{ s}^{-1}$. In the second paper[95] Bouges-Bocquet shows that the reduction of PC^+ and Cyt f^+ in the 10 ms range is correlated with an increase in the electric field, named phase b. In the dark, electron transfer involving a carrier of electrons across the membranes, a proton carrier, R^1 as terminal reductant, and PC^+ and Cyt f^+ as terminal oxidants would account for this field generation. These conclusions are summarized in the hypothetical scheme in Figure 17, which depicts the arrangement of the carriers within the membrane of the photosynthetic apparatus. Using spinach

[92] D. D. Smith, B. R. Selman, K. K. Voegili, G. Johnson, and R. A. Dilley, *Biochim. Biophys. Acta*, 1977, **459**, 468.

[93] H. Hardt and B. Kok, *Plant Physiol.*, 1977, **60**, 225.

[94] B. Bouges-Bocquet, *Biochim. Biophys. Acta*, 1977, **462**, 362.

[95] B. Bouges-Bocquet, *Biochim. Biophys. Acta*, 1977, **462**, 371.

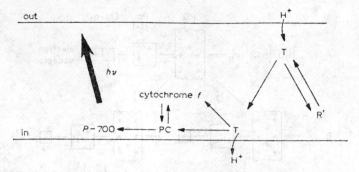

Figure 17 *A representation of the carriers within the membrane of the photosynthetic apparatus of* Chlorella pyrenoidosa
(Reproduced by permission from *Biochim. Biophys. Acta*, 1977, **462**, 371)

chloroplasts, Hachnel[96] has performed flash titrations to look at the electron transport between plastoquinone and chlorophyll a_1 and at the reaction kinetics and function of plastocyanin *in situ*. By means of a balance sheet of electron equivalents it was established that electron equivalents from reduced plastoquinone are accepted by oxidized plastocyanin, cytochrome f, and chlorophyll a_1, thus excluding the possibility of the involvement of any additional carriers. The course of the change of absorbance of plastocyanin measured at 584 nm indicated an initial lag in the reduction followed by a rise which had a t_{50} of 20 ms. With two successive groups of saturating flashes after far-red illumination, simultaneous oxidation and reduction kinetics of plastoquinone, cytochrome f, plastocyanin, and chlorophyll a_1 were observed. The fast reduction of chlorophyll a_1 by plastocyanin showed no effect of inhibitors such as 3-(3′,4′-dichlorophenyl)-1,1-dimethylurea or reduced phenazine methosulphate, but was completely blocked by potassium cyanide. The observed kinetics have been interpreted as giving evidence for cytochrome f functioning between plastoquinone and chlorophyll a_1, in parallel with plastocyanin, with the greater part of the electron flow by-passing this carrier. Figure 18 shows in a simplified form the proposed scheme of linear electron transport between the two light reactions.

Cyclic electron flow in cell-free preparations from the blue-green alga *Nostoc muscorum* has been studied by Knalf.[97] Cytochrome b_6 can be photo-oxidized and photoreduced by PS I and appears to have a redox potential near to zero. Addition of ADP to the preparation in the presence of ferridoxin and light results in the oxidation of cytochrome b_6 and the reduction of cytochrome f, suggesting the existence of a coupling site between the two cytochromes. The acceleration of the dark reduction of photo-oxidized cytochrome b_6 observed on addition of ADP raises the possibility of a second coupling site on the reducing side of cytochrome b_6. The question of whether cytochrome b-559 is a functional carrier between the photosystems in well-coupled spinach chloroplasts is the subject of a study by Whitmarsh and Cramer,[98] who find that the t_{50} for reduction of

[96] W. Hachnel, *Biochim. Biophys. Acta*, 1977, **459**, 418.
[97] D. B. Knaff, *Arch. Biochem. Biophys.*, 1977, **182**, 540.
[98] J. Whitmarsh and W. A. Cramer, *Biochim. Biophys. Acta*, 1977, **460**, 280.

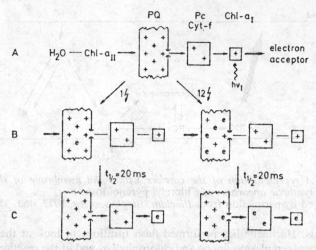

Figure 18 *Simplified scheme of linear electron transport between the two light reactions. Redox states of the electron carriers are shown. A: after preillumination with System I light; B, C (left) after one flash; (right) after a group of twelve flashes*
(Reproduced by permission from *Biochim. Biophys. Acta*, 1977, **459**, 418)

cytochrome b-559 is 100 ± 10 ms, compared to 6—10 ms for plastoquinone. This slow reduction of b-559 compared to the plastoquinone pool implies that electrons are transferred efficiently from PS II to plastoquinone without involvement of cytochrome b-559. This report contrasts markedly with the observations of Leach *et al.*[99] on cytochrome b-559 isolated from spinach and the alga *Burnilleriopsis filiformis*, a member of the Xanthophyceae. The cytochrome appeared to be active as an electron acceptor in a diaphorase system using NADPH as donor and ferredoxin plus ferredoxin–NADP reductase as redox partners. It was photo-oxidized with PS I particles illuminated with red light (707 nm) and also by PS II particles illuminated with 652 nm light.

Horton and Croze[100] have made potentiometric measurements on cytochrome b-559 after treatment with hydroxylamine and find two components of $E_m = 7.8$ at $+240$ mV in addition to a $+90$ mV species. In control chloroplasts, b-559 is typically high-potential, at $+383$ mV. The role of cytochrome b-559 in the functioning of PS II suggests a dependence of the photo-oxidizability of PS II on the redox properties of this cytochrome.

Purification and Properties of Photosynthetic Components.—Gray[101] has reported a purification procedure for cytochrome f from higher plants based on extraction with organic solvents followed by chromatography on a variety of supports such as DEAE-cellulose, Sephadex G100, hydroxyapatite, and Bio gel A. Interestingly, the purified cytochrome exhibits no band at 695 nm in the oxidized form.

[99] H. J. Leach, H. Boehme, and P. Boger, *Biochim. Biophys. Acta*, 1977, **462**, 12.
[100] P. Horton and E. Croze, *Biochim. Biophys. Acta*, 1977, **462**, 86.
[101] J. C. Gray, *Biochem. Soc. Trans.*, 1977, **5**, 326.

The ratio of cytochrome *f* to chlorophyll and of chlorophyll to *P*-700 has been measured for *Chlorella fusca* and a chlorophyll-deficient mutant of the same organism by Wild and Feildner.[102] Somewhat surprisingly, the mutants appear to have a high photosynthetic efficiency, based on chlorophyll content, with Hill reaction activities of PS I and PS II four to five times higher than those of the normal green form. The higher capacities for CO_2 fixation and electron transport are thought to be due to the formation of photosynthetic units which are four- to seven-fold smaller than those of the normal strain because of the deficiency in the light-harvesting chlorophyll–protein complex.

An interesting observation on the atypical cytochrome *c*-558 from dark-grown cultures of *Euglena gracilis* has been made by Miller and Rapoport.[103] They report that the haem appears to be bound to the protein by a single thioether bond, and as a result of the reductive cleavage (using sodium amalgam) of this bond followed by esterification have isolated a porphyrin which is identical with synthetic 2-vinyl-4-ethyldeuteroporphyrin-IX dimethyl ester. This establishes the *Euglena* porphyrin *c*-558 as 2-vinyl-4-(α-S-cysteinylethyl)deuteroporphyrin-IX (2) and the mono-thioether linkage as being at the 4α-ethyl position. Workers in

(2)

this same laboratory[104] have compared the porphyrins from *R. rubrum* cytochrome c_2 and yeast and horse heart cytochromes *c*, using [1]H n.m.r. and c.d. The identity of the spectra indicates that, chemically and stereochemically, the three porphyrins are identical, although differences in the [1]H n.m.r. spectra of *R. rubrum* and horse heart cytochrome *c* suggest an opposite stereochemistry at the porphyrin and thioether bond.

[102] A. Wild and K. H. Feildner, *Planta*, 1977, **136**, 281.
[103] M. J. Miller and H. Rapoport, *J. Amer. Chem. Soc.*, 1977, **99**, 3479.
[104] J. T. Slama, C. G. Wilson, C. E. Grimshaw, and H. Rapoport, *Biochemistry*, 1977, **16**, 1750.

4 Iron–Sulphur Proteins

Despite a comparatively short history, the study of the iron–sulphur proteins has been an extremely active research area in recent years. It therefore seems desirable to begin by mentioning some recent reviews on the subject. A relatively brief introduction is given by Palmer[105] and another fairly short review, dealing mainly with synthetic analogues, has appeared this year.[106] However, perhaps the most comprehensive treatise is to be found in the three volumes edited by Lovenberg.[107] Volume III of this series has been published during the past year, and work by contributors will be referred to where appropriate in this section. The above three references should provide ready access to the past literature on iron–sulphur proteins.

The broad arrangement of this part of the Report will be to use the natural division of the three types of iron–sulphur protein, *i.e.* the rubredoxins, the 2Fe:2S ferredoxins, and the 4Fe:4S iron–sulphur proteins, as sub-headings. Nevertheless, some overlapping between these divisions is difficult to avoid because some papers deal with proteins from more than one class.

Rubredoxins.—The rubredoxins are the simplest class of iron–sulphur proteins, typically containing only one iron atom per molecule and having no labile sulphur atoms. The prototype protein of the class has been the rubredoxin from *Clostridium pasteurianum*, and X-ray diffraction studies[108] have shown that the iron is liganded to four cysteine sulphur atoms arranged in approximately tetrahedral shape. The initial X-ray work suggested that one of the Fe—S bonds was shorter than the other three. However, recently refined X-ray structures (to 1.2 Å resolution) have shown this not to be the case. A similar conclusion was also reached independently by Bunker and Stearn[109] on the basis of an extended X-ray absorption fine structure (EXAFS) technique. They found that the average Fe—S bond length was 2.267 ± 0.003 Å.

The rubredoxin from *C. pasteurianum* has also been the subject of two different types of detailed spectroscopic examination recently. Rivoal *et al.*[110] have investigated the low-temperature m.c.d. spectra of the oxidized protein. Their results, illustrated in Figure 19, were interpreted as indicating the presence of two one-electron charge-transfer transitions $(S \rightarrow Fe^{3+})$ in the region 15000—28000 cm^{-1}. A first-moments analysis of the lower energy band between 15000 and 23000 cm^{-1} was consistent with it being the orbital transition $t_1 \rightarrow e$. It was concluded that axial distortion of the lowest excited charge-transfer state was responsible for the splitting of the low-energy band of the absorption spectrum, whilst spin–orbit coupling was responsible for the temperature dependence of the m.c.d. spectrum.

[105] G. Palmer, in 'The Enzymes', ed. P. D. Boyer, Academic Press, New York, 1975, Vol. 12, 1.
[106] R. H. Holm, in 'Biological Aspects of Inorganic Chemistry', ed. A. W. Addison, W. R. Cullen, D. Dolphin and J. James, Wiley Interscience, New York, 1977, 71.
[107] 'Iron–Sulphur Proteins', ed. W. Lovenberg, Academic Press, New York, 1973, 1976, Vols. I, II, and III.
[108] L. Jensen, *Ann. Rev. Biochem.*, 1974, **43**, 461.
[109] B. Bunker and E. A. Stearn, *Biophys. J.*, 1977, **19**, 253.
[110] J. C. Rivoal, B. Briat, R. Cammack, D. O. Hall, K. K. Rao, I. N. Douglas, and A. J. Thomson, *Biochim. Biophys. Acta*, 1977, **493**, 122.

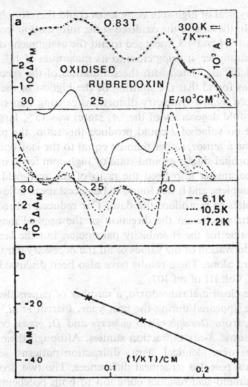

Figure 19 (a) The absorption ($\times \cdot \times \cdot$) and m.c.d. spectra of oxidized C. pasteurianum rubredoxin in Tris buffer. Concentration is 10^{-5} M, path length
1 mm, $B = 0.83$ T. Temperatures at which m.c.d. spectra were
measured, ———, 300 K; –·–·, 17.2 K; ····, 10.5 K; and – – – –,
6.1 K. (b) Plot of the first moment, ΔM_1, measured from 15 000 to
23 500 cm^{-1}, against $1/kT$
(Reproduced by permission from Biochim. Biophys. Acta, 1977, **493**, 122)

The second spectroscopic study was that of Schulz and Debrunner[111] on the
Mössbauer characteristics of both oxidized and reduced rubredoxin. Having
carried out the necessary replacement of ^{56}Fe by ^{57}Fe, they obtained spectra
under a number of different conditions of temperature and applied magnetic
field, and then parametrized them using the Hamiltonian equations (1) and (2)
for the spin and nuclear contributions.

$$\mathcal{H}_S = D[S_z{}^2 - S(S+1)/3 + \lambda(S_x{}^2 - S_y{}^2)] + \beta S \cdot \tilde{g} \cdot H \qquad (S = 2 \text{ or } \tfrac{5}{2}) \qquad (1)$$

$$\mathcal{H}_N = \langle S_i \rangle \cdot \tilde{A} \cdot I - \beta_n g_n H \cdot I + (eQV_{zz}/12)[3\,I_z{}^2 - \tfrac{15}{4} + \eta(I_x{}^2 - I_y{}^2)] \qquad (2)$$

Both the redox states of rubredoxin were found to offer favourable conditions
for Mössbauer spectroscopy, because at low temperatures the spin relaxation
in frozen solution is unusually slow, and the large magnetic hyperfine inter-

111 C. Schulz and P. G. Debrunner, *J. Phys. (Paris) Colloq.*, 1976, **6**, 153.

actions that are typical of high-spin complexes could therefore be observed with high resolution. In the case of the oxidized state, information from e.p.r., including the g values of 9.4 and 4.3, was used to aid the assignment of the value of λ, the rhombicity parameter, it being close to its maximum of $1/3$. However, there were some problems associated with the simulation of the spectra for oxidized rubredoxin. It was found that the splitting of the highest of the three Kramer's doublets (produced by low-symmetry distortions causing spin–orbit interactions which lift the six-fold degeneracy of the $^6A_1$ state) was 15% larger than that of the lowest. Since no value of λ could produce this ratio, the parameter A was allowed to become a tensor, rather than be equal to the isotropic Fermi-contact term A_0 for the orbital singlet ground state of high-spin ferric iron. In addition, the rhombicity parameter also fixed the ratio of the zero-field energies of the three Kramers doublets, and it was found to be necessary to adjust the energies of the highest doublets to reproduce the data. For reduced rubredoxin, the small dependence of the spectra on the direction of the applied magnetic field also implied a high value for the rhombicity parameter, but the lack of e.p.r. data necessitated the derivation of the values of all the necessary parameters from the Mössbauer spectra alone. These results have also been discussed at some length in Chapter 10 of Vol. III of ref 107.

Aside from the clostridial rubredoxin, a number of papers dealing with other rubredoxins have appeared during the past year. Pierrot *et al.*[112] have isolated two rubredoxins, from *Desulphovibrio vulgaris* and *D. gigas*, crystallized them, and carried out some X-ray diffraction studies. Although their cell parameters and space groups were identical, X-ray diffraction patterns, sequence studies, and reactivity data revealed structural differences. The two proteins had 37 of their 52 and 54 amino-acid residues common to both polypeptide chains, with the N-terminal methionine of the *D. gigas* rubredoxin being blocked. But, despite this, an NADH + H$^+$: rubredoxin oxidoreductase from *D. gigas* showed ten times more activity with its native substrate than with that from *D. vulgaris*. In a later paper, Adman *et al.*[113] reported the structural determination of the *D. vulgaris* rubredoxin to 2 Å resolution, using X-ray data in conjunction with a molecular replacement method. The starting model was derived from that of *C. pasteurianum* rubredoxin, using atoms common to both proteins according to sequence. Difference Fourier methods were used to obtain a partial refinement and the two rubredoxins were found to differ mainly in groups on the surface of the molecule. No major differences in conformation or hydrogen bonding were detected at the present level of analysis. The Fe—S bonds ranged in length from 2.15 to 2.35 Å, but did not differ significantly from the mean value of 2.29 Å.

The phylogenetic aspects of the amino-acid sequences of the *D. vulgaris* and *D. gigas* rubredoxins have been discussed by Vogel *et al.*[114] Their primary structures, when compared with rubredoxins from other anaerobic bacteria, revealed

[112] M. Pierrot, R. Haser, M. Frey, M. Bruschi, J. Le Gall, L. C. Siecker, and L. H. Jensen, *J. Mol. Biol.*, 1976, **107**, 179.
[113] E. T. Adman, J. C. Sieker, L. H. Jensen, M. Bruschi, and J. Le Gall, *J. Mol. Biol.*, 1977, **112**, 113.
[114] H. Vogel, M. Bruschi, and J. Le Gall, *J. Mol. Evol.*, 1977, **9**, 111.

only 12 identical residues, these being mainly concerned with the two groups of residues that each contain two iron-binding cysteines. A phylogenetic tree based on the primary structures was presented.

The isolation of a rubredoxin from another *Desulphovibrio* species, *D. desulphuricans* (Norway), was reported by Bruschi *et al.*[55] This protein was found to be acidic and to possess one iron atom and four cysteine residues; its molecular weight was approximately 6000.

The rather atypical rubredoxin from *Pseudomonas oleovorans* was the subject of a report by May and Kuo.[115] They have succeeded in immobilizing the protein on CNBr-treated Sepharose 4B, to yield a preparation with all the spectral features of the unbound oxidized protein. This bound form was capable of accepting electrons from dithionite, and from NADPH in the presence of spinach ferredoxin: NADP reductase. The reduction potential of the immobilized rubredoxin was similar to that of the soluble protein, and it was found that the iron atom could be both removed and replaced. The bound rubredoxin was, however, much more resistant to denaturation by guanidine hydrochloride.

The properties of another iron–sulphur protein that contains two iron atoms per molecule, bound in rubredoxin fashion, have been described by Moura *et al.*[116] Isolated from *Desulphovibrio gigas*, this protein appeared to be an example of a new type of iron–sulphur protein and was named desulphoredoxin. The molecule had a molecular weight of 7900 and contained no labile sulphur, but possessed eight cysteine residues. In all, the protein consisted of 73 amino-acids, although six types of residue, namely histidine, arginine, proline, isoleucine, phenylalanine, and tryptophan, were not present in the molecule. The *N*-terminal sequence was determined up to 35 residues and gave no indications of any homology with other non-haem iron proteins. The electronic spectrum of the protein contrasted with that of the two-iron rubredoxin from *Pseudomonas oleovorans* in that it was not simply the addition of two spectra of one-iron rubredoxins. This result led to the suggestion that two iron centres interact, and, by analogy with haemerythrin and other model compounds containing two interacting iron atoms, a weak shoulder at 720 nm was assigned as a 'magnon' *d–d* band. Preliminary e.p.r. measurements were reported to be more complex than the spectrum of the *P. oleovorans* rubredoxin.

2Fe:2S Ferredoxins.—Although an *X*-ray structure for one of the proteins in this class has yet to be published, a wealth of spectroscopic evidence both on the proteins themselves and on synthesized chemical analogues (where *X*-ray structures are known) leaves little doubt that the iron atoms in molecules of this type are each bound, in tetrahedral fashion, to two cysteinyl sulphur atoms and two common bridging sulphide ions. Even so, an *X*-ray structure of a two-iron ferredoxin remains a desirable objective, and Ogawa *et al.*[117] have published preliminary *X*-ray data on a chloroplast-type ferredoxin prepared from *Spirulina platensis*. This protein crystallized in the orthorhombic system, with cell dimen-

[115] S. W. May and J. Y. Kuo, *J. Biol. Chem.*, 1977, **252**, 2390.
[116] I. Moura, M. Bruschi, J. Le Gall, J. J. G. Moura, and A. V. Xavier, *Biochem. Biophys. Res. Comm.*, 1977, **75**, 1037.
[117] K. Ogawa, T. Tsukihara, H. Tahara, Y. Katsube, Y. Matsura, N. Tanaka, M. Kukudo, K. Wada, and H. Matsubara, *J. Biochem. (Japan)*, 1977, **81**, 529.

sions $a = 63.32$, $b = 28.51$, and $c = 108.09$ Å. The space group was $C222_1$ and each asymmetric unit contained one molecule. Electron-density maps at 5 and 3.5 Å resolution were synthesized, using the best phase angles calculated by the single isomorphous method coupled with the anomalous dispersion method. The difference Fourier synthesis, with the anomalous scattering difference of the native data, showed the locations of the iron atoms, and it was concluded that the active centre was close to the surface of the molecule.

Perhaps the prototype plant ferredoxin is that extracted from spinach, and this particular protein has continued to attract the interest of a number of workers. Davis and San Pietro[118, 119] have chemically modified spinach ferredoxin with trinitrobenzenesulphonic acid to produce a trinitrophenylated species that is modified at a single amino-group. This modified ferredoxin could accept electrons from photosystem I and pass them on to cytochrome c, but was incapable of forming a complex with the flavoprotein ferredoxin: NADP+ oxidoreductase and thus could not participate in the photoreduction of NADP. The fact that the modified protein was not able to inhibit the reduction of dichlorophenol indophenol by the flavoprotein led to the conclusion that the site of interaction between the ferredoxin and the flavoprotein, resulting in inhibition of diaphorase activity, was the same site as that responsible for the spectrally observable complex between the two proteins. Furthermore, this site was identical to the site of involvement of ferredoxin in NADP reduction. Data obtained on the reduction of cytochrome c supported a mechanism whereby ferredoxin could bind to the flavoprotein or to cytochrome c and therefore supported the suggestion that it was the complex between ferredoxin and cytochrome which served as the true substrate for reduction by the flavoprotein. Chromatographic evidence was obtained for a complex between the cytochrome and the iron–sulphur protein.

Another study connected with the involvement of spinach ferredoxin in photosynthesis was that of Boehme.[120] He prepared antibodies against spinach ferredoxin and ferredoxin: NADP oxidoreductase and used them as specific inhibitors with osmotically shocked and washed chloroplasts. The object of Boehme's work was to gain information on the cyclic electron-transport system in chloroplasts, the existence of which has been known for some time, but which nevertheless remains incompletely defined. One component known to be involved in the system, cytochrome b_6, was used to monitor the effects of the antibodies. Photoreduction of cytochrome b_6 was found to be unaffected by the reductase antibody, but was decreased by the ferredoxin antibody. Although the preparation of chloroplasts resulted in the release of ferredoxin, a portion of the protein appeared to remain bound, and it was this fraction that was responsible for the reduction of cytochrome b_6. The fact that the reductase antibody did not affect the photoreduction of cytochrome b_6, despite being able to inhibit the photoreduction of NADP, argued against the involvement of the reductase in the cyclic electron-transport system.

[118] D. J. Davis and A. San Pietro, *Biochem. Biophys. Res. Comm.*, 1977, **74**, 33.
[119] D. J. Davis and A. San Pietro, *Arch. Biochem. Biophys.*, 1977, **182**, 266.
[120] H. Boehme, *European J. Biochem.*, 1977, **72**, 283.

The electron-transfer reactivity of spinach ferredoxin with a number of reagents has been examined in a kinetic study by Rawlings *et al.*[121] The oxidation of the reduced protein by $[Fe(edta)]^-$, $[Fe(hedta)]$,* horse heart cytochrome *c*, and horse metmyoglobin followed second-order behaviour in every case; rate constants, activation enthalpies, and activation entropies were reported. Application of the Marcus theory, using data from the reaction with $[Fe(edta)]$, yielded a value of $1.7 \times 10^{-3} \, l \, mol^{-1} \, s^{-1}$ for the electrostatically corrected self-exchange rate constant of spinach ferredoxin. This low value was taken to imply an extreme inaccessibility of the redox centre of the protein. A Marcus-type analysis also suggested that electron transfer from ferredoxin to cytochrome *c* was particularly inefficient, whilst data on the dependence on pH and on ionic strength of the $[Fe(edta)]$–ferredoxin reaction, when fitted to the Marcus ionic strength equation, indicated a charge of -9.4 on ferredoxin at 25.8 °C and pH 7.8. It is worth mentioning at this point that a review of the theory of electron-transfer reactions, dealing particularly with applications of the Marcus theory to iron–sulphur proteins, has been provided by Bennett as Chapter 9 of Vol. III of Ref. 107.

Publications have appeared over the year which have dealt with the isolation of a number of other plant-type ferredoxins. Thus, Altosaar *et al.*[122] have purified a ferredoxin from leaves of *Sambucus racemosa* L. They have also characterized its spectral properties and determined its amino-acid composition. The c.p.r. spectrum was centred around $g = 1.957$ and the protein had a molecular weight of 10 700.

Hase *et al.*[123, 124] have isolated and sequenced two ferredoxins each from *Equisetum telmateia* and *E. arvense*. The *E. telmateia* ferredoxins, types I and II, each had only four cysteine residues in a total of 95 and 93 residues respectively, and although the *N*-terminal residues of both proteins were heterogeneous, alanine was concluded to be their genuine terminal residue. The proteins exhibited 29 differences in amino-acid residues, with 3 inverted replacements. It was necessary to insert one gap at position 32 of ferredoxin II in order to align the two sequences with greatest homology. The two proteins from *E. arvense* also had 95 and 93 residues and their sequences differed by only one residue each from the corresponding *E. telmateia* ferredoxins, although there were 31 sites of difference between the *E. arvense* sequences. Taking these facts together, the authors suggested that duplication of the ferredoxin gene occurred at an evolutionary stage long before the divergence of the two *Equisetum* species.

Two plant-type ferredoxins have also been isolated and purified from the blue-green alga *Nostoc verrucosum*.[125] Both had similar absorption spectra to plant ferredoxins, but their molecular weights, as determined by gel filtration, were approximately 18 000. These proteins were active in the photoreduction of NADP by broken spinach chloroplasts, although to different extents. There was also some difference in their ability to transfer electrons from NADPH to cytochrome

* hedta is *N*-hydroxyethyl-ethylenediaminetriacetate.

121 J. Rawlings, S. Wherland, and H. B. Gray, *J. Amer. Chem. Soc.*, 1977, **99**, 1968.
122 J. Altosaar, B. A. Bohm, and I. E. P. Taylor, *Canad. J. Biochem.*, 1977, **55**, 159.
123 T. Hase, K. Wada, and H. Matsubara, *J. Biochem. (Japan)*, 1977, **82**, 267.
124 T. Hase, K. Wada, and H. Matsubara, *J. Biochem. (Japan)*, 1977, **82**, 277.
125 M. Shin, M. Sukenoku, R. Oshino, and Y. Kitazume, *Biochim. Biophys. Acta*, 1977, **460**, 85.

c, using the *Nostoc* ferredoxin:NADP oxidoreductase. The involvement of ferredoxins in the photosynthetic process of another *Nostoc* species, *Nostoc mucosum*, has been investigated by Arnon *et al.*[126] These workers have used e.p.r. to examine directly the iron–sulphur centres found in membrane fragments. A number of chloroplast-bound centres had previously been identified by this means as being associated with photosystem I. These were centre A (bound ferredoxin), with *g* values of 1.86, 1.94, and 2.05 and a potential of −530 mV, centre B, with *g* values of 1.89, 1.92, and 2.05 and a potential of −580 mV, and centre X, with *g* values of 1.78, 1.88, and 2.08. Centre X could only be observed after prior reduction of centres A and B with dithionite. In contrast to previous studies, Arnon *et al.* illuminated their preparations prior to freezing for e.p.r. measurements. They found that when water was the reductant the addition and photoreduction of soluble ferredoxin occurred without a notable decrease in the signals of centres A and B, but that when NADP was also present the signals of centres A and B and of soluble ferredoxin were all diminished. This was proposed to be the first direct evidence of light-induced electron transport between membrane-bound iron–sulphur centres and ferredoxin:NADP oxidoreductase.

Three papers have appeared recently[127–129] concerning a two-iron ferredoxin from the halophilic bacterium *Halobacterium halobium*. The ferredoxin was found to represent approximately 1 % of the total soluble protein of the organism, and a molecular weight of approximately 15000 was calculated on the basis of its amino-acid composition. The mid-point potential was −345 mV and the e.p.r. spectrum of the reduced form had *g* values of $g_x = 1.9$, $g_y = 1.97$, and $g_z = 2.07$, which are typical of plant and algal ferredoxins. The e.p.r. signal was only clearly defined below 60 K, which indicated the spin relaxation to be more rapid than that of plant ferredoxins and considerably more rapid than that of adrenodoxin-type proteins. Since the slowness of relaxation is linked with antiferromagnetic coupling, this implied that the coupling in the halobacterial ferredoxin was less than in other two-iron ferredoxins. The protein was not able to mediate electron transport in the NADP photoreduction system of chloroplasts, but extracts of the bacterial cells did catalyse its reduction by NADH. Sequence determination showed the protein to consist of 128 amino-acids, of which four were cysteines whose relative positions in the sequence were the same as those of the four cysteines involved in iron binding in plant ferredoxins. The sequence had a high similarity with the ferredoxin from *Nostoc mucosum*, there being 39 identical residues in 96 sites compared and 27 additional conservative residues. There was, however, a long extra *N*-terminal region, containing 41 % of the acidic residues, which would appear to be part of a commonly found adaptation to life in concentrated salt solutions. The ferredoxin was shown to function physiologically as an electron acceptor in the enzymic oxidation of α-ketoglutarate, pyruvate, and α-ketobutyrate.

[126] D. I. Arnon, H. Y. Tsujimoto, and T. Hiyama, *Proc. Natl. Acad. Sci., U.S.A.*, 1977, **74**, 3826.
[127] L. Kerscher and D. Oesterhelt, *European J. Biochem.*, 1976, **71**, 101.
[128] T. Hase, S. Wakabayashi, H. Matsubara, L. Kerscher, D. Oesterhelt, K. K. Rao, and D. O. Hall, *F.E.B.S. Letters*, 1977, **77**, 308.
[129] L. Kerscher and D. Oesterhelt, *F.E.B.S. Letters*, 1977, **83**, 197.

Some further reports on the properties of adrenodoxin have been published during the past year. Katagiri and Takitawa[130] have monitored the formation of a 1:1 complex between adrenodoxin and cytochrome P-450$_{sec}$ by use of absorption spectroscopy, gel filtration, and density-gradient centrifugation techniques. The binding could only be induced by the presence of cholesterol bound to the cytochrome. Williams-Smith and Cammack[131] have used e.p.r. to determine the reduction potential of membrane-bound adrenodoxin as -248 ± 15 mV, whilst Lambeth and Kamin[132] have investigated (kinetically) the involvement of adenodoxin in the reduction of cytochrome c by the flavoprotein–adrenodoxin reductase. A complex between adrenodoxin and its reductase was found to be the catalytically active species.

The reminder of this section on two-iron ferredoxins will be devoted to those papers which are connected with the rationalization of the unusual spectroscopic properties of these systems. The initially odd observation that e.p.r. signals were only observed from reduced proteins of this class has been explained in terms of antiferromagnetic coupling between the magnetic moments of the two high-spin iron atoms and of the fact that the system accepts only a single electron on reduction. Antiferromagnetic coupling has also been shown to account for the observation of g_{av} values less than 2 in these proteins. Further studies have been reported this year which have been directed at enhancing our understanding of this antiferromagnetic coupling. For reviews of past work, Sands and Durham[133] and Cammack *et al.* in Vol. III of ref. 107 provide considerable information on the important contributions of e.p.r. and Mössbauer spectroscopy.

Blum *et al.*[134] have determined the value of the exchange coupling constant, J, for both oxidized and reduced spinach ferredoxin, by resonance Raman techniques. Fitting the observed Raman band positions to those expected on the basis of the Gibson model[135] led to values for J_{ox} and J_{red} of -172 and -74 cm^{-1} respectively. Several other bands associated with Fe—S and C—H stretching were also observed, aside from those assigned to antiferromagnetic coupling. As these authors point out, resonance Raman spectroscopy is unique in its ability to measure the positions of the higher levels in a ladder of exchange-coupled states. However, the fact that the observed levels did not follow the model exactly suggested that the exchange coupling constants fall off with increasing temperature, due to increases in the average distance between the two iron atoms.

E.p.r. methods have also been used to measure the exchange coupling constants of a number of systems. Salerno *et al.*[136] have monitored variations in the e.p.r. linewidth and signal intensity with temperature and have arrived at values for J_{red} of 90 cm^{-1} for the S-1 centre of succinic dehydrogenase, 65 cm^{-1} for

[130] M. Katagiri and O. Takitawa, *Biochem. Biophys. Res. Comm.*, 1977, **77**, 804.

[131] D. L. Williams-Smith and R. Cammack, *Biochim. Biophys. Acta*, 1977, **499**, 432.

[132] J. D. Lambeth and H. Kamin, *J. Biol. Chem.*, 1977, **252**, 2908.

[133] R. H. Sands and W. R. Durham, *Quart. Rev. Biophys.*, 1975, **7**, 443.

[134] H. Blum, R. Adar, J. C. Salerno, and J. S. Leigh, *Biochem. Biophys. Res. Comm.*, 1977, **77**, 650.

[135] J. F. Gibson, D. O. Hall, J. H. M. Thornley, and F. R. Whatley, *Proc. Natl. Acad. Sci., U.S.A.*, 1966, **56**, 987.

[136] J. C. Salerno, T. Ohnishi, H. Blum, and J. S. Leigh, *Biochim. Biophys. Acta*, 1977, **494**, 191.

Rieske's iron–sulphur centre, and 270 cm^{-1} for adrenodoxin. The similar behaviour of the N-1a centre of NADH–ubiquinone reductase to that of the S-1 centre was taken as indicating that the former was probably a two-iron ferredoxin with a J_{red} value of 90 cm^{-1}. These authors noted an interesting association of the value of J to the character of the e.p.r. spectrum of the 2Fe:2S centre proteins, in that those with the most axial e.p.r. spectra had the highest values of J, and *vice versa*.

Another aspect of the exchange coupling of two-iron ferredoxins has been reported by Blumenfeld *et al.*[137] The dependencies of the e.p.r. signals of pea ferredoxin, the membrane-bound ferredoxin from bean, and the N-2 centre of bovine heart mitochondria with temperature differed depending on whether dithionite or thermalized electrons produced by γ-irradiation was used to reduce the proteins. The latter method apparently produced states with an increased spin–lattice interaction, which was rationalized in terms of a change in the exchange integral of antiferromagnetic coupling.

Finally, low-temperature m.c.d. spectra of three 2Fe:2S proteins have been published by Thomson *et al.*[138] The fully oxidized ferredoxins from spinach and *Spirulina maxima* gave temperature-independent m.c.d. spectra between room temperature and 18 K. They therefore showed there to be no contribution to the m.c.d. spectra at room temperature from any population of low-lying excited states originating from exchange coupling. The reduced proteins and reduced adrenodoxin did, however, exhibit temperature-dependent m.c.d. spectra. A partial interpretation of the oxidized m.c.d. spectrum was made on the assumption that 'valence localization' of the d-electrons on the two iron atoms took place, thus justifying the use of the simpler rubredoxin system[110] as a model. Even so, it was concluded that unambiguous spectral assignments could not be made.

4Fe:4S Iron–Sulphur Proteins.—As was the case for the two-iron ferredoxins, spectroscopy, both of proteins and of model analogues, has made a considerable contribution to our knowledge of 4Fe:4S cluster systems. Cammack,[139] in a short review, has recently detailed some of the observations that have led up to our present interpretation of the apparently divergent properties of some of these proteins. The basic concepts are the antiferromagnetic coupling of the high-spin iron atoms making up the cluster and the three-state hypothesis of Carter *et al.*[140] which has allowed the redox properties and the observation of two different paramagnetic states to be rationalized.

Recent work directed at a theoretical understanding of the properties of 4Fe:4S clusters has included that of Bogner *et al.*[141] who have reported a model for the interpretation of Mössbauer spectra of reduced *Clostridium pasteurianum* ferredoxin obtained over a wide range of temperatures and magnetic fields.

[137] L. A. Blumenfeld, D. S. H. Burbaev, A. V. Lebanidze, and A. F. Vanin, *Studia Biophys.*, 1977, **63**, 143.
[138] A. J. Thomson, R. Cammack, D. O. Hall, K. K. Rao, B. Briat, J. C. Rivoal, and J. Badoz, *Biochim. Biophys. Acta*, 1977, **493**, 132.
[139] R. Cammack, *J. Phys. (Paris) Colloq.*, 1976, **6**, 137.
[140] C. W. Carter, J. Kraut, S. T. Freer, R. A. Alden, L. C. Siecker, E. Adman, and L. H. Jensen, *Proc. Natl. Acad. Sci., U.S.A.*, 1972, **69**, 3526.
[141] L. Bogner, F. Parak, and K. Gersonde, *J. Phys. (Paris) Colloq.*, 1976, **6**, 177.

Blum *et al.*[142] have carried out a parallel study to that of Salerno *et al.*[136] using the variation of e.p.r. signal intensity and linewidth with temperature as a means of determining the energy of a low-lying excited state in the oxidized HiPiP from *Chromatium vinosum*. These authors justify the use of a Gibson model[135] for the antiferromagnetic coupling on the basis of Mössbauer and n.m.r. evidence which indicates that the four iron atoms are magnetically inequivalent. Their results showed that the $S=\frac{1}{2}$ ground state and the $S=\frac{3}{2}$ excited state differed in energy by 160 ± 10 cm^{-1}. However, the observed ratio of the multiplicities of the excited to ground states was, at 15, neither predicted by their model nor by other molecular orbital models. They concluded that there were apparently sources of degeneracy other than those due to spin multiplicity, in the first excited states of *Chromatium* HiPIP.

One of the many applications of spectroscopic techniques, and particularly of e.p.r., in the field of iron–sulphur proteins has been the characterization of the iron–sulphur compounds of the mitochondrial electron-transfer chain. Beinert, in a chapter of Vol. III of ref. 107, has reviewed, in some detail, the character of the iron–sulphur centres that have been discovered so far, and has discussed their distribution amongst the mitochondrial complexes I, II, and III.

An interesting approach to this problem has been published recently by Albracht and Subramanian.[143] These workers have investigated the hyperfine interactions of ^{57}Fe atoms in the e.p.r. spectra of submitochondrial particles prepared from the yeast *Candida utilis*, grown in a medium enriched with ^{57}Fe. Information was gained on both the 4Fe:4S and 2Fe:2S clusters found in mitochondria. Thus, in the oxidized state, line broadening due to ^{57}Fe was clearly seen in the 4Fe:4S cluster centre 3 of the succinate dehydrogenase, complex II. However, in reduced mitochondria, e.p.r. signals from 4Fe:4S clusters were only visible below 20 K; these were for centres 2, 3, and 4 of NADH dehydrogenase, complex I. Between 50 and 80 K it was concluded that only the spectra of the 2Fe:2S proteins were detectable, and hyperfine splitting constants were determined for the centres 1 of NADH and succinate dehydrogenases, the Rieske centre of complex III, and the centre 2 of succinate dehydrogenase. The results are summarized in Table 5. All these 2Fe:2S centres showed hyperfine interaction with two iron nuclei in the z direction, but for the centres 1 of NADH and succinate dehydrogenases the effective interaction with the two nuclei was very different: Furthermore, the strength of the greatest interaction in the x–y direction was much greater than in the z direction, and the Fe atoms were quite unequal with respect to free electron spin, the ferric iron having four times the interaction of the ferrous. This behaviour was not, however, exhibited by the Rieske centre, where there was no detectable hyperfine interaction in the x direction and only a small interaction in the y direction.

Another mitochondrial iron–sulphur centre was the subject of an investigation by Ruzicka and Beinert.[144] This protein, isolated from beef heart mitochondria and purified to electrophoretic homogeneity, contained one FAD group and a single 4Fe:4S cluster. The cluster had g values of $g_x=1.886$, $g_y=1.939$, and

[142] H. Blum, J. C. Salerno, R. C. Prince, and J. S. Leigh, *Biophys. J.*, 1977, **20**, 23.
[143] S. P. J. Albracht and J. Subramanian, *Biochim. Biophys. Acta*, 1977, **462**, 36.
[144] F. J. Ruzicka and H. Beinert, *J. Biol. Chem.*, 1977, **252**, 8440.

Table 5 Properties of the Fe—S centres in ^{57}Fe-containing submitochondrial particles

Centre	Complex I			Complex II				Complex III
	1	2	3	4	3	2	1	3
Cluster type	[2Fe:2S]	[4Fe:4S]	[4Fe:4S]	[4Fe:4S]	[4Fe:4S]	[2Fe:2S]	[2Fe:2S]	[2Fe:2S]
Charge	2-(2-;3-)	2-(2-;3-)	2-(2-;3-)	2-(2-;3-)	1-(1-;2-)	2-	2-(2-;3-)	2-(2-;3-)
$A_{z1,2}$/mT	1.2	0.85	0	—	—	1.3	1.3	1.25
A_{y1}/mT	2.0	—	—	—	—	2.1	2.1	~0.55
A_{y2}/mT	0	—	—	—	—	0.3	0.3	~0.55
A_{x1}/mT	2.0	—	0	0	—	~1.6	~1.6	0
A_{x2}/mT	0	—	0	0	—	small	small	0

$g_z = 2.086$ in the reduced state. The protein could take up 3 electrons and the reduction potential of the flavin was approximately 55 mV lower than that of the cluster. The fact that the iron–sulphur group was reduced by the electron-transferring flavoprotein of the β oxidation pathway of fatty acids and could be re-oxidized by ubiquinone led to the conclusion that the protein functions *in vivo* as an electron carrier in fatty acid oxidation. Beinert, in the review mentioned above, has suggested that this protein is in fact the 'centre 5' originally classified as part of complex I.

The apparent similarity of some of the iron–sulphur centres in mammalian succinate dehydrogenase to centres found in *Rhodopseudomonas sphaeroides* led Ingledew and Prince[145] to examine the possibility that the bacterial proteins might make up a succinate dehydrogenase. Using chromatophores prepared from this organism, three iron–sulphur centres were found to be removed by alkaline washing, a treatment which also caused the loss of succinate dehydrogenase activity. Two of these three components were ferredoxin-like centres with reduction potentials of $+50$ and -250 mV at pH 7.0 and g_y values of 1.94. The third was a HiPIP-type centre, of potential $+80$ mV at pH 7.0, with e.p.r. signals centred at $g = 2.01$. Their ease of removal, coupled with the fact that the washing did not interfere with the kinetics of light-induced electron transport, or allow cytochrome c_2 to escape from the vesicular interior, suggested that the three iron sulphur components were located on the outer surface of the chromatophore membrane. It also implied that they were not involved in the cyclic electron-transfer pathway. The washing did not, however, remove a Rieske-type centre with $g_y = 1.90$ from the chromatophore, and may also leave a third centre (of $g_y = 1.94$ type and potential -350 mV) still bound.

Unlike the two-iron ferredoxins, the study of proteins containing 4Fe:4S clusters has been able to benefit from the availability of X-ray structures. The tertiary structures of the HiPIP from *Chromatium vinosum*, which has a single cluster, and the ferredoxin from *Peptococcus aerogenes*, which contains two clusters, are known. Carter, in a chapter of Vol. III of ref. 107, has discussed at some length the structural features of both these proteins, which are illustrated in Figure 20. The iron–sulphur clusters of the two proteins are immersed in cavities bounded principally by non-polar amino-acid side-chains. Nevertheless, in spite of the expected low dielectric constant of such environments, the protein redox potentials are different from those of model analogues measured in more homogeneous solvent environments of similar type. Carter suggested that the uniqueness of the protein potentials arises from the fact that backbone peptide groups penetrate the 'oil droplet' at the core of the protein with geometries appropriate for hydrogen bonding to specific sulphur atoms. Three of the four cysteine groups linked to each cluster (46, 63, and 77 in HiPIP and 8, 11, and 45 and 35, 38, and 18 in ferredoxin) are followed by hairpin turns in the polypeptide chain, so that a hydrogen bond may be made with the amide nitrogen of residue $(n+2)$. It appears that both HiPIP and ferredoxin form similar hydrogen bonds to cysteinyl sulphur, but different hydrogen bonds to inorganic sulphur.

---

[145] W. J. Ingledew and R. C. Prince, *Arch. Biochem. Biophys.*, 1977, **178**, 303

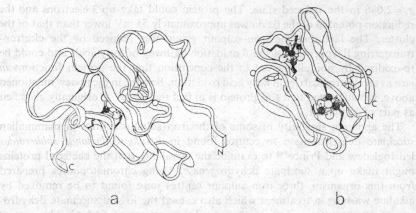

a   b

Figure 20 *Chain tracings of (a)* Chromatium *HiPIP and (a)* P. aerogenes *ferredoxin. Both molecules are drawn to the same scale and with one of the
ferredoxin* Fe$_4$S$_4$* *clusters (cluster II) and the HiPIP cluster in the
same relative orientation with respect to the two tyrosine residues
associated with them*
(Reproduced by permission from 'Iron–Sulphur Proteins', ed. W. Lovenberg,
Academic Press, New York, Vol. III)

Particular attention has also been drawn to tyrosine residues which are to be
found close to both clusters in ferredoxin and to the single cluster of HiPIP.
In the latter protein there is an expansion of the cluster on reduction, and the
Tyr-19 residue moves closer. Recently, Carter[146] has reported that the active
sites of HiPIP and ferredoxin can be brought into equivalent orientations by
assuming that the clusters belong to the symmetry point group C_s. In this way
polypeptide segments connecting cysteines 46 and 63 in HiPIP and cysteines 18
and 35 in ferredoxin become analogous, in the sense that they are of the same
length, connect equivalent cysteinyl sulphur atoms, and have similar twisted
antiparallel β conformations. Furthermore, the tyrosine residues 19 (in HiPIP)
and 2 (in ferredoxin) also become analogous in that they interact with equivalent
sulphur atoms. Carter has stated that these interactions with polypeptide backbone and tyrosine side-chains place the iron–sulphur clusters into diastereomeric
environments. He has invoked evidence from circular dichroism to support this
theory (Figure 21) and suggests that such a relationship may underlie the reason
for the contrasting redox behaviour of the two proteins.

A theoretical approach to the understanding of the effects of the environment
on the reduction potentials of iron–sulphur clusters has been made by Kassner
and Yang.[147] They have calculated the contribution that a difference in dielectric
constant might make to the overall free-energy change of redox reactions by
means of the Born equation. Furthermore, they have also allowed for the effect
that the dielectric constant of the solvent has on the microscopic dielectric
constant of the cluster. Calculations combining these two effects suggested that

[146] C. W. Carter, *J. Biol. Chem.*, 1977, **252**, 7802.
[147] R. J. Kassner and W. Yang, *J. Amer. Chem. Soc.*, 1977, **99**, 4351.

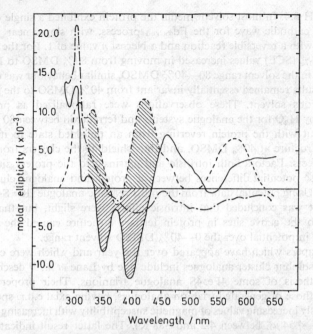

Figure 21 *Circular dichroism spectra of native reduced HiPIP at pH 7.8 (———)*
and super-reducible HiPIP at pH 9.85 in 85% DMSO (–·–·–·), both at
4 °C, and of oxidized ferredoxin (– – –) at pH 9.5 and 5 °C. Molar
ellipticity, $[\theta]$, is defined by $[\theta] = (\theta/10)(M/lC)$, where θ is the measured
ellipticity (in degrees), M is the molecular weight, l is the path length
(in cm), and C is the concentration (in g ml^{-1}). All ellipticities are
expressed per dmol of $Fe_4S_4^*S\gamma_4$ cluster. Cross-hatched regions show
transitions with opposite rotation in the two proteins
(Reproduced by permission from *J. Biol. Chem.*, 1977, **252**, 7802)

they may represent a significant contribution to the observed differences in
potential between protein and analogue iron–sulphur clusters, as well as predict-
ing the effects of different solvents on the latter. However, as the authors point
out, it may be difficult to assign an appropriate value to the dielectric constant
of the environment surrounding a cluster buried in a protein, but their calculations nevertheless give some indication of how the relative polarity of the cluster
binding site might affect the observed redox potential.

Hill *et al.*[148] have also noted that previous comparisons of reduction potentials
of iron–sulphur proteins in aqueous solution and their model analogues in non-
aqueous solution showed an apparent substantial negative shift of the latter to
the former. They have investigated this effect by d.c. polarographic measurement
of the half-wave potentials of the isoelectronic redox couples of *Clostridium
pasteurianum* ferredoxin and $[Fe_4S_4(SR)_4]^{2-, \ 3-}$ models at the dropping mercury
electrode, in solvent mixtures ranging from 80% DMSO–20% H_2O to 100%

[148] C. L. Hill, J. Renaud, R. H. Holm, and L. E. Mortenson, *J. Amer. Chem. Soc.*, 1977, **99**, 2549.

H_2O, at pH 8.4. In most solvent media the protein exhibited a single diffusion-controlled cathodic wave for the $Fd_{ox/red}$ process, with slopes near -59 mV, consistent with a reversible reaction and a Nernst n value of 1. For the analogue couples, $E_{1/2}$ (SCE) values increased in moving from 80% DMSO to H_2O. For ferredoxin in the solvent range 80—40% DMSO, similar behaviour was observed, but the results remained essentially invariant from 40% DMSO to the limit of a pure aqueous solvent. These observations were rationalized as progressive solvation by H_2O for the analogue systems and ferredoxin between 80 and 40% DMSO, but with the protein reverting from an unfolded state to its normal aqueous structure at 40% DMSO, and thus shielding the clusters from further solvent effects. Factors both intrinsic and extrinsic to the protein sites which might cause potential differences between a protein and analogue cluster were discussed. Using potential data from the most realistic analogue [R = S-Cys(Ac)-NHMe], it was concluded that intrinsic effects were slight, but that factors extrinsic to the active sites in protein tertiary structure caused the observed differences in potential over the 0—40% DMSO solvent range.

Other papers which have appeared over the year and which were concerned with iron–sulphur cluster analogues include one by Lane *et al.*[149] describing the direct synthesis of some 4Fe:4S analogue trianions. Their properties were similar to those of generated trianion analogues, having axial e.p.r. spectra and continuously increasing values of magnetic susceptibility with increasing temperature (2.05—4.54 μ_B between 4.2 and 388 K). The latter result indicates a spin doublet ground state and interactions within the cluster core. Mössbauer features were also comparable with several reduced ferredoxins, and suggested that iron sub-site inequivalence in the proteins was an intrinsic property of the clusters and was only secondarily influenced by protein structure. Holm and Ibers have published a review of the properties of synthetic iron–sulphur clusters in a chapter of Vol. III of ref. 107.

The formation of analogues by extrusion of the iron–sulphur core from proteins was the subject of a report by Gillum *et al.*[150] They have produced both $[Fe_4S_4(SPh)_4]$ and $[Fe_2S_2(SPh)_2]$ analogues in essentially quantitative yield by treating proteins with benzthiol in a medium of 4:1 v/v hexamethylphosphoramide:H_2O (aqueous component pH 8.5). The method was successful with both oxidized *C. pasteurianum* ferredoxin (and redox mixtures), oxidized *Bacillus stearothermophilus* ferredoxin, reduced *Chromatium* HiPIP, and the iron protein of *C. pasteurianum* nitrogenase, all of which contain 4Fe:4S sites. The problems associated with core extrusions of multi-site systems and any possible difficulties in identifying 2Fe:2S units due to dimer–tetramer conversion were also considered.

A kinetic study of the dissolution of 4Fe:4S clusters from five ferredoxins and *Chromatium* HiPIP has been undertaken by Maskiewicz and Thomas,[151] with particular attention to the effects of pH. In the acid pH range, log k_{obs} *vs.* pH

[149] R. W. Lane, A. G. Wedd, W. O. Gillum, E. J. Laskowski, R. J. Holm, R. B. Frankel, and G. C. Papaefthymiou, *J. Amer. Chem. Soc.*, 1977, **99**, 2350.
[150] W. O. Gillum, L. E. Mortenson, J. S. Chen, and R. H. Holm, *J. Amer. Chem. Soc.*, 1977, **99**, 584.
[151] R. Maskiewicz and C. B. Thomas, *Biochemistry*, 1977, **16**, 3024.

profiles for the dissolution of the ferredoxin clusters followed the same rate law as for the hydrolysis of synthetic $[Fe_4S_4(SR)_4]^{2-}$ cluster ions. By analogy, it was therefore proposed that the dissolution at acid pH occurred by hydrolysis. At neutral and alkaline pH values, dissolution was found to be related to initial ligand exchange and possible extrusion of the core, which may then be hydrolysed. Interestingly, the log k_{obs} *vs.* pH profile for the dissolution of *Bacillus polymyxa* ferredoxin II (which contains only a single cluster) followed the same rate law as other ferredoxins containing two clusters, which suggested that for the latter the rates, and acid–base equilibrium constants, were very similar for both clusters. No correlation between the redox potentials of the ferredoxins and their dissolution behaviour was detected.

The kinetics of the electron-transfer reactions of a number of proteins, including *Chromatium* HiPIP, with penta-aminepyridineruthenium(III) have been reported by Cummings and Gray.[152] The kinetic parameters determined for HiPIP were $k_{obs}=1.1 \times 10^3$ l mol^{-1} s^{-1}, $\Delta H^{\ddagger}=9.4$ kcal mol^{-1} and $\Delta S^{\ddagger}=13$ e.u. (at 25 °C, pH 6.5, $\mu=0.5$ mol l^{-1}). The electrostatically corrected protein self-exchange constants (k_{11}^{corr}), when compared with data from analogous reactions with the other small redox reagents $[Ru(NH_3)_6]$ and $[Fe(edta)]^{2-}$, indicated that $[Ru(NH_3)_5py]^{3+}$ facilitates electron transfer from metalloproteins. This was attributed to the penetration of the protruding edge of the π-conjugated pyridine ligand into the protein interior, thus allowing direct protein–reagent redox centre overlap in the transition state for electron transfer. Figure 22 shows a plot of the ratio between the calculated value of k_{11}^{corr} for a variety of reagents over k_{11}^{corr} for $[Fe(edta)]$ *vs.* k_{11}^{corr} for $[Fe(edta)]$. The values of

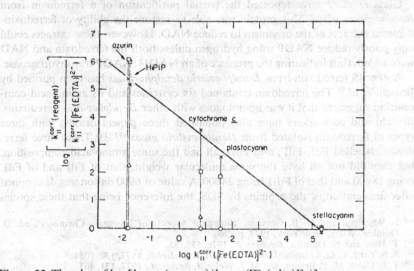

Figure 22 *The plot of* $\log[k_{11}^{corr}(reagent)/k_{11}^{corr}([Fe(edta)]^{2-})]$ *versus* $\log k_{11}^{corr}([Fe(edta)]^{2-})$; ($\square$) $[Co(phen)_3]^{3+}$; ($\times$) $[Ru(NH_3)_5py]^{3+}$; ($\triangle$) $[Ru(NH_3)_6]^{2+}$; ($\bigcirc$) $[Fe(edta)]^{2-}$
(Reproduced by permission from *J. Amer. Chem. Soc.*, 1977, **99**, 5158)

[152] D. Cummings and H. B. Gray, *J. Amer. Chem. Soc.*, 1977, **99**, 5158.

$k_{11}{}^{corr}$ for HiPIP may be seen to show a wide variation with reagent. The reasons for such variations have been discussed recently by Wherland and Gray[153] in an extensive review of theoretical and experimental observations on protein electron-transfer reactions. By their arguments, HiPIP would seem to possess a variety of mechanisms available for electron transfer.

The concluding part of this section will be devoted to recent reports connected with the isolation and characterization of proteins containing 4Fe:4S clusters. Hase and Ohmiya[154] have determined the amino-acid sequence of a ferredoxin isolated from *Bacillus stearothermophilus*. The polypeptide chain consisted of 81 amino-acid residues, of which only four were cysteines. Three of these were located near the N-terminus as a Cys^{11}-xx-Cys^{14}-xx-Cys^{17} segment, whilst the fourth, Cys-61, was followed by a proline residue and lay in the C-terminal half of the molecule. The sequence was homologous, in many segments, with those of other bacterial ferredoxins. The presence of a relatively high number of glutamic acid residues (a good helix former) and the relatively low number of cysteines was put forward as a possible explanation of the good thermal stability and the insensitivity of this protein to oxygen.

Yang *et al.*[155] have described the isolation of another ferredoxin of high thermal stability. Their protein, from *Clostridium thermoacetium*, had a molecular weight of approximately 7300, a partial specific volume of 0.67, and a pI of 3.25. The extinction coefficient of the absorbance maximum at 390 nm ($E = 16.8 \times 10^3$ l mol^{-1} cm^{-1}) was more in line with those of single-cluster ferredoxins than those with two. The protein retained 50% of its activity after treatment at 80 °C and functioned in the transfer of electrons from pyruvate to NADP.

Glass *et al.*[156] have reported the partial purification of a ferredoxin from *Ruminococcus albus*. This protein was able to restore the ability of ferredoxin-depleted extracts of the organism to reduce NAD. However, these extracts could only poorly reduce NADP using hydrogen unless both the ferredoxin and NAD were added, thus indicating the presence of an NADH:NADP transhydrogenase.

A 4Fe:4S ferredoxin from *Desulphovibrio desulphuricans* has been purified by Bruschi *et al.*[55] The ferredoxin contained six cysteines and its amino-acid composition suggested that it was homologous with other *Desulphovibrio* ferredoxins. Bruschi and co-workers have also published three papers dealing with three types of ferredoxin isolated from *Desulphovibrio gigas*.[157-159] These three ferredoxins, labelled FdI, FdI', and FdII, all had the same amino-acid composition, but they did not all have the same molecular weight, that of FdI and of FdI' being 18000 and that of FdII being 24000. A value of 6000 dalton was determined after dissociation of the proteins by SDS, the inference being that these species

153 S. Wherland and H. B. Gray, in 'Biological Aspects of Inorganic Chemistry', ed. D. Dolphin, Wiley, New York, 1977, p. 289.
154 T. Hase and N. Ohmiya, *Biochem. J.*, 1976, **159**, 55.
155 S. S. Yang, L. G. Llungdahl, and J. Le Gall, *J. Bacteriol.*, 1977, **130**, 1084.
156 T. L. Glass, M. P. Bryant, and M. J. Wolin, *J. Bacteriol.*, 1977, **131**, 463.
157 M. Bruschi, E. C. Hatchikian, J. Le Gall, J. J. G. Moura, and A. V. Xavier, *Biochim. Biophys. Acta*, 1976, **449**, 275.
158 J. J. G. Moura, A. V. Xavier, M. Bruschi, and J. Le Gall, *Biochim. Biophys. Acta*, 1977, **459**, 278.
159 R. Cammack, K. K. Rao, D. O. Hall, J. J. G. Moura, A. V. Xavier, M. Bruschi, J. Le Gall, A. Deville, and J. P. Gayda, *Biochim. Biophys. Acta*, 1977, **490**, 311.

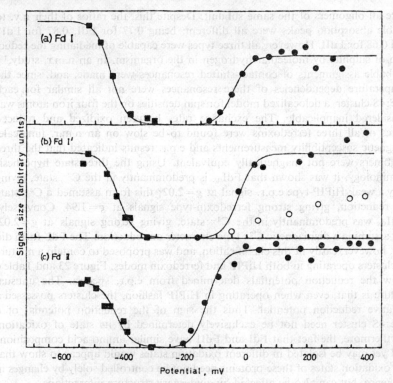

Figure 23 *E.p.r. signal intensities of (a) FdI; (b) FdI'; (c) FdII as a function of redox potential.* ■, *g = 1.94 signal;* ●, *g = 2.02 signal. In (b), open circles represent the intensity of the g = 2.02 signal after reduction to the lowest potential and re-oxidation with ferricyanide. Signal sizes were measured by the peak-to-dip size of the principal feature. E.p.r. spectra were recorded under varying conditions of gain setting, and the signal intensities of the different proteins are not comparable with each other*

(Reproduced by permission from *Biochim. Biophys. Acta*, 1977, **490**, 311)

Table 6 *Relative e.p.r. signal intensities and redox potentials for the three forms of* D. gigas *ferredoxins*

Form	Signal intensity (spins per 6456 mol. wt.)		Mid-point potential /mV	
	$g = 2.02$	$g = 1.94$	C^-/C^{2-}	C^{2-}/C^{3-}
FdI	0.07	0.68	−50	−455
FdI'	0.28	0.33	−30	−430
FdII	0.73	0.02	−130	−437

The $g = 2.02$ signals in the oxidized (C^-) state were measured in the proteins as prepared. The intensities of the $g = 1.94$ signals in the reduced (C^{3-}) state were measured in samples reduced with sufficient dithionite solution to give maximum signal intensities. Spectra were recorded at 15 K at a microwave power of 1 mW. In each case, the intensity of the whole spectrum was integrated; for the reduced signal, between 0.31 and 0.37 T, and for the oxidized signal, between 0.31 and 0.38 T. Values are in electron spins per subunit molecular weight of 6456. Mid-point potentials are expressed relative to the standard hydrogen electrode.

were all oligomers of the same subunit. Despite this, the ratios of their u.v. to visible absorption peaks were all different, being 0.77 for FdI, 0.87 for FdI', and 0.68 for FdII. However, all three types were capable of mediating the reduction of sulphite by molecular hydrogen in the organism. In an n.m.r. study,[158] probable assignments of contact-shifted resonances were made, and, since the temperature dependencies of these resonances were not all similar for each 4Fe:4S cluster, a delocalized model for spin densities on the four iron atoms was considered inapplicable. The exchange rates between oxidized and reduced states of all three ferredoxins were found to be slow on an n.m.r. timescale. Magnetic susceptibility measurements and e.p.r. results indicated that the three oligomers were not magnetically equivalent. Using the three-state hypothesis terminology, it was shown that FdI_{ox} is predominantly in the C^{2-} state, giving only a weak HiPIP-type e.p.r. signal at $g = 2.02$; this form assumed a C^{3-} state on reduction, giving strong ferredoxin-type signals at $g = 1.94$. Conversely, $FdII_{ox}$ was predominantly in the C^{1-} state, giving strong signals at $g = 2.02$, but assuming a diamagnetic C^{2-} type character on reduction. The FdI' form did not, however, easily fit this classification, and was proposed to contain a mixture of clusters operating in both HiPIP and ferredoxin modes. Figure 23 and Table 6 show the reduction potentials determined from e.p.r. spectra. The unusual feature is that, even when operating in HiPIP fashion, the clusters possessed a negative reduction potential. Thus the sign of the reduction potential of a 4Fe:4S cluster need not be exclusively determined by its state of oxidation. Furthermore, the fact that FdI and FdII have similar amino-acid compositions and yet may be isolated in different oxidation states would appear to show that the oxidation states of these proteins need not be controlled solely by changes in sequence, but can also be affected by quaternary structure interactions.

Finally, the primary structure of a HiPIP from a halophilic denitrifying coccus, provisionally classified as a *Paracoccus*, has been determined by Tedro *et al.*[160] The protein contained 71 residues and was as similar to HiPIP's from photosynthetic bacteria as they are to each other. The protein was highly acidic, as is generally the case for proteins isolated from halophiles.

[160] S. M. Tedro, T. E. Meyer, and M. D. Kamen, *J. Biol. Chem.*, 1977, **252**, 7826.

7

Oxidases and Reductases

BY B. E. SMITH & P. F. KNOWLES

In this chapter we have reviewed papers from the literature of 1977 on the chemical and biochemical properties of the copper-, iron-, and molybdenum-containing oxidases and reductases. Some enzymes or aspects of them were not discussed in the literature during that year, and therefore are not included in this review. Wherever possible, at the beginning of each section we have referred to more comprehensive reviews in order to mitigate this problem. Nevertheless, some will no doubt feel that their favourite topic has been neglected or will not concur with our judgement of which topics are of chemical and biochemical interest. To these readers we extend our apologies.

1 Oxidases and Hydroxylases containing Copper; Superoxide Dismutases

The copper centres in this group of oxidative enzymes may be classified as type I (intensely blue, with a Cu^{2+} e.p.r. signal having small $A_{\parallel}$), type II (similar visible absorption spectrum and e.p.r. parameters to simple Cu^{2+} complexes), or type III (considered to be spin-paired Cu^{2+} ions and therefore diamagnetic). There are no known examples of oxidative enzymes having type I copper alone.

Type II Copper.—*Amine Oxidases.* A brief review of the literature to 1975 has been written.[1] The amine-binding site of the amine oxidase from pig plasma has been probed, using the fluorescent substrate kynuramine;[2] only 1 mole of the amine was found to bind ($K_a = 1.8 \times 10^5$ l mol^{-1}) under anaerobic conditions despite there being two identical subunits in the enzyme. The data also indicate that hydrophobic forces as well as covalent attachment to the carbonyl grouping present in the enzyme are involved in the substrate binding. For other important studies on the reaction mechanism of pig plasma monoamine oxidase, the work of Pettersson and co-workers[3] should be consulted. Buffoni *et al.*[4] have shown that 3-hydrazino-pyridazine derivatives, which are used as anti-hypertensive agents, are non-competitive inhibitors against amine binding to amine oxidases containing copper but fail to show this inhibition with mitochondrial amine oxidases which contain flavin nucleotide as cofactor.

Proton-relaxation studies have been carried out on the diamine oxidase from pig kidney.[5] The results suggest that (i) water that is accessible to solvent is

[1] K. T. Yasunobu, H. Ishizaki, and N. Ninamura, *Mol. Cell. Biochem.*, 1976, **13**, 3.
[2] J. D. Massey and J. E. Churchich, *J. Biol. Chem.*, 1977, **252**, 8082.
[3] A. Lindstrom and G. Pettersson, *European J. Biochem.*, 1978, **83**, 131.
[4] F. Buffoni, G. Ignesti and R. Pirisano, *Farmaco, Ed. Sci.*, 1977, **32**, 404.
[5] M. D. Kluetz and P. G. Schmidt, *Biochemistry*, 1977, **16**, 5191.

bound to Cu^{2+} in the enzyme; (ii) Cu^{2+} is not located near the oxidizing site in the enzyme; (iii) oxygen does not compete with water for a binding site on the Cu^{2+}; (iv) histamine binds such that an imidazole ring nitrogen is proximal to the Cu^{2+}; (v) ammonia does not bind to the Cu^{2+}.

Galactose Oxidase. Galactose oxidase has a single polypeptide chain of molecular weight 68 000 and one e.p.r.-detectable Cu^{2+} ion. It is thus the simplest type II copper enzyme. Low-temperature e.p.r. studies[6] at 9 and 35 GHz reveal nitrogen hyperfine structure on both $g_{\parallel}$ and $g_{\perp}$. Further hyperfine structure is observed when fluoride binds. It has been concluded that the Cu^{2+} site in galactose oxidase has pseudo-square-planar geometry, with one water molecule and two histidines as ligands. Weiner *et al.*[7] have presented evidence that the Cu^{2+} site in galactose oxidase quenches the fluorescence from some of the tryptophan residues in the polypeptide chain, and calculated that the Cu^{2+} must be $\leqslant 1.2$ nm from these tryptophans. The decrease in fluorescence intensity when certain of the tryptophans are chemically modified supports the postulate that the substrate, *i.e.* galactose, is proximal to Cu^{2+} and tryptophan in the enzyme–substrate complex. Other studies[8] show that the activity of galactose oxidase is reduced to 2% of the activity of the native enzyme when 2 out of the 18 tryptophan residues are oxidized by *N*-bromosuccinimide; this chemical modification also causes a marked decrease in the optical rotation at 314 nm (assigned to a charge-transfer transition involving Cu^{2+}), modifies the e.p.r. spectrum, and increases the chemical stability of the copper site. These data are discussed in terms of a model in which oxidation of the tryptophans causes the Cu^{2+} site to change geometry from pseudo-square-planar towards tetrahedral. Kwiatkowski *et al.*[9] found that the activity of galactose oxidase is lost when one moiety of iodoacetamide reacts; it has been implied from these and other results that one polypeptide histidine has a role in the active site of galactose oxidase.

Tyrosine Hydroxylase. Tyrosine hydroxylase from some sources has been reported to be a copper protein.[10] The toxic amino-acid L-mimosine, found in tropical legumes, has been shown[11] to be a competitive inhibitor against L-dopa for tyrosine hydroxylase from mouse myelomas, though no evidence was presented on whether or not the copper site is affected. Hoeldtke and Kaufman[12] have purified tyrosine hydroxylase from bovine adrenals to near homogeneity. Analytical ultracentrifuge and gel-electrophoresis data suggest that the molecular weight of the enzyme is 34 000. Analysis for metals revealed the presence of 0.5—0.75 gram atom of Fe per mole of enzyme; no copper was detected. At present, the role (if any) of iron in the hydroxylation mechanism is unknown.

[6] R. D. Bereman and D. J. Kosman, *J. Amer. Chem. Soc.*, 1977, **99**, 7322.
[7] R. E. Weiner, M. J. Ettinger, and D. J. Kosman, *Biochemistry*, 1977, **16**, 1602.
[8] D. J. Kosman, M. J. Ettinger, R. D. Bereman, and R. S. Giordano, *Biochemistry*, 1977, **16**, 1597.
[9] L. D. Kwiatkowski, L. Siconolfi, R. E. Weiner, R. S. Giordano, R. D. Bereman, M. J. Ettinger, and D. J. Kosman, *Arch. Biochem. Biophys.*, 1977, **182**, 712.
[10] S. Gutteridge and D. Robb, *European J. Biochem.*, 1975, **54**, 107.
[11] H. Hashiguchi and H. Takahashi, *Mol. Pharmacol.*, 1977, **13**, 362.
[12] R. Hoeldtke and S. Kaufman, *J. Biol. Chem.*, 1977, **252**, 3160.

Dopamine β-Hydroxylase. The enzyme from bovine adrenals is reported[13] to be a tetramer, with four bound copper atoms; approximately 66% of these coppers are detectable by e.p.r. Reduction of the enzyme with ascorbate eliminates the e.p.r. signal whilst oxidation with $K_3Fe(CN)_6$ or by oxygen during substrate-turnover conditions yields all coppers in e.p.r.-detectable form. The 4-hydroxyl grouping in tyramine and octopamine cause hyperfine splitting in the $g_\perp$ component of the e.p.r. spectrum. Deoxygenation of the enzyme leads to a decrease in the copper e.p.r. signal, which suggests the existence of an oxygenated complex. Kinetic studies reveal that the toxic amino-acid L-mimosine is a non-competitive inhibitor against binding of tyramine to the enzyme.[11]

L-*Tryptophan-2,3-dioxygenase.* Contrary to the work of Brady *et al.*,[14] Makino and Ishimura[15] found iron but no copper in the enzyme purified from liver.

Superoxide Dismutases. For convenience, the forms of superoxide dismutase containing type II copper and zinc, iron, or manganese will be reported in this section.

A short review covering the evolution of copper- and iron-containing forms of superoxide dismutase as well as other metalloproteins is available.[16] Classification of different photosynthetic organisms according to whether or not they produce Cu/Zn, Fe, or Mn superoxide dismutases has revealed clear evolutionary patterns.[17, 18]

Most assays for superoxide dismutase are based upon inhibition of some reaction potentiated by superoxide radicals, and can be termed 'negative assays'. A new 'negative assay' has been reported by Chunakov *et al.*[19] which utilizes the effect of superoxide dismutase on the reduction of tetrazolium salts by superoxide radicals. Rigo and Rotilio[20] have described a polarographic procedure for assay of both superoxide dismutase and catalase in samples of whole blood; superoxide dismutase at concentrations as low as 2×10^{-11} mol l^{-1} has been detected. Misra and Fridovich[21] have reported a sensitive and convenient 'positive assay' for superoxide dismutase. The assay is based upon the inhibition by superoxide radicals of the peroxidation of dianisidine that is catalysed by horseradish peroxidase; addition of superoxide dismutase relieves this inhibition and allows a coupled spectrophotometric assay for the dismutase. Catalase interferes with the assay, but this problem can be overcome by addition of a catalase inhibitor, namely 3-amino-1,2,4-triazole.

Superoxide Dismutases containing Cu/Zn. A review covering physical and chemical studies on Cu/Zn superoxide dismutase from bovine erythrocytes has been

13 G. A. Walker, H. Kon, and W. Lovenberg, *Biochim. Biophys. Acta*, 1977, **482**, 309.
14 F. O. Brady, M. E. Monaco, H. J. Forman, G. Schutz, and P. Feigelson, *J. Biol. Chem.*, 1972, **247**, 7915.
15 R. Makino and Y. Ishimura, *J. Biol. Chem.*, 1976, **251**, 7722.
16 E. Frieden, *Trends Biochem. Sci.*, 1976, **1**, 273.
17 L. E. A. Henry and D. O. Hall, *Plant Cell Physiol.*, Special issue No. 3, on 'Photosynthetic organelles' 1977, p. 377.
18 K. Asada, S. Kanematsu, and K. Uchida, *Arch. Biochem. Biophys.*, 1977, **179**, 243.
19 V. M. Chunakov and L. F. Osinskaya, *Voprosy Med. Khim.*, 1977, **23**, 712.
20 A. Rigo and G. Rotilio, *Analyt. Biochem.*, 1977, **81**, 157.
21 H. P. Misra and I. Fridovich, *Analyt. Biochem.*, 1977, **79**, 553.

published.[22] Lippard *et al.*[23] report the close similarity between the n.m.r. spectra of native superoxide dismutase from bovine erythrocytes and the apo-enzyme to which one zinc atom per subunit had been re-added. It is suggested that zinc organizes the structure of the active site. The 270 MHz ^1H n.m.r. spectra of Cu^{2+}/Zn^{2+} and apo superoxide dismutase have been analysed by Cass *et al.*[24] The data require that at least four and probably six histidine residues serve as ligands to the metals in each subunit of the native enzyme; this conclusion is consistent with the *X*-ray crystallographic data.[25] Other spectral lines which have been assigned originate from His-19, His-41, Tyr-108, and the *N*-terminal acetyl group. Some pH titration measurements show that the proton attached to C-2 of His-41 exchanges rapidly at pH > 8.

Beem *et al.*[26] have described their e.p.r. studies on different metal-substituted forms of superoxide dismutase from bovine erythrocytes, in which Ag^+ resides in the native Cu^{2+} sites and either Cu^{2+} or Co^{2+} is in the native Zn^{2+} sites; the effect of azide, an inhibitor of enzyme activity, on these substituted forms of the enzyme is also reported. It is concluded that azide binds to the Cu^{2+} in the native enzyme; this conclusion is contrary to that reached by Fee and Gaber,[27] *i.e.* that azide binds to Zn^{2+} in the native enzyme.

The new e.p.r. data have been considered in a detailed examination of the active site of superoxide dismutase, utilizing interactive computer graphics model-building.[26] The results indicate that His-61, which binds to both copper and zinc, does not lie in the plane of the copper and its three other histidine ligands but occupies a position intermediate between planar and axial. This feature might well account for the rhombicity of the e.p.r. spectra and also could relate to the catalytic activity of the enzyme.

Two papers by Rigo *et al.*[28, 29] describe detailed studies on the properties of superoxide dismutase samples prepared by adding controlled amounts of Cu^{2+} to the apoenzyme. A striking result is that the specific activity of enzyme having only one bound copper is twice that of enzyme having two bound coppers, and favours the concept of anti-co-operative interactions between the two copper sites during the catalytic cycle.

McAdam,[30] using pulsed radiolysis, has shown that the effects of buffer concentration and addition of salt on the activity of superoxide dismutase can be interpreted in terms of general ionic-strength phenomena rather than a specific effect. However, the effects of the univalent anions azide and halide, which are

[22] S. J. Lippard, A. R. Burger, K. Ugurbil, J. S. Valentine, and M. W. Pantoliano, *Adv. Chem. Series*, 1977, **162**, 251.
[23] S. J. Lippard, A. R. Burger, K. Ugurbil, M. W. Pantoliano, and J. S. Valentine, *Biochemistry*, 1977, **16**, 1136.
[24] A. E. G. Cass, H. A. O. Hill, B. E. Smith, J. V. Bannister, and W. H. Bannister, *Biochemistry*, 1977, **16**, 3061.
[25] J. S. Richardson, K. A. Thomas, D. H. Rubin, and D. C. Richardson, *Proc. Nat. Acad. Sci. U.S.A.*, 1975, **72**, 1349.
[26] K. M. Beem, D. C. Richardson, and K. V. Rajagopalan, *Biochemistry*, 1977, **16**, 1930.
[27] J. A. Fee and B. P. Gaber, *J. Biol. Chem.*, 1972, **247**, 60.
[28] A. Rigo, P. Viglino, L. Calabrese, D. Cocco, and G. Rotilio, *Biochem. J.*, 1977, **161**, 27.
[29] A. Rigo, M. Terezi, P. Viglino, L. Calabrese, D. Cocco, and G. Rotilio, *Biochem. J.*, 1977, **161**, 31.
[30] M. E. McAdam, *Biochem J.*, 1977, **161**, 697.

competitive inhibitors against binding of superoxide radical to the enzyme, are specific and can be understood in terms of binding to Cu^{2+} in the enzyme.[31]

Human and bovine Cu/Zn superoxide dismutase differ in their content of the aromatic amino-acids tryptophan and tyrosine. Finazzi-Agro *et al.*[32] have compared the fluorescence and the catalytic properties of the two forms of the enzyme, and concluded that the aromatic residues have little effect on enzyme activity. A similar experimental approach has been used by Permyakov *et al.*,[33] who report that the fluorescence quantum yield from phenylalanine residues in superoxide dismutase from green peas has an anomalously low value of 1.1% and is constant over the pH range 5—10. It is speculated that the low value for the quantum yield results from interactions between certain phenylalanine residues and the copper sites. Since thermal denaturation of the enzyme occurs only at 70 °C, it is concluded[33] that the protein structure has high rigidity.

There have been several reports on the presence and distribution of superoxide dismutases in brain. Sharoyan *et al.*[34, 35] claim that there are three isozymes of superoxide dismutase in bovine brain; these have been purified and examined by optical and e.p.r. spectroscopy. The results suggest that the isozymes have closely similar structure. Other workers[36] have found no evidence for any regional or subcellular distribution of superoxide dismutase in the brain. Van Berkel *et al.*[37, 38] have analysed different cell populations in rat liver for superoxide dismutase, and presented evidence that parenchymal cells have both the Mn and Cu/Zn forms of the enzyme whilst non-parenchymal cells have only the Cu/Zn enzyme. The physiological reason behind this observation is unknown.

Superoxide Dismutases containing Mn. The manganese-containing superoxide dismutase from *Bacillus stearothermophilus* is being studied in detail. There are two subunits, each of molecular weight 20 000, and 2 gram atom of Mn. Smit *et al.*[39] report the unit-cell dimensions of three orthorhombic crystal forms of the enzyme. The apoenzyme can be prepared by exposure to 8M-urea at low pH;[40] the dimeric structure is maintained during this treatment. Full activity can be restored to the apoenzyme by adding Mn^{2+} in the presence of 8M-urea at acid pH, followed by dialysis to remove the urea.

McAdam *et al.*[41] have studied the catalytic mechanism of the manganese enzyme from *Bacillus stearothermophilus*, using pulse radiolysis. Previous studies

[31] A. Rigo, R. Stevanato, R. Viglino, and G. Rotilio, *Biochem. Biophys. Res. Comm.*, 1977, **79**, 776.
[32] A. Finazzi-Agro, D. Cocco, L. Calabrese, W. H. Bannister, and F. Bossa, *Internat. J. Biochem.*, 1977, **8**, 389.
[33] E. A. Permyakov, E. A. Burstein, Y. Sawada, and I. Yamazaki, *Biochim. Biophys. Acta*, 1977, **491**, 149.
[34] S. G. Sharoyan, A. A. Shaldzhian, R. M. Nalbandyan, and G. Kh. Buntyatyan, *Voprosy Biokhim. Mozga*, 1976, **11**, 105.
[35] S. G. Sharoyan, A. A. Shaljian, R. M. Nalbandyan, and H. Ch. Buniatian, *Biochim. Biophys. Acta*, 1977, **493**, 478.
[36] T. C. Loomis, G. Yee, and W. L. Stahl, *Experientia*, 1977, **32**, 1374.
[37] T. J. C. Van Berkel, J. K. Kruijt, R. G. Slee, and J. F. Koster, *Arch. Biochem. Biophys.*, 1977, **179**, 1.
[38] T. J. C. Van Berkel and J. K. Kruijt, *European J. Biochem.*, 1977, **73**, 223.
[39] J. D. G. Smit, J. Pulver-Sladek, and J. M. Jansonius, *J. Mol. Biol.*, 1977, **112**, 491.
[40] C. J. Broke, J. I. Harris, and S. Sata, *J. Mol. Biol.*, 1976, **107**, 175.
[41] M. E. McAdam, F. Lavelle, R. A. Fox, and E. N. Fielden, *Biochem. J.*, 1977, **165**, 71 and 81.

led to the suggestion that the Cu/Zn- and Fe-containing superoxide dismutases operate *via* a simple two-step mechanism:

$$E_A + O_2^- \xrightarrow{k_1} E_B + O_2 \qquad \text{(Step 1)}$$

$$2H^+ + E_B + O_2^- \xrightarrow{k_2} E_A + H_2O_2 \qquad \text{(Step 2)}$$

For the manganese enzyme, two extra reaction steps are needed to explain the pulse radiolysis results:

$$E_B + O_2^- \xrightarrow{k_3} E_C \qquad \text{(Step 3)}$$

$$E_C \xrightarrow{k_4} E_A \qquad \text{(Step 4)}$$

Steps 1 and 2 constitute a fast oxidation/reduction cycle whilst steps 3 and 4 provide a slow cycle. Another difference between the Mn and Cu/Zn forms of the enzyme is that the activity of former is dependent on pH over the range 6.5—10.2 whereas that of the latter is independent of pH. The Mn enzyme also seems less sensitive to inhibition by H_2O_2, N_3^-, and CN^-. Moustafa-Hassan and Fridovich[42] have reported that production of the manganese enzyme can be induced by addition of methyl viologen during the anaerobic growth of *Escherichia coli*. It is suggested that methyl viologen diverts electron flow from a cyanide-sensitive to a cyanide-insensitive pathway, resulting in increased levels of superoxide radical, which derepresses the biosynthesis of the enzyme.

It is interesting to note that the manganese enzyme from the red alga *Porphyridium cruentum* appears to have only one gram atom of manganese per mole of enzyme;[43] the enzyme has a molecular weight of 40 000 and two subunits. Some properties of the enzyme, including the effects of inhibitors and the preparation of apoenzyme, have been studied. The relationship between this form of superoxide dismutase and those from other organisms with respect to evolution has been discussed.

Superoxide Dismutases containing Fe. Superoxide dismutase from *Mycobacterium tuberculosis* has been purified to homogeneity[44] and shown to have a molecular weight of 88 000. There are four equal subunits and 4 gram atom of Fe per mole of enzyme. The amino-acid composition is closely similar to that of other forms of superoxide dismutase containing iron. It has also been observed that an antibody to the superoxide dismutase that contains manganese is only partially effective against the iron enzyme.

Bruschi *et al.*[45] have studied the *N*-terminal sequence of amino-acids in the iron-containing superoxide dismutase from the strict anaerobe *Desulphovibrio desulphuricans*. A striking feature is that more than 50% of the amino-acids are hydrophobic, suggesting that the *N*-terminal sequence may be oriented towards the centre of the protein. There is sequence homology between the *Desulphovibrio* enzyme and superoxide dismutase from prokaryotes and mitochondria, but not

[42] H. Moustafa-Hassan and I. Fridovich, *J. Biol. Chem.*, 1977, **252**, 7667.
[43] H. P. Misra and I. Fridovich, *J. Biol. Chem.*, 1977, **252**, 6421.
[44] E. Kusonose, K. Ichihara, Y. Noda, and M. Kusonose, *J. Biochem. (Japan)*, 1977, **80**, 1343.
[45] M. Bruschi, E. C. Hatchikian, J. Bonicel, G. Bovier-Lapierre, and P. Couchoud, *F.E.B.S. Letters*, 1977, **76**, 121.

with the Cu/Zn form of the enzyme found in bovine erythrocytes. In another report from this same group,[46] the enzyme is shown to have between 1.4 and 1.9 gram ions of high-spin iron(III) per mole of enzyme. The significance of superoxide dismutases in strict anaerobes is discussed. Superoxide dismutase from *Photobacterium leiognathi* has been purified and its catalytic mechanism studied by pulse radiolysis.[47] Two samples of the enzyme containing 1.6 and 1.15 gram atom of iron per mole of enzyme were studied separately. The results indicate that only part of the iron in the sample containing 1.6 gram atom of iron participates in the catalytic mechanism. Taken with the observation that two distinct forms of iron are detected by e.p.r. spectroscopy, it is concluded that the enzyme has one gram atom of iron per mole of enzyme and variable amounts of extraneous iron.

Types I, II, and III Copper.—*Laccase and other Phenol Oxidases.* Malmström[48] has reviewed the mechanism of oxygen reduction in oxidases, quoting laccase and cytochrome oxidase as specific examples. From rapid-quench e.p.r. experiments using reduced laccase and $^{17}O_2$, Branden and Deinum[49] have obtained evidence for $H_2^{17}O$ bound equatorially to the type II copper; this indicates that type II copper must be close to the dioxygen-reducing site in the enzyme. Graziani *et al.*[50] have shown that type II copper can be reversibly removed from *Rhus vernicifera* laccase by dialysis against dimethylglyoxime plus potassium ferricyanide at pH 5.0 followed by dialysis against EDTA. No properties of this simplified form of the enzyme have yet been reported. Pulsed e.p.r. has been used to show that types I and II copper in *Rhus vernicifera* laccase and Type II copper in porcine caeruloplasmin have imidazole as a ligand.[51] Linear field-effect measurements further indicate that the ligands to the type I copper in laccase are arranged tetrahedrally.

Laccase from *Rhus vernicifera* reacts specifically with hydrogen peroxide to give a stable product which might be a functionally significant intermediate.[52] An earlier report from the same group[53] showed that the complex with hydrogen peroxide has an absorption band at 325 nm, which has been attributed to reaction with the type III site; it is proposed that the reduced form of type III copper in laccase binds oxygen in a manner analogous to that in the haemocyanins.

A purification procedure and some properties of the extracellular laccase from *Botrytis cinerea* have been described.[54] The enzyme is shown to have an exceptionally low isoelectric point and great stability to low pH. Mayer *et al.*[55] claim that the enzyme is a glycoprotein, of molecular weight 56 000 and containing a single

[46] E. C. Hatchikian and Y. A. Henry, *Biochemie*, 1977, **59**, 153.
[47] F. Lavelle, M. E. McAdam, E. M. Fielden, P. B. Roberts, K. Paget, and A. M. Michelson, *Biochem. J.*, 1977, **161**, 3.
[48] B. G. Malmström, *Adv. Chem. Series*, 1977, **162**, 173.
[49] R. Branden and J. Deinum, *F.E.B.S. Letters*, 1977, **73**, 144.
[50] M. T. Graziani, L. Morpurgo, G. Rotilio, and B. Mondovi, *F.E.B.S. Letters*, 1976, **70**, 87.
[51] B. Mondovi, M. T. Graziani, W. B. Mims, R. Oltzik, and J. Peisach, *Biochemistry*, 1977, **16**, 4198.
[52] I. Pecht, O. Farver, and M. Goldberg, 1977, *Adv. Chem. Series*, 1977, **162**, 179.
[53] O. Farver, M. Goldberg, D. Lancet, and I. Pecht, *Biochem. Biophys. Res. Comm.*, 1976, **73**, 494.
[54] M. Dubernet, P. Ribereau-Gayon, H. R. Lerner, H. Eitan, and A. M. Mayer, *Phytochemistry*, 1977, **16**, 191.
[55] A. M. Mayer, I. Marbach, A. Marbach, and A. Sharon, *Phytochemistry*, 1977, **16**, 1051.

gram atom of copper per mole of enzyme. It has also been reported[56] that catechol oxidase, purified from grapes, has a copper content of 0.5 gram atom per mole of enzyme (mol. wt. 80 000). Esterbauer *et al.*[57] have developed a new procedure for the assay of laccase and other polyphenol oxidases in which the quinone reaction product is converted into a colourless adduct, by reaction with 2-nitro-5-thiobenzoic acid. The new procedure is claimed to be as sensitive as established procedures, yet more convenient.

Ascorbate Oxidase. The properties of ascorbate oxidase purified from two plant sources have been reported.[58] E.p.r. and optical absorption spectral studies suggest that fluoride binds to the type II copper, and that azide 'modifies the environment of the copper' whilst cyanide completely reduces both type I and type II copper. These and other results for binding of anions to the enzyme are compared with similar binding to laccase and caeruloplasmin. Krul and Dawson[59] report on the properties of apo-ascorbate oxidase. The quaternary structure is similar to that of the native enzyme though the apoenzyme is more readily dissociated into subunits. It is suggested that one role of copper in the native enzyme is to maintain the quaternary and tertiary structures.

Cytochrome Oxidase.—Cytochrome oxidase cannot be described by the Type I, Type II, Type III categories for copper enzymes, and it requires a separate classification. A recent literature survey is available.[60]

Studies on the Metal Sites in Oxidized and Reduced Cytochrome Oxidases. Magnetic susceptibility measurements on purified cytochrome oxidase from beef have been made at room temperature, using a 270 MHz n.m.r. spectrometer.[61] The results suggest that both the oxidized and reduced forms of the enzyme have one of the haems (designated cytochrome a_3) in the high-spin state and that this haem is antiferromagnetically coupled to one of the two cupric ions. The other copper and the low-spin haem (designated cytochrome a) are magnetically isolated. These observations are in broad agreement with recent e.p.r. and other physical measurements[62–64]

Two groups have reported on their magnetic c.d. studies with beef heart cytochrome oxidase.[65, 66] Fortunately, these reports are in close agreement. The oxidized enzyme is suggested to have one haem (the cytochrome a) in the low-spin ferric state, the other haem (cytochrome a_3) being in the high-spin ferric state. Thomson *et al.*[66] further observed that the signal from cytochrome a is unaffected by the addition of anions whereas the signal from cytochrome a_3 is high-spin in the presence of fluoride ions and low-spin in the presence of cyanide

56 M. Kidrom, E. Harel, and A. M. Mayer, *Phytochemistry*, 1977, **16**, 1050.
57 H. Esterbauer, E. Schwazl, and M. Hayn, *Analyt. Biochem.*, 1977, **77**, 486
58 V.Ts. Aikazyan and R. M. Nalbandyan, *Biokhimiya*, 1977, **42**, 2027.
59 K. G. Krul and C. R. Dawson, *Bioinorg. Chem.*, 1977, **7**, 71.
60 R. A. Capaldi and M. Briggs, in 'The Enzymes of Biological Membranes', Vol. 4, ed. A. Martinosi, Plenum Press, New York, 1976, 87.
61 K.-E. Falk, T. Vänngård, and J. Angström, *F.E.B.S. Letters*, 1977, **75**, 23.
62 H. Beinert, R. E. Hansen, and C. R. Hartzell, *Biochim. Biophys. Acta*, 1976, **423**, 339.
63 R. Aasa, S. P. J. Albracht, K.-E. Falk, D. Lanne, and T. Vänngård, *Biochim. Biophys. Acta*, 1976, **422**, 260.
64 G. Palmer, G. T. Babcock, and L. E. Vickery, *Proc. Nat. Acad. Sci. U.S.A.*, 1976, **73**, 2206.
65 G. T. Babcock, L. E. Vickery, and G. Palmer, *J. Biol. Chem.*, 1976, **251**, 7907.
66 A. J. Thomson, T. Brittain, C. Greenwood, and J. P. Springall, *Biochem. J.*, 1977, **165**, 327.

ions. The reduced enzyme has cytochrome a in the form of low-spin iron(II) and the cytochrome a_3 in the form of high-spin iron(III). These general conclusions have been confirmed[66] by other studies on the carbon monoxide derivative of the enzyme.

The biological significance of changes in spin state of the cytochrome a_3 following the binding of certain anions has been discussed by Nicholls *et al.*[67] in terms of the proposal by Perutz and co-workers[68] that changes in the spin state of the iron in haemoglobin are linked to changes of conformation in the protein. Wikstrom[69] has presented further evidence to support his postulate that this conformational change is associated with vectorial translocation of protons across the mitochondrial membrane, which is the basis for the energy-conservation mechanism. Hu *et al.*[70] have carried out X-ray absorption-edge studies of oxidized and reduced cytochrome. Reduction with dithionite results in increased amounts of a cuprous species, presumed to arise from the other copper centre. The heretical suggestions were made that neither of the copper ions in the native enzyme gives rise to an e.p.r. signal and that a protein grouping, in addition to cytochrome a, might be an acceptor of electrons from the physiological substrate, *i.e.* ferrocytochrome c. These results should be treated with some caution since it has recently been pointed out[63] that cytochrome oxidase often contains extraneous as well as catalytically active copper.

These paramagnetic copper species could be resolved by e.p.r. spectral-subtraction procedures.[63] Similar procedures have been described by Greenaway *et al.*,[71] who give the e.p.r. parameters for intrinsic e.p.r.-detectable copper in the enzyme. The e.p.r. spectrum is interpreted in terms of flattened tetrahedral symmetry about the Cu^{2+} with an additional rhombic distortion. Anisotropy in the spectrum is suggested to originate from weak dipolar interaction with a low-spin iron(III) species (cytochrome a) located at a distance of 7 Å. These investigators[71] also observed a broad e.p.r. signal at temperatures $\leqslant 5$ K which they suggest might originate from the 'e.p.r. undetectable' copper and cytochrome a_3.

Schroedl and Hartzell[72] report oxidative titrations on reduced cytochrome oxidase, using ferricyanide, 1,1'-bis(hydroxymethyl)ferricinium ion, or oxygen as oxidant. The results indicate that four electrons are donated to the oxidants and suggest that the metal components of the oxidase behave as haem–copper pairs having mid-point potentials of 240 and 340 mV. However, the above procedure for determining mid-point potentials in metalloenzymes has been criticized by Lanne *et al.*,[73] who show that the e.p.r. spectrum of oxidized cytochrome oxidase differs from that of the cytochrome oxide/ferricyanide complex; this suggests that the redox mediator might modify the true mid-point potential for metal centres in the enzyme.

[67] P. Nicholls, L. C. Peterson, M. Miller, and F. B. Hansen, *Biochim. Biophys. Acta*, 1976, **449**, 188.
[68] M. F. Perutz, E. J. Heidner, J. E. Ladner, J. G. Beetlestone, C. Ho, and E. F. Slade, *Biochemistry*, 1974, **13**, 2187.
[69] M. K. F. Wikstrom, *Nature*, 1977, **266**, 271.
[70] V. W. Hu, S. I. Chan, and G. S. Brown, *Proc. Nat. Acad. Sci. U.S.A.*, 1977, **74**, 3821.
[71] F. T. Greenaway, S. H. P. Chan, and G. Vincow, *Biochim. Biophys. Acta*, 1977, **490**, 62.
[72] M. A. Shroedl and C. R. Hartzell, *Biochemistry*, 1977, **16**, 1327.
[73] B. Lanne, B. G. Malmström, and T. Vänngård, *F.E.B.S. Letters*, 1977, **77**, 146.

Studies on the Metal Sites in Inhibited Forms of Cytochrome Oxidase. The number of reducing equivalents required to form the partially reduced carbon-monoxide-inhibited form of cytochrome oxidase has been determined as 2.0 ± 0.3 per mole of enzyme.[74] It is suggested that both the 'e.p.r.-silent' copper and the cytochrome a_3 sites must be reduced for carbon monoxide to bind. Magnetic circular dichroism studies at 293 K show[75] that the cytochrome a_3 and cytochrome a species are in their low-spin ferrous and low-spin ferric forms respectively; it is further suggested that, at temperatures < 100 K, the bond between carbon monoxide and cytochrome a_3 is readily photolysed, allowing the haem a_3 to revert to the high-spin form, which in turn triggers a change at the haem a site from low- to high-spin.

From studies on the CO stretch band at 1963.5 cm^{-1} in the partially reduced complex of carbon monoxide with cytochrome oxidase, Yoshikawa et al.[76] concluded that carbon monoxide binds only to cytochrome $a_3{}^{2+}$, and it was postulated that this is the site of oxygen binding in the reduced native enzyme. The width of this band in the i.r. spectrum suggests that the carbon monoxide is located in a non-aqueous site, which might well facilitate reduction of oxygen. Unfortunately, no information can yet be drawn from these studies on whether a second metal centre is involved in stabilization of the postulated μ-peroxo intermediate during reduction of dioxygen.

Wever et al.[77] have studied the binding of carbon monoxide to reduced cytochrome oxidase, using optical absorption spectroscopy. The results of these and e.p.r. studies indicate that formation of the partially reduced carbon monoxide derivative results in re-oxidation of both the e.p.r.-detectable copper and possibly also the cytochrome a. Redox-titration experiments also suggest that both the 'invisible' copper and cytochrome a_3 must be reduced for carbon monoxide to bind with high affinity. From a detailed analysis of redox-titration data on cytochrome complexes with fully or partially reduced cytochrome oxidase, Shroedl and Hartzell[78, 79] concluded that cytochrome c can only bind to the enzyme after the low-potential cytochrome a^{2+} had been oxidized. Thus all the reports from different laboratories on the properties of partially reduced cytochrome oxidase are in broad agreement.

It has been shown[80] that sulphite is both an inhibitor and a slow reductant of oxidized cytochrome oxidase. When limited reaction with sulphite is allowed, the following species are detected by low-temperature e.p.r.: low-spin cytochrome a^{3+}, low-spin cytochrome $a_3{}^{3+}$, e.p.r.-detectable copper ($g_{\parallel} = 2.17$, $g_{\perp} = 2.03$), and a new Cu^{2+} e.p.r. signal with $g_{\parallel} = 2.19$ and $g_{\perp} = 2.05$. This new signal is claimed to result because sulphite converts cytochrome a_3 into its low-

[74] D. F. Wilson and Y. Miyata, *Biochim. Biophys. Acta*, 1977, **461**, 218.
[75] T. Brittain, J. Springall, C. Greenwood, and A. J. Thomson, *Biochem. J.*, 1976, **159**, 811.
[76] S. Yoshikawa, M. C. Choc, M. C. O'Toole, and W. S. Caughey, *J. Biol. Chem.*, 1977, **252**, 5498.
[77] R. Wever, J. H. Van Drooge, A. O. Muijsers, E. A. Barber, and B. F. van Gelder, *European J. Biochem*, 1977, **73**, 149.
[78] M. A. Schroedl and C. R. Hartzell, *Biochemistry*, 1977, **16**, 4961.
[79] M. A. Schroedl and C. R. Hartzell, *Biochemistry*, 1977, **16**, 4966.
[80] C. H. A. Seiter, S. G. Angelos, and R. A. Perreault, *Biochem. Biophys. Res. Comm.*, **78**, 1977, 761.

spin state, which disrupts the antiferromagnetic coupling between cytochrome a_3 and copper.

Kinetic Studies on Cytochrome Oxidase. Stopped-flow studies by Orii and King[81] on the reaction between reduced cytochrome oxidase and oxygen indicate (at least) three intermediates, designated I, II, and III according to the order of their appearance. Compound I shows the characteristics of a true intermediate in the catalytic cycle. These observations have been confirmed and extended in other laboratories. Antonini *et al.*,[82] using stopped-flow spectrophotometry, have shown that the products of the reaction of fully reduced oxidase with oxygen (termed 'oxygen-pulsed oxidase') are functionally distinct from the oxidase as isolated. The differences can be accounted for by assuming different rates of intramolecular electron transfer in the oxygen-pulsed and resting oxidase. The oxygen-pulsed oxidase is postulated to involve a complex between O_2^{2-}, cytochrome a_3, and the 'invisible' copper in its cuprous valence state; cytochrome a and the other copper are suggested to be in their oxidized states. Quench–freeze e.p.r. studies have revealed[83] that a high-spin ferric signal appears rapidly following the reaction of reduced oxidase with ferrocytochrome c and oxygen; stopped-flow experiments reveal that this 'oxygen-pulsed' form of the enzyme has optical absorption maxima at 423 and 601 nm relative to bands at 427 and 598 nm in the native enzyme. Similar experiments have been carried out by Beinert and Shaw,[84] who observed that intense e.p.r. signals from high- and low-spin haem were produced rapidly during the anaerobic re-oxidation of reduced cytochrome oxidase by ferricyanide. These species were relatively short-lived (0.1—2 s) and accounted for 70—90% of the total haem, *i.e.* the major proportion of both haem centres can be detected by e.p.r. under these reaction conditions. If the generally accepted assignment of the low-spin signal to cytochrome a is adopted, the high-spin signal must originate from cytochrome a_3. Key questions about the molecular mechanism of action of cytochrome oxidase are raised in the discussion.

Kornblatt[85] has observed that ferrocytochrome c (but not ferricytochrome c) increases the rate at which cytochrome oxidase (λ_{max} 428 nm) is converted into its conformational isomer (λ_{max} 418—423 nm), which would correspond to the oxygen-pulsed oxidase. Reaction conditions have been established which allow the study of these different forms of cytochrome oxidase. Yoshikawa *et al.*[86] claim that co-addition of ascorbate during the reaction of reduced cytochrome oxidase with oxygen stabilizes the oxygenated intermediate I[82] and allows its reaction with cyanide to be studied. The results confirm Orii and King's[81] conclusion that intermediate I is catalytically significant.

Three distinct oxygenated forms of cytochrome oxidase, produced by the reaction of reduced and partially reduced enzyme with oxygen, have been studied

[81] Y. Orii and T. E. King, *J. Biol. Chem.*, 1976, **251**, 7487.
[82] E. Antonini, M. Brunori, A. Colosimo, C. Greenwood, and M. T. Wilson, *Proc. Nat. Acad. Sci. U.S.A.*, 1977, **74**, 3128.
[83] S. Rosen, R. Branden, T. Vänngård, and D. G. Malmström, *F.E.B.S. Letters*, 1977, **74**, 25.
[84] H. Beinert and R. W. Shaw, *Biochim. Biophys. Acta*, 1977, **462**, 121.
[85] J. A. Kornblatt, *Canad. J. Biochem.*, 1977, **55**, 458.
[86] S. Yoshikawa, T. Ueno, and T. Sai, *J. Biochem. (Japan)*, 1977, **82**, 1361.

by i.r. spectroscopy.[87] An intense absorption band at 606—609 nm has been assigned to a charge-transfer complex between ferrocytochrome a_3 and its associated copper, in the oxidized state, with ferricytochrome a and its associated copper in the oxidized state. The relationship between these three oxygenated forms and intermediates I, II, and III of Orii and King[82] has not yet been established.

Greenwood *et al.*[88] have carried out stopped-flow studies on the reduction by ferrocytochrome c of the partially reduced carbon-monoxide-inhibited form of cytochrome oxidase. Simultaneous fast reduction of cytochrome a and oxidation of ferrocytochrome c is observed; however, more ferrocytochrome c is oxidized than cytochrome a is reduced, which suggests that there is intramolecular migration of electrons within the oxidase. The reaction:

$$\text{cytochrome } c^{2+} + \text{cytochrome } a^{3+} \underset{k_{-1}}{\overset{k_1}{\rightleftharpoons}} \text{cytochrome } c^{3+} + \text{cytochrome } a^{2+}$$

has been studied by temperature-jump relaxation spectrophotometry and shown[89] to be close to equilibrium with $k_1 = 9 \times 10^6 \ \text{l mol}^{-1} \text{s}^{-1}$ and $k_{-1} = 10^6 \ \text{l mol}^{-1}$ s^{-1}. A slower relaxation process has been assigned to electron transfer from cytochrome a^{2+} to another redox centre in the enzyme, which is speculated to be the visible Cu^{2+}.

Petersen[89, 90] has carried out careful steady-state kinetic studies on cytochrome oxidase, particularly on the effect of inhibitors. The results are critically discussed in terms of different conformational forms of the enzyme. Electrochemically generated oxidants and reductants have been used[91] to oxidize/reduce cytochrome oxidase; the rates for electron transfer were compared with those calculated according to Marcus theory. Sharrock and Yonetani[92] have studied the kinetics of binding of carbon monoxide to cytochrome a_3 in purified cytochrome oxidase over a wide range of temperatures and concentrations of carbon monoxide. The results suggest biphasic binding due to environments of different fluidity.

Finally in this section, the reduction of cytochrome oxidase$_{551}$ from *Pseudomonas aeruginosa* by chromous ion[93] and by the copper protein azurin[94] has been studied, using stopped-flow spectrophotometry. Reaction with the reducing agent appears to be diffusion-controlled. The authors comment that it is premature to draw conclusions on the mechanism of electron transfer within the enzyme from the present evidence.

Subunit Structure of Cytochrome Oxidase and its Location in the Mitochondrial Membrane. Considerable effort is being directed towards isolating the subunits of cytochrome oxidase, using procedures which (hopefully) retain their native

[87] D. Chance and J. S. Leigh, *Proc. Nat. Acad. Sci. U.S.A.*, 1977, **74**, 4777.
[88] C. Greenwood, T. Brittain, M. Wilson, and M. Brunori, *Biochem. J.*, 1976, **157**, 591.
[89] L. C. Petersen, *Canad. J. Biochem.*, 1977, **55**, 706.
[90] L. C. Petersen, *Biochim. Biophys. Acta*, 1977, **460**, 299.
[91] L. N. Mackey and T. Kuwana, *Bioelectrochem. Bioenerg.*, 1976, **3**, 596.
[92] M. Sharrock and T. Yonetani, *Biochim. Biophys. Acta*, 1977, **462**, 718.
[93] D. Barber, S. R. Parr, and C. Greenwood, *Biochem. J.*, 1977, **163**, 629.
[94] S. R. Parr, D. Barber, C. Greenwood, and M. Brunori, *Biochem. J.*, 1977, **167**, 447.

conformations. In a careful paper, Werner[95] describes an isolation procedure for the seven subunits of the enzyme from *Neurospora crassa*; the data do not allow a definitive statement of which subunits have bound haem or copper. Two groups of investigators report their findings on the metal content of isolated subunits from beef heart cytochrome oxidase. Yu and Yu[96] have extended an earlier report[97] and concluded that the haem-binding subunits have molecular weights of 11.6×10^3 and 40×10^3 whereas the copper-binding subunits have molecular weights of 21×10^3 and 40×10^3. Further studies[98] on the subunit of molecular weight 11.6×10^3 have revealed a high degree of primary sequence homology with the β-chain of haemoglobin. This observation suggests that the isolated subunit may have oxygen-binding capacity and that the oxygen reductase activity represents some form of co-operativity resulting from assembly of the separate subunits. Gutteridge *et al.*[99] have found haem to be located in subunits with molecular weights 35×10^3, 21×10^3, and 12.1×10^3, whilst copper is found in subunits with molecular weights of 12.1×10^3 and 3.4×10^3. The differences between these reports perhaps originate from dissociation of the metals from their native subunits and extraneous re-association. Evidence has also been presented to suggest that cytochrome *a* 'anchors' the polypeptide chains into a functional unit that is essential for the oxidation of ferrocytochrome *c*.[100]

Affinity-labelling experiments[101, 102] indicate that the binding site for cytochrome *c* is probably located on the subunit of the oxidase that has molecular weight 21×10^3. Other studies[103, 104] suggest that the haem cleft of the ferrocytochrome *c* is oriented towards the oxidase in the enzyme–substrate complex.

Cross-linking and other experiments are establishing the arrangement of cytochrome oxidase subunits in the membrane. The results of Briggs and Capaldi[105] are in reasonable accord with the earlier findings of Eytan *et al.*[106] It has been found that[105] the limiting cross-linked product is a species of molecular weight 140 000, which is suggested to be the active unit of cytochrome oxidase. Other studies[107] suggest that this may be a dimer composed of two cytochrome *a–a₃* complexes, and this is also the picture emerging from the image-reconstruction studies based on electron micrographs of cytochrome oxidase in two-dimensional vesicle crystals,[108] which also clearly establish that the protein spans the membrane bilayer. It is interesting to note that e.p.r. studies on oriented multilayers containing cytochrome oxidase indicate that the haem planes in

95 S. Werner, *European J. Biochem.*, 1977, **79**, 103.
96 C.-A. Yu and L. Yu, *Biochim. Biophys. Acta*, 1977, **495**, 248.
97 C.-A. Yu, L. Yu, and T. E. King, *Biochem. Biophys. Res. Comm.*, 1977, **74**, 670.
98 T. Tanaka, M. Hanui, K. Yasunobu, C. A. Yu, L. Yu, Y. H. Wei and T. S. King, *Biochem. Biophys. Res. Comm.*, 1977, **76**, 1014.
99 S. Gutteridge, D. B. Winter, X. Bruyninck, and H. S. Mason, *Biochem. Biophys. Res. Comm.*, 1977, **78**, 945.
100 Y. Orii, M. Nanabe, and M. Yoneda, *J. Biochem. (Japan)*, 1977, **81**, 505.
101 W. Birchmeier, C. E. Kohler, and G. Schatz, *Proc. Nat. Acad. Sci. U.S.A.*, 1976, **73**, 4334.
102 W. Birchmeier, *Mol. Cell Biochem.*, 1977, **14**, 81.
103 N. Staudenmayer, S. Ng, M. D. Smith, and F. Millett, *Biochemistry*, 1977, **16**, 600.
104 H. T. Smith, N. Staudenmayer, and F. Millett, *Biochemistry*, 1977, **16**, 4971.
105 M. M. Briggs and R. A. Capaldi, *Biochemistry*, 1977, **16**, 73.
106 G. D. Eytan, R. C. Carroll, G. Schatz, and E. Racker, *J. Biol. Chem.*, 1975, **250**, 8598.
107 N. C. Robinson and R. A. Capaldi, *Biochemistry*, 1977, **16**, 375.
108 R. Henderson, R. A. Capaldi, and J. S. Leigh, *J. Mol. Biol.*, 1977, **112**, 631.

cytochrome a and cytochrome a_3 are roughly perpendicular to the plane of the bilayer.[109]

The Cytochrome P-450 Complex.—The cytochrome P-450 complex, which catalyses the hydroxylation of a wide variety of endogenous and foreign compounds, appears to have three components: (i) cytochrome P-450; (ii) reduced pyridine nucleotide–cytochrome P-450 reductase, and (iii) an iron–sulphur protein. A lipid requirement is also suggested for membrane-bound forms of the complex.

An excellent recent review on the structure and function of cytochrome P-450 complexes has been written by Dus.[110] The following comments might act as a useful introduction to this diverse and rapidly developing field. The designation 'P-450' derives from the wavelength maximum of its complex with carbon monoxide; inactive forms of cytochrome P-450, whether produced by extraction or by the action of inhibitors, are characterized by a carbon monoxide complex having λ_{max} 420 nm, and these are consequently designated as cytochrome P-420. Cytochrome P-450 complexes from mammalian sources are tightly bound to membranous structures and are consequently difficult to extract and purify without inactivation; by contrast, some cytochrome P-450 species of microbial origin, notably the one from *Pseudomonas putida* involved in the hydroxylation of camphor, and termed P-450$_{cam}$, are soluble, and hence more readily purified.

Cytochrome P-450 Complexes from Micro-organisms. Champion and Gunsalus[111] have studied the properties of purified cytochrome P-450$_{cam}$ in the presence of saturating concentrations of D-($+$)-camphor by resonance Raman spectroscopy; the haem centre would be high-spin ferric under these reaction conditions. A strong band at 351 cm^{-1}, assigned to an Fe—N vibration, was observed. This band and another at 691 cm^{-1} disappeared when the cytochrome P-450 was converted into cytochrome P-420; the authors consider this result as indicating that cysteine forms one axial ligand and conclude that resonance Raman spectroscopy is a very promising technique for structural studies on cytochrome P-450. Polarization absorption spectroscopy has also been used to examine single crystals of camphor-bound cytochrome P-450 in its oxidized, reduced, and carbon-monoxide-reduced states.[112] The high-spin oxidized species shows two broad z-polarized bands at 567 and 323 nm which are suggested to arise from cysteine mercaptide charge-transfer transitions to iron; perturbation of the d-orbitals is postulated to account for the high rhombicity. The high-spin reduced species has no z-polarized bands, possibly owing to protonation of the cysteine sulphur, which consequently prevents orbital mixing with the d-orbitals on the iron. The low-spin carbon-monoxide-reduced species has an intense x, y-polarized band at 363 nm, assigned to a mercaptide sulphur–porphyrin charge-transfer transition; this result indicates that a similar structure might hold for the low-spin ferrous

[109] M. Ericinska, J. K. Blaisie, and D. F. Wilson, *F.E.B.S. Letters*, 1977, **76**, 235.
[110] K. Dus, in 'The Enzymes of Biological Membranes' Vol. 4, ed. A. Martinosi, Plenum Press, New York, 1976, p. 199.
[111] P. M. Champion and I. C. Gunsalus, *J. Amer. Chem. Soc.*, 1977, **99**, 2000.
[112] K. K. Hanson, S. G. Sligor, and I. C. Gunsalus, *Croat. Chem. Acta*, 1977, **49**, 237.

P-450–substrate dioxygen complex. Ullrich *et al.*,[113] from comparison of the electronic absorption spectra of high-spin ferric cytochrome *P*-450 (substrate-free form) and free haem derivatives of known structure, also detected a mercaptide–haem linkage. These authors have further suggested[113] that a hydroxyl might be the ligand co-ordinated at the sixth position on haem which becomes substituted when substrate binds. Contrary to this conclusion, crystal-field analyses of e.p.r. signals from various low-spin ferric forms of cytochrome *P*-450 suggest[114] that the two axial ligands are cysteine and histidine. The histidine ligand can be displaced by cyanide, guanidine, or an amine.

The binding of camphor to cytochrome *P*-450 from *Pseudomonas putida* and spin conversions in the camphor-bound complex have been studied by low-temperature stopped-flow spectrophotometry.[115] In particular, the effects of temperature, pH, and the nature of the solvent have been considered. The major conclusion drawn is that both the camphor-binding mechanism and the spin conversions depend on the protonic activity of the medium. Spin-state equilibria for the camphor-bound species have been studied in more detail, using the same technique,[116, 117] which has also been employed to follow the reaction between the natural electron donor putidaredoxin and cytochrome *P*-450 in the presence of the substrate, camphor.[118] For this latter reaction, the rate-limiting step is suggested to be binding of putidaredoxin to camphor; the rate constant and activation parameters for subsequent electron-transfer steps have also been determined. Cytochrome *P*-450 from *Pseudomonas putida* and from liver microsomes have been compared in[119] their substrate-free forms by e.p.r. spectroscopy. The spectra are quite similar, though the heterogeneity of the liver protein presents difficulties in interpretation.

A relatively simple procedure for preparation of cytochrome *P*-450 in 33% overall yield from microsomes of anaerobically grown yeast has been described.[120] SDS polyacrylamide gel electrophoresis indicates a single major band of mol. wt. 55 000 and the preparation is free from cytochrome b_5, NADH–cytochrome b_5 reductase, and NADPH–cytochrome *P*-450 reductase. The haem centre in the native protein is low-spin ferric and is converted into high-spin ferric by the addition of lanosterol. In these and other properties, the protein behaves similarly to that isolated from liver, and the yeast system shows considerable promise for studies in the future.

Cytochrome P-450 *Complexes from Mammalian Systems.* Forms of cytochrome *P*-450 from rabbit liver microsomes are probably the best characterized species from mammalian sources. A partial amino-acid sequence has been carried out[121]

113 V. Ullrich, H. H. Ruf, and P. Wende, *Croat. Chem. Acta*, 1977, **49**, 213.
114 M. Chevion, J. Peisach, and W. E. Blumberg, *J. Biol. Chem.*, 1977, **252**, 3637.
115 R. Lange, G. Hui Bon Hoa, P. Debey, and I. C. Gunsalus, *European J. Biochem.*, 1977, **77**, 479.
116 R. Lange, P. Debey, and P. Douzou, *Croat. Chem. Acta*, 1977, **49**, 279.
117 R. Lange, C. Bonfils, and P. Debey, *European J. Biochem.*, 1977, **79**, 623.
118 P. Debey, G. Hui Bon Hoa, and I. C. Gunsalus, 1977, *Croat. Chem. Acta*, 1977, **49**, 309.
119 R. E. Ebel, D. H. O'Keefe, and J. A. Peterson, *Arch. Biochem. Biophys.*, 1977, **183**, 317.
120 Y. Yoshida, Y. Aoyama, H. Kumaoka, and S. Kubota, *Biochem. Biophys. Res. Comm.*, 1977, **78**, 1005.
121 D. A. Haugen, L. G. Armes, K. T. Yasunobu, and M. J. Coon, *Biochem. Biophys. Res. Comm.*, 1977, **77**, 967.

on two highly purified forms of cytochrome P-450, designated P-450$_{LM2}$ and P-450$_{LM3}$. A noteworthy observation is that P-450$_{LM2}$ contains a high concentration of hydrophobic amino-acids in the N-terminal region.

Johnson and Muller-Eberhard[122] have described the properties of two forms of cytochrome P-450 isolated from rabbit liver microsomes following induction with 2,3,7,8-tetrachlorodibenzo-p-dioxin. The two forms are immunologically distinct, and their difference spectra in the presence and absence of n-octylamine are also distinct. There is also different substrate specificity when the cytochrome P-450 complex is reconstituted.

The purification and properties of cytochrome P-450[123] from bovine adrenocortical mitochondria have been described; the protein has a subunit molecular weight of 47 500, and the possible molecular weight of the active species is 185 000. It is claimed that the same protein species is responsible for the hydroxylation of steroid substrates at the 11- and 18-positions. However, using difference spectrophotometry, two forms of cytochrome P-450 involved in the cleavage of the side-chain of cholesterol have been isolated from adrenal mitochondria;[124] these forms of the protein differ in their binding specificity for steroids. The relationship between these two forms of the protein has not yet been established.

There are two reports indicating that cytochrome P-450 can only accommodate a one-electron redox process per haem. Peterson *et al.*[125] obtained this result from reductive titrations on two purified forms of liver microsomal P-450; it was found that re-oxidation of the reduced P-450 is accompanied by transfer of one electron to oxidizing agents such as ferricytochrome c, ferricytochrome b_5, and potassium ferricyanide. Cooper *et al.*[126] obtained the same result from reductive titrations on intact rat liver microsomes under strictly anaerobic conditions.

The mid-point potentials of cytochrome P-450 in forms of bovine adrenal cytochrome P-450 have been determined.[127] There appear to be differences in these potentials, depending on the presence or absence of substrates or inhibitors.

The equilibrium between low- and high-spin forms of cytochrome P-450 has been studied by e.p.r.,[128] spectrophotometry,[129] and stopped-flow spectrophotometry at low temperatures.[130] The e.p.r. studies[128] carried out on rabbit liver microsomes confirm the established shift from low to high spin state on binding substrate: it is speculated that this shift might facilitate reduction of the P-450. Further evidence supporting this conclusion has been obtained by spectrophotometric studies on purified cytochrome P-450.[129] The stopped-flow studies[130] demonstrate the formation of kinetically distinct species following the

122 E. F. Johnson and U. Muller-Eberhard, *J. Biol. Chem.*, 1977, **252**, 2839
123 M. Watanuki, B. E. Tilley, and P. F. Hall, *Biochim. Biophys. Acta*, 1977, **483**, 236.
124 C. R. Jefcoate, *J. Biol. Chem.*, 1977, **252**, 8788.
125 J. A. Peterson, R. E. White, Y. Yasukochi, M. L. Coones, D. H. O'Keefe, R. E. Ebel, B. S. S. Masters, D. P. Ballou, and M. J. Coon, *J. Biol. Chem.*, 1977, **252**, 4431.
126 D. Y. Cooper, M. D. Cannon, H. Schleyer, and O. Rosenthal, *J. Biol. Chem.*, 1977, **252**, 4755.
127 D. L. Williams-Smith and R. Cammack *Biochim. Biophys. Acta*, 1977, **499**, 432
128 H. Rein, O. Ristan, J. Friedrich, G. R. Janig, and K. Ruckpaul, *Croat. Chem. Acta*, 1977, **49**, 251.
129 H. Rein, O. Ristan, J. Friedrich, G. R. Janig, and K. Ruckpaul, *F.E.B.S. Letters*, 1977, **75**, 19.
130 E. Begard, P. Debey, and P. Douzou, *F.E.B.S. Letters*, 1977, **75**, 52.

reaction of reduced liver microsomes with oxygen. These species have tentatively been identified as oxyferro-complexes of cytochrome *P*-450.

Finally, in this section, mention should be made of two reports[131, 132] that suggest that cytochrome b_5 might mediate electron transfer between NADPH and cytochrome *P*-450 in liver microsomes, and that adrenodoxin might play a similar role in mitochondria from adrenal cortex.[133] There are therefore clear parallels to the cytochrome *P*-450 system in *Pseudomonas putida*, where putidaredoxin plays a similar role of electron mediator.[118]

2 Haem-containing Enzymes

Peroxidase.—The physicochemical properties and reaction mechanism of peroxidase[134] and the nature of the products of its reaction with many organic substrates[135] have been reviewed recently. Peroxidases utilize H_2O_2 or related compounds to oxidize a large number of substrates. They are widely distributed and have been isolated from many higher plants, animal tissues, and yeasts and other micro-organisms. All plant peroxidases have ferriprotoporphyrin IX (1) as the prosthetic group. Extensive work on the enzyme from the roots of the horseradish (HRP) has recently been supplemented by work on the enzyme from Japanese radish (JRP) and turnip (TP).

(1) Ferriprotoporphyrin IX

The characterized peroxidase enzymes from animal tissues are more variable. Thyroid peroxidase (ThP) is most like HRP; glutathione peroxidase (GluP) contains one Se atom per subunit and lactoperoxidase (LacP) and myeloperoxidase (MP) have a more unsaturated (possibly formyl) type of porphyrin in their prosthetic group. MP contains two atoms of porphyrin-bound iron per molecule, and the two haem environments may not be the same. Chloroperoxidase has been isolated from a mould and is similar to HRP, but it has the unique ability to catalyse the oxidation of Cl^- to Cl^+.

131 Y. Inai and R. Sata, *Biochem. Biophys. Res. Comm.*, 1977, **75**, 420.
132 S. Fujita and J. Peisach, *Biochem. Biophys. Res. Comm.*, 1977, **78**, 328.
133 N. Katagiri, O. Takukawa, H. Sato, and K. Suhara, *Biochem. Biophys. Res. Comm.*, 1977, **77**, 804.
134 H. B. Dunford and J. S. Stillman, *Coordination Chem. Rev.*, 1976, **19**, 187.
135 B. C. Saunders, in 'Inorganic Biochemistry,' ed. G. L. Eichhorn, Elsevier, Amsterdam, London, and New York, 1973, Vol. 2, p. 988.

Cytochrome *c* peroxidase (C*c*P) has been purified from bakers yeast and from *Pseudomonas aeruginosa*. Most peroxidases are glycoproteins, but C*c*P and MP are not. Many of the purified enzymes consist of mixtures of isoenzymes with similar catalytic properties but differing chemical consititution and physical properties.

Horseradish Peroxidase (HRP) and other Plant Peroxidases.—*Haem Structure.* The native enzyme has generally been considered to have iron(III) bound to the porphyrin in an octahedral complex, with a nitrogen-containing amino-acid, probably histidine (proximal), from the polypeptide chain as the fifth ligand and an H_2O molecule as the sixth.[134] Another histidine residue (distal) may be hydrogen-bonded to the co-ordinated water molecule (Figure 1). The spectra

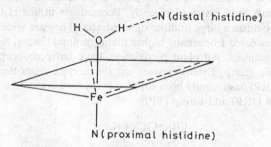

Figure 1 *Proposed ligand environment of iron*(III) *in protoporphyrin IX in peroxidase*

are pH-dependent, and e.p.r. studies[136] indicated that high-spin ferric complexes were formed at low pH and low-spin ones at high pH. At neutral pH, a complex that was in thermal equilibrium between high- and low-spin forms and that had different *g* values from those found at the extremes of pH was observed. It was concluded that three species were involved in the acid → base transition:

HRP (acid)   ⟷   HRP (neutral)   ⟷   HRP (alkaline)
high-spin   high-spin ⇌ low-spin   low-spin
$Fe^{III}(H_2O)$ $Fe^{III}(OH^-)$

However, when $H_2{}^{16}O$ was substituted by $H_2{}^{17}O$ at pH 6 in solutions of HRP, the e.p.r. spectra showed no broadening.[137] Similar experiments with ferric myoglobin did induce broadening. Thus under these conditions either HPR contained pentaco-ordinated iron or the sixth co-ordination position was occupied by an oxygen atom from an amino-acid of the polypeptide chain. In addition, the magnetic circular dichroism (m.c.d.) spectrum of the acid form of native HRP was similar to that of acetatoiron(III) protoporphyrin IX dimethyl ester, also suggesting high-spin ferric iron with an electronic structure similar to that in a pentaco-ordinated haem complex. No change in spectrum was observed between pH 5.2 and pH 9, and the acid → alkali transition had $pK_a = 11.0$. The alkaline form had the m.c.d. spectrum of a low-spin complex,

[136] M. Tamura and H. Horii, *Biochim. Biophys. Acta*, 1972, **284**, 20.
[137] S. Vuk-Pavlovic and Y. Siderer, *Biochem. Biophys. Res. Comm.*, 1977, **79**, 885.

and the near-i.r. m.c.d. spectrum suggested that the overall strength of the ligand field was between that of the low-spin component of metmyoglobin hydroxide and that of metmyoglobin azide. Whilst this was consistent with an OH group having its basicity strengthened by hydrogen bonding (Figure 1) as the sixth ligand, other alternatives were considered possible.[138]

^{1}H n.m.r. studies of the protons on or near the haem indicated that the alkaline form of HPR was not formed by a protolytic dissociation of an iron-bound H_2O molecule but by direct co-ordination to an amino-acid, possibly histidine, from the polypeptide chain. N.m.r. studies also indicated that access of azide to the sixth co-ordination position of the iron was modulated by ionization of an amino-acid residue, which, from its $pK_a = 5.9$, was suggested to be histidine.[139] Visible spectral studies indicated that the high- to low-spin ferric transition could also be effected by high pressures.[140] A stopped-flow study on the transition between the neutral and alkaline forms of the P_1 and P_7 TP isoenzymes showed one fast and two slow steps for P_1 but just one fast step for P_7. The pH dependence of the fast steps in both cases was interpreted in terms of deprotonation of the H_2O molecule in the sixth co-ordination position of the haem Fe.[141]

The c.d. and optical absorption spectra of artificial peroxidases prepared from apoHRP and haems with a range of substituents in place of the normal vinyl groups indicated that the unsaturated side-chains interacted with aromatic amino-acids of the polypeptide chain. These unsaturated complexes also gave a higher rate of formation of compound I (see below) with H_2O_2.[142] The visible and Soret bands in the c.d. spectra of two of the TP isoenzymes, *i.e.* P_1 and P_7, had similar wavelengths but different ellipticities. When compared with the spectra of the HRP and JRP enzymes, the most common property of plant peroxidases appeared to be a negative c.d. band in the Soret region for all the reduced [Fe^{II}] enzymes, which became positive upon binding cyanide as the sixth ligand.[143]

Polypeptide Structure. The complete amino-acid sequence of the dominant cathodic isoenzyme of HRP has been determined.[144] It consisted of 308 amino-acid residues, including three histidines (giving a polypeptide molecular weight of 33 890), one haem prosthetic group, four internal disulphide bridges, and eight neutral carbohydrate side-chains attached to aspartate residues.

The amino-acid sequences around the proximal and (probably) distal histidine residues in the TP isoenzymes P_1, P_2, P_3, and P_7 were determined and compared with the sequence of HRP isoenzyme C. Substitutions were rare close to the histidine residues, but became more frequent at greater distances. The sequences of P_1, P_2, P_3, and HRP C all contained two histidine residues at positions 40 and 42 in their distal sequences; however, in P_7, which has abnormal properties,

[138] N. Kobayashi, T. Nozawa, and M. Hatano, *Biochim. Biophys. Acta*, 1977, **493**, 340.
[139] 1. Morishima, S. Ogawa, T. Inubushi, T. Yonezawa, and T. Iizuka, *Biochemistry*, 1977, **16**, 5109.
[140] G. B. Ogunmola, A. Zipp, F. Chen, and W. Kauzmann, *Proc. Nat. Acad. Sci. U.S.A.*, 1977, **74**, 1.
[141] D. Job, J. Ricard, and H. B. Dunford, *Arch. Biochem. Biophys.*, 1977, **179**, 95.
[142] P. I. Ohlsson, K. G. Paul, and I. Sjoholm, *J. Biol. Chem.*, 1977, **252**, 8222.
[143] D. Job and H. B. Dunford, *Canad. J. Biochem.*, 1977, **55**, 804.
[144] K. G. Welinder, *F.E.B.S. Letters*, 1976, **72**, 119.

residue 40 was phenylalanine. The forms P_1, P_2, and P_3 all had several sites of carbohydrate attachment whereas P_7 had only one.[145]

The paramagnetic high-spin iron in native HRP apparently protected the enzyme from photodynamic action. However, several amino-acids, including L-histidine and L-tyrosine residues in the low-spin CN^- or OH^- complexes, were damaged by illumination with visible light in the presence of sensitizers such as eosin Y.[146]

Three glycoprotein isoenzyme groups, G_1 (mol. wt. 27 000), G_2 (43 000), and G_3 (47 000), were separated from tobacco peroxidase. The group G_1 was inactivated at 50 °C whereas G_2 and G_3 were stable at temperatures up to 70 °C.[147]

Tryptophan fluorescence was used as an indicator of protein conformation in native HRP, apoHRP, and the protoporphyrin–apoHRP complex. Coupled with c.d. data, these studies provided evidence for the similarity of the tertiary structure of the protoporphyrin–apoHRP complex and the native holoenzyme but the holo- and apo-enzymes were greatly different.[148]

The Active Species. It is generally accepted that the enzymatic cycle for HRP and most other peroxidases is as shown in Scheme 1. The two-equivalent oxidation of the native enzyme to Compound I does not yield Fe^V but Fe^{IV}, with the extra electron abstracted from the protein or porphyrin resulting in the formation of a radical cation (R·). Compound II is not a radical. The m.c.d. spectra of HRP Compounds I and II indicated that the chromophores of both are electronically similar.[149]

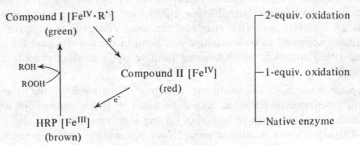

The enzymatic cycle for horseradish peroxidase

Scheme 1

The electrons for the reductive steps of the cycle are abstracted from the oxidizable substrates. Kinetic studies on the oxidation of *o*-dianisidine by HRP–H_2O_2 showed that deprotonation of a group on the enzyme with pK_a = 6.5 markedly decreased the enzymic activity. Above pH 9, an additional ionizing group, with pK_a = 8.9, controlled the enzyme's activity.[150]

145 K. G. Welinder and G. Mazza, *European J. Biochem.*, 1977, **73**, 353.
146 Y.-J. Kang and J. D. Spikes, *Biochem. Biophys. Res. Comm.*, 1977, **74**, 1160.
147 A. Nessel and M. Maebler, *Z. Pflanzenphysiol.*, 1977, **83**, 235.
148 A. P. Savitskii, N. N. Ugarova, and I. V. Berezin, *Bioorg. Khim.*, 1977, **3**, 1242.
149 M. J. Stillman, B. R. Hollebone, and J. S. Stillman, *Biochem. Biophys. Res. Comm.*, 1976, **72**, 554.
150 O. V. Lebedeva, N. N. Ugarova, and I. V. Berezin, *Biokhimiya*, 1977, **42**, 1372.

The Mössbauer spectra of Compound II of JRP, in the presence of an applied magnetic field, indicated that the Fe in Compound II had a low-spin $(t_g)^4$, $S = 1$ paramagnetic configuration,[151] consistent with earlier results and with e.p.r., magnetic susceptibility, and resonance Raman data.

In aqueous glycerol solutions the rate of formation of Compound I of HRP by reaction with *m*-chloroperbenzoic acid, but not with H_2O_2, was diffusion-controlled.[152] The interpretation of earlier results where ethanol and methanol had been used to decrease the dielectric constant of the medium might be in error, since the dominant effect with HRP was binding of the alcohols to the sixth co-ordination position.

Optical spectral results had indicated that a low-spin compound was formed on photoreduction of Compound I. Photoreduction of Compound I from HRP in frozen glycerol–H_2O solution at 77 K,[153] 10 K, or 100 K[154] led to the intensification of a radical species in the e.p.r. spectra. A similar species was formed on irradiation of Compound II.[153] It is thought that the sixth ligand position in Compound I is occupied by [O]. Reduction to Compound II involves addition of a proton and thus its sixth ligand may be [OH]. The above photolysis experiments were therefore rationalized[153] in terms of the formation of Intermediate Y, a low-spin type of Compound II:

$$\underset{\text{Compound I}}{\text{OFe}^{IV} \text{ R·}} \xrightarrow[+\text{H}^+ + \text{e}^-]{h\nu} \underset{\text{Intermediate Y}}{\text{HOFe}^{III} \text{ R·}}$$

$$\underset{\text{Compound II}}{\text{HOFe}^{IV}} \xrightarrow{h\nu} \underset{\text{Intermediate Y}}{\text{HOFe}^{III}\text{R·}}$$

An alternative interpretation was in terms of haem-photosensitized oxidation of the peptide groups close to the radical site of Compound I.[154]

HRP may be reduced by dithionite to ferroperoxidase, which can react with O_2 to form Compound III, a substance also formed directly from HRP in the presence of a large excess of H_2O_2.[134] When HRP was reduced by γ-irradiation in water–glycerol solution at 77 K, a non-equilibrium mixture of low- and high-spin forms of ferroperoxidase was formed which on thawing relaxed to the normal high-spin form.[155] Similar relaxation effects were observed on reduction of the HRP–CN complex. The reduced complex dissociated in two consecutive stages, the first stage corresponding to dissociation of the CN^- followed by a slow conformational change of the enzyme to its normal spin state.[156] The reduced forms of the enzyme have also been studied *via* the cobalt-substituted enzyme, which can be reduced to an e.p.r.-active cobaltous form.[157] The ^{14}N superhyperfine structure of this e.p.r. spectrum unambiguously showed that the proximal axial ligand was a nitrogenous base.

[151] T. Harami, Y. Maeda, Y. Morita, A. Trautwein, and U. Gonser, *J. Chem. Phys.*, 1977, **67**, 1164.
[152] H. B. Dunford and W. D. Hewson, *Biochemistry*, 1977, **16**, 2949.
[153] M. Chu, H. B. Dunford, and D. Job, *Biochem. Biophys. Res. Comm.*, 1977, **74**, 159.
[154] A. R. McIntosh and M. J. Stillman, *Biochem. J.*, 1977, **167**, 31.
[155] S. N. Maganov, A. M. Arutyunyan, L. A. Blyumenfel'd, R. M. Davydov, and Yu. A. Sharonov, *Doklady Akad. Nauk S.S.S.R.*, 1977, **232**, 695.
[156] S. P. Kuprin, *Biofizika*, 1977, **22**, 929.
[157] M.-Y. R. Wang, B. M. Hoffman. and P. F. Hollenberg, *J. Biol. Chem.*, 1977, **252**, 6268.

Reactions with Oxidizable Substrates. Not all of the enzymic reactions are associated with the haem. The apoenzyme of peroxidase prepared from the medium of cultured peanut cells possessed polyphenol oxidase and indoleacetic acid oxidase activities but not peroxidase activity, which was regained on reconstitution of the holoenzyme.[158] Studies on the inhibition of HRP peroxidation of scopoletin by indoleacetic acid also indicated that oxidation of the inhibitor did not occur at the peroxidation site.[159]

Although Compounds I and II are major reactive species of HRP, oxidizable substrates will also bind to the native protein. Conformational changes on binding indolepropionic acid to both the high-spin and low-spin ferric (CN⁻ complex) enzymes have been observed from shifts in the ^1H n.m.r. resonances that are due to the peripheral methyl groups of haem.[160]

Many peroxidase reactions result in the formation of free radicals or electronically excited states.[161-163] In the aerobic, HRP-catalysed oxidation of dihydroxyfumarate the substrate was probably first oxidized to a radical, which then reacted with O_2 to form superoxide ion. This in turn reduced much of the enzyme to inactive Compound III, and also took part in the non-enzymic oxidation of substrate. Compounds which reacted with Compound III enhanced the enzymic reaction and became hydroxylated, probably by the OH⁺ radical.[164]

Oscillatory kinetics were observed in the reaction between HRP, H_2O_2, and NADH. The reaction in the presence of limiting O_2 concentration occurred in four phases:[165] initial burst, induction, steady state, and termination. A total of eleven reactions and nine rate equations were used to simulate the kinetics. The major reactions were as follows: the initial burst was due to trace amounts of H_2O_2 in the NADH preparations, leading to the formation and reaction of Compound I; the induction phase corresponded to the build-up of Compound II and H_2O_2 levels; the main reaction in the steady state was between NADH and Compound II. However, NADH also reacted with H_2O_2 to form NAD⁺ radical, which in turn reacted with O_2 to form O_2^- and NAD^+. The O_2^- could react with NADH to re-form NAD⁺ and H_2O_2, or disproportionate to $H_2O_2 + O_2$, or react with HRP to form Compound III. The terminating reaction was probably the reaction between Compound III and NAD⁺ radical. Thyroxine acted as a catalyst in the above reaction. The initial step in this catalysis was probably the reaction of thyroxine with Compound I to form a phenoxy radical, which then attacked NADH to form the NAD⁺ radical.[166]

In the peroxidase-catalysed reaction of morphine hydrochloride with human serum albumin, model reactions with synthetic polypeptides indicated that

158 O. P. Srivastava and R. B. van Huystee, *Canad. J. Bot.*, 1977, **55**, 2630.
159 D. L. Reigh and E. C. Smith, *Experientia*, 1977, **33**, 1451.
160 I. Morishima, S. Ogawa, T. Inubushi, and T. Yonezawa, *F.E.B.S. Letters*, 1977, **80**, 177.
161 K. Zinner, C. C. C. Vidigal, N. Duran, and G. Cilento, *Arch. Biochem. Biophys.*, 1977, **180**, 452.
162 B. W. Griffin, *F.E.B.S. Letters*, 1977, **74**, 139.
163 N. Duran, K. Zinner, C. C. C. Vidigal, and G. Cilento, *Biochem. Biophys. Res. Comm.*, 1977, **74**, 1146.
164 B. Halliwell, *Biochem. J.*, 1977, **163**, 441.
165 K. Yokota and I. Yamazaki, *Biochemistry*, 1977, **16**, 1913.
166 K. Takayama and M. Nakano, *Biochemistry*, 1977, **16**, 1921.

tyrosine and histidine (but not glutamic acid) residues might be involved.[167] A decrease in the lysine content of proteins incubated in the presence of HRP, catechol or other H donor, and H_2O_2 probably occurred through oxidation of some residues to lysyl aldehyde by the quinone formed from the oxidation of catechol. Covalently linked dimers, trimers, and higher polymers of the proteins were observed, indicating that cross-linking between protein molecules was facilitated by the free radicals or quinones formed in the peroxidase-catalysed reaction.[168, 169] Sodium iodide, at concentrations greater than 1 μmol l^{-1}, activated the oxidation of protein sulphydryl groups by H_2O_2 and HRP, lacto-peroxidase, or myeloperoxidase. The iodide was apparently oxidized to I_2, which reacted with the sulphydryl groups to form the sulphenyl iodide derivative, but this derivative could dissociate to sulphenic acid and iodide at low [I$^-$], and thus each iodide ion could participate in the oxidation of several sulphydryl groups.[170]

Cytochrome *c* Peroxidase (C*c*P).—This enzyme oxidizes ferrocytochrome *c* at a much higher rate than other substrates. Its properties have been reviewed recently.[171] In many ways it is similar to HRP; however, the equivalent of Compound I, *i.e.* Compound ES, is much more stable, and red rather than green, although still formulated as [Fe^{IV}—R·]. Compound II of C*c*P can apparently exist in two forms, either [Fe^{IV}] or [Fe^{III}·R·].

C*c*P from yeast has been the most studied, but the enzyme isolated from *Thiobacillus thiooxidans* is very similar, with protohaem as the prosthetic group, a molecular weight of 32 000, and virtually the same activity towards ferro-cytochrome *c* from *T. thiooxidans* or horse heart.[172] C*c*P from *Pseudomonas* was stable between pH 6 and pH 7.4. At low pH, the gross tertiary structure changed simultaneously with the loss of enzymic activity and changes in the secondary structure. However, at high pH, the secondary structure and the protein environ-ment of the haem remained stable after the tertiary structure had changed, and the activity decreased.[173] The bimolecular rate constant for the formation of Compound ES from C*c*P and H_2O_2 was not diffusion-controlled.[174]

The Mössbauer spectra of Compound ES were measured over a range of temperatures and applied magnetic fields, and the properties reproduced with the choice of a single parameter, the axial crystal field, within a crystal-field model which assumed that the iron was Fe^{IV}, with unpaired spin $S = 1$. The model predicted the observed positive sign for the electric field gradient but over-estimated its magnitude, possibly because of the expected covalent charge compensation of the extreme oxidation state. No evidence for interaction be-

[167] M. J. Deutsch, D. L. Roerig, and R. I. H. Wang, *Biochem. Pharmacol.*, 1977, **26**, 1267.
[168] M. A. Stahmann and A. K.Spencer, *Biopolymer*, 1977, **16**, 1299.
[169] M. A. Stahmann, *Adv. Exp. Med. Biol.*, 1977, **86B**, 285.
[170] E. L. Thomas and T. M. Aune, *Biochemistry*, 1977, **16**, 3581.
[171] T. Yonetani, in 'The Enzymes', ed. P. Boyer 3rd edn., Academic Press, New York, 1976, Vol. XIII, p. 345.
[172] T. Tano, K. Sakai, T. Sugio, and K. Imai, *Agric. and Biol. Chem. (Japan)*, 1977, **41**, 323.
[173] R. Soininen and N. Kalkkinen, *Acta. Chem. Scand. (B)*, 1977, **31**, 604.
[174] S. J. Loo and J. E. Ermann, *Biochim. Biophys. Acta*, 1977, **481**, 279.

tween the iron and the e.p.r.-observable free radical was found, indicating that they were well separated.[175]

Biphasic steady-state kinetics were observed for the oxidation of a range of ferrocytochromes c by yeast CcP. Binding studies indicated two binding sites for cytochrome c on the enzyme. It was suggested that there were two kinetically active sites on the enzyme; one of high and one of low affinity, which might correspond to the free radical and Fe^{IV} haem of Compound ES respectively.[176]

Thyroid Peroxidase (ThP).—Bovine ThP was separated into five active components with molecular weights between 73 000 and 340 000. The most active component (mol. wt. 93 000) was purified further to a purity ratio ($\varepsilon_{soret} : \varepsilon_{280}$) of 0.55 and specific activities (for the oxidation of I^- and guaiacol) of 300 and 460 μmol min^{-1} (mg protein)$^{-1}$ respectively. The other components were less stable on purification.[177] Binding studies on a concanavalin A–agarose column confirmed that pig ThP was a glycoprotein.[178]

ThP catalyses both the iodination of tyrosine residues of thyroglobulin (TG) and the coupling of di-iodotyrosine residues of TG to form the hormone thyroxine. Two independent investigations[179, 180] showed that the TG structure was important in determining the efficiency of the hormone formation reaction. Kinetic investigations demonstrated a lag between the formation and the coupling of the iodotyrosine residues which form thyroxine. The coupling reaction required enzymic oxidation of the iodotyrosine residues and the presence of iodide ions.[181] Iodination of TG *in vivo* resulted in a highly heterogenous population of iodinated TGs, but iodination *in vitro* did not. Thus the results *in vivo* must be associated with the localization of the protein and enzyme in the thyroid gland.[182]

Myeloperoxidase (MP) and Lactoperoxidase (LacP).—Canine MP has been purified to homogeneity and crystallized. The molecular weight was 142 000, containing 2 gram atom of Fe. Sodium dodecyl sulphate (SDS) polyacrylamide gel electrophoresis demonstrated the presence of components of mol. wt. 57 500 and 10 500.

The haem and carbohydrate were attached to the larger subunit. Treatment of MP with dimethyl suberimidate yielded a species of mol. wt. 22 000, *i.e.* in the holoenzyme the two light subunits were adjacent.[183]

The NAD(P)H oxidase activity of granules from resting leukocytes was shown to be due to MP,[184] and the visible emission spectra of the MP–H_2O_2–Cl^- system and phagocytizing polymorphonuclear leukocytes were the same, sug-

[175] G. Lang, K. Spartalian, and T. Yonetani, *Biochim. Biophys. Acta*, 1976, **451**, 250; *J. Phys. (Paris), Colloq.*, 1976, 217.
[176] C. H. Kang, S. Ferguson-Miller, and E. Margoliash, *J. Biol. Chem.*, 1977, **252**, 919.
[177] N. M. Alexander, *Endocrinology (Philadelphia)*, 1977, **100**, 1610.
[178] J. T. Neary, D. Koepsell, B. Davidson, A. Armstrong, H. V. Strout, M. Soodak, and F. Maloof, *J. Biol. Chem.*, 1977, **252**, 1264.
[179] D. Deme, J. M. Gaveret, J. Pommier, and J. Nunez, *European J. Biochem.*, 1976, **70**, 7.
[180] L. Lamas and A. Taurog, *Endocrinology (Philadelphia)* 1977, **100**, 1129.
[181] D. Deme, J. Pommier, and J. Nunez, *European J. Biochem.*, 1976, **70**, 435.
[182] J. M. Gavaret, D. Deme, J. Nunez, and G. Salvatore, *J. Biol. Chem.*, 1977, **252**, 3281.
[183] J. E. Harrison, S. Pabalan, and J. Schultz, *Biochim. Biophys. Acta*, 1977, **493**, 247.
[184] K. Kakinuma and B. Chance, *Biochim. Biophys. Acta*, 1977, **480**, 96.

gesting that the light-producing processes were the same.[185] The apparent affinities of Cl^- and H_2O_2 for MP varied with pH, and they were mutually competitive inhibitors for MP activity.[186] During the chlorination of taurine by the $MP-H_2O_2-Cl^-$ system, the enzyme was inactivated. The inactivation rate increased with increasing H_2O_2 and decreased with increasing Cl^- concentrations. The results implied that MP may react with substrate only a limited number of times.[187]

The toxic effects of SCN^- in the $LacP-H_2O_2-CNS^-$ system were attributed to its oxidation to thiocyanogen (SCN), which could then modify either the —SH groups or, when H_2O_2 was present in excess, the tyrosine, tryptophan, and histidine residues of proteins.[188] An alternative interpretation was based on the observation that an oxidized form of SCN^- accumulated during incubations of LacP or MP with SCN^- at low concentrations of H_2O_2. The same product was formed by alkaline hydrolysis of $(SCN)_2$, and it was proposed that it is hypo-thiocyanate ion ($OSCN^-$) which is the antimicrobial agent generated by these systems.[189] Singlet oxygen was produced by, and has been proposed as the antimicrobial agent of, the $LacP-H_2O_2-Br^-$ system.[190]

Glutathione Peroxidase (GluP).—GluP isolated from rat liver was purified to a specific activity of 278 μmol of NADPH oxidized per minute per mg of protein. Polyacrylamide disc and SDS disc gel electrophoresis indicated that the enzyme existed in multiple forms, with a subunit of molecular weight 19 000. The native enzyme was probably a tetramer, with mol. wt. 76 000, although GluP existed as a large aggregate after homogenization of the liver.[191] X-Ray photoelectron spectroscopy on GluP from rat liver showed Se 3d-electron signals in the 55 cV region, but no signals from Fe 2p-electrons. The Se was probably not bound to oxygen.[192] Contrary to earlier reports, the Se of GluP was removed, and the enzyme was inactivated, by reaction with KCN, but only after oxidation of the enzyme with cumene hydroperoxide.[193]

Uterine Peroxidase (UP).—UP activity increased considerably when animals were treated with oestradiol.[194, 195] The enzyme was solubilized, with a mol. wt. of $\simeq 50\,000$, by the treatment of membranous material with aqueous solutions of bivalent ions, Ca^{2+} being particularly effective.[194] When UP, H_2O_2, oestradiol, and tyrosine were allowed to react together, the steroid and tyrosine associated in ratios of 1 : 2 and 1 : 3.[196] The reaction product of the oxidation of oestradiol

[185] B. R. Andersen, A. M. Brendzel, and T. F. Lint, *Infect. Immun.*, 1977, **17**, 62.
[186] J. M. Zgliczynski, R. J. Selvaraj, B. B. Paul, T. Stelmaszynska, P. K. F. Poskitt, and A. J. Sbarra, *Proc. Soc. Exp. Biol. Med.*, 1977, **154**, 418.
[187] J. W. Naskalski, *Biochim. Biophys. Acta*, 1977, **485**, 291.
[188] T. M. Aune and E. C. Thomas, *Biochemistry*, 1977, **16**, 4611.
[189] T. M. Aune and E. C. Thomas, *European J. Biochem.*, 1977, **80**, 209.
[190] J. F. Piatt, A. S. Cheema, and P. J. O'Brien, *F.E.B.S. Letters*, 1977, **74**, 251.
[191] F. H. Stults, J. W. Forstrom, D. T. Y. Chiu, and A. L. Tappel, *Arch. Biochem. Biophys.*, 1977, **183**, 490.
[192] D. Chiu, A. L. Tappel, and M. M. Millard, *Arch. Biochem. Biophys.*, 1977, **184**, 209.
[193] J. R. Prohaska, S.-H. Oh, W. G. Hoekstra, and H. E. Ganther, *Biochem. Biophys. Res. Comm.*, 1977, **74**, 64.
[194] C. R. Lyttle and E. R. DeSombre, *Proc. Nat. Acad. Sci. U.S.A.*, 1977, **74**, 3162.
[195] C. R. Lyttle and E. R. DeSombre, *Nature*, 1977, **268**, 337.
[196] P. H. Jellinck and T. McNabb, *Steroids*, 1977, **29**, 525.

by $K_3Fe(CN)_6$ or HRP has been characterized as a tetramer, with the monomers probably linked through C—O bonds in the 2- and 4-positions.[197]

Catalase.—The prosthetic group of the catalases is thought to be similar to that found in the plant peroxidases, *i.e.* a ferric protoporphyrin IX molecule bound to the polypeptide through co-ordination by an amino-acid in the fifth iron co-ordination site and with an oxygen-containing ligand in the sixth site. The nature of the fifth ligand is still controversial, and a distal amino-acid is probably involved in some reactions.[198] The c.d. and m.c.d. spectra of catalase and of its Compounds I and II indicated that the ligand at the fifth co-ordination position of iron was probably not histidine.[199] Investigation of the dependence of the ^1H solvent H_2O relaxation rate on temperature and frequency showed that the enhancement of the rate caused by formate and by formate with fluoride did not involve the presumed H_2O molecule at the sixth co-ordination position of iron, but was attributable to rapidly exchanging protons outside the axial co-ordination positions of the iron.[200]

A major function of catalase is the decomposition of peroxides by the overall reaction:

$$2ROOH \longrightarrow 2ROH + O_2$$

$$Ferricatalase \underset{ROH + O_2 \quad ROOH}{\overset{ROOH \quad ROH}{\rightleftharpoons}} Compound\ I$$

In addition, the enzyme can participate in the peroxide-dependent oxidation of various substrates (Scheme 2).[198]

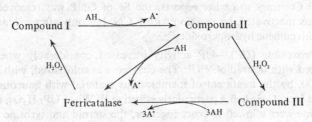

The peroxide-dependent oxidation of substrates (*AH*) by catalase

Scheme 2

Catalase from *Bacteroides distasonis* was purified and, in common with other catalases, shown to be a tetramer of identical subunits, each of molecular weight about 60 000 and each containing one prosthetic group.[201] Two, genetically distinct, catalase isoenzymes were purified from *Nassarius obsoleta*.[202]

[197] J. K. Norymberski, *F.E.B.S. Letters*, 1977, **76**, 321.
[198] G. R. Schonbaum and B. Chance, in 'The Enzymes' ed. P. Boyer, 3rd edn., Academic Press, New York, 1976, Vol. XIII, p. 363.
[199] M. Kajiyoshi and F. K. Anan, *J. Biochem. (Japan)*, 1977, **81**, 1319.
[200] S. Vuk-Pavlovic and D. L. Williams-Smith, *Biochemistry*, 1977, **16**, 5465.
[201] E. M. Gregory, J. B. Kowalski, and L. V. Holdeman, *J. Bacteriol.*, 1977, **129**, 1298.
[202] M. S. Nelson and J. G. Scandalios. *J. Exp. Zool.*, 1977, **199**, 257.

The enzyme from the liver of the American white tail deer crystallized in two forms, one of which appeared to be virtually isomorphous with crystals of bovine liver catalase.[203] However, 5—7% of purified mouse hepatic catalase was of higher molecular weight than the native form, although containing subunits of the same size.[204] Porcine erythrocyte catalase associates to form dimers, trimers, and tetramers when stored at 4 °C for more than a week.[205] The monomer and dimer were purified and shown to have molecular weights of 280 000 and 550 000, with frictional ratios of 1.33 and 1.51 respectively.[206] The association was due to the formation of an intermolecular disulphide link by the aerial oxidation of a single —SH group on the surface of each molecule.[205, 206] The specific activity of the enzyme was almost unchanged by dimerization but the m.c.d. spectrum was significantly changed, and the dimer was more stable.[206] Intermolecular cross-linking of catalase with glutaraldehyde or the immobilization of the enzyme on a gel had no effect on its K_m.[207] Cross-linking of human erythrocyte catalases from normal and Swiss acatalasemic heterozygotes resulted in inter-subunit and intermolecular cross-linking. The former stabilized the enzyme and markedly prevented development of the peroxidase activity which is produced from catalase by urea.[208]

Catalase from baboon catalysed the synthesis of cinnabarinate from 3-hydroxy-anthranilate.[209] The enzyme from human red cell haemolysate acted as 3-(3′,4′-dihydroxyphenyl)-L-alanine (L-dopa) peroxidase.[210] Catalase also acted as a peroxidase when it oxidized NADH in the presence of Mn^{2+} and phenols.[211] Catalase was inhibited by 4-hydroxypyrazole, which is a major product of pyrazole metabolism in rats, thus explaining the observation of the inhibition of catalase activity caused by pyrazole in experiments *in vivo* but not in those *in vitro*.[212] Catalase was also significantly inhibited at low pH, in the presence of Cl⁻, owing to the formation of a stable catalase–chloride compound that has one chloride ion per haem atom by its reaction with undissociated HCl.[213]

The tritium isotope effect on the peroxidation of ethanol by rat- and ox-liver catalase was shown to be close to the true isotope effect on the rate constant in the catalytic step involving the scission of the C—H bond.[214] The low value of this isotope effect (2.5) compared with that for methanol[198] was attributed to the non-linear arrangement of the transient complex with ethanol, which leads to greater conservation of the C—H bond energy.[215]

203 D. J. Burkey and A. McPherson, *Experientia*, 1977, **33**, 880.
204 M. B. Baird, H. R. Massie, and L. S. Birnbaum, *Biochem. J.*, 1977, **163**, 449.
205 A. Takeda and T. Samejima, *Biochim. Biophys. Acta*, 1977, **481**, 420.
206 A. Takeda and T. Samejima, *J. Biochem. (Japan)*, 1977, **82**, 1025.
207 M. T. Atallah and H. O. Hultin, *J. Food Sci.*, 1977, **42**, 7.
208 J. Hajdu, S. R. Wyss, and H. Aebi, *European J. Biochem.*, 1977, **80**, 199.
209 N. Savage and W. Prinz, *Biochem. J.*, 1977, **161**, 551.
210 Y. C. Awasthi, S. K. Srivastava, L. M. Snyder, L. Edelstein, and N. L. Fortier, *J. Lab. Clin. Med.*, 1977, **89**, 763.
211 B. Halliwell, *F.E.B.S. Letters*, 1977, **80**, 291.
212 F. H. Deis, G. W. J. Lin, and D. Lester, *F.E.B.S. Letters*, 1977, **79**, 81.
213 W. J. Litchfield, *F.E.B.S. Letters*, 1977, **83**, 281.
214 S. E. Damgaard, *Biochem. J.*, 1977, **167**, 77.
215 W. P. Jencks, 'Catalysis in Chemistry and Enzymology', McGraw-Hill, New York, 1969, Ch. 4.

L-Lactate–Cytochrome *c* Reductase (Cytochrome b_2).—This enzyme, from the yeasts *Saccharomyces cerevisiae* and *Hansenula anomala*, has a molecular weight near 235 000, and it consists of four equivalent subunits, each containing one protohaem and one FMN as the prosthetic groups. Cytochrome b_2 is stereospecific for (+)-L-lactate. Cytochrome *c* is considered to be the physiological electron acceptor, but other oxidants may be used.[216]

The *Saccharomyces* enzyme was found to bind only one molecule of cytochrome *c* to the tetrameric enzyme,[217, 218] indicating negative co-operativity between the subunits,[219] whereby the binding of a molecule of cytochrome *c* to one subunit decreased the affinity of the binding sites on the other subunits. However, the *Hansenula* enzyme bound four molecules of cytochrome *c* per tetramer, and thus it was suggested that perhaps the *Saccharomyces* enzyme had somehow been modified on purification.[218]

Limited proteolysis with trypsin cleaved the enzyme into haem and flavin-containing moieties,[220, 221] the haem core from the *Hansenula* enzyme having a molecular weight of $15\,000 \pm 1000$.[220] The flavin-containing protein was a tetramer of molecular weight 140 000[220] to 160 000,[221] of which each protomer could be cleaved into fragments of molecular weights 18 300 and 21 600.[221] The tetrameric flavoprotein was still reducible by substrate, but could not form a complex with cytochrome *c*.[220]

A different form of the haem-deficient enzyme was prepared by removal of both prosthetic groups (by dialysis in the presence of guanidinium hydrochloride, followed by reconstitution of the polypeptides with flavin). The resultant flavoprotein had the same kinetic properties, in its interaction with L-lactate, as the holoenzyme, but it was 200 times slower in its interaction with ferricyanide. Thus the haem might act as the electron mediator between the flavin and cytochrome *c*.[222]

Indolyl-3-alkane α-Hydroxylase (Tryptophan Hydroxylase).—A new haem-containing enzyme that catalyses the hydroxylation of the side-chain of a variety of 3-substituted indoles has been isolated from a soil isolate, *Pseudomonas* XA, by two independent groups.[223, 224] The enzyme had a molecular weight of between 250 000[223] and 280 000,[224] could be crystallized, and contained one haem residue per molecule.[223] One molecule of L-tryptophan was hydroxylated to one molecule of 3-indolylglycoaldehyde with the consumption of one molecule of oxygen and the formation of one molecule each of ammonia and carbon dioxide. No hydrogen peroxide was detected.[224]

[216] Y. Hatefi and D. L. Stiggall, in 'The Enzymes', ed. P. Boyer, Academic Press, New York, 1976, Vol. XIII, p. 175.
[217] T. Yoshimura, A. Matsushima, K. Aki, and K. Kakiuchi, *Biochim. Biophys. Acta*, 1977, **492**, 331.
[218] M. Prats, *Biochimie*, 1977, **59**, 621.
[219] A. Baudras, M. Krupa, and F. Labeyrie, *European J. Biochem.*, 1971, **20**, 58.
[220] M. Prats, *European J. Biochem.*, 1977, **75**, 619.
[221] M. Gervais, O. Groudinsky, Y. Risler, and F. Labeyrie, *Biochem. Biophys. Res. Comm.*, 1977, **77**, 1543.
[222] M. Iwatsubo, M. Mevel-Ninio, and F. Labeyrie, *Biochemistry*, 1977, **16**, 3558.
[223] J. Roberts and H. J. Rosenfeld, *J. Biol. Chem.*, 1977, **252**, 2640.
[224] K. Takai, H. Ushiro, Y. Noda, S. Narumiya, T. Tokayama, and O. Hayaishi. *J. Biol. Chem.*, 1977, **252**, 2648.

3 Iron–Sulphur Enzymes

NADH–Ubiquinone Reductase (Complex I).—Complex I represents, in a highly purified form, the segment of the mitochondrial electron-transport system from NADH to ubiquinone. It is reported to contain 1.4—1.5 nmol of FMN, 16–28 nanogram atom of both non-haem iron and acid-labile sulphur, 4.2—4.5 nmol of ubiquinone-10, and 0.22 mg of lipid per mg of protein.[216, 225]

The iron–sulphur centres in the NADH-reduced enzyme give rise to an overlapping complex of e.p.r. resonances at low temperatures. The resolution of these into signals from individual centres is still the subject of some controversy. Orme-Johnson *et al.* concluded that the resonances arise from four centres with approximately equal intensities[226] (Table 1). Ohnishi attempted the resolution of the centres by potentiometric titration (Table 1). She found mid-point potentials (E_m) at pH 7.2 of -380 and -240 mV associated with Centre 1, and she thus resolved it into Centres 1a and 1b respectively. Centres 2, 3, and 4 had values of E_m of -20, -240, and -405 mV respectively. She also found a multiplicity of peaks (Table 1), with $E_m = -260$ mV, at temperatures near 4.2 K, and tentatively assigned these to Centres 5 and 6.[227] Albracht *et al.*[228] have further investigated the shape and intensity of the e.p.r. resonances and concluded that five centres (Table 1) were enough to account for all the observations, but that Centres 1a and 1b had only about 25% of the integrated intensity of each of Centres 2, 3, and 4.

A study of the ^{57}Fe hyperfine interaction in the e.p.r. spectra of reduced submitochondrial particles from the yeast *Candida utilis* indicated that Centres 1a and 1b were 2Fe–2S centres and that Centres 2, 3, and 4 were 4Fe–4S centres.[229] Interaction between Centre 3 and another paramagnetic centre in Complex 1 has been detected. At -240 mV, the intensity of Centre 3 was about half developed. When the potential was lowered further, the intensity of the line near $g = 1.86$ decreased anomalously, reaching a minimum at -380 mV and then increasing to a maximum. Concurrently, a line at $g = 3.93$ appeared, reaching a maximum at -380 mV and then disappearing. It was suggested that the above phenomena were due to an (otherwise undetected) two-electron carrier in its intermediate redox state. The line at $g = 3.93$ was interpreted as arising from $\Delta M_S - 2$ transitions of interacting free-radical and Fe–S centres or from two strongly coupled Fe–S centres.[230]

The e.p.r. spectra of reduced membrane particles from *Paracoccus denitrificans* were similar to those of Complex I from mitochondria. The behaviour of a signal similar to Centre 2, under sulphate-limited growth of this organism, led to the suggestion that there was a close correlation between Site I phosphorylation, rotenone sensitivity, and the presence of the e.p.r. signal like that of Centre 2.[231]

[225] H. Beinert, in 'Iron–Sulphur Proteins', ed. W. Lovenberg, Academic Press, New York, San Francisco, and London, 1977, Vol. III, p. 61.
[226] N. R. Orme-Johnson, R. E. Hansen, and H. Beinert, *J. Biol. Chem.*, 1974, **249**, 1922.
[227] T. Ohnishi, *Biochim. Biophys. Acta*, 1975, **387**, 475.
[228] S. P. J. Albracht, G. Dooijewaard, F. J. Leeuwerik, and B. van Swol, *Biochim. Biophys. Acta*, 1977, **459**, 300.
[229] S. P. J. Albracht and J. Subramanian, *Biochim. Biophys. Acta*, 1977, **462**, 36.
[230] J. C. Salerno, T. Ohnishi, J. Lim, W. R. Widger, and T. E. King, *Biochem. Biophys. Res. Comm.*, 1977, **75**, 618.
[231] E. M. Meijer, R. Wever, and A. H. Stouthamer, *European J. Biochem.*, 1977, **81**, 267.

NADH-reduced Complex I generated O_2^- and H_2O_2 from oxygen, the latter even in the presence of superoxide dismutase, apparently *via* a reduced form of ubiquinone. Similar conclusions were reached from experiments on Complex III[232] (see below).

Succinate Dehydrogenase.—The succinate dehydrogenase that has been studied most is that purified from beef heart mitochondria as part of Complex II of the respiratory chain. It contains one mole of covalently bound FAD and eight non-haem iron and acid-labile sulphur atoms per molecule of molecular weight 97 000, consisting of two subunits, of molecular weights 70 000 and 27 000. Both subunits contain iron, and the larger also contains flavin,[216] but the assignment of the iron and sulphur into types of cluster was not well established.[225] A comparison of an iron-counting method[233] (in which clusters were progressively dissociated with NO and cysteine to form e.p.r.-detectable complexes containing one iron atom each) with Cammack's method[234] indicated that succinate dehydrogenase contained two [2Fe–2S] and one [4Fe–4S] cluster per unit of molecular weight 97 000.[233] This assignment was consistent with the observation of two e.p.r. signals for which $g = 1.94$, *i.e.* S-1 and S-2, in the reduced enzyme and one 'HiPIP'-type of signal, *i.e.* S-3, at $g = 2.01$ in the oxidized enzyme[225] and with studies on the ^{57}Fe hyperfine interaction in the e.p.r. spectra of reduced submitochondrial particles from *Candida utilis*.[229]

Centres S-1 and S-2 have e.p.r. spectra with the same g values, ($g_z = 2.03$, $g_y = 1.93$, and $g_x = 1.91$) but with different temperature dependences and redox potentials. It was deduced that spin coupling occurs between these centres from the observed desaturation of the S-1 signal when S-2 was reduced, from the broadening and splitting of the signal between 16 and 6.5 K, and from the appearance of a half-field $\Delta M_s = 2$ signal at $g = 3.88$ (arising from the anti-ferromagnetically coupled system) under these conditions.[230]

Purified succinate dehydrogenase can oxidize succinate with the use of artificial electron acceptors but it cannot utilize the (presumed) natural electron acceptor, *i.e.* ubiquinone. Studies on the inter-relationships between the stoicheiometry of the flavin and the Fe–S centres and the catalytic activity of purified succinate dehydrogenase, compared with the properties of the enzyme reconstituted with membranes and the original membrane-bound enzyme, did not support the idea that the properties of Centres S-1 and S-2 in the reconstituted enzyme differed from those found in other preparations.[235] However, purified succinate dehydrogenase had a lower turnover number with PMS as oxidant than had the reconstituted enzyme. This observation was attributed by Ackrell *et al.*[236] to positive modulation of the enzyme activity by the membrane rather than to deactivation on extraction. This hypothesis was challenged by Vinogradov *et al.*, when it was observed that, by using an alternative oxidant, *i.e.* TMPD

[232] E. Cadenas, A. Boveris, C. I. Ragan, and A. O. M. Stoppani, *Arch. Biochem. Biophys.*, 1977, **180**, 248.

[233] J. C. Salerno, T. Ohnishi, J. Lim, and T. E. King, *Biochem. Biophys. Res. Comm.*, 1976, **73**, 833.

[234] R. Cammack, *Biochem. Soc. Trans.*, 1975, **3**, 482.

[235] H. Beinert, B. A. C. Ackrell, A. D. Vinogradov, E. B. Kearney, and T. P. Singer, *Arch. Biochem. Biophys.*, 1977, **182**, 95.

[236] B. A. C. Ackrell, E. B. Kearney, and T. P. Singer, *J. Biol. Chem.*, 1977, **252**, 1582.

Table 1 *E.p.r. signals from NADH–ubiquinone reductase (Complex I): g factors (peaks)*

	Ref. a			Ref. b			Ref. c		
Centre 1	2.022	1.938	1.923	—			—		
Centre 1a	—			2.03	1.94		2.021	1.938	
Centre 1b	—			2.03	1.94	1.91	2.021	1.928	
Centre 2	2.054	1.922		2.05	1.93		2.054	1.922	
Centre 3	2.100	1.866	1.862	2.10	(1.93?)	1.87	2.103	1.93—1.94	1.884
Centre 4	2.103	?	1.864	2.11	(1.93?)	1.88	2.04	1.92—1.93	1.863
Centres 5 & 6				2.11,	2.06,	2.03			
				1.93,	1.90,	1.88			

(a) N. R. Orme-Johnson, R. E. Hansen, and H. Beinert, *J. Biol. Chem.*, 1974, **249**, 1922.
(b) T. Ohnishi, *Biochim. Biophys. Acta*, 1975, **387**, 475.
(c) S. P. J. Albracht, G. Dooijewaard, F. J. Leeuwerik, and B. van Swol, *Biochim. Biophys. Acta*, 1977, **459**, 300.

(the radical form of $NNN'N'$-tetramethyl-p-phenylenediamine), the activity of the purified enzyme was double that found with PMS. The TMPD activity was not enhanced on re-incorporation of the enzyme into the membrane, and thus Vinogradov *et al.* concluded that the re-activation phenomenon was associated with the artificial oxidant rather than with modification of the enzyme.[237] Ackrell *et al.* re-examined the two oxidants but found no difference in activity with the purified enzyme.[238] On this evidence, and on the basis of the observation that PMS re-oxidized the succinate-reduced enzyme in a few milliseconds,[239] well within the enzyme turnover time, they concluded that the rate of oxidation by PMS was not rate-limiting, and therefore that the enhancement of activity on reconstitution was associated with a modification of the enzyme by the membrane.[238]

The presumed natural activity of succinate dehydrogenase, *i.e.* the ability to reduce ubiquinone, was recovered[240] by mixing the purified enzyme with a small protein, of molecular weight 15 000, that was isolated from Complex III[241] (see below).

Carbonamides[239, 242] and thenoyltrifluoroacetone (TTF) strongly inhibited the re-oxidation of Centres S-1 and S-3 by PMS but not their reduction by succinate, suggesting that these inhibitors block electron transport from succinate dehydrogenase to the ubiquinone pool at some point on the oxidizing side of Centre S-3.[239] This conclusion was supported by the observation that the free-radical e.p.r. signal from ubiquinone and the e.p.r. signal from S-3 were similarly inhibited by TTF, ethanol, and ferricyanide.[243]

Succinate dehydrogenases from aerobes usually have covalently bound flavin whereas it is usually non-covalently bound in those from anaerobes. An exception to this rule is apparently the enzyme from the anaerobe *Vibrio succinogenes*, which has covalently bound flavin.[244]

The enzymes isolated from chromatophores of *Rhodopseudomonas sphaeroides*[245] and *Rhodospirillum rubrum*[246] were similar to the mammalian enzymes in many of their properties. The enzymes from *R. sphaeroides* was bound to the outer surface of the chromatophores and had Fe–S centres similar to centres S-1, S-2, and S-3, with mid-point potentials of $+50$, -250, and $+80$ mV respectively at pH 7.[245] In addition, the enzyme from *R. rubrum* appeared to have a third centre, with g values of 2.03, 1.93, and 1.91. The three Fe–S centres with $g_{av} = 1.95$ were distinguished by having mid-point potentials at $+50$, -160, and -380 mV respectively. It was suggested that the last centre might

[237] A. Vinogradrov, V. G. Goloveshkina, and E. V. Gavrikova, *F.E.B.S. Letters*, 1977, **73**, 235.
[238] B. A. C. Ackrell, C. J. Coles, and T. P. Singer, *F.E.B.S. Letters*, 1977, **75**, 249.
[239] B. A. C. Ackrell, E. B. Kearney, C. J. Coles, T. P. Singer, H. Beinert, Y.-P. Wan, and K. Folkers, *Arch. Biochem. Biophys.*, 1977, **182**, 107.
[240] C. A. Yu, L. Yu, and T. E. King, *Biochem. Biophys. Res. Comm.*, 1977, **79**, 939.
[241] C. A. Yu, L. Yu, and T. E. King, *Biochem. Biophys. Res. Comm.*, 1977 **78**, 259.
[242] P. C. Mowery, D. J. Steenkamp, B. A. C. Ackrell, T. P. Singer, and G. A. White, *Arch. Biochem. Biophys.*, 1977, **178**, 495.
[243] A. N. Tikhonov, D. Sh. Burbaev, I. V. Grigolava, A. A. Konstantinov, M. Yu. Ksenzenko, and E. Ruuge, *Biofizika*, 1977, **22**, 734.
[244] W. C. Kenney and A. Kröger, *F.E.B.S. Letters*, 1977, **73**, 239.
[245] W. J. Ingledew and R. C. Prince, *Arch. Biochem. Biophys.*, 1977, **178**, 303.
[246] R. P. Carithers, D. C. Yoch, and D. I. Arnon, *J. Biol. Chem.*, 1977, **252**, 7461.

be associated with super-reduced Centre S-3, or with a portion of the enzyme that had retained the properties of the membrane-bound species.[246]

A study of the e.p.r. spectra of mitochondria from a variety of species of higher plant indicated that the point of interaction of ubiquinone and Centre S-3 of succinate dehydrogenase might be the branching point of the main and alternative respiratory pathways in higher plants.[247]

Ubiquinone–Cytochrome c Oxidoreductase (Complex III).—Complex III (or the cytochrome b–c_1 complex) from the mitochondrial respiratory chain is a complex enzyme, containing one [2Fe–2S] centre, one cytochrome c_1, and two cytochrome b moieties, plus several polypeptides that contain no prosthetic group.

Evidence for interactions between the b- and c-type haems in Complex III was obtained from the anomalous behaviour of various haem b marker bands in the resonance Raman spectra of membranous preparations (from pigeon heart mitochondria) that had been frozen at redox potentials between -105 and $+276$ mV.[248]

Electrophoresis of the enzyme that had been purified from beef heart mitochondria on SDS polyacrylamide gels gave eight polypeptide bands, in the ratios $1 : 1 : 2 : 1 : 1 : 2 : 2 : 2$. In the presence of 8M-urea, or with a more cross-linked gel, the smallest band split into three. The molecular weights of these ten bands were estimated as 43 000, 40 000, 28 000, 29 000, 24 000, 12 000, 8000, 6000, 5000, and 4000, giving a minimal molecular weight of 242 000, in agreement with the value of 250 000 obtained from ultracentrifuge studies. Bands 3 and 7 were cytochrome b haemoproteins, band 4 was the cytochrome c_1 haemoprotein, but band 6 was also associated with cytochrome c_1 and was identified as the so-called oxidation factor; band 5 was the Rieske Fe–S protein.[249]

In a separate study, seven subunit bands, with molecular weights of 53 000, 50 000, 37 000, 30 000, 28 000, 17 000, and 15 000, in the ratios $2 : 2 : 2 : 3 : 2 : 2 : 5$, were found.[250] In experiments with cross-linking reagents, followed by two-dimensional gel electrophoresis, the most prominent subunit pairs observed (under mild conditions) were I and II, II and IV, I and V, and VI and VII. At higher concentrations of reagent, larger aggregates were produced, until an aggregate of molecular weight 310 000, equivalent to the enzyme monomer, was dominant.[251]

The enzyme from *Saccharomyces cerevisiae* YF gave seven bands on SDS polyacrylamide gel electrophoresis, with molecular weights of 43 000, 40 000, 32 000, 24 000, 22 000, 20 000, and 18 000, in the ratios $1 : 1 : 3 : 1 : 1 : 1 : 1$. It has a minimal molecular weight of 255 000 and it contains four non-haem iron atoms and two each of the cytochromes b and c_1.[252]

Mild digestion of the Complex III from beef heart with trypsin was followed by SDS polyacrylamide gel electrophoresis, and the data correlated with the loss of electron-transfer activity. The non-haem iron–protein cleavage occurred at

[247] P. R. Rich, A. L. Moore, W. J. Ingledew, and W. D. Bonmer, Jr., *Biochim. Biophys. Acta*, 1977, **462**, 501.
[248] F. Adar and M. Erecinska, *F.E.B.S. Letters*, 1977, **80**, 195.
[249] C. A. M. Marres and E. C. Slater, *Biochim. Biophys. Acta*, 1977, **462**, 531.
[250] L. Yu, C.-A. Yu, and T. E. King, *Biochim. Biophys. Acta*, 1977, **495**, 232.
[251] R. J. Smith and R. A. Capaldi, *Biochemistry*, 1977, **16**, 2629.
[252] J. Reed and B. Hess, *Z. physiol. Chem.*, 1977, **358**, 1119.

a rate comparable with the rate of loss of electron-transfer activity. Only the rate of cleavage of this protein was altered by changing the redox state of the enzyme, indicating that redox-linked conformational changes were small.[253]

A small protein of molecular weight 15 000, isolated from active cytochrome b-c_1 complex, was identified as being responsible for electron transfer between succinate dehydrogenase and ubiquinone (see above). The protein was tentatively assigned the role of a ubiquinone carrier.[241] A protein–quinone complex, having a mid-point potential of $+150$ mV, and being capable of reducing ferricyto-chrome c_2 of *R. sphaeroides* chromatophores with a τ of 1—2 ms, required two electrons and two protons for its reduction. The state of reduction of this complex seemed to govern the rate at which the cyclic, photosynthetic electron-transport system could operate, and thus it might play a central role in electron and proton movements in Complex III.[254]

Electron-transferring Flavoprotein (ETF) Dehydrogenase–Ubiquinone Reductase from Beef Heart Mitochondria.—This iron–sulphur flavoprotein[225] was purified to apparent homogeneity, and it contained one FAD and one [4Fe–4S] cluster in a subunit of molecular weight 66 000. It could accept three electrons per flavin group, the Fe–S cluster being reduced to a species with an e.p.r. spectrum that has g values of 2.086, 1.939, and 1.886. The mid-point potential of the Fe–S cluster was 55 mV higher than that of the flavin. In mitochondria, the Fe–S cluster potential was $+40$ mV. Reduction of the Fe–S cluster, by ETF from the β-oxidation pathway of fatty acids, occurred with a half-time of < 10 ms, and when in its reduced form, it was oxidized by ubiquinone-1 with a half-time of 2 ms. Direct electron transfer from ETF to ubiquinone-1 only occurred readily when the Fe–S flavoprotein was present. Thus the protein was an efficient ETF dehydrogenase and ubiquinone reductase *in vitro*. Since the e.p.r. signal of the Fe–S cluster was enhanced in the brown adipose tissue of cold-acclimatized guinea-pigs, it was postulated that the protein might also function *in vivo* as an electron carrier in the oxidation of fatty acids.[255]

Nitrite Reductase.—The six-electron reduction of nitrite to ammonia can be carried out by enzymes of at least three types. In bacteria and fungi, a flavo-iron protein is found.[216, 256] The FAD moiety accepts electrons from NAD(P)H and transfers them to an unusual haem, sirohaem[257] (2), which binds and reduces nitrite. Sirohaem is also found in sulphite reductase. The enzyme from algae and higher plants utilizes ferrodoxin as reductant, and contains sirohaem and a single [2Fe–2S] centre in place of the FAD found in the bacterial enzyme.[258] The sirohaem had a mid-point potential of -50 mV, with $n=1$, but the mid-point potential of the [2Fe–2S] centre was too low (approximately -550 mV) for accurate determination.[259]

253 M. B. Ball, R. L. Bell, and R. A. Capaldi, *F.E.B.S. Letters*, 1977, **83**, 99.
254 R. C. Prince and P. L. Dutton, *Biochim. Biophys. Acta*, 1977, **462**, 731.
255 F. J. Ruzicka and H. Beinert, *J. Biol. Chem.*, 1977, **252**, 8440.
256 M. Losada, *J. Mol. Catalysis*, 1975–6, **1**, 245.
257 M. J. Murphy, L. M. Siegel, S. R. Tove, and H. Kamin, *Proc. Nat. Acad. Sci. U.S.A.*, 1974, **71**, 612.
258 P. J. Aparicio, D. B. Knaff, and R. Malkin, *Arch. Biochem. Biophys.*, 1975, **169**, 102.
259 M. L. Stoller and R. Malkin, *F.E.B.S. Letters*, 1977, **81**, 271.

HO$_2$CCH$_2$CH$_2$ CH$_2$CO$_2$H

HO$_2$CCH$_2$—   N   N   —CH$_2$CH$_2$CO$_2$H

H$_3$C   Fe   CH$_3$

     N   N

HO$_2$CCH$_2$ CH$_2$CO$_2$H

HO$_2$CCH$_2$CH$_2$ H   H   CH$_2$CH$_2$CO$_2$H

(2) Sirohaem

The enzyme from spinach leaves was recently purified to specific activities of 140[260] and 110[261] μmol of NO$_2^-$ reduced per minute per mg of protein. The latter preparation stoicheiometrically reduced nitrite to ammonia, with no detectable, free, nitrogen-containing intermediates. However, hydroxylamine could act as a substrate, and optical spectroscopy indicated the presence of enzyme-bound intermediates when nearly all the nitrite was depleted. Nitrite, hydroxylamine, and the inhibitor cyanide formed complexes with both the oxidized and reduced enzymes, but the inhibitor CO bound only to the reduced enzyme. Nitrite formed a 1 : 1 complex with the enzyme, with $K_{diss} = 3.2 \times 10^{-6}$ mol l^{-1}.[261]

Nitrite is utilized as a terminal electron acceptor in respiration. Here the nitrite reductase contains two haem moieties. Studies on the respiratory particle from *Streptomyces griseus* showed that azide specifically inhibited nitrite reductase (but not cytochrome oxidase) activity and that haem *b* apparently acted as an intermediate electron donor to haem *d*, which acted as the nitrite reductase.[262]

The enzyme from *Pseudomonas* contained one haem *c* and one haem *d* in a molecule of molecular weight 70 000. Ferrocytochrome *c* acted as electron donor. The enzyme could also act as a cytochrome oxidase with oxygen as oxidant. Both haems were reported to react simultaneously with NO$^+$, which is in equilibrium with NO$_2^-$ in aqueous solution. M.c.d. and c.d. spectroscopy indicated that haem–haem interaction only occurred in the reduced enzyme, and could be perturbed by changes of pH or by the binding of inhibitor to the haem.[263]

Hydrogenase.—Hydrogenase is known to be an iron–sulphur protein, but its metal content has been the subject of some controversy. The enzyme from *Clostridium pasteurianum* has been reported to contain four[264] or twelve[265] non-haem iron atoms per molecule of molecular weight 60 000. The preparations apparently had very similar specific activities. Cluster-extrusion experiments have apparently confirmed both of these analytical compositions, indicating the

[260] S. Ida, *J. Biochem. (Japan)*, 1977, **82**, 915.
[261] J. M. Vega and H. Kamin, *J. Biol. Chem.*, 1977, **252**, 896.
[262] Y. Inoue, *Biochim. Biophys. Acta*, 1977, **459**, 88.
[263] Y. Orii, H. Shimada, T. Nozawa, and M. Hatano, *Biochem. Biophys. Res. Comm.*, 1977, **76**, 983.
[264] D. L. Erbes, R. H. Burris, and W. H. Orme-Johnson, *Proc. Nat. Acad. Sci. U.S.A.*, 1975, **72**, 4795.
[265] J.-S. Chen and L. E. Mortenson, *Biochim. Biophys. Acta*, 1974, **371**, 283.

11

presence of one[264] and three[266] [4Fe–4S] clusters respectively. Both preparations exhibited e.p.r. spectra characteristic of $[4Fe–4S]^{3-}$ and $[4Fe–4S]^{1-}$ centres, but the integrations of the spectra differed, being 0.7—1.0 spins per mole of $(3-)$ and 0.07 spins per mole of $(1-)^{264}$ and 1.6—1.8 spins per mole of $(3-)$ and 0.7 spins per mole of $(1-)^{267}$ for the four-iron and twelve-iron preparations respectively. Carbon monoxide is a competitive inhibitor of hydrogenase, and ^{13}CO broadened both types of e.p.r. spectra, indicating that there is direct binding to the [4Fe–4S] clusters.[264]

The above data have been interpreted in terms of the alternative hypotheses (a) that one [4Fe–4S] centre cycles between the $(1-)$ and $(3-)$ oxidation levels when catalysing the reaction $2e + 2H^+ \rightleftharpoons H_2$,[264] or (b) that one [4Fe–4S] centre acts as the catalytic site and two other centres act as electron-transferring sites to and from the catalytic site.[266] In the latter scheme, each centre need only cycle between $(1-)$ and $(2-)$ or $(2-)$ and $(3-)$ oxidation levels.

Preparations of hydrogenase from other sources are less well studied. The most recent preparations of the enzyme from *Desulphovibrio vulgaris* were reported to contain 7—9 atoms of Fe per molecule[268] and the enzyme from *Chromatium* contains 4 Fe atoms and 4 acid-labile sulphur atoms per molecule.[269] The latter enzyme had a molecular weight of $\simeq 70\,000$ and dissociated into active subunits of molecular weight $\simeq 35\,000$ in the presence of 1% SDS or deoxycholate. In 6% SDS, further dissociation into species of molecular weight $\simeq 20\,000$ was observed, indicating that the enzyme was a tetramer.[270]

Adenylyl Sulphate Reductase.—This enzyme is an iron–sulphur flavoprotein, on the sulphate-reduction pathway, which catalyses the reduction of adenosine 5'-phosphosulphate (APS) to SO_3^{2-} and AMP or, in the presence of a suitable acceptor such as cytochrome *c* or ferricyanide, the reverse reaction:

$$APS + 2e \rightleftharpoons AMP + SO_3^{2-}$$

The enzyme has a molecular weight of 170 000—219 000 and contains one FAD molecule and 4—13 atoms of non-haem iron and 4—12 atoms of acid-labile sulphur per molecule, depending on the source.[216]

The mechanism of the reverse reaction was investigated by stopped-flow and difference spectrophotometry. The enzyme-bound FAD was rapidly reduced by SO_3^{2-}. On subsequent addition of AMP, the $FADH_2$ was oxidized to FADH and the Fe–S clusters were reduced. Cytochrome *c* was reduced directly by the enzyme, without the involvement of O_2^-.[271]

Trimethylamine Dehydrogenase (TMADH).—TMADH catalyses the oxidative demethylation of trimethylamine to dimethylamine and formaldehyde. The enzyme has a molecular weight of 147 000 and contains two types of prosthetic

266 W. O. Gillum, L. E. Mortenson, J.-S. Chen, and R. H. Holm, *J. Amer. Chem. Soc.*, 1977, **99**, 584.
267 J.-S. Chen, L. E. Mortenson, and G. Palmer, in 'Iron and Copper Proteins', ed. K. T. Yasunobu, H. F. Mower, and O. Hayaishi, Plenum Press, New York, 1976, p. 68.
268 T. Yagi, K. Kimura, H. Daidoji, F. Sakai, S. Tamura, and H. Inokuchi, *J. Biochem (Japan)*, 1976, **79**, 661.
269 P. H. Gitlitz and A. I. Krasna, *Biochemistry*, 1975, **14**, 2561.
270 T. Kakuno, N. O. Kaplan, and M. D. Kamen, *Proc. Nat. Acad. Sci. U.S.A.*, 1977, **74**, 861.
271 K. Adachi and I. Suzuki, *Canad. J. Biochem.*, 1977, **55**, 91.

groups, one being organic and of unknown structure and the other containing iron and acid-labile sulphur. Direct analysis showed the presence of 4 gram atom each of iron and acid-labile sulphur, and cluster-extrusion experiments indicated that these formed a single [4Fe–4S] cluster.[272]

Pyrazon Dioxygenase.—This enzyme has been isolated from bacteria capable of degrading pyrazon, the active ingredient of the herbicide Pyramin. The enzyme consisted of three components: Component B, a ferredoxin of molecular weight 12 000; Component A_2, a flavoprotein of mol. wt. 67 000, containing FAD, and Component A_1, a [2Fe–2S]-containing protein of molecular weight 180 000.[273] The authors postulated the mechanism shown in Scheme 3.

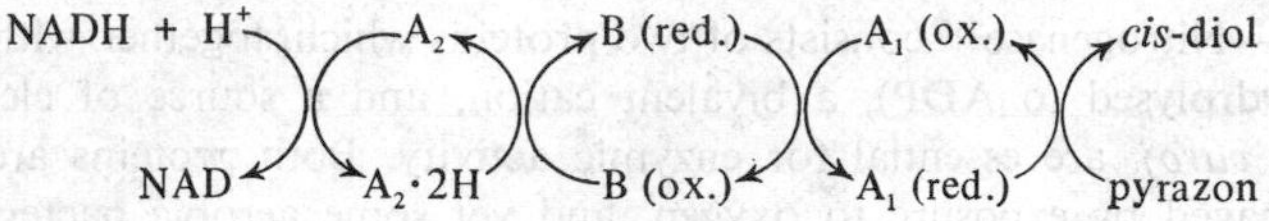

NADH + H⁺   A₂   B (red.)   A₁ (ox.)   cis-diol

NAD   A₂·2H   B (ox.)   A₁ (red.)   pyrazon + O₂

Scheme 3

4 Other Non-haem Iron-containing Enzymes

Lipoxygenase.—E.p.r. and optical studies have indicated the involvement of non-haem iron in catalysis by lipoxygenase.[274, 275] The pea enzyme contained one atom of iron in a molecule of molecular weight 98 000. Very high g values (near 7.2, 6.1, and 5.7) were observed in the e.p.r. spectra of the iron in enzyme–substrate and enzyme–product complexes,[274] indicating that the active iron was in the high-spin ferric state.[275]

The enzyme catalyses the oxygenation (by molecular oxygen) of polyunsaturated acids containing 1,4-*cis,cis*-pentadiene systems to conjugated hydroperoxy fatty acids.[276] The exact nature of the product seems to be dependent on the source of the enzyme, since arachidonic acid was oxidized to 8,15-dihydroperoxy-5,9,11,13-eicosatetraenoic acid by soybean lipoxygenase-1,[277] and to 12L-hydroxy-5,8,10,14-eicosatetraenoic acid by the enzyme from human platelets.[278]

ω-Hydroxylase.—The ω-hydroxylase from *Pseudomonas oleovorans* catalyses the hydroxylation of fatty acids and alkanes and the epoxidation of alkanes. NADH, a reductase, rubredoxin, and oxygen are also required. The hydroxylase was characterized as a non-haem iron protein that has one iron atom and one cysteine residue per polypeptide chain of molecular weight 40 800. The enzymic activity was reversibly inhibited by cyanide. The iron was removed from the reduced

[272] C. L. Hill, D. J. Steenkamp. R. H. Holm, and T. P. Singer, *Proc. Nat. Acad. Sci. U.S.A.*, 1977, **74**, 547.

[273] K. Sauber, C. Fröhner, G. Rosenberg, J. Eberpächer, and F. Lingens, *European J. Biochem.*, 1977, **74**, 89.

[274] L. J. M. Spaapen, J. F. G. Vliegenthart, and J. Boldingh, *Biochim. Biophys. Acta*, 1977, **488**, 517.

[275] E. K. Pistorius, B. Axelrod, and G. Palmer, *J. Biol. Chem.*, 1976, **251**, 7144.

[276] J. Verhagen, A. A. Bouman, J. F. G. Vliegenthart, and J. Boldingh, *Biochim. Biophys. Acta*, 1977, **486**, 114; H. Aoshima, T. Kajiwara, A. Hatamaka, H. Nakatani, and K. Hironi, *ibid.*, p. 121.

[277] G. S. Bild, C. S. Ramadoss, S. Lim, and B. Axelrod, *Biochem. Biophys. Res. Comm.*, 1977, **74**, 949.

[278] P. K. Ho, C. P. Walters, and H. R. Sullivan, *Biochem. Biophys. Res. Comm.*, 1977, **76**, 398.

enzyme by dialysis against EDTA and its activity was restored by incubation of the apoprotein with Fe^{2+}.[279]

Ribonucleotide Reductase.—This enzyme, from *Escherichia coli*, consists of two non-identical subunits, B1 and B2, both of which are required for activity. The subunit B2 has been shown to contain two non-equivalent high-spin Fe^{III} atoms in an antiferromagnetically coupled complex, and it also exhibited an e.p.r. signal with $g = 2.0047$, which was assigned to a free radical on the protein.[280] This radical was shown to be associated with a tyrosine residue of the polypeptide chain.[281]

5 Molybdenum-containing Enzymes

Nitrogenase.—Nitrogenase[282] consists of two proteins which, together with ATP (which is hydrolysed to ADP), a bivalent cation, and a source of electrons ($Na_2S_2O_4$ *in vitro*), are essential for enzymic activity. Both proteins are irreversibly damaged by exposure to oxygen, and yet some aerobic bacteria can fix nitrogen. These bacteria have developed complex mechanisms to protect the nitrogenase under conditions of high oxygen tensions. One of these mechanisms, in *Azotobacter vinelandii*, seems to involve the formation of an inactive complex with an iron–sulphur protein. Reduction of the resultant complex results in its dissociation and the reappearance of nitrogenase activity.[283]

Nitrogenase has only been purified from free-living bacteria or from *Rhizobia* in symbiosis with legumes, but nitrogenase activity has been observed in the endophyte suspensions derived from the root nodules of several non-legumes.[284]

Structural Studies. Both nitrogenase proteins contain iron and acid-labile sulphur in approximately equivalent amounts, and the larger protein, the MoFe protein, also contains molybdenum. This protein has a molecular weight of $220\,000 \pm 20\,000$[282, 285, 286] and consists of four subunits. Reports differ as to whether these subunits are identical or are of two types. The controversy is at least partially attributable to the use of sodium dodecyl sulphate (SDS) polyacrylamide gel electrophoresis as a diagnostic tool, since the use of different brands of SDS gave either one or two polypeptide bands on the gel, depending on the source of MoFe protein. Tryptic digestion, followed by two-dimensional electrophoresis of the separated polypeptide bands, showed that the MoFe protein of *Klebsiella pneumoniae* definitely contained two each of two types of subunit.[287]

[279] R. T. Ruettinger, G. R. Griffith, and M. J. Coon, *Arch. Biochem. Biophys.*, 1977, **183**, 528.
[280] C. L. Atkin, L. Thelander, and P. Reichard, *J. Biol. Chem.*, 1973, **248**, 7464.
[281] B.-M. Sjoberg, P. Reichard, A. Graslund, and A. Ehrenberg, *J. Biol. Chem.*, 1977, **252**, 536.
[282] R. R. Eady and B. E. Smith, in 'Dinitrogen Fixation', 1978, Section 2, 399, ed. R. W. F. Hardy and R. C. Burns, Wiley Interscience, New York, 1978, Section 2, p. 399; R. R. Eady and J. R. Postgate, *Nature*, 1974, **249**, 805; W. G. Zumft and L. E. Mortenson, *Biochim. Biophys. Acta*, 1975, **416**, 1.
[283] H. Haaker and C. Veeger, *European J. Biochem.*, 1977, **77**, 1; G. Scherings, H. Haaker, and C. Veeger, *ibid.*, p. 621.
[284] J. Van Straten, A. D. L. Akkermans, and W. Roelofsen, *Nature*, 1977, **266**, 257.
[285] R. H. Swisher, M. L. Landt, and F. J. Reithel, *Biochem. J.*, 1977, **163**, 427.
[286] E. P. Starchenkov, N. A. Protsenko, and E. D. Krugova, *Fiziol. Biokhim. Kul't Rast.*, 1977, **9**, 167.
[287] C. Kennedy, R. R. Eady, E. Kondorosi, and D. K. Rekosh, *Biochem. J.*, 1976, **155**, 383.

The most active preparations of the MoFe protein contain two Mo atoms and 24–33 Fe atoms per unit of molecular weight 220 000, depending on the scource.[282] The reported acid-labile sulphur contents are generally lower than the iron contents, but they have been assumed to be equivalent. However, this assumption has recently been challenged.[288]

Mössbauer spectroscopic studies of the MoFe protein showed the existence of three or four types of iron atom arranged in clusters.[289, 290] One of the Mössbauer-active species was associated with a distinctive e.p.r. signal, with g values near 4.3, 3.7, and 2.01, observed in the protein as isolated in the presence of $Na_2S_2O_4$.[282]

It is generally assumed that molybdenum is at the enzyme's substrate-binding and -reducing site, but the evidence for this assumption is all circumstantial, and the function of the Mo is shrouded in mystery.[291] A considerable step towards an understanding of this function was the isolation and characterization of a molybdenum-containing cofactor, FeMoco,[288] which was apparently distinct from the molybdenum cofactors isolated from xanthine oxidase and nitrate reductase,[292] and which contained molybdenum, iron, and acid-labile sulphur in the ratios 1 : 8 : 6.[288] At least some of the iron in the cofactor was associated with the centre with the $g = 4.3$, 3.7, and 2.01 e.p.r. signal, but the centre was found to be very unusual when compared with the Fe–S centres found in the ferredoxins since it was resistant to attack by thiophenol, acid, and 2,2′-bipyridyl. However, the centre was destroyed by mercurials.[293]

The smaller nitrogenase protein (Fe protein) has a molecular weight of 57 000—68 000, depending on its source, with two (apparently equivalent) subunits[282] and a single Fe_4S_4 cluster.[266, 289] The first sequence of a nitrogenase protein, the Fe protein from *Clostridium pasteurianum*, was determined, and it was confirmed that the subunits were identical but that no sequence homology existed between the Fe protein and other iron-containing proteins. The six cysteine residues per monomer were not 'grouped' as in the ferredoxins.[294]

The Fe protein in extracts from some organisms was apparently inactive, but was activated by incubation with a trypsin-sensitive, oxygen-sensitive membrane component in the presence of Mg^{2+} and ATP. Mn^{2+} enhanced the activation.[295]

Mechanistic Studies. ATP as its monomagnesium salt binds to the Fe protein and activates electron transfer from this protein to the MoFe protein.[282]

Water proton n.m.r. relaxation studies on the binding of Mn^{2+} and Mg^{2+} to the Fe protein showed that both metal ions were bound to four, apparently

[288] V. K. Shah and W. J. Brill, *Proc. Nat. Acad. Sci. U.S.A.*, 1977, **74**, 3249.

[289] B. E. Smith and G. Lang, *Biochem. J.*, 1974, **137**, 169.

[290] E. Münck, H. Rhodes, W. H. Orme-Johnson, L. C. Davis, W. J. Brill, and V. K. Shah, *Biochim. Biophys. Acta*, 1975, **400**, 32.

[291] B. E. Smith, *J. Less-Common Metals*, 1977, **54**, 465.

[292] P. T. Peinkos, V. K. Shah, and W. J. Brill, *Proc. Nat. Acad. Sci. U.S.A.*, 1977, **74**, 5468.

[293] J. Rawlings, V. K. Shah, J. R. Chisnell, W. J. Brill, R. Zimmermann, E. Münck, and W. Orme-Johnson, *J. Biol. Chem.*, 1978, **253**, 1001.

[294] M. Tanaka, M. Haniu, K. T. Yasunobu, and L. E. Mortenson, *J. Biol. Chem.*, 1977, **252**, 7081, 7089, 7093.

[295] P. W. Ludden and R. H. Burris, *Science*, 1976, **194**, 424; Y. Okon, J. P. Houchins, S. L. Albrecht, and R. H. Burris, *J. Gen. Microbiol.*, 1977, **98**, 87; S. Nordland, U. Erickson, and H. Baltscheffsky, *Biochim. Biophys. Acta*, 1977, **462**, 187.

equivalent, sites with dissociation constants of 0.35 ± 0.05 and 1.7 ± 0.3 mmol l^{-1} respectively.[296] Earlier gel-equilibration data indicated two binding sites for MgATP,[297] but more recent studies using the same technique have suggested 3—5 binding sites.[298] Unfortunately, it was not possible to determine whether the metal nucleotide (ATP or ADP) complexes were also bound at four sites from the proton-relaxation data, but the data did indicate that ternary complexes

with the configuration EMS or $E\begin{smallmatrix} M \\ \diagdown \\ \diagup \\ S \end{smallmatrix}$ were formed with the Fe protein.[296] A

similar, independent study indicated that the structure of these ternary complexes altered in the presence of the MoFe protein.[299]

There is increasing evidence that ATP, in addition to activating electron transfer from the Fe protein to the MoFe protein, has another role more intimately associated with substrate reduction.[300] Further evidence for this hypothesis came from studies on inhibition by the product MgADP. MgADP was a competitive inhibitor of the MgATP-activated, pre-steady-state, protein–protein electron transfer, but no heterotropic interactions were observed. However, the steady-states kinetics of H_2 evolution by the enzyme did show heterotropic interactions between MgATP and MgADP. The data suggested that the enzyme could bind up to two molecules of either MgATP or MgADP but was unable to bind both nucleotides together.[301] Kinetic studies of dithionite ion utilization and ATP hydrolysis by nitrogenase yielded a rate law which was interpreted as indicating two binding sites for MgATP.[302] In accordance with earlier studies,[303] $SO_2^{\cdot-}$, formed by dissociation of $S_2O_4^{2-}$, was the actual electron donor to nitrogenase.[302] An Arrhenius plot for ATP hydrolysis by the enzyme did not show the characteristic break[282] near 20 °C that was observed for substrate reduction and $S_2O_4^{2-}$ utilization.[302] The break in the Arrhenius plot for substrate reduction was eliminated by treating the enzyme from *Azotobacter vinelandii* with phospholipase A, but it reappeared after incubating the enzyme with specific phospholipids for one hour.[304]

The reaction of the *A. vinelandii* nitrogenase complex with *p*-chloromercuribenzoate resulted in inactivation of the Fe protein, indicating that the Fe protein had accessible sulphydryl groups which were necessary for enzymic activity.[305]

296 E. O. Bishop, M. D. Lambert, D. Orchard, and B. E. Smith, *Biochim. Biophys. Acta*, 1977, **482**, 286.
297 M. Y. Tso and R. H. Burris, *Biochim. Biophys. Acta*, 1973, **309**, 263.
298 P. Wyeth and W. H. Orme-Johnson, personal communication.
299 L. A. Syrtsova, T. N. Pisarskaya, I. I. Nazarova, A. M. Uzenskaya, and G. I. Likhtenstein, *Bioorg. Khim.*, 1977, **3**, 1251.
300 B. E. Smith, R. N. F. Thorneley, R. R. Eady, and L. E. Mortenson, *Biochem. J.*, 1976, **157**, 439.
301 R. N. F. Thorneley and A. Cornish-Bowden, *Biochem. J.*, 1977, **165**, 255.
302 G. D. Watt and A. Burns, *Biochemistry*, 1977, **16**, 264.
303 R. N. F. Thorneley, M. G. Yates, and D. J. Lowe, *Biochem. J.*, 1976, **155**, 137.
304 F. Ceuterick, K. Heremans, and H. DeSmedt, *Arch. Int. Physiol. Biochim.*, 1977, **85**, 394.
305 A. P. Sadkov, A. I. Kotel'nikov, and R. I. Gvozdev, *Izvest. Akad. Nauk S.S.S.R.*, *Ser. biol.*, 1977, 610.

These groups were used for electron-density labelling in an electron-microscopic study of the nitrogenase and its metal clusters.[306]

The stoicheiometry of the active nitrogenase complex has not yet been resolved. Studies with the proteins from *Klebsiella pneumoniae* and *Azotobacter chroococcum* were interpreted in terms of a 1 : 1 Fe protein–MoFe protein complex.[307] However, a catalytically inactive 2 : 1 complex was formed from the Fe protein of *Clostridium pasteurianum* and the MoFe protein of *A. vinelandii*. The 1 : 1 complex was still active in the presence of *A. vinelandii* Fe protein, and kinetic data indicated competitive binding of the Fe proteins, thus implying that a fully active complex had 2 : 1 Fe protein : MoFe protein stoicheiometry.[308]

When the Fe protein of *C. pasteurianum* and the MoFe protein of *K. pneumoniae* were combined and assayed, the hydrogen-evolving activity was linear with time, whereas the acetylene-[300] and dinitrogen[309]-reducing activities showed lags of 8 min and 35 min respectively before becoming linear with time. These data were interpreted in terms of different enzymic species of this heterologous nitrogenase being required for the reduction of protons, acetylene, and dinitrogen, activation steps being required for the formation of the species reducing the last two substrates. Similar results have now been observed for the proton- and acetylene-reducing activities of the homologous *K. pneumoniae* nitrogenase at low temperature (10 °C) or high MoFe protein : Fe protein ratio. The lag phase for acetylene reduction was accompanied by a complementary burst phase for concomitant hydrogen evolution.[310]

An e.p.r.-active transient species, associated with the MoFe protein, was observed during steady-state catalysis by *A. chroococcum* nitrogenase. The e.p.r. signal was rhombic, with *g* values at 2.14, 2.001, and 1.976. The reducible substrates C_2H_2, HN_3, and HCN (but not N_2) eliminated the signal.[311] Additional transient species, also associated with the MoFe protein, were observed in the presence of the inhibitor CO. Their *g* values were 2.08, 1.97, and 1.92 (0.01 atm CO) and 2.16, 2.07, and 2.04 (0.2 atm CO).[311, 312] The significance of these transient species is not yet clear.

Xanthine Oxidase and Xanthine Dehydrogenase.—Milk xanthine oxidase is by far the best understood of the molybdenum-containing enzymes, and it is not possible here to do justice to the extensive earlier work recently reviewed by Bray.[313] Xanthine dehydrogenase and aldehyde oxidase are closely similar to

[306] L. A. Levchenko, A. V. Raevskii, G. I. Likhtenstein, A. P. Sadkno, and T. S. Pivovarova, *Biokhimiya (Moscow)*, 1977, **42**, 1755.
[307] R. N. F. Thorneley, R. R. Eady, and M. G. Yates, *Biochim. Biophys. Acta*, 1975, **403**, 269.
[308] D. W. Emerich and R. H. Burris, *Proc. Nat. Acad. Sci. U.S.A.*, 1976, **73**, 4369.
[309] B. E. Smith, R. R. Eady, R. N. F. Thorneley, M. G. Yates, and J. R. Postgate, in 'Recent Developments in Nitrogen Fixation'; Proceedings of the 2nd International Symposium at Salamanca, 1976, ed. W. Newton, J. R. Postgate, and C. Rodriguez-Barrueco, Academic Press, London, New York, and San Francisco, 1977, 191.
[310] R. N. F. Thorneley and R. R. Eady, *Biochem. J.*, 1977, **167**, 457.
[311] M. G. Yates and D. J. Lowe, *F.E.B.S. Letters*, 1976, **72**, 121.
[312] R. H. Burris and W. Orme-Johnson, in 'Proceedings of the 1st International Symposium on Nitrogen Fixation, 1974', ed. W. E. Newton and C. J. Nyman, Washington State University Press, 1976, Vol. 1, p. 208.
[313] R. C. Bray, in 'The Enzymes', ed. P. Boyer, 3rd edn., Academic Press, New York, 1975, Vol. XII, p. 299.

xanthine oxidase in structure and substrate specificity. The reactions catalysed by this group of enzymes are generally of the form:

$$RH \xrightarrow{-2e, -H^+, + OH^-} ROH$$

where RH is the reducing substrate. The oxygen introduced into RH is derived from water.

Reactions. Xanthine oxidase has a wide substrate specificity. It was found to oxidize a number of uncharged and anionic methyl derivatives of pteridin-4-one at a high rate. Pteridin-4-one (3) was oxidized to the 4,7-dione. Attack at C-2 in the pyrimidine ring was enhanced by methylation at C-6, or by the presence of a double bond between C-2 and N-3 ,whereas attack at C-7 was enhanced by a double bond between C-6 and C-7.[314] An 8-phenyl group markedly reduced the rate of oxidation of hypoxanthine (4), but increased the rate of oxidation of 6-thioxopurine (5).[315] The bismuth (but not the palladium) complex of 6-thioxo-purine could be oxidized to 6-thiouric acid.[316] The slow rate of enzymic oxidation of 3-hydroxyxanthine to 2-hydroxyuric acid was doubled by the formation of the 1-methyl derivative.[317]

    (3) Pteridin-4-one   (4) Hypoxanthine   (5) 6-Thioxopurine

Structure and Function. The enzyme contains one molybdenum atom, one FAD molecule, and one each of two (2Fe–2S) centres (FeSI and FeSII) in each half of a dimeric molecule of molecular weight 300 000.[313] Unless prepared by affinity chromatography,[318] the preparations always contain one or more non-functional forms. The most prominent of these are the naturally occurring demolybdo-form and the desulpho-form. The desulpho-enzyme might be prepared by the reaction of the native enzyme with cyanide, with the formation of free thiocyanate, or it may occur spontaneously, by loss of sulphur, during purification or storage. It was proposed that the sulphur atom in question originated from a persulphide group (R—S—S⁻) that is present at the active centre.[319] Recently, the hypothesis that it is derived from a cysteine residue, bound through the sulphur atom to molybdenum, has been postulated.[320] Central to the new theory is the formation of cyanoalanine from the proposed cysteine residue when the enzyme is attacked by cyanide. It should be possible to check this by using ¹⁴CN-labelling. Unfortunately, reports of experiments of this type are contradictory.[320]

314 F. Bergmann, L. Levene, I. Tamir, and M. Rahat, *Biochim. Biophys. Acta*, 1977, **480**, 21.
315 F. Bergmann, L. Levene, H. Govrin, and A. Frank, *Biochim. Biophys. Acta*, 1977, **480**, 39.
316 R. N. Lewis and G. Olivier, *Res. Comm. Chem. Pathol. Pharmacol.*, 1977, **18**, 377.
317 F. Bergmann and L. Levene, *Biochim. Biophys. Acta*, 1977, **481**, 359.
318 D. Edmondson, V. Massey, G. Palmer, L. M. Beacham, and G. B. Elion, *J. Biol. Chem.*, 1972, **247**, 1597.
319 V. Massey and D. Edmondson, *J. Biol. Chem.*, 1970, **245**, 6595.
320 M. P. Coughlan, *F.E.B.S. Letters*, 1977, **81**, 1.

Formaldehyde was also found to inactivate xanthine oxidase, possibly by formylation of a nitrogen ligand of molybdenum.[321] Studies with turkey liver xanthine dehydrogenase showed that, while the molybdenum environment was certainly affected, possibly as postulated, formaldehyde probably also bound to other sites on the enzyme, and affected the flavin loci.[322]

A new non-functional form of the enzyme, containing stable Mo^V, was prepared by the reaction of the desulpho-form with ethylene glycol. The e.p.r. signal of this species had g values at 1.980, 1.973, and 1.967, and was very similar to signals previously reported from turkey liver xanthine dehydrogenase and rabbit liver aldehyde oxidase. It was suggested that the species derived from reaction with ethylene glycol might contain a $-COCH_2OH$ residue bound to a nitrogen ligand of the molybdenum,[323] as postulated for formaldehyde.[321]

It is believed that, during enzyme turnover, reducing equivalents from a substrate such as xanthine are transferred to molybdenum, which cycles between Mo^{VI}, Mo^V, and Mo^{IV} oxidation states, and are then rapidly distributed between all the reducible centres of each half-molecule before being transferred to oxygen or other oxidizing substrates. The relative extents of reduction of the various centres depend on their relative redox potentials. The redox potentials of all of the centres of milk xanthine[324] oxidase and turkey liver xanthine dehydrogenase[325] were measured. The major difference between the two enzymes was in the $FADH \cdot \rightleftharpoons FADH_2$ redox potential, which was -236 ± 20 mV for the oxidase and -366 ± 20 mV for the dehydrogenase. Very little $FADH_2$ was formed during turnover of the dehydrogenase, and since only $FADH_2$ was expected to react with O_2, the lack of oxidase activity of the dehydrogenase was explained.

The molybdenum cofactor of xanthine oxidase is discussed below.

Nitrate Reductase.—Nitrate ion is used by living organisms as a source of nitrogen for growth and as a terminal electron acceptor. These processes are mediated by assimilatory and respiratory nitrate reductases respectively. The structures of nitrate reductases are very variable, almost the only unifying factor being the presence of molybdenum. Some enzymes contain a b-type cytochrome and others contain non-haem iron and acid-labile sulphur groups. NAD(P)H is frequently used as an electron donor, but ferredoxins are also used. FAD is important in the functioning of some enzymes. An assay of the total enzymic activity commonly measures the enzyme-mediated reduction of nitrate to nitrite by NAD(P)H. However, partial activities of the enzyme can also be measured, *viz.* the reduction of cytochrome c by NAD(P)H (this dehydrogenase activity does not involve molybdenum) and the reduction of nitrate by reduced methyl or benzyl viologen; this activity involves molybdenum but is independent of NAD(P)H and flavin. For a description of earlier work, the reader is referred to the reviews of Losada,[256] Hewitt,[326] and Stouthamer.[327]

[321] F. M. Pick, M. A. McGartoll, and R. C. Bray, *European J. Biochem.*, 1971, **18**, 65.
[322] I. N. Fhaoláin and M. P. Coughlan, *Internat. J. Biochem.*, 1977, **8**, 441.
[323] D. J. Lowe, M. J. Barber, R. T. Pawlik, and R. C. Bray, *Biochem. J.*, 1976, **155**, 81.
[324] R. Cammack, M. J. Barber, and R. C. Bray, *Biochem. J.*, 1976, **157**, 469.
[325] M. J. Barber, R. C. Bray, R. Cammack, and M. P. Coughlan, *Biochem. J.*, 1977, **163**, 279.
[326] E. J. Hewitt, *Ann. Rev. Plant Physiol.*, 1975, **26**, 73.
[327] A. H. Stouthamer, *Adv. Microbiol. Physiol.*, 1976, **14**, 315.

Spectral studies on the purified nitrate reductase from spinach indicated the presence of a b-type cytochrome.[328] The nitrate-induced nitrate reductase in the cytoplasmic membrane of *Azotobacter chroococcum* may also contain cytochrome b_1 as a cofactor, since this moiety was reduced by $Na_2S_2O_4$ and then re-oxidized by nitrate.[329]

The electron-transport pathway of the assimilatory nitrate reductase of *Neurospora crassa* was determined as:

$$NADPH \longrightarrow FAD \longrightarrow cytochrome\ b_{557} \longrightarrow Mo \longrightarrow NO_3^-$$

NADPH dehydrogenase activity was lost when the NADP analogue 3-aminopyridine–adenine dinucleotide was covalently linked to the enzyme, which, however, could still use reduced methyl viologen or $FADH_2$ as electron donors for the reduction of nitrate. Sulphydryl group activity was also lost in the inactivated enzyme, but the inactivation could be prevented by oxidizing the enzyme with FAD. The results suggested the presence of a functional sulphydryl group in the electron-transport chain,[330] *i.e.*

$$NADPH \longrightarrow -SH \longrightarrow FAD \longrightarrow cytochrome\ b_{557} \longrightarrow Mo \longrightarrow NO_3^-$$

The soluble assimilatory nitrate reductase from *Acinetobacter calcoaceticus* was characterized as a molybdenum enzyme of molecular weight 96 000 that could use reduced viologen [but not NAD(P)H] as the electron donor. Nitrate reductase activity was lost in the presence of $Na_2S_2O_4$ at temperatures above 10 °C. This inactivation could be prevented or reversed by 2 mM-KCNO.[331]

There is evidence that the respiratory enzyme from *Escherichia coli* is associated with the membrane phospholipids *via* —SH groups.[332] This enzyme has been reported to contain 20 atoms of non-haem iron and acid-labile sulphur, with 1—2 molybdenum atoms, in a unit that has a molecular weight of about 320 000.[333] When released from the membrane fraction by heat treatment, it was electrophoretically inhomogeneous, owing to attack by a proteolytic enzyme.[334] After treatment with trypsin, the enzyme had a molecular weight of $\simeq 200\ 000$ and retained full nitrate reductase activity with methyl viologen as reductant. However, it had lost the ability to aggregate into oligomers.[334]

Two e.p.r. signals from molybdenum(V),[335] but, contrary to an earlier report,[336] none from molybdenum(III), were observed with the *E. coli* enzyme. One of the molybdenum(V) signals showed interaction with a proton that was exchangeable with the solvent. Both molybdenum(V) signals could be reduced by dithionite and re-oxidized by nitrate, but it was not clear which, if either, was associated with an intermediate in the enzyme-catalysed reaction.[335]

[328] B. A. Notton, R. J. Fido, and E. J. Hewitt, *Plant Sci. Letters*, 1977, **8**, 165.
[329] R. Vila, J. A. Bárcena, A. Llobell, and A. Paneque, *Biochem. Biophys. Res. Comm.*, 1977, **75**, 682.
[330] N. K. Amay, R. H. Garett, and B. M. Anderson, *Biochim. Biophys. Acta*, 1977, **480**, 83.
[331] A. Villalobo, J. M. Roldán, J. Rivas, and J. Cardenas, *Arch. Microbiol.*, 1977, **112**, 127.
[332] E. Azoulay, C. Rivière, G. Giordano, J. Pommier, M. Denis, and G. Ducet, *F.E.B.S. Letters*, 1977, **79**, 321.
[333] P. Forget, *European J. Biochem.*, 1974, **42**, 325.
[334] J. A. DeMoss, *J. Biol. Chem.*, 1977, **252**, 1696.
[335] R. C. Bray, S. P. Vincent, D. J. Lowe, R. A. Clegg, and P. B. Garland, *Biochem. J.*, 1976, **155**, 201.
[336] D. V. DerVartanian and P. Forget, *Biochim. Biophys. Acta*, 1975, **379**, 74.

The molybdenum cofactor of nitrate reductase is discussed below.

Sulphite Oxidase.—Sulphite oxidase is a molybdo-enzyme containing haem. It catalyses the oxidation of the sulphite that is generated in the oxidative degradation of sulphur-containing amino-acids. The enzyme is a dimer of subunits that each have a molecular weight of 57 000, and it can utilize various electron acceptors, including cytochrome *c*, ferricyanide, and molecular oxygen.[337]

Tryptic digestion of the enzyme from rat liver resulted in loss of the ability to oxidize sulphite with cytochrome *c* as the electron acceptor, whereas sulphite → O_2 activity was partially and sulphite → ferricyanide activity was completely retained. Two fragments were separated from the digest. One of molecular weight 9500, and derived from the *N*-terminus of the enzyme, contained the haem (cytochrome b_5). It mediated electron transfer between NADH–cytochrome b_5 reductase and cytochrome *c*, thus demonstrating its function. The other fragment was a dimer of subunits of molecular weight 47 000, containing all the molybdenum of the enzyme and retaining its e.p.r. properties. Reduction by sulphite generated a spectrum attributable to Mo^V.[338]

Chymotryptic digestion of the chicken liver enzyme led to similar observations. A haem-binding fragment of molecular weight about 11 000, derived from the *N*-terminus, was isolated and characterized. The *N*-terminal sequence of sulphite oxidase showed a strong similarity to those of liver microsomal cytochrome b_5 and baker's yeast cytochrome b_2 core. A common evolutionary origin for the three *b*-type cytochromes was suggested.[339]

The molybdenum cofactor of sulphite oxidase is discussed below.

The Molybdenum Cofactors (Moco) from Nitrate Reductase, Xanthine Oxidase, and Sulphite Oxidase.—Early genetic work with the fungus *Aspergillus nidulans* indicated that there was a common cofactor, *cnx*, possibly containing molybdenum, for nitrate reductase and xanthine dehydrogenase.[340] Later work with the nitrate-reductase-deficient mutant strains *nit*-1, *nit*-2, and *nit*-3 of the fungus *Neurospora crassa* showed that an active NADPH–nitrate reductase could be obtained by incubation *in vitro* of *nit*-1 crude extracts with extracts from *nit*-2, *nit*-3, or uninduced wild-type.[341] Acid-treated molybdenum enzymes, including nitrogenase, were also reported to complement *nit*-1 extracts. These results led to the hypothesis that all molybdenum-containing enzymes had a common cofactor, *i.e.* Moco.[342] This cofactor was dialysable (*i.e.* with a molecular weight < 5000) and very unstable, resisting all attempts to isolate and characterize it.

Shah and co-workers have now isolated both FeMoco from nitrogenase[288] (see above) and Moco from nitrate reductase and xanthine oxidase.[292] FeMoco and Moco are distinct, Moco being the smaller. Moco did not complement *nit B* nitrogenase mutants that were deficient in FeMoco, and FeMoco did not

[337] J. L. Johnson and K. V. Rajagopalan, *J. Clin. Invest.*, 1976, **58**, 543.
[338] J. L. Johnson and K. V. Rajagopalan, *J. Biol. Chem.*, 1977, **252**, 2017.
[339] B. Guiard and F. Lederer, *European J. Biochem.*, 1977, **74**, 181.
[340] J. A. Pateman, D. J. Cove, B. M. Reirer, and D. B. Roberts, *Nature*, 1964, **201**, 58.
[341] A. Nason, A. D. Antoine, P. A. Ketchum, W. A. Frazier, and K.-Y. Lee, *Proc. Nat. Acad. Sci. U.S.A.*, 1970, **65**, 137.
[342] A. Nason, K.-Y. Lee, S.-S. Pan, P. A. Ketchum, A. Lamberti, and J. de Vries, *Proc. Nat. Acad. Sci. U.S.A.*, 1971, **68**, 3242.

complement *nit*-1; earlier results to the contrary[342] were due to the use of impure nitrogenase preparations as the source of the cofactor. However, Moco from xanthine oxidase did complement *nit*-1, indicating that the cofactors from xanthine oxidase and nitrate reductase were identical.[292]

The genetics of Moco synthesis are extremely complex, with the products of five loci on the *Aspergillus nidulans* chromosome being involved in concerted action to form Moco.[343]

A molybdenum-containing complex (MCC), obtained from spinach nitrate reductase by treatment at pH 2.5, activated an apoprotein derived from molybdenum-deficient plants grown with tungsten. The apoprotein retained cytochrome *c* reductase activity associated with two species, these sedimenting with values of 3.75 or 8.15 on sucrose-density centrifugation. An apoprotein activating factor of molecular weight between 10 000 and 30 000 could also be extracted from nitrate reductase absorbed on AMP-Sepharose. This factor could be further fractionated into two fragments of different sizes, but with the loss of activating ability. The relationships between this factor, MCC, and Moco are not known.[344]

Demolybdo sulphite oxidase from rat liver was prepared from molybdenum-deficient rats through the use of tungsten as a competitive molybdenum antagonist. Some of the molybdo-enzyme contained tungsten, and this fraction could be activated by incubation with molybdate.[345] The apoprotein with vacant molybdenum sites was not activated by molybdate, but it could be activated by an unstable molybdenum cofactor that was isolated from mitochondrial outer membranes. This apoprotein fraction was also activated by molybdenum cofactors from a variety of rat tissues, *E. coli*, *N. crassa*, and human tissue and from acidified milk xanthine oxidase.[346]

The above data indicate that xanthine oxidase, nitrate reductase, and sulphite oxidase have interchangeable, and possibly identical, molybdenum cofactors, but detailed evidence on the structure of these cofactors is still lacking.

[343] R. H. Garrett and D. J. Cove, *Molec. Gen. Genetics*, 1976, **149**, 179.
[344] E. J. Hewitt, B. A. Notton, and G. J. Rucklidge, *J. Less-Common Metals*, 1977, **54**, 537.
[345] H. P. Jones, J. L. Johnson, and K. V. Rajagopalan, *J. Biol. Chem.*, 1977, **252**, 4988.
[346] J. L. Johnson, H. P. Jones, and K. V. Rajagopalan, *J. Biol. Chem.*, 1977, **252**, 4994.

8
Metalloenzymes

A. GALDES & H. A. O. HILL

We are concerned in this chapter with those metalloenzymes in which the metal ion does not itself undergo a redox reaction during catalysis. The ability of metal ions to provide electrophilic centres in enzymes and in most cases to do so under conditions (*e.g.* pH) relevant to biological processes lends itself to a number of processes, *e.g.* the activation of amino-acid residues in proteins, the co-ordination and polarization of substrates, coenzymes, solvent, *etc.*, or any combination of these. Most metal ions can function in this manner, albeit less effectively, in simple co-ordination compounds. However, in biology, though all may be called, few are chosen. For example, the Lewis acidity (assuming this is relevant) of copper(II) ion is at least the equal of any other bivalent cation of the first transition series. Though copper plays a major role in many electron-transfer processes, it does not appear to function in any non-redox metalloprotein. The simplest rationale of the selection of metal ions for metalloenzymes would be that those metal ions which can readily take part in redox reactions under normal biological conditions are *excluded* from those proteins in which they would be required to have a role related to their Lewis acidity. Obviously there are exceptions to this generalization, such as cobalt(II) in oxaloacetate transcarboxylases from *P. shermanii*.[1] Nevertheless it appears that biology has chosen to use magnesium(II), calcium(II), manganese(II), nickel(II), and first and foremost zinc(II) to fulfill the role of Lewis acids in metalloenzymes. [The redox chemistry of manganese(II) and nickel(II) is not readily expressed under 'normal' biological conditions, though it is possible to alter the redox chemistry by co-ordination. Thus the role of manganese in superoxide dismutase or in the chloroplast is presumably related to its redox chemistry.] Of these ions, zinc(II) plays the most important role *in metalloenzymes*, as opposed[2] to *metal-activated enzymes*. There are many reasons why zinc(II) can play such a role: it is a reasonable Lewis acid, yet it does not readily hydrolyse (form hydroxo-derivatives at pH 7); it is fairly catholic in its choice of ligand atoms, being found in complexes in which oxygen, nitrogen, or sulphur atoms co-ordinate to it; there are no special constraints, apart from those controlling the stereochemistry of all metal complexes, determining one particular co-ordination number or geometry; the rates of substitution reactions of zinc (II) are fast; and, of course, it does not take part in redox reactions under conditions pertinent to biology. It is therefore not surprising that the literature is dominated in this area by the study of zinc metalloenzymes.

[1] D. B. Northrop and H. G. Wood, *J. Biol. Chem.*, 1969, **244**, 5801.
[2] B. L. Vallee, *Adv. Protein. Chem.*, 1955, **10**, 317; *ibid.*, 1961, **16**, 401.

Magnesium-dependent enzymes are also much studied, and particularly relevant to this Report are investigations in which magnesium is replaced by manganese(II).

1 Zinc Metalloenzymes

The first recorded indication[3] that zinc was a required element for a living organism was made in 1869 when Raulin showed that zinc deficiency retarded the growth of *Aspergillus niger*. Hard on the heels of that report came the observation that zinc was a constituent of plants,[4] vertebrates,[4] and animals.[5] No more was heard for over half a century until Bertrand and Bhattacherjee provided[6] conclusive evidence that zinc is essential for the normal growth of rodents. Today, a minimum hypothesis requires that zinc is essential[7-9] for all forms of life.

The second period in the development of zinc in biology dates from the identification,[10] in 1940, of this element in carbonic anhydrase. Over a decade followed before a second zinc metalloenzyme was identified,[11] *viz.* carboxypeptidase. Now over eighty, zinc-containing metalloenzymes have been identified, representing each of the six categories of enzymes designated by the International Union on Biochemistry (I.U.B.) Commission on enzymes (Table 1). Carbohydrates, lipids, proteins, and nucleic acids are synthesized or degraded by processes which require[12] (Figure 1) zinc metalloenzymes as catalysts. Two penetrating articles by B. L. Vallee, who has provided much of the momentum behind the advance of the subject, have been published.[12, 13] Other useful articles include a discussion[14] in depth of the best-characterized zinc metalloenzymes, a timely and brief summary[15] of recent developments, a more extensive discussion[16] of the biochemistry of zinc metalloenzymes relevant to biological problems, an article[17] particularly concerned with zinc and micro-organisms, and a detailed discussion[18] of zinc chemistry and biochemistry. A detailed discussion of publications relevant to the mechanisms of well-characterized zinc metalloenzymes is given in Chapter 2 of these Reports.

³ J. Raulin, *Ann. Sci. Natl. Botan. et Biol. Vegetale*, 1869, **11**, 93.
⁴ G. Lechartier and G. Bellamy, *Compt. rend.*, 1877, **84**, 687.
⁵ F. Raoult and H. Breton, *Compt. rend.*, 1877, **85**, 40.
⁶ G. Bertaud and R. C. Bhattacherjee, *Compt. rend.*, 1934, **198**, 1823.
⁷ 'Trace Elements in Human Health and Disease', ed. A. S. Prasad, Academic Press, New York, 1976.
⁸ E. J. Hewitt and T. A. Smith, 'Plant Mineral Nutrition', The English Universities Press, London, 1975.
⁹ E. J. Underwood, 'Trace Elements in Human and Animal Nutrition', Academic Press, New York, 4th edn., 1977.
¹⁰ D. Keilen and T. Mann, *Biochem. J.*, 1940, **34**, 1163.
¹¹ B. L. Vallee and H. Neurath, *J. Biol. Chem.*, 1955, **217**, 253.
¹² B. L. Vallee, *Trends Biochem. Sci.*, 1976, **1**, 88.
¹³ B. L. Vallee, in 'Biological Aspects of Inorganic Chemistry', ed. A. W. Addison, W. R. Cullen, D. Dolphin, and B. R. James, Wiley-Interscience, New York, 1977.
¹⁴ J. F. Chlebowski and J. E. Coleman in 'Metal Ions in Biological Systems', ed. H. Sigel, Vol. 6. Marcel Dekker, New York, 1976, Vol. 6, p. 77.
¹⁵ J. F. Riordan, *Ann. Clin. Lab. Sci.*, 1977, 7, 119.
¹⁶ J. F. Riordan and B. L. Vallee in 'Trace Elements in Human Health and Disease', ed A. S. Prasad, Academic Press, New York, 1976.
¹⁷ M. L. Failla, in 'Microorganisms and Minerals', ed. E. D. Weinberg, Marcel Dekker, New York, 1977, p. 151.
¹⁸ M. R. Dunn, *Structure and Bonding*, 1975, **23**, 61.

Table 1 *Zinc-containing metalloenzymes*

Enzyme	E.C. Number	Source
Class 1: *Oxidoreductases*		
Alcohol dehydrogenase	1.1.1.1	Plants, yeast, vertebrates
D-Lactate cytochrome reductase	1.1.2.4	Yeast
Superoxide dismutase	1.15.1.1	Plants, yeast, vertebrates
Class 2: *Transferases*		
Transcarboxylase	2.1.3.1	*P. shermanii*
Aspartate transcarbamalase	2.1.3.2	*E. coli*
Phosphoglucomutase	2.7.5.1	Yeast
RNA polymerase I and II	2.7.7.6	*E. gracilis*, yeast, plants
DNA polymerase I and II	2.7.7.7	*E. coli*, invertebrates
Deoxynucleotidyl transferase	2.7.7	Vertebrates
Reverse transcriptase	2.7.7	Oncogenic viruses
Mercaptopyruvate sulphur transferase	2.8.1.2	*E. coli*
Class 3: *Hydrolases*		
Alkaline phosphatase	3.1.3.1	*E. coli*, vertebrates
Phospholipase C	3.1.4.3	*B. cereus*
α-D-Mannosidase	3.2.1.24	Plants, vertebrates
Leucine aminopeptidase	3.4.11.1	Yeast, vertebrates
Carboxypeptidase A	3.4.12.2	Vertebrates
Carboxypeptidase B	3.4.12.3	Vertebrates
Carboxypeptidase G1	3.4.12	*P. stutzeri*
Dipeptidase	3.4.13.11	Vertebrates
Neutral protease	3.4.24.4	Micro-organisms
Collagenase	3.4.24.3	*Clostridium histolyticum*, vertebrates
β-Lactamase II	3.5.2.8	*B. cereus*
Class 4: *Lyases*		
Aldolase	4.1.2.13	Bacteria, fungi
L-Rhamnulose-1-phosphate aldolase	4.1.2.9	*E. coli*
Carbonic anhydrase	4.2.1.1	Vertebrates, plants
δ-Aminolaevulinic acid dehydratase	4.2.1.24	Vertebrates
Glyoxalase I	4.4.1.5	Vertebrates, yeast
Class 5: *Isomerases*		
Phosphamannose isomerase	5.3.1.8	Yeast
Class 6: *Ligases*		
tRNA synthetase	6.1.1	*E. coli*
Pyruvate carboxylase	6.4.1.1	Yeast

Carbonic Anhydrase —Carbonic anhydrase is a zinc metalloenzyme present in animals, plants, and certain micro-organisms which catalyses the reversible hydration of carbon dioxide ($CO_2 + H_2O \rightleftharpoons HCO_3^- + H^+$). In addition to this physiological reaction, carbonic anhydrase also catalyses the hydrolysis of many esters and several aldehydes. Those best characterized, the bovine and human enzymes, are monomeric, containing 1 gram atom of tightly bound zinc per 30000 molecular weight.[19] X-Ray crystallographic studies[20] of the human

[19] S. Lindskog, L. E. Henderson, K. K. Kannan, A. Liljas, P. O. Nyman and B. Strandberg, in 'The Enzymes', ed. P. D. Boyer, Academic Press, New York, 1971, Vol. 5, p. 587.
[20] K. K. Kannan, B. Notstrand, K. Fridborg, S. Lougren, A. Ohlessen, and M. Petef, *Proc. Nat. Acad. Sci., U.S.A.*, 1975, **72**, 51.

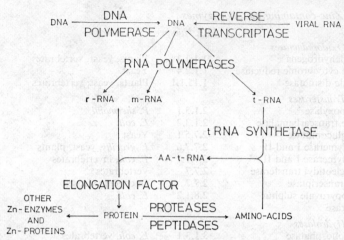

Figure 1 *Zinc-dependent processes in nucleic acid and protein metabolism*

enzyme reveal that the zinc ion lies near the bottom of a deep cleft, co-ordinated to three histidyl residues in a distorted tetrahedral geometry, with the fourth co-ordination site presumably occupied by a water molecule.

Despite the fact that carbonic anhydrase was the first zinc metalloenzyme identified,[10] there is still controversy about the nature of the various active-site species and the detailed mechanisms of their action (see Chapter 2). In particular, the identity of the group with a $pK_a \approx 7$ that is involved in the mechanism, and the stereochemistry around the zinc ion during catalysis, are still in dispute. The various mechanisms proposed assume either ionization of a histidine group (bound or not to the zinc) and nucleophilic attack on CO_2 by the co-ordinated imidazolate anion,[21,22] or ionization of the Zn^{II}-co-ordinated water and nucleophilic attack of CO_2 by OH^-.[19] Several papers relating to this problem have been published during 1977.

The crystal structure of the human carbonic anhydrase B–imidazole complex has been determined.[23] The authors conclude that the imidazole is bound weakly to the Zn^{II} in the active site, being located in a possible fifth co-ordination site, without displacing the water molecule (Figure 2). This implies that the zinc ion has four nearly tetrahedral ligand directions and a somewhat distant fifth ligand. It has in fact been shown[24] that imidazole is a unique competitive inhibitor of human carbonic anhydrase B in that its binding to the enzyme is independent of the ionizations of active-site residues. This is consistent with the proposal[23] that this inhibitor, unlike all other known inhibitors, does not displace the solvent molecule in the fourth ligand position. By analogy with this situation, Kannan and colleagues[23] propose a mechanism of action whereby the CO_2 binds weakly

[21] J. M. Pesando, *Biochemistry*, 1975, **14**, 675.
[22] R. K. Gupta and J. M. Pesando, *J. Biol. Chem.*, 1975, **250**, 2630.
[23] K. K. Kannan, M. Petef, K. Fridborg, H. Cid-Dresdner, and S. Lougren, *F.E.B.S. Letters*, 1977, **73**, 115.
[24] C. D. Strader and R. G. Khalifah, *Biochemistry*, 1977, **16**, 5717.

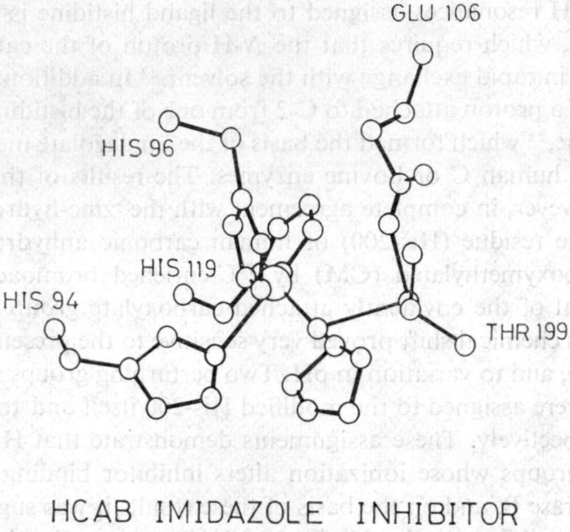

GLU 106

HIS 96

HIS 119

HIS 94

THR 199

HCAB IMIDAZOLE INHIBITOR

Figure 2 *The imidazole-binding site in human carbonic anhydrase B–imidazole complex*
(Redrawn from *F.E.B.S. Letters*, 1977, **73**, 115, by permission)

to the fifth co-ordination site of the metal ion in the hydrophobic region of the active site of the enzyme.

Some [13]C n.m.r. longitudinal relaxation studies on Co[II]-substituted human carbonic anhydrase B,[25] using ^{13}C-enriched HCO_3^- and CO_2 at several pH values, have shown that the distance of these species from the Co[II] ion is essentially invariant with pH, and is approximately 3.6 ± 0.2 Å. These results therefore indicate that the CO_2 and HCO_3^- are bound close to the metal ion, and this fits the mechanism proposed by Kannan.[23]

Ab initio molecular orbital calculations,[26] no less, on the binding of Zn[II] with imidazole (ImH), water, and carbon dioxide as well as imidazolate (Im⁻) and hydroxide have been performed. It was calculated that the bond strength increases in the order $CO_2 < OH_2 < ImH < Im^- < OH^-$. The computed bond lengths fall near 1.85—1.90 Å. The authors[26] concluded that Zn[II] can bind CO_2, though less strongly than the other ligands studied. Their results suggest that even when a close approach of CO_2 to Zn[II] in the active site of carbonic anhydrase is prevented, its interaction with the metal ion as a fifth, distant, ligand may still be favourable. As to the effect of Zn[II] on the ease of deprotonation of water and imidazole, it was concluded that whilst binding to Zn[II] greatly facilitates the process for both species, the ionization of a co-ordinated water molecule is intrinsically much more favoured than ionization of imidazole. These results therefore lend support to the 'zinc-hydroxide' mechanism.

Recent [1]H n.m.r. studies[27] on the *C*-2-H and *N*-H resonances of the histidine residues in human, rhesus monkey, and bovine carbonic anhydrases further tend to rule out the imidazolate anion model. In particular, the observation in this

[25] P. J. Stein, S. P. Merrill, and R. W. Henkens, *J. Amer. Chem. Soc.*, 1977, **99**, 3194.
[26] D. Demoulin, A. Pullman, and B. Sarkar, *J. Amer. Chem. Soc.*, 1977, **99**, 8498.
[27] I. D. Campbell, S. Lindskog, and A. I. White, *Biochim. Biophys. Acta*, 1977, **484**, 443.

study of the N-H resonances assigned to the ligand histidine is not compatible with this model, which requires that the N-H proton of the catalytically functional ligand be in rapid exchange with the solvent.[22] In addition, the downward shift with pH of a proton attached to C-2 from one of the histidine ligands in the human B enzyme,[21] which formed the basis of the imidazolate mechanism, is not observed in the human C or bovine enzymes. The results of this recent n.m.r. study[27] are, however, in complete agreement with the 'zinc-hydroxide' model.

The active-site residue (His-200) of human carbonic anhydrase B has been specifically carboxymethylated (CM) by ^{13}C-enriched bromoacetate.[28, 29] The ^{13}C n.m.r. signal of the covalently attached carboxylate group could be easily detected, and its chemical shift proved very sensitive to the presence of inhibitors in the active site, and to variation in pH. Two perturbing groups with pK_a values of 6.0 and 9.2 were assigned to the modified His-200 itself and to the Zn-bound H_2O ligand respectively. These assignments demonstrate that His-200 is one of the active-site groups whose ionization alters inhibitor binding in CM-human carbonic anhydrase B, and on the basis of these results it was suggested that this residue may also influence the binding of HCO_3^- to native human carbonic anhydrase B. It was also proposed that the carboxylate group of the CM-enzyme can interact with the Zn or its H_2O ligand and that this interaction may be responsible for the much higher pK_a of the H_2O ligand in CM-carbonic anhydrase B compound compared to that in the native enzyme (9.2 *vs.* 7.6).

Detailed spectroscopic studies[30-32] on Co^{II} bovine carbonic anhydrase alone, and in the presence of inhibitors, have led to the observation of a previously unrecorded weak absorption band ($\varepsilon \approx 10$) at 13 500 cm^{-1} in the presence of halide, acetate, and benzoate but not in their absence. This band is considered indicative of the presence of five-co-ordinate species. It was concluded that the native enzyme is four-co-ordinate at every pH value but that the halide, acetate, and benzoate complexes exist in equilibrium between four- and five-co-ordinate species.

Previous proton relaxation enhancement (PRE) studies[33] on Co^{II} carbonic anhydrase had failed to show any protons exchangeable between water bound to the Co^{II} ion and the bulk solution at low pH values, in spite of the assumed presence of a cobalt-bound water molecule. It has been demonstrated[34] that this was due to the presence of the Tris–sulphate buffer used during these studies, and if this buffer is omitted the expected PRE is readily seen. Similarly, proton relaxation enhancement of solvent was observed for the copper(II)-substituted enzyme[35] over the pH range 5.0—10.0, indicating that, even in this enzymatically inactive derivative, a metal-bound water molecule is present. Furthermore, it was found that acetate and iodide can bind to this Cu^{II} enzyme without affecting the

[28] R. G. Khalifah, *Biochemistry*, 1977, **16**, 2236.
[29] R. G. Khalifah, D. J. Strader, S. H. Bryant, and S. M. Gibson, *Biochemistry*, 1977, **16**, 2241.
[30] I. Bertini, C. Luchinat, and A. Scozzafava, *Inorg. Chim. Acta*, 1977, **22**, L23.
[31] I. Bertini, C. Luchinat, and A. Scozzafava, *J. Amer. Chem. Soc.*, 1977, **99**, 581.
[32] I. Bertini, C. Luchinat, and A. Scozzafava, *Bioinorg. Chem.*, 1977, **7**, 225.
[33] M. E. Fabry, S. H. Koenig, and W. E. Schillinger, *J. Biol. Chem.*, 1970, **245**, 4256.
[34] I. Bertini, G. Cauti, C. Luchinat, and A. Scozzafava, *Biochem. Biophys. Res. Comm.*, 1977, **78**, 158.
[35] I. Bertini, G. Cauti, C. Luchinat, and A. Scozzafava, *Inorg. Chim. Acta*, 1977, **23**, L14.

enhancement, and hence these anions presumably form five-co-ordinate complexes. Carbon-13 n.m.r. relaxation studies[36] have confirmed that acetate binds to carbonic anhydrase at sites other than the metal ion.

From a comparison of the exchange of ^{18}O between $H_2{}^{18}O$ and CO_2 it was shown[37] that, though buffer-facilitated proton transfer involving free carbonic anhydrase is compatible with the observed data for the bovine enzyme, such a buffer-facilitated proton-transfer pathway alone cannot account for the exchange that is catalysed by the human B isoenzyme.

The second-order rate constants for the removal of the zinc(II) from bovine carbonic anhydrase B have been measured,[38] using various chelating agents, at pH 5.0. The following order was found: pyridine-2,6-dicarboxylic acid $\gg$ pyridine-2-carboxylic acid > pyridine-2,3-dicarboxylic acid $\geqslant$ 1,10-phenanthroline $\gg$ 2,2'-bipyridyl ($\gg$ EDTA). Two possible mechanisms for this process were proposed: S_N1, involving a slow spontaneous dissociation of the zinc enzyme followed by a rapid reaction of free zinc ions with chelating agent, and S_N2, involving rapid formation of a ternary complex which is followed by a slow dissociation of the complex. It was concluded that 1,10-phenanthroline and the pyridinecarboxylic acid chelating agents follow an S_N2 pathway, 2,2'-bipyridyl a mixed S_N1/S_N2 pathway, while the EDTA-type chelating agents followed an S_N1 pathway. In fact, no evidence for the formation of an EDTA–enzyme ternary complex could be found. In a further paper,[39] this fact was exploited to measure the dissociation constant for Co^{II} carbonic anhydrase. This was calculated to be approx. 1.2×10^{-6} s^{-1} at pH 5.0. The formation constant for Co^{II} carbonic anhydrase at this pH was kinetically estimated to be $0.72 \, l \, mol^{-1} \, s^{-1}$, and hence a binding (association) constant of $\log K = 5.8 \pm 0.1 \, l \, mol^{-1}$ was derived. This compares very well with the value of $6.0 \pm 0.3 \, l \, mol^{-1}$ calculated from equilibrium dialysis studies. These results have led[40] to a rapid preparation of apo-carbonic anhydrase (2 hours, *vs.* the usual 10 days).

In an interesting study[41] of the folding of bovine carbonic anhydrase from the disorganized to the native structure, an azosulphonamide inhibitor was used as a probe to follow the dynamics of formation of the region of the active site. It was concluded that a tight binding site for the inhibitor forms at an intermediate stage in the folding process, prior to the recovery of the active conformation. A subsequent conformational change results in even tighter inhibitor binding and completes the formation of the active site.

A number of reports utilizing novel techniques in the study of carbonic anhydrase have appeared. Laser Raman scattering spectra for human carbonic anhydrase B have been obtained[42] and are reported to be in good quantitative agreement with *X*-ray data in the prediction of the secondary structure and the state (buried/exposed) of the tyrosine residues. No change in conformation

---

[36] I. Bertini, C. Luchinat, and A. Scozzafava, *J.C.S. Dalton*, 1977, 1962.
[37] C. Tu and D. N. Silvermann, *J. Biol. Chem.*, 1977, **252**, 3332.
[38] Y. Kidani and J. Hirose, *J. Biochem. (Japan)*, 1977, **81**, 1383.
[39] Y. Kidani and J. Hirose, *Chem. Letters*, 1977, 475.
[40] J. B. Hunt, M.-J. Rhee, and C. B. Storm, *Analyt. Biochem.*, 1977, **79**, 614.
[41] B. P. N. Ko, A. Yazgan, P. L. Yeagle, S. C. Lottich, and R. W. Henkens, *Biochemistry*, 1977, **16**, 1720.
[42] W. S. Craig and B. P. Gaber, *J. Amer. Chem. Soc.*, 1977, **99**, 4130.

between the holo- and apo-enzymes was evident. The latest weapon in the arsenal of spectroscopic methods, EXAFS, has been applied[43] to bovine carbonic anhydrase. It was concluded that iodide binds directly to the Zn in the enzyme, with a Zn—I distance of 2.65 ± 0.06 Å (compared to 2.62 Å in ZnI_2). Some [113]Cd n.m.r. studies[44] on [113]Cd^{II}-substituted human carbonic anhydrase B have been performed. A single, broad (~ 300 Hz) line, centered at about 228 p.p.m., could be observed. On addition of two equivalents of NaCl this resonance sharpens to 60 Hz and shifts to 238.6 p.p.m., while the addition of one equivalent of K[13]CN causes the resonance to split into a doublet ($J_{Cd-C} = 1060$ Hz), centered at 410 p.p.m. (Figure 3). The coupling constant observed is the largest known for

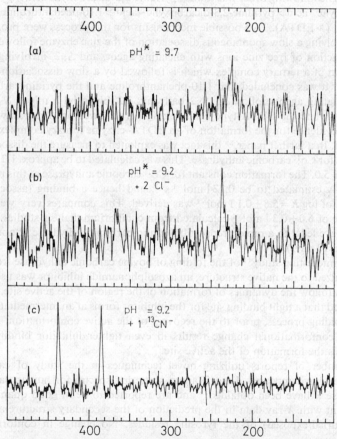

Figure 3 *The* [113]Cd *n.m.r. spectra of* [113]Cd *substituted human carbonic anhydrase B. (a) in the absence of inhibitors; (b) in the presence of two equivalents of NaCl; (c) in the presence of 1 equivalent of* K[13]CN. *The pH* values are uncorrected for the presence of* D_2O
(Reproduced by permission from *J. Amer. Chem. Soc.*, 1977, **99**, 4499)

[43] G. S. Brown, G. Navon, and R. G. Shulmann, *Proc. Nat. Acad. Sci. U.S.A.*, 1977, **74**, 1794
[44] J. L. Sudmeier and S. J. Bell, *J. Amer. Chem. Soc.*, 1977, **99**, 4499.

Cd, and indicates a lifetime for the Cd—C bond of 10^{-2} s. From n.m.r. relaxation studies[45] of frozen solutions of carbonic anhydrase with or without Mn^{II} at the active site, it was concluded that, even in these frozen solutions, exchange of water in the immediate vicinity of the protein with the Mn^{II} is rapid ($< 10^{-6}$ s). Raman spectroscopy has been used[46] to study the ionization state of a sulphonamide inhibitor bound to carbonic anhydrase. It was determined that the sulphonamide was bound to the enzyme as SO_2NH_2 rather than SO_2NH^-, and hence protonated sulphonamide must bind directly to the basic form of the enzyme. As with other sulphonamides, a one-step binding mechanism is suggested from the rate data for association and dissociation of the enzyme inhibitor. These results indicate that sulphonamides form outer-sphere complexes with the metal at the active site of the enzyme.

Several papers relating to kinetic studies[47-49] on carbonic anhydrase have been published. These are reviewed in Chapter 2.

Finally, we were pleased to note that an excellent article[50] intended to introduce undergraduate students to inorganic biochemistry is illustrated by reactions of carbonic anhydrase.

Peptidases.—*Carboxypeptidases.* The zinc-containing carboxypeptidases are secreted as an inactive zymogen (procarboxypeptidase) in the pancreatic juice of animals.[51] As their name implies, these enzymes are capable of cleaving the C-terminal amino-acid from proteins and peptides, and they also catalyse the hydrolysis of certain esters.[52] Bovine carboxypeptidase A is one of the most intensively studied zinc metalloenzymes. The enzyme, as isolated, contains 1 gram atom of zinc per protein molecular weight of 34500,[53] and removal of the metal results in a totally inactive apoenzyme.[54] The activity can be restored by re-addition of zinc or one of a number of other bivalent metal ions.[55] These metal-substitution studies provide an elegant demonstration of the role of the metal ion during catalysis. Table 2 shows a comparison of the kinetic parameters for the Zn^{II}-, Co^{II}-, Mn^{II}-, and Cd^{II}-enzyme-catalysed hydrolysis of a peptide substrate and its structural analogue ester substrate. These data indicate that the primary role of the metal in peptide hydrolysis is to function in the catalytic process, and that it has little to do with peptide binding, while the reverse is true for ester hydrolysis.[56] The X-ray structure[57, 58] shows the molecule to be ellipsoidal, with a cleft (containing the Zn) associated with the active site. The zinc is bound to two

[45] E. Hsi and R. G. Bryant, *J. Phys. Chem.*, 1977, **81**, 462.
[46] R. L. Petersen, T.-Y. Li, J. T. McFarland, and K. L. Watters, *Biochemistry*, 1977, **16**, 726.
[47] Y. Pocher, L. Bjorkquist, and D. W. Bjorkquist, *Biochemistry*, 1977, **16**, 3967.
[48] Y. Pocher and D. W. Bjorkquist, *Biochemistry*, 1977, **16**, 5698.
[49] Y. Pocher and D. W. Bjorkquist, *J. Amer. Chem. Soc.*, 1977, **99**, 6537.
[50] R. S. McQuate, *J. Chem. Educ.*, 1977, **54**, 645.
[51] E. Walschmidt-Leitz and A. Purr, *Chem. Ber.*, 1929, **62B**, 2217.
[52] J. F. Riordan and B. L. Vallee, *Biochemistry*, 1963, **2**, 1460.
[53] B. L. Vallee and H. Neurath, *J. Amer. Chem. Soc.*, 1954, **76**, 5006.
[54] J. P. Felber, T. L. Coombs, and B. L. Vallee, *Biochemistry*, 1962, **1**, 231.
[55] J. E. Coleman and B. L. Vallee, *J. Biol. Chem.*, 1960, **235**, 390.
[56] D. S. Auld and B. Holmquist, *Biochemistry*, 1974, **13**, 4355.
[57] G. N. Reeke, J. A. Hartsuck, M. L. Ludwig, F. A. Quiocho, T. A. Steiz, and W. N. Lipscomb, *Proc. Nat. Acad. Sci. U.S.A.*, 1967, **58**, 2220.
[58] W. N. Lipscomb, J. A. Hartsuck, G. N. Reeke, F. A. Quiocho, P. H. Bethge, M. L. Ludwig, T. A. Steitz, H. Muirhead, and J. C. Coppola, *Brookhaven Symp. Biol.*, 1968, **21**, 24.

Table 2 *Metallocarboxypeptidase-catalysed hydrolysis of* Bz-(Gly)₂-L-Phe *and* Bz-(Gly)₂-L-OPhe[a]

Metal	Bz-(Gly)₂-L-Phe		Bz-(Gly)₂-L-OPhe	
	k_{cat}/min^{-1}	$10^{-3} K_M^{-1}/l\ mol^{-1}$	$10^{-4} k_{cat}/min^{-1}$	$K_M^{-1}/l\ mol^{-1}$
Cobalt	6000	1.5	3.9	3300
Zinc	1200	1.0	3.0	3000
Manganese	230	2.8	3.6	660
Cadmium	41	1.3	3.4	120

(a) Assays performed at 25 °C, pH 7.5, 1.0M-NaCl, and a buffer concentration of 0.05M-Tris for peptide hydrolysis and 10⁻⁴M-Tris for ester hydrolysis.[56]

histidine residues (residues 69 and 196) and one glutamate residue (residue 72). The fourth position is presumed to be occupied by a water molecule which is replaced by substrate when the substrate is bound. The co-ordination is distorted tetrahedral. Other catalytic residues in the active site include Glu-270, Tyr-248, and Arg-145. Based on the crystallographic data, detailed mechanisms of action for carboxypeptidase A have been proposed.[59] These mechanisms, and recent work pertaining to them,[60, 61] are reviewed in Chapter 2. Although undoubtedly correct in their general features, it must be borne in mind that these mechanistic conclusions are based on the assumption that the kinetic and chemical properties are conserved on crystallization. In fact it has been shown[62, 63] that the detailed kinetics of carboxypeptidase A solutions differ from those of enzyme crystals, and it has been suggested that different conformations of the active site exist in the two physical states.[64] Detailed kinetic studies on crystals over a range of enzyme concentrations, substrate concentrations, and crystal sizes have been carried out.[65] The results were interpreted according to a recent theory for insolubilized enzymes.[66] The marked differences between the kinetic behaviour of crystals and solutions were confirmed. These differences were shown to be genuine, and not due to artifacts arising from diffusion limitations or surface phenomena. It was found that, for all substrates examined, crystallization of the enzymes markedly reduces catalytic efficiency (k_{cat} is reduced 20- to 1000-fold). In addition, substrate inhibition, apparent in solution for some di- and depsi-peptides, was abolished with the crystals. Larger substrates with normal kinetics in solution could exhibit activation with the crystals. The physical state of the enzyme also affected the mode of action of known modifiers of the peptidase activity of the enzyme. Similar differences between the solution and crystal kinetics were observed for carboxypeptidase B.[67] It has been concluded[65] that these differences raise serious questions about the capability of crystallography to delineate binding modes, as well as to visualize possible mechanisms of action for carboxypeptidase. Dynamic

[59] W. N. Lipscomb, *Chem. Soc. Rev.*, 1972, **1**, 319.
[60] S. Scheiner and W. N. Lipscomb, *J. Amer. Chem. Soc.*, 1977, **99**, 3466.
[61] R. Breslow and D. L. Wernick, *Proc. Nat. Acad. Sci. U.S.A.*, 1977, **74**, 1303.
[62] C. A. Spilburg, *Fed. Proc. Fed. Am. Soc. Exp. Biol.*, 1974, **33**, 1529.
[63] C. A. Spilburg, J. L. Bethune, and B. L. Vallee, *Proc. Nat. Acad. Sci. U.S.A.*, 1974, **71**, 3922
[64] J. F. Riordan and G. Muszynska, *Biochem. Biophys. Res. Comm.*, 1974, **57**, 447.
[65] C. A. Spilburg, H. L. Bethune, and B. L. Vallee, *Biochemistry*, 1977, **16**, 1142.
[66] E. Katchalski, I. Silman, and R. Goldman, *Adv. Enzymol.*, 1971, **34**, 445.
[67] G. M. Alter, D. L. Leussing, H. Neurath, and B. L. Vallee, *Biochemistry*, 1977, **16**, 3663

studies have shown the existence of rapidly interconvertible structures of carboxypeptidase A in solution. For example, recent resonance Raman spectra[68] of arsanilazotyrosine-248 carboxypeptidase A in solution contain multiple, discrete bands which change as a function of pH, demonstrating the existence of interconvertible species, and similar conclusions have been reached on the basis of c.d. measurements.[69] Crystallization, however, might single out a particular conformation which does not permit the most efficient catalysis. This is suggested to be a possible reason for the observed kinetic differences between the crystals and the solutions of carboxypeptidase.[65, 67]

The substitution of the essential Zn^{II} ion in carboxypeptidase A with Co^{II} results in an enzyme which retains its peptidase and esterase activities. It has been reported that oxidation of the Co^{II} carboxypeptidase A to Co^{III} carboxypeptidase A with H_2O_2 yields a product with negligible peptidase activity, but which retains esterase activity.[70, 71] In a recent paper[72] it is claimed that this retention of esterase activity is dependent on the mode of preparation of the Co^{II} enzyme. These authors claim that, when the Co^{II} is introduced by dialysing the Zn^{II} carboxypeptidase A against a large molar excess of Co^{II}, the oxidation product retains the esterase activity. If the Co^{II} is introduced into the metal-free apoenzyme, the oxidation product loses the esterase activity. Other authors,[73] however, report that oxidation of Co^{II} carboxypeptidase A to Co^{III} carboxypeptidase A with *m*-chloroperbenzoate results in a totally inactive enzyme. Attempts by these latter authors to prepare the Co^{III} enzyme using the peroxide[70, 71] method failed, as the e.p.r. spectrum of the product revealed that a substantial fraction of the enzyme was still in the Co^{II} state. In the Co^{III} carboxypeptidase A produced by Van Wart and Vallee,[73] no e.p.r. spectrum could be detected, as is expected for Co^{III}, which is usually diamagnetic. The reaction of *m*-chloroperbenzoate with Co^{II} carboxypeptidase A followed saturation kinetics and was prevented by the inhibitor β-phenylpropionate. Furthermore, under the conditions employed, no oxidation of Co^{II} *E. coli* alkaline phosphatase occurred. It has therefore been suggested[73] that *m*-chloroperbenzoate is acting as an active-site-directing oxidizing agent for Co^{II} carboxypeptidase.

A systematic m.c.d. study[74] of Co^{II} carboxypeptidase A has been published. This is in agreement with earlier studies[75] and shows that the addition of substrates and inhibitors neither alters the overall shape, nor the sign, of the m.c.d. signal, but affects the fine structure of the Co^{II} carboxypeptidase spectrum.

A report of a two-step purification scheme[76] based on affinity chromatography for bovine carboxypeptidase A and B has appeared. The purification of carboxy-

[68] R. K. Scheule, H. E Van Wart, B. L. Vallee, and H. A. Scheraga, *Proc. Nat. Acad. Sci. U.S.A.*, 1977, **74**, 3272.
[69] A. A. Klesov and B. L. Vallee, *Bioorg. Khim.*, 1977, **3**, 964.
[70] E. P. Kang, C. B. Storm, and F. W. Carson, *Biochem. Biophys. Res. Comm.*, 1972, **49**, 621.
[71] E. P. Kang, C. B. Storm, and F. W. Carson, *J. Amer. Chem. Soc.*, 1975, **97**, 6723.
[72] M. M. Jones, J. B. Hunt, C. B. Storm, P. S. Evans, F. W. Carson, and W. J. Pauli, *Biochem. Biophys. Res. Comm.*, 1977, **75**, 253.
[73] H. E. Van Wart and B. L. Vallee, *Biochem. Biophys. Res. Comm.*, 1977, **75**, 732.
[74] A. A. Klesov and B. L. Vallee, *Bioorg. Khim.*, 1977, **3**, 958.
[75] B. Holmquist, T. A. Kaden, and B. L. Vallee, *Biochemistry*, 1975, **14**, 1454.
[76] S. P. Ager and G. M. Hass, *Analyt. Biochem.*, 1977, **83**, 285.

peptidase B from human pancreas[77] has also been reported. Two forms of the enzyme (named B_1 and B_2) could be separated. These enzymes appear to be very similar to the corresponding bovine enzyme, and the N-terminal amino-acid sequence of the B_1 enzyme differs in only two places (out of the twenty residues sequenced) from the bovine enzyme.

A substantial catalytic activity has been found for bovine procarboxypeptidase A acting on halogenated acyl-amino-acids.[78] The most surprising result of this study is that apo-procarboxypeptidase A will hydrolyse trifluoroacetyl-L-phenyl-alanine, albeit at a lower rate than the holo- or Mn^{II}-enzyme. It has been sug-gested[78] that with this substrate the polarization of the carbonyl carbon is accomplished by the fluorine atoms, making the presence of the metal ion at the active site unnecessary.

Leucine Aminopeptidase. The essential Zn^{II} atoms (two per subunit) in the hexa-meric enzyme from bovine lens have been substituted with Mn^{II}, Co^{II}, Cd^{II}, and Mg^{II} by dialysis of the native enzyme against solutions of these cations.[79] The amount of metal bound was reported to be in the range 6—12 gram atom per mole of protein. This substitution results in retained activity for the Mn^{II}- and Co^{II}-substituted enzymes; the Mg^{II}- and Cd^{II}-substituted enzymes are inactive. The native bovine enzyme is activated by Mg^{II} and Mn^{II}.[80] The stability of the hexameric structure of the bovine enzyme in guanidine hydrochloride has been investigated.[81] The enzyme retains its hexameric structure in 0.5M-guanidine hydrochloride. From 0.75M- to 2.5M-guanidine–HCl, dissociation into trimers occurs, while above 2.75M-guanidine–HCl monomers dominate. Partial unfold-ing of the subunits accompanies this dissociation.[81]

A number of papers relating to the quaternary structure of leucine amino-peptidases have appeared,[82,83] including a preliminary X-ray study.[84] In the latter work[84] the diffraction pattern is reported to extend to approximately 2.2 Å resolution and shows a $P6_322$ space group, the dimensions of the unit cell being $a = 132$ and $c = 122$ Å. The asymmetric unit consists of one protomer of molecular weight 54000.

The isolation of leucine aminopeptidase from *Aspergillus oryzae* has been reported.[85] The molecular weight (37500) of this enzyme is significantly smaller than that from mammalian sources, but, like the latter, it is a metalloenzyme.

[77] D. V. Marinkovic, J. N. Marinkovic, E. G. Erdos, and C. J. G. Robinson, *Biochem. J.*, 1977, **163**, 253.

[78] P. C. Plese and W. D. Behnke, *Biochim. Biophys. Acta*, 1977, **483**, 172.

[79] U. Kettmann, *Ergeb. Exp. Med.*, 1977, **24**, 103.

[80] M. Ludwig, *Ergeb. Exp. Med.*, 1977, **24**, 83.

[81] G. Lassmann, W. Damerau, D. Schwarz, R. Kleine, and M. Frohne, *Studia Biophys.*, 1977 **63**, 149.

[82] M. Ludwig, H. Hanson, N. A. Kiselev, V. Ya. Stel'mashcuh, and V. L. Tsuprun, *Acta Biol. Med. Ger.*, 1977, **36**, 157.

[83] G. Wangermann, I. M. Edintsov, G. R. Ivanitskii, A. S. Kuniskii, R. Reichelt, and M. A. Tsyganov, *Biofizika*, 1977, **22**, 599.

[84] F. Jurnak, A. Rich, L. van Loon-Klassen, H. Bloemendal, A. Taylor, and F. H. Carpenter, *J. Mol. Biol.*, 1977, **112**, 149.

[85] N. M. Ivanova, T. L. Vaganova. A. Ya. Strongin, and V. M. Stepanov, *Biokhimiya*, 1977, **42**, 843.

Other Proteases. The active site of thermolysin, a zinc endopeptidase isolated from *Bacillus thermoproteolyticus*, is reported[86] to be similar to that of carboxypeptidase, and these two enzymes seem to be the product of convergent evolution. A n.m.r. investigation[87] of the ionizable residues at the active site of the Mn^{II}-substituted enzyme shows two perturbing groups with pK_a's of approximately 8.5 and 9.5. These were tentatively assigned to Tyr-157 and the metal-bound water molecule. The binding behaviour of the inhibitor *N*-trifluoroacetyl-phenylalanine to this Mn^{II}-substituted enzyme strongly suggests that two binding sites are found on the enzyme, the tighter of which leads to non-productive binding.

A few years ago, collagenase from the micro-organism *Clostridium histolyticum* was shown to be a zinc metalloenzyme.[88] In a recent paper[89] it is reported that mammalian collagenases are also zinc metalloenzymes. The collagenase-like peptidase of rat testis[90] has also independently been shown to be a zinc metalloenzyme.

In an elegant piece of work Holmquist[91] demonstrates that, in contradiction to earlier reports,[92] *Bacillus cereus* 'microprotease' is a monomeric enzyme containing 1 gram atom of Zn^{II} per molecular weight of 34000, and hence is a typical zinc-containing neutral protease. The earlier reports[92] claiming that this enzyme was oligomeric, with a subunit of mol. wt. 2700, seem to have been due to autocatalytic digestion of the enzyme. Several papers[93, 94] relating to the isolation and characterization of new metallo-neutral proteases have appeared.

Alcohol Dehydrogenases.–Of those pyridine-nucleotide-dependent enzymes which catalyse the reversible oxidation of alcohols to aldehydes, that from horse liver has been the subject of the most detailed studies. It consists[95] of two identical subunits of molecular weight 40000, contains 4 gram atom per mole of zinc(II), and binds two moles of NAD$^+$ per mole of enzyme. X-Ray diffraction studies confirmed suggestions made[96] on the basis of kinetic, equilibrium, and thermodynamic investigations which indicated that there were two types of zinc ion. That at the active site is situated at the bottom of a hydrophobic pocket, in a deep cleft about 25 Å from the protein surface, and is co-ordinated to two cysteinyl sulphurs and an imidazole group of a histidine. The distorted tetrahedral configuration is completed by a water molecule. In contrast, the other type of zinc(II) ion is close to the surface of the enzyme yet not exposed to solvent. It is co-ordinated to four cysteinyl sulphurs. The active-site zinc is considered to act as an electrophilic centre, promoting hydride transfer to NAD$^+$ *via* the formation of a zinc(II) alcoholate complex.

[86] W. R. Kester and B. W. Matthews, *J. Biol. Chem.*, 1977, **252**, 7704.
[87] W. L. Bigbee and F. W. Dahlquist, *Biochemistry*, 1977, **16**, 3798.
[88] E. Marper and S Seifter, *Israel J. Chem.*, 1974, **12**, 515.
[89] J. L. Seltzer, J. J. Jeffrey, and A. Z. Eisen, *Biochim. Biophys. Acta*, 1977, **485**, 179.
[90] J. Lukac and E. Koren, *J. Reprod. Fertil.*, 1977, **49**, 95.
[91] B. Holmquist, *Biochemistry*, 1977, **16**, 4591.
[92] R. U. Schenk and J. Bjorksten, *Fin. Kemistsamf. Medd*, 1974, **82**, 26.
[93] P. T. Varandani and L. A. Shroyer, *Arch. Biochem. Biophys.*, 1977, **181**, 82.
[94] A. T. H. Abdelal, E. H. Kennedy, and D. G. Ahearn, *J. Bacteriol.*, 1977, **130**, 1125.
[95] C. I. Bränden, H. Jornvall, H. Eklund, and B. Furugren, in 'The Enzymes', ed. P. Boyer Academic Press, New York, 1975, 3rd edn., Vol. 11, p. 103.
[96] D. E. Drum and B. L. Vallee, *Biochemistry*, 1970, **9**, 4078.

In an interesting study[97] of the binding of coenzyme analogues, by *X*-ray diffraction, it was shown that whereas the adenosine moiety of analogues (*e.g.* 3-iodopyridine–adenine dinucleotide) is bound in a fashion similar to ADP-ribose or NAD$^+$, the pyridine ring is oriented quite differently, in particular away from the catalytically active zinc ion, as shown in Figure 4. It has been concluded that

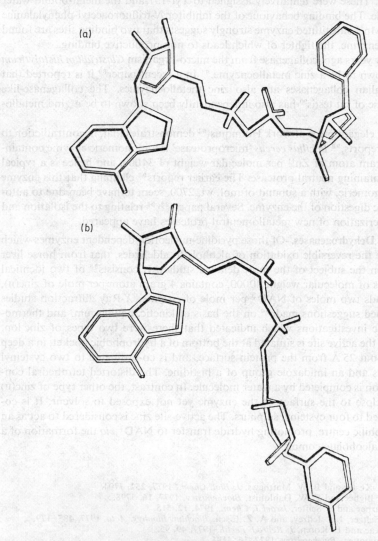

Figure 4 *Schematic diagram of the conformations of* (a) *3-iodopyridine– and pyridine–adenine dinucleotide and* (b) *of* NAD$^+$ *and NADH bound to liver alcohol dehydrogenase*
(Reproduced by permission from *Biochem. Soc. Trans.*, 1977, **5**, 612)

97 C.-I. Bränden, *Biochem. Soc. Trans.*, 1977, **5**, 612.

the hydrogen bond between Thr-178 and the carboxamide is essential for the proper positioning of the nicotinamide moiety of the coenzyme.

The binding of two inhibitor molecules, *i.e.* imidazole and 1,10-phenanthroline, has also been studied[98] by X-ray diffraction. Both inhibitors bind to the catalytic zinc ion, and in doing so they displace the water molecule which was tightly bound in the holoenzyme. Interestingly, in binding to the zinc, no consequent structural changes were observed in the rest of the protein. Figure 5 shows the electron density of 1,10-phenanthroline superimposed on the electron density of the apoenzyme. It is concluded that NAD$^+$ binding causes a pK shift of the zinc-bound water molecule from 9.6 to 7.6. The resulting hydroxide ion, still bound to the zinc, may then facilitate binding of alcohols. A proton may then be removed from the alcohol, with the formation of H_2O and alcoholate bound to the zinc, or there may be a transition state in which both hydroxyl ion and neutral alcohol are bound to the zinc.

Most of the fluorescence of the protein is provided by Trp-314. In an interesting investigation[99] of the quenching of this fluorescence it was concluded that there

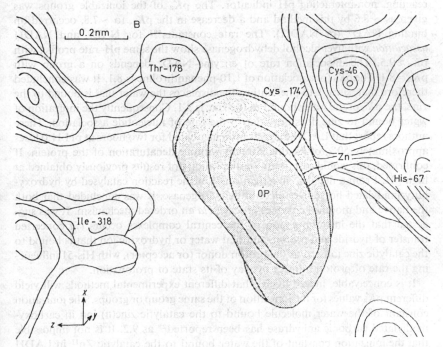

Figure 5 *The electron density of 1,10-phenanthroline (OP) bound to horse liver alcohol dehydrogenase superimposed on the apoenzyme electron density projected down the y-axis*
(Reproduced by permission from *European J. Biochem.*, 1977, **77**, 173)

98 T. Boiwe and C.-I. Brändén, *European J. Biochem.*, 1977, **77**, 173.
99 J. K. Wolfe, C. F. Weiding, H. R. Halvorson, J. D. Shore, D. M. Parker, and J. J. Holbrook, *J. Biol. Chem.*, 1977, **252**, 433.

is a pH-dependent conformational change which leads to an altered environment of this residue. The intrinsic pK_a is 9.8, shifted to 10.6 in D_2O. In the presence of NAD^+ the limiting pK_a is 7.6, implying that NAD^+ binds to the unprotonated form of the enzyme $10^{2.2}$ times more tightly than to the protonated form. It was suggested that the single most plausible explanation for the common pH dependence of protein fluorescence, ADP–ribose-binding, and NAD^+-binding rate is a pH-dependent conformational change which simultaneously exposes Trp-314 to solvent and covers the hydrophobic coenzyme-binding site. The scheme proposed is such that the functional group with a pK_a of 9.8 controls the conformational state of the enzyme. NAD^+ binds to the acid conformation and subsequently causes another conformational change, resulting in the perturbation of the pK_a to 7.6. Alcohol then binds to the unprotonated form of the functional group with a pK_a of 7.6 in the binary enzyme–NAD^+ complex and converts the enzyme into the alkaline conformation. Thus, at neutral pH, liver alcohol dehydrogenase undergoes two conformational changes *en route* to the ternary complex in which hydride transfer occurs. Similar results were obtained[100] by an independent temperature-jump study of relaxation effects, using Cresol Red as a rapidly reacting, non-interacting pH indicator. The pK_a of the ionizable groups was given as ~ 8.6 by this method and a decrease in the pK_a, to ~ 7.8, occurred on binding NAD^+ (or NADH). The rate constants[101] for NAD^+ and NADH *association* with liver alcohol dehydrogenase show the same pH–rate profile, with $pK_a \approx 9.5$. The dissociation rate of enzyme–NAD^+ depends on a group with $pK_a = 8.1$. The pK_a for association of 1,10-phenanthroline is 8.1. It was concluded therefore that, since 1,10-phenanthroline displaces the water that is bound to the zinc, the pK_a of the water bound to the zinc is 8.1. Consequently the ionization of water bound to zinc does not control the rate of nucleotide association. These authors report that LADH that has been incubated for two hours at pH 11.0 loses approximately 50% of its zinc content without denaturation of the protein. If confirmed, this would cast doubt on the validity of results previously obtained at high pH. From a detailed investigation[102] of the reaction catalysed by hydroxy-butyrimidylated horse liver alcohol dehydrogenase, it was concluded that both the native and modified enzymes proceed *via* an ordered mechanism. It was suggested that the interconversion of the central complexes occurs *via* concerted transfer of hydride and proton, and that water or hydroxide ion that is bound to the catalytic zinc ion acts as the proton donor (or acceptor), with His-51 influencing the rate of proton transfer by way of its state of protonation.

It is conceivable, indeed likely, that different experimental methods will yield different pK_a values for the ionization of the same group or groups. The ionization constant of the water molecule bound to the catalytic zinc(II) ion in carboxy-methylated carbonic anhydrase has been reported[29] as 9.2. It is not impossible that the ionization constant of the water bound to the catalytic Zn^{II} in LADH should have a value not far different. However, it is probable that such equilibria do not represent the ionization of a water molecule *in isolation*, *i.e.* the ionization

[100] I. Giannini, V. Baroncelli, G. Boccalon, and P. Renzi, *J. Mol. Catal.*, 1977, **2**, 39.

[101] M. C. De Traglia, J. Schmidt, M. F. Dunn, and J. T. McFarland, *J. Biol. Chem.*, 1977, **252**, 3493.

[102] R. T. Dworschack and B. V. Plapp, *Biochemistry*, 1977, **16**, 2716.

is of water interacting not only with the Zn^{II} but with other polar residues in its immediate environment. The different experimental methods used will perturb this environment to different extents and lead to altered apparent pK_a values. The observed effect of the coenzyme must result from a dramatic reorientation of the zinc(II) and/or the water molecule with respect to the rest of the protein.

In a spirited defence[103] of a proposal made[104] previously that the two enzyme subunits in LADH are not kinetically equivalent,[105] a thorough investigation of the transient kinetics, stoicheiometry, and rate parameters of the oxidation of benzyl alcohol is described. The principal conclusion is that 'half-of-the-sites' reactivity *is* a characteristic feature of LADH during oxidation of benzyl alcohol, but only when the concentration of the substrate is non-inhibitory. In contrast, in a study[106] of the reaction of LADH with benzyl alcohol and benzyldehyde substrates, *no* evidence was found for non-equivalence of sites, and the results were consistent with independently functioning active sites. The results at non-inhibitory concentrations of substrate are compatible with the known rate constants for aldehyde dissociation and the reverse hydride transfer. An interesting investigation[107] of the pressure dependence of the rates of reaction of ethanol with yeast and liver ADH has been described. With the yeast enzyme there is an increase in volume following the formation of the activated complex in the rate-determining step. The results are consistent with a random two-substrate mechanism. The liver enzyme shows a more complicated behaviour if ethanol is present at concentrations that are inhibitory. A random mechanism has been proposed[108] for the reaction of rat liver alcohol dehydrogenase. In a detailed study[109] of the binding of alkyl-cyclohexanols it was possible, in conjunction with the results derived from the X-ray diffraction studies, to show that the cyclohexanol rings of all substrates studied have the same orientation relative to the nicotinamide moiety of the coenzyme. 2-Mercaptoethanol has been found[110] to act as a substrate for both calf and horse liver alcohol dehydrogenase, and acts as an apparently competitive inhibitor with respect to ethanol.

The arginyl-specific reagents butane-2,3-dione and phenylglyoxal inhibit[111] horse liver alcohol dehydrogenase. Two arginines are modified per subunit, and it is presumed that one of the arginines so modified is Arg-47, which binds to the pyrophosphate bridge of the coenzyme. Consequently, coenzyme binding is abolished on modification, and inhibition results. It was shown that the other arginine that is modified, (out of the 12 arginyl residues per subunit) is Arg-84. This is on the surface on the opposite side of the catalytic domain in relation to the active site. Presumably the local environment leads to this enhanced reactivity; a

103 A. Baici and P. L. Luisi, *J. Mol. Biol.*, 1977, **114**, 267.
104 P. L. Luisi and E. Bignetti, *J. Mol. Biol.*, 1974, **88**, 653.
105 S. A. Bernhard, M. F. Dunn, P. L. Luisi, and P. Schack, *Biochemistry*, 1970, **9**, 185.
106 C. F. Weidig, H. R. Halvorson, and J. D. Shore, *Biochemistry*, 1977, **16**, 2916.
107 E. Morild, *Biophys. Chem.*, 1977, **6**, 351.
108 M. Feraudi, M. Kohlmeier, and G. Schmolz, *J. Mol. Catalysis*, 1977, **2**, 171.
109 H. Dutler, *Biochem. Soc. Trans.*, 1977, **5**, 617.
110 C. R. Geren, C. M. Olomon, T. T. Jones, and K. E. Ebner, *Arch. Biochem. Biophys.*, 1977, **179**, 415.
111 H. Jörnvall, L. G. Lane, J. F. Riordan, and B. L. Vallee, *Biochem. Biophys. Res. Comm.*, 1977, **77**, 73.

chance consequence of the tertiary structure of the protein, and unrelated to function.

There is considerable interest in the different kinetic behaviour and substrate specificity of the different isozymes of liver alcohol dehydrogenases. The major isozymes of the horse liver enzyme were compared[112] in a study which led to the conclusion the subunits may not act independently. The major isozyme from the human liver displays[113] non-linear kinetics in the oxidation of ethanol, and it was suggested that either the subunits have different kinetic behaviour or that there exists negative co-operativity between the subunits, reminiscent of the well-documented properties of alkaline phosphatase (see below). In the same study, the human and horse liver enzymes were compared in their response to chemical modification. Acetimidylation increases the activities of both horse and human enzymes, though a variant subunit isolated from atypical liver is not activated. The purpose of such investigations is well illustrated by the discovery[114] that, whilst the physical properties of an isozyme of human liver alcohol dehydrogenase (characterized by its electrophoretic behaviour) did not differ from others, it did not catalyse the oxidation of methanol, and its K_m for ethanol was one hundred times greater than that of other forms. It was suggested that this particular alcohol dehydrogenase could make a significant contribution to the elimination of alcohol in Man, especially when other isozymes are saturated. The isoenzyme from horse liver alcohol dehydrogenase, composed of two identical subunits, SS, (where S indicates steroid-active, distinguishing it from E, ethanol-active) has been purified and characterized.[115] It contains 4 gram atom of Zn per molecule (dimer). Kinetic data have been reported for an extensive range of substrates. Important differences between human and horse liver alcohol dehydrogenases with respect to alcohol oxidation and aldehyde reduction in substrates containing non-polar groups have been revealed.[116] Whereas the human enzyme is markedly dependent on the hydrophobicity of these non-polar portions of the substrate, the horse enzyme is not.

In two magistral papers on yeast alcohol dehydrogenase the sequence is reported[117] and compared[118] with that of the liver enzyme. The latter paper provides an illuminating account of the insight to be gained from such comparisons. It is suggested that the entire yeast and horse liver alcohol dehydrogenases are distantly homologous, and though they have a common ancestor, estensive evolutionary changes have occurred. The general folding of the catalytic domains of the horse and yeast enzymes is similar. The most conserved regions in the entire protein are those containing the structures that build up or surround the active site in the horse liver enzyme. Significantly, the three zinc ligands at the active site (Cys-46, His-67, and Cys-174) of this enzyme are common to both proteins (Cys-43, His-66, and Cys-153 in the yeast enzyme).

112 R. T. Dworschack and B. V. Plapp, *Biochemistry*, 1977, **16**, 111.
113 A. Dubied, J. P. Von Wartburg, D. P. Bohlken, and B. V. Plapp, *J. Biol. Chem.*, 1977, **252**, 1464.
114 W. F. Bosron, T.-K. Li, L. G. Lange, W. P. DaFeldecker, and B. L. Vallee, *Biochem. Biophys. Res. Comm.*, 1977, **74**, 85.
115 C. N. Ryzewski and R. Pietruszko, *Arch. Biochem. Biophys.*, 1977, **183**, 73.
116 A. A. Klesov, L. G. Lange, A. J. Sytkowski, and B. L. Vallee, *Bioorg. Khim.*, 1977, **3** 1141.
117 H. Jörnvall, *European J. Biochem.*, 1977, **72**, 425.
118 H. Jörnvall, *European J. Biochem.*, 1977, **72**, 443.

A fascinating result is that the four cysteinyl residues which bind the second zinc in the horse liver enzyme are strictly conserved in the yeast enzyme, yet in a region in which there are only random similarities. It is suggested that 'potential zinc ligands need not be conserved in space although strictly homologous in primary structure'. Thus, although the residues which bind the second, structural zinc ion are conserved, such are the other differences in the two sequences in this region that it is possible that these cysteinyl residues in the yeast enzyme are not able to adopt the conformation which would allow them to act as ligands. This would be consistent with the first[119] and most recent work[120] on the zinc content of the yeast enzyme. The enzyme from baker's yeast has a molecular weight of 149 000 and consists of four subunits. Vallee and Hoch had found 4.1 gram atom per mole of catalytically active yeast, but subsequent work, perhaps influenced by the stoicheiometry of the horse liver enzyme, especially once the crystal structure of the latter had been determined, had suggested up to 8 gram atom per mole. The problem of additional and extraneous metal ions binding to enzymes, let alone metalloenzymes, is not unknown. Indeed, with the yeast enzyme, it had been shown in a careful study[121, 122] that yeast alcohol dehydrogenase could bind up to 35 gram atom of Zn per mole! The most recent work,[120] which is exemplary in its protocol, thoroughness, and clarity, shows that there are indeed 4 gram atom of Zn per mole of enzyme. The enzyme also binds 4 mole of NADH per mole of enzyme, consistent with the binding of 1 gram atom of zinc and 1 mole of NADH per subunit. Four zinc ions are required for activity, and they can be replaced by ^{65}Zn or Co to give $(YADH)^{65}Zn_4$ and $(YADH)Co_4$. The latter enzyme has 17% of the activity of the native enzyme. Thus the evidence suggests that the four zinc ions are analogous to those which occupy the active site in the corresponding liver enzymes and that there are no 'structural' zinc ions in the yeast enzyme. The evolutionary relevance will no doubt be the subject of much interesting speculation.

The residue Arg-47 in liver alcohol dehydrogenase is replaced by a histidine (residue 44) in the yeast enzyme. On the basis of a detailed investigation[123] of the yeast enzyme, mainly involving the reaction of the enzyme with diethyl pyro-carbonate, it has been concluded that this histidine is an essential and reactive residue. If this is the case, the mechanisms of the two enzymes would appear to differ. From an analysis of the reaction kinetics of yeast alcohol dehydrogenase, it has been suggested[124] that there are two kinetically distinct active sites, which bind substrate and coenzyme in a defined sequence, ethanol binding first, followed by NAD^+. A brief report[125] has appeared of the inhibition by ethanol of the maximum rates of reduction of acetaldehyde for both yeast and horse liver alcohol dehydrogenases. It was concluded that, for both enzymes, the inhibition observed is consistent with the formation of an enzyme–ethanol complex. The inhibition by 1,10-phenanthroline has also been investigated.[126] A preliminary report sug-

[119] B. L. Vallee and F. L. Hoch, *Proc. Nat. Acad. Sci. U.S.A.*, 1955, **41**, 327.
[120] A. J. Sytkowski, *Arch. Biochem. Biophys.*, 1977, **184**, 505.
[121] K. Wallenfels and H. Sund, *Biochem. Z.*, 1957, **329**, 31.
[122] K. Wallenfels and H. Sund, *Biochem. Z.*, 1957, **329**, 59.
[123] C. J. Dickenson and F. M. Dickinson, *Biochem. J.*, 1977, **161**, 73.
[124] M. Feraudi, M. Kohlmeier, and G. Schmolz, *Ital. J. Biochem.*, 1977, **26**, 12.
[125] F. M. Dickinson and C. J. Dickenson, *Biochem. Soc. Trans.*, 1977, **5**, 767.
[126] F. M. Dickinson and S. Berrieman, *Biochem. J.*, 1977, **167**, 237.

gests[127] that the alcohol dehydrogenase from tea seeds contains 2 gram atom of Zn per mole, *i.e.* one Zn per subunit. Alcohol dehydrogenase from peas has been shown[128] to be inactivated by iodoacetate, the inactivation rate being decreased by 1,10-phenanthroline and by ATP and other phosphates. Finally, a useful publication describes[129] the preparation of various NAD analogues.

Alkaline Phosphatases.—Alkaline phosphatases catalyse[14, 130] the hydrolysis of alkyl and aryl phosphates and, as the name implies, show a maximum activity in the pH range 7.5—10.0. Zinc has been shown to be an integral component of those enzymes which have been adequately characterized, that from *Micrococcus sodonensis* being[131] the only exception to date. The enzyme from *E. coli* has been most intensively investigated. It has a molecular weight of 80000, is dimeric, and consists of two identical subunits. As usual, there was considerable disagreement about the zinc content. It now seems agreed that, when isolated with due precautions against the loss of zinc, *E. coli* alkaline phosphatase contains 4 gram atom per mole, *i.e.* per dimer. The addition of the first two zinc ions to the apoenzyme is sufficient to restore activity, the second two further stabilizing the protein structure. The early report[132, 133] of the magnesium content was largely overlooked until recently, when it was found[134, 135] that the binding of magnesium depends on the pH and indeed on the zinc content. The presence of magnesium increases the activity of the enzyme containing 2 gram atom per mole and that containing 4 gram atom per mole. In the most recent paper[136] it was shown that, though the zinc ion content is clearly 4 gram atom per mole, the magnesium content is closely related to the conditions chosen for optimal growth of the organism and purification of the enzyme. These are *not* the conditions which optimize the binding of magnesium ion. As isolated, the enzyme often contains 2 gram atom of Zn per mole, but the evidence is pretty conclusive that there are a maximum of six metal sites available, four of which normally bind Zn^{II} and two Mg^{II}. The relationship between the metal ions (Zn^{II} and Mg^{II}) required for expression of full activity and the structure of the protein has been highlighted in a study[137] using differential scanning calorimetry. The latter technique allows an assessment of the thermodynamic parameters which accompany the unfolding of proteins. Binding of the first two Zn^{II} ions markedly increases (see Figure 6) the stability of the enzyme by 290 kJ mol^{-1}, and there is some indication of co-operativity in binding. The next pair of Zn^{II} ions added further stabilizes the protein by ~ 125 kJ mol^{-1}, and the suspected role of the Mg^{II} ion in further stabilizing the protein is confirmed by this study. As expected, Mg^{II} is without

[127] J. Sekiya, T. Kajiwara, T. Miura, and A. Hatanaka, *Agric. and Biol. Chem. (Japan)*, 1977, **41**, 713.
[128] R. Lapka and S. Leblova, *Physiol. Plant.*, 1977, **39**, 86.
[129] C. Woenckhous and R. Jeck, *Methods Enzymol.*, 1977, **45**, 249.
[130] H. N. Fernley, *Enzymes*, 1971, **4**, 417.
[131] R. H. Colew and E. C. Heath, *J. Biol. Chem.*, 1971, **246**, 1556.
[132] D. J. Plocke and B. L. Vallee, *Biochemistry*, 1962, **1**, 1039.
[133] R. T. Simpson and B. L. Vallee, *Biochemistry*, 1968, **7**, 4343.
[134] W. F. Bosron, F. S. Kennedy, and B. L. Vallee, *Biochemistry*, 1975, **14**, 2275.
[135] R. A. Anderson, W. F. Bosron, F. S. Kennedy, and B. L. Vallee, *Proc. Nat. Acad. Sci. U.S.A.*, 1975, **72**, 2989.
[136] W. F. Bosron, R. A. Anderson, M. C. Falk, F. S. Kennedy, and B. L. Vallee, *Biochemistry*, 1977, **16**, 610.
[137] J. F. Chlebowski and S. Mabrey, *J. Biol. Chem.*, 1977, **252**, 7042.

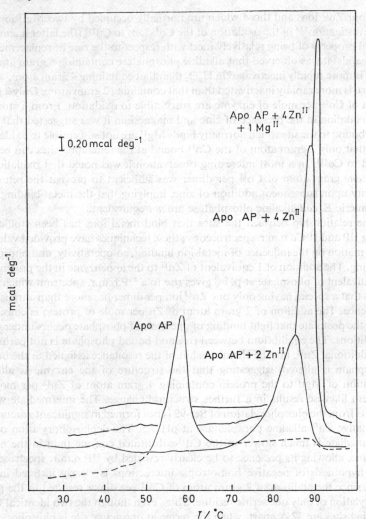

Figure 6 *The effects of metal ion composition on thermally induced transitions of alkaline phosphatase*
(Reproduced by permission from *J. Biol. Chem.*, 1977, **252**, 7042)

effect on the stability of the protein when Zn^{II} is absent. Consideration of the entropic contribution to the relative free energy of the metalloproteins suggests that the internal order in the metalloproteins is greater than in the apoenzyme.

It has been shown[138] that the enzyme can be reconstituted with 6 gram atom of Co^{II} per mole. The Co^{II} can therefore occupy both the sites which accommodate

[138] R. A. Anderson and B. L. Vallee, *Biochemistry*, 1977, **16**, 4388.

the four Zn^{II} ions and those which are normally occupied by two Mg^{II} ions. In an investigation[138] of the oxidation of the Co^{II} ions to Co^{III} (the latter having the useful property of being relatively inert with respect to the rate of replacement of its ligands) it was observed that alkaline phosphatase containing 6 gram atom of Co^{II} is more rapidly inactivated in H_2O_2 than that containing 4 gram atom, which in turn is more rapidly inactivated than that containing 2 gram atom. Only 2 gram atom of Co^{II} per mole of enzyme are susceptible to oxidation. From a study of the oxidation in the presence of zinc and magnesium it was suggested that Co^{II} ions bound to the site which normally binds Mg^{II} are not susceptible to oxidation, and that only 2 gram atom of the Co^{II} bound at the remaining sites can be oxidized to Co^{III}. In a most interesting observation it was noted that oxidation of only *one* gram atom of Co^{II} per dimer was sufficient to prevent the return of activity upon subsequent addition of zinc, implying that the metal-binding sites of dimeric *E. coli* alkaline phosphatase are *not* equivalent.

The relationship between the sites that bind metal ions has been studied,[139] using ^{31}P and ^{113}Cd n.m.r. spectroscopy; these techniques have provided valuable information on the influence of metal-ion binding, co-operativity, and phosphate binding. The addition of 1 equivalent of Zn^{II} to the apoenzyme in the presence of 1 equivalent of phosphate at pH 8 gives rise to a ^{31}P n.m.r. spectrum which indicates that a species having only one Zn^{II} ion per dimer has more than a transitory existence. The addition of 2 gram atom of Zn per mole of protein is consistent with the postulate that tight binding of one mole of phosphate occurs under these conditions. The equilibrium between free and bound phosphate is not perturbed by additional Zn^{II}, but the chemical shift of the resonance assigned to the bound phosphate is altered, suggesting that the structure of the enzyme is altered. Addition of Mg^{II} to the protein containing 4 gram atom of Zn^{II} per mole of protein likewise results in a further structural change. The intermediate which results from the phosphorylation of Ser-99 is not formed in significant amounts in the native Zn^{II} alkaline phosphatase at pH 7.5. The dephosphorylation of the intermediate is much slower in the Cd^{II}-substituted enzyme than in the native enzyme, allowing its presence to be clearly revealed by ^{31}P n.m.r. spectroscopy. The hypothesis of negative homotropic interactions is clearly justified in this case, since the addition of 2 gram atom of Cd^{II} per dimer resulted in the phosphorylation of only one serine residue. Thus, even though the two identical Cd^{II}-binding sites are 32 Å apart, when the serine at one active site is phosphorylated the other is rendered unfit for reaction. The same phenomenon of negative co-operativity is shown in the formation of the phosphate complex of the cobalt-substituted enzyme. The ^{113}Cd spectra provide a dramatic demonstration of these effects. The enzyme containing 2 gram atom of ^{113}Cd per mole gives rise to a ^{113}Cd n.m.r. spectrum shown in Figure 7. Only one resonance is observed, showing that the two ^{113}Cd atoms must occupy identical environments. The addition of 1 equivalent of phosphate gives rise to a spectrum containing *two* ^{113}Cd resonances. The addition of further phosphate *does not* affect the ^{113}Cd resonances. Thus one phosphate bound per dimer not only changes the chemical environment of both cadmium ions but renders one different from the other. The role of

[139] J. F. Chlebowski, I. M. Armitage, and J. E. Coleman, *J. Biol. Chem.*, 1977, **252**, 7053.

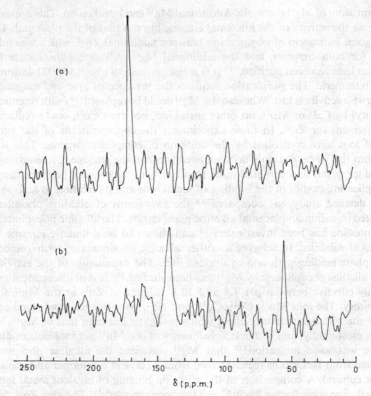

δ (p.p.m.)

Figure 7 *The* ^{113}*Cd n.m.r. spectra of (a)* $[^{113}Cd^{II}]_2$*alkaline phosphatase and (b) phosphoryl-*$[^{113}Cd^{II}]_2$*alkaline phosphatase, both at pH 6.5*
(Reproduced by permission from *J. Biol. Chem.*, 1977, **252**, 7053)

negative co-operativity in the function of the protein is, at present, unclear. The relationship between Mg^{II} and Zn^{II} has been further explored[140] in a study of rat placenta alkaline phosphatase. Though the binding of Zn^{II} at the active site of the partially purified soluble enzyme gives rise to the active enzyme, the binding of additional Zn^{II} at the site which normally binds the activator, *i.e.* Mg^{II}, leads to an inactive enzyme at alkaline pH. In one of the few studies of its kind, the investigation was extended, to discover whether this phenomenon was also exhibited *in vivo*. Using tissue sections, it was shown that, indeed, Zn^{II} displaces and replaces Mg^{II} at the sites normally occupied by the latter. The activities of the two enzymes, *i.e.* that with Zn^{II} at the 'activator' site and that with Mg^{II}, have the same catalytic activity at pH 7.0, but the Zn^{II} 'activated' enzyme has little activity at alkaline pH. It has been proposed that Zn^{II} and Mg^{II} concentrations regulate the activity of alkaline phosphatase in rat placenta. The alkaline phosphatases of human intestine and placenta likewise contain[141] 4 gram atom of Zn^{II} per mole and

[140] C. Petitclerc and C. Fecteau, *Canad. J. Biochem.*, 1977, **55**, 474.
[141] M. Sugiura, K. Hirano, S. Iino, H. Suzuki, and T. Oda, *Chem. and Pharm. Bull.* (*Japan*), 1977, **25**, 653.

3 gram atom of Mg^{II} per mole. Additional Mg^{II} can be taken up. This appears to increase the activity of the intestinal enzyme but not that of the placental. There was some indication of competition between additional Zn^{II}, which was inhibitory for both enzymes, and the additional Mg^{II}. Alkaline phosphatase from human liver has been purified.[142] It is a glycoprotein of about 136000 dalton and it is tetrameric. The purification requires the presence of zinc *and* magnesium, otherwise activity is lost. Whereas the Mg^{II} could be replaced[143] (with retention of activity) by Ca^{II} or Mn^{II}, no other metal ion, not even Co^{II}, could replace the requirement for Zn^{II}. In these experiments the concentrations of the various metal ions were controlled by the addition of complexing agents. This is not without risk since, not only do such agents inhibit the enzyme (by removing the metal ions), but they can bind quite tenaciously to the protein, perhaps, thereby, complicating studies of the binding of metal ions and its relation to activity.

A detailed study has compared[144] the two forms of alkaline phosphatase isolated from human placental alkaline phosphatase. The alkaline phosphatase of rat intestine has been investigated[145] and shown to be a dimeric enzyme with identical subunits. It behaves as other alkaline phosphatases with respect to phosphate binding with non-equivalent sites. The stimulation of the activity of milk alkaline phosphatase by Mg^{II} has been studied.[146] It is pH-dependent and is mainly effective between pH 7.8 and 10.2. Binding of Zn^{II} at the Mg^{II} site is inhibitory. The ions Mn^{II}, Co^{II}, Cd^{II}, and Ni^{II}, on the other hand, bind at the Mg^{II} site with a stimulatory effect which is slightly less than that of Mg^{II}. Mn^{II} binds most tightly, and with a stoicheiometry of one Mn^{II} per molecule of dimer. There was some indication[147] that Mg^{II} stabilized the alkaline phosphatase isolated from HeLa cells (cells derived from a cervical carcinoma and grown in tissue culture). A comparison of the effect on binding of bivalent metal ions to apo-alkaline phosphatase from *E. coli* has been reported.[148] The ions Zn^{II}, Mn^{II}, Co^{II}, and Cd^{II}, which induce tight binding of one phosphate ion per dimer, have a different effect on the u.v. spectrum of the protein from Ni^{II} and Hg^{II}, which do not induce phosphate binding. The addition of 2 gram atom of either Mn^{II} or Cd^{II} per mole of enzyme was sufficient to produce the maximum effect. With Zn^{II} and Co^{II} the spectra were affected up until the addition of 4 gram atom of the bivalent metal ion per mole of enzyme. It was claimed that the addition of Mg^{II} had *no* effect, unlike the reports mentioned above. The inhibition of *E. coli* alkaline phosphatase by a number of phenylarsonic acids was investigated.[149] For example, 4-(4-aminophenylazo)phenylarsonic acid has a single strong binding site. It can, however, be displaced by inorganic phosphate, and it was concluded that the arsonic acids bind in the anion-binding site of the enzyme. Whilst the spectrum of this azo dye was perturbed by the holoenzyme, it was unperturbed by

[142] K.-D. Gerbitz, H. J. Kolb, and O. H. Wieland, *Z. physiol. Chem.*, 1977, **358**, 435.
[143] K.-D. Gerbitz, *Z. physiol. Chem.*, 1977, **358**, 1491.
[144] G. J. Doellgast, J. Spiegel, G. R. A. Guenther, and W. H. F. Fishman, *Biochim. Biophys. Acta*, 1977, **481**, 59.
[145] N. Malik and P. J. Butterworth, *Arch. Biochem. Biophys.*, 1977, **179**, 113.
[146] G. Linden, D. Chappelet-Tordo, and N. Lazdunski, *Biochim. Biophys. Acta*, 1977, **483**, 100.
[147] S. C. Hung and G. Melnykovych, *Enzyme*, 1977, **22**, 28.
[148] H. Szajn and H. Csopak, *Biochim. Biophys. Acta*, 1977, **480**, 143.
[149] H. Szajn, H. Csopak, and G. Folsch, *Biochim. Biophys. Acta*, 1977, **480**, 154.

the apoenzyme. The alkaline phosphatases of human kidney have been compared[150] with those of human liver and intestine. Marked similarities were noted. Finally, from a study of the activation of rat intestinal alkaline phosphatase by sorbitol and carbohydrates and its inhibition by phosphates, it was suggested[151] that alkaline phosphatase may be involved[152] in calcium transport, since the latter is influenced similarly by these entities.

RNA and DNA Polymerases.—Of all zinc-dependent enzymes currently under investigation none arouse[12] more interest than RNA and DNA polymerases, since they offer the first insight into possible causal relationships between the dramatic consequences of zinc deprivation or deficiency and the biochemical role of zinc. For example, in studies[153] on *Euglena gracilis*, grown in a medium seriously deficient in zinc, defects were observed[154] in nucleic acid and protein synthesis, and even in cellular division. All stages of cellular development in this organism were affected, *i.e.* those analogous to the first growth period, G_1, the period which follows (characterized by DNA synthesis), S, and the second growth period, G_2, to mitosis require zinc. Figure 1 indicates the key roles played by zinc metalloenzymes in these essential processes. During the year there has been a flurry of activity in the investigation of these nucleotidyl transferases, in particular, DNA polymerases[155-157] and RNA polymerases.[158-160] It had previously been shown that some RNA-dependent DNA polymerases[161-163] (reverse transcriptases) from tumour viruses contain zinc. In addition, Elongation Factor 1 from rat liver, which catalyses the binding of amino-acyl-tRNA to an RNA–ribosome site through the formation of an amino-acyl-tRNA–EF1–GTP ternary complex, requires zinc.[164] In a most important recent report the s-Met-tRNA synthetase from *E. coli* has been shown[165] to contain zinc. These and related results serve to emphasize the likely importance of zinc to the growth and development of all organisms.

There has been much concern with the location of the zinc ions, the interaction of template or substrate with the enzymes, and the relationship of zinc and the 'activator' metal ion (usually magnesium) to the fidelity of transcription. The

150 L. Korngold, *Int. Arch. Allergy Appl. Immunol.*, 1977, **54**, 300.
151 Y. Dupuis, A. Digaud, and N. Fontaine, *Calcif. Tissue Res.*, 1977, **22**, 556.
152 G. H. Bourne, in 'The Biochemistry and Physiology of Bone', ed. G. H. Bourne, Academic Press, New York, 1972, Vol. II, p. 79.
153 K. H. Falchuk, D. W. Fawcett, and B. L. Vallee, *J. Cell. Sci.*, 1975, **17**, 57.
154 K. H. Falchuk, A. Krishan, and B. L. Vallee, *Biochemistry*, 1975, **14**, 3439.
155 J. P. Slater, A. S. Mildvan, and L. A. Loeb, *Biochem. Biophys. Res. Comm.*, 1971, **44**, 37.
156 C. F. Springgate, A. S. Mildvan, R. Abramson, J. L. Engle, and L. A. Loeb, *J. Biol. Chem.*, 1973, **248**, 5987.
157 J. P. Slater, A. S. Mildvan, and L. A. Loeb, *Biochem. Biophys. Res. Comm.*, 1971, **44**, 37.
158 M. C. Scrutton, W. C. Wu, and D. A. Goldthwait, *Proc. Nat. Acad. Sci. U.S.A.*, 1971, **68**, 2497.
159 D. S. Auld, I. Atsuya, C. Campino, and P. Valenzuela, *Biochem. Biophys. Res. Comm.*, 1976, **69**, 548.
160 H. Lattke and U. Weser, *F.E.B.S. Letters*, 1976, **65**, 288.
161 D. S. Auld, H. Kawaguchi, D. M. Livingstone, and B. L. Vallee, *Proc. Nat. Acad. Sci. U.S.A.*, 1974, **71**, 2091.
162 D. S. Auld, H. Kawaguchi, D. M. Livingstone, and B. L. Vallee, *Biochem. Biophys. Res. Comm.*, 1975, **62**, 296.
163 B. J. Poiesz, G. Seal, and L. A. Loeb, *Proc. Nat. Acad. Sci. U.S.A.*, 1974, **71**, 4892.
164 S. Kotsiopoulos and S. C. Mohr, *Biochem. Biophys. Res. Comm.*, 1975, **67**, 979.
165 L. M. Psorske, D. S. Auld, and M. Cohn, *Fed. Proc.*, 1977, **36**, 706.

DNA-dependent RNA polymerase from *E. coli*, which contains 2 gram atom of zinc ions per mole of enzyme, has[166] the subunit structure $\alpha_2\beta\beta'\sigma$. The separated subunits obtained in the presence of 7M-urea do not contain zinc. However, it was found that both the β- and β'-subunits would take up zinc ions to give 0.6 ± 0.3 and 1.4 ± 0.5 gram atom of tightly bound zinc per mole of subunit respectively. It was suggested that at least one of the two tightly bound zinc ions in the RNA polymerase is located in the β'-subunit, whilst the other Zn^{II} ion may be in β'-, or β-, or at the contact domain of these two subunits. Since there does not yet exist a method for the removal of the zinc and its replacement by other metal ions, the reported isolation[167] of the DNA-dependent RNA polymerase from *E. coli* grown in a medium that is deficient in Zn^{II} and supplemented with Co^{II} will be of great value in delineating the role of the metal ion. After a slight lag, upon the introduction of the Co^{II} ions, the growth of the organism recovered. This provides a dramatic illustration of the close similarity between the properties, both structural and functional, of the zinc proteins and of their Co^{II} analogues. The purified enzyme contains 1.8—2.2 gram atom of cobalt per mole of enzyme, is enzymatically as active as the Zn-RNA polymerase, and the physical properties are all but identical. Naturally, the Co-RNA polymerase is coloured, with a spectrum as shown in Figure 8. The two major peaks are presumably associated with two different Co^{II} ions since they are differently affected by the addition of substrates or templates. Oxidation of the Co^{II} to Co^{III} results[166] in at least one Co^{III} ion remaining tightly bound to the isolated β'-subunit, even in the presence

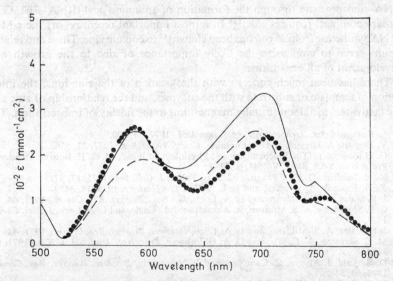

Figure 8 *Difference spectra between Co^{II}- and Zn^{II}-RNA polymerase;* (——) *no substrate or template added;* (.....) *0.4 mM-ATP or GTP added;* (– – – –) *10^{-5} M-d(pT)$_{10}$ added*
(Reproduced by permission from *European J. Biochem.*, 1977, **72**, 425)

[166] C.-W. Wu, F. Y.-H. Wu, and D. C. Speckhard, *Biochemistry*, 1977, **16**, 5449.
[167] D. C. Speckhard, F. Y.-H. Wu, and C.-W. Wu, *Biochemistry*, 1977, **16**, 5228.

of 6.5M-urea. This is consistent with other work on the same enzyme, which suggested[167] that one of the two intrinsic metal ions is located in the β'-subunit that is responsible for DNA binding and the other metal ion may be in β'- or β-subunit, or at the contact domain of these two subunits. Unfortunately, in view of the new[168] interpretation of the mechanism of inhibition, the effects of 1,10-phenanthroline upon the spectrum were not reported. An investigation of the kinetic studies showed that the two enzymes had V_{max} the same for both; K_m values for UTP and ATP were also similar, though the apparent K_m for T7 DNA was lower for the cobalt enzyme. The accuracy of transcription was the same for both enzymes, but the Co-RNA polymerase is less efficient in initiating RNA chains, as demonstrated by studying the rates of GTP/ATP incorporation into the 5'-terminal of RNA products. The important conclusion from the work is that the metal ion of RNA polymerase is involved in recognition of the promotor and specific initiation in RNA synthesis.

These studies have a direct bearing on the role(s) of the zinc. The zinc ion may co-ordinate the 3'-OH of the terminal nucleotide in the growing RNA chain, be involved in recognition of specific promotor sites on the DNA template, or be required for the maintenance of the proper subunit arrangement.

The RNA polymerase from *B. subtilis* has many features in common with that from *E. coli*. The most recent investigation[169] of the *B. subtilis* enzyme was concerned with the chromatography of the enzyme after dissociation into subunits using 6.5M-urea. One subunit was shown to contain both zinc ions, and, assuming that there had not been a 'scrambling' of the zinc ions during the experiment, it was suggested that in the holoenzyme both zinc ions were contained in the second largest subunit (designated β) and the re-designated β' (to be consistent with the nomenclature given[166] to the *E. coli* polymerase). Thus the location of zinc ions in the two polymerases may be related.

In an interesting study it was shown[170] that the RNA polymerase isolated from *Euglena gracilis* grown on a zinc-deficient medium contains 2 gram atom of Zn per 700000 gram of protein and differs from RNA polymerases isolated from zinc-replete *E. gracilis*. The messenger RNA isolated from the zinc-deficient cells had a different base composition, giving rise to a two-fold increase in the $(G + C)/(A + U)$ ratio. Since the intracellular concentration of manganese increases in zinc-deficient cells, the effect of Mn^{II} concentration on the activity of the RNA polymerase from zinc-deficient cells was investigated. It was shown that increasing the concentration of the Mn^{II} leads to an increased incorporation of GMP relative to UMP. Such studies should allow some insight to be gained into the biochemical mechanism of the interplay between metal-ion content in cells. The yeast RNA polymerase III has been isolated[171] and shown[172] to contain 2 gram atom of zinc per mole of enzyme, with a molecular weight of 380000. The enzyme was inhibited by 1,10-phenanthroline and, as expected, the enzyme-

168 V. d'Aurora, A. M. Stern, and D. S. Sigman, *Biochem. Biophys. Res. Comm.*, 1977, **78**, 170.
169 S. M. Halling, F J. Sanchez-Anzaldo, R. Fukuda, R. H. Doi, and C. F. Meares, *Biochemistry*, 1977, **16**, 2880.
170 K. H. Falchuk, C. Hardy, L. Ulpino, and B. L. Vallee, *Biochem. Biophys. Res. Comm.*, 1977, **77**, 314.
171 T. M. Wandzilak and R. W. Benson, *Biochem. Biophys. Res. Comm.*, 1977, **76**, 247.
172 T. M. Wandzilak and R. W. Benson, *Biochemistry*, 1978, **17**, 426.

bound zinc did not exchange with free zinc. The DNA-dependent RNA polymerase II from wheat-germ has been shown[173] to have a surprisingly high zinc content (7 gram atom per mole), which was removed on prolonged dialysis against 1,10-phenanthroline. The DNA-dependent RNA polymerase I, so classified because α-amanitin does not affect its activity, has been isolated[174] from *E. gracilis*. It contains 2.2 gram atom of Zn per molecular weight of 624000 and is inhibited by 1,10-phenanthroline. Four DNA polymerases have been isolated[175, 176] from extra-mitochondrial fractions of the marine photosynthetic diatom *Cylindrotheca fusiformis*. The enzymes were inhibited by 1,10-phenanthroline and presumably are zinc metalloenzymes. As with other DNA polymerases, they have an absolute requirement for Mg^{II} or Mn^{II}, the former being preferred. In addition, they are activated by high concentrations of K^+. Though, in the organism, DNA replication is uniquely dependent on silicate, no effect of silicate on the activity of the DNA polymerases was observed. The relationship between each polymerase and their functions in cell growth is discussed. From studies performed on an *E. coli* DNA polymerase it was suggested[168] that the ubiquitous inhibition of both RNA and DNA polymerases by 1,10-phenanthroline may be due to inhibition by the bis-1,10-phenanthroline-copper(I) cation, formed from adventitious copper(II) impurities with subsequent reduction by thiols, and not by co-ordination of the 1,10-phenanthroline to the enzyme-bound zinc ions. The reason offered for the supposed inhibition by $[Cu^I(1,10\text{-phen})_2]^+$ was due to its fortuitous complementarity to the active site. With the availability of the cobalt-containing polymerases it should be straightforward to test whether or not 1,10-phenanthroline co-ordinates to the cobalt(II) ion. It has been suggested,[177] on the basis of inhibition by 1,10-phenanthroline, that the zinc in RNA polymerase is not involved in template DNA binding. However, 1,10-phenanthroline does inhibit the pyrophosphate exchange, and this was adduced as evidence for the involvement of enzyme-bound zinc in the activation of the 3'-OH of the sugar moiety, thereby facilitating nucleophilic attack of the growing chain towards the α-phosphate of the entering nucleotide. Though a variety of ligands, such as bathophenanthroline disodium disulphonate (1), are[178] effective inhibitors of the RNA-dependent RNA polymerase from influenza virus, they were unfortunately without effect on viral polypeptide synthesis on infected cells. However, this is an approach which deserves further investigation.

The Role of the Activator. The interaction of Mn^{II}, substrates, and initiators with the RNA polymerase from *E. coli* has[179] been investigated. It was shown that Mn^{II} binds tightly to the enzyme and weakly at 6 ± 1 additional sites. After a thorough study, it was concluded that the tightly bound Mn^{II} is at the site responsible for chain elongation and that the Mn^{II} interacts with the leaving pyrophosphate group. In studies on the activity of human DNA polymerase

173 P. Petranyi, J. J. Jendrisak, and R. R. Burgess, *Biochem. Biophys. Res. Comm.*, 1977, **74**, 1031
174 K. H. Falchuk, L. Ulpino, B. Magus, and B. L. Vallee, *Biochem. Biophys. Res. Comm.*, 1977, **74**, 1206.
175 T. W. Okita and B. E. Volcani, *Biochem. J.*, 1977, **167**, 601.
176 T. W. Okita and B. E. Volcani, *Biochem. J.*, 1977, **167**, 611.
177 H. Lattke and U. Weser, *F.E.B.S. Letters*, 1977, **83**, 297.
178 J. S. Oxford and D. D. Perrin, *Ann. New York Acad. Sci.*, 1977, **284**, 613.
179 R. Koren and A. S. Mildvan, *Biochemistry*, 1977, **16**, 241.

(1)

in vitro it was observed[180] that the substitution of Mn^{II} for Mg^{II} had profound effects on the kinetic parameters of the polymerase reaction. Depending on the reaction conditions, Mn^{II} may be as effective as Mg^{II}, strongly preferred, or absolutely required. This influence of the metal 'activator', in contrast to the role of the tightly bound zinc, appears to be concerned with the binding, alignment, and conformational alteration of the incoming deoxynucleoside triphosphate within the active site of the polymerase. An interesting investigation has been concerned[181] with the effect of different metal activators on the fidelity of DNA synthesis, using the DNA polymerase from avian myeloblastosis virus. With poly[d(A–T)] as a template, the error frequencies for dCMP incorporation were $1:1400$, $1:1100$, and $1:600$ for Mg^{II}, Co^{II}, and Mn^{II} respectively. Increasing the concentration of Mg^{II} above that required for maximum activity had no effect on the error frequency, whilst increasing the concentration of Co^{II} and Mn^{II} resulted in a greater frequency of errors, which was due to differential rates of complementary and non-complementary nucleotide incorporation. The effect of Ni^{II} on DNA synthesis was much the same as that of Mg^{II}. On the other hand, addition of Ni^{II} (or Co^{II}, or Mn^{II}) to the Mg^{II}-activated polymerase led to a decrease in the fidelity of DNA synthesis. It is possible that metal ions such as Mn^{II}, Co^{II}, and Ni^{II}, in addition to binding at the Mg^{II} site, also bind to additional sites in the enzyme, with consistent structural changes. The relationship to the carcinogenicity of the metals is briefly considered.

In an attempt to understand the mutagenicity of metals, the influence of the metal ions on the activity of *E. coli* DNA polymerases I, II, and III was investigated.[182] It was shown that Be, Ca, and Cu weakly and Zn, Cd, and Hg strongly inhibited the polymerase reaction, presumably by replacing magnesium. However, the results derived from the investigation of the misincorporation at concentrations of metal ions which were sufficient to activate the enzyme; at higher concentrations, misincorporation occurred, suggesting that binding of Mn^{II} at additional inhibitory sites is necessary for misincorporation to occur. Whilst Be, Fe, Co, and Ni ions apparently increased misincorporation, Cd, Hg, and Zn had little effect. Studies on the affinity of the polymerase for the template also revealed the difference between the effects of Mn and Mg.

[180] T. S.-F. Wang, D. C. Eichler, and D. Korn, *Biochemistry*, 1977, **22**, 4927.
[181] M. A. Sirover and L. A. Loeb, *J. Biol. Chem.*, 1977, **252**, 3605.
[182] M. Miyaki, I. Murata, M. Osabe, and T. Ono, *Biochem. Biophys. Res. Comm.*, 1977, **77**. 845.

Other Zinc Metalloenzymes.—The effects of metal ions on the activity of δ-aminolaevulinic acid dehydratase continue to attract great attention. It is now well established that Zn^{II} activates this enzyme, while Pb^{II} is a strong inhibitor. Recent work[183-185] extends these results and demonstrates that Al^{III} is a good activator of this enzyme, while Cu^{II} is a powerful inhibitor.

Two forms of α-D-mannosidase have been isolated from *Phaseolus vulgaris*[186] and both are reported to contain two mole of Zn^{II} per mole of protein. The presence of four different types of α-D-mannosidase in rat tissue has been reported.[187] One of these (active at acidic pH) is a Zn-dependent enzyme, while the other three (active near to neutral pH) are claimed to be zinc-independent but activated by Fe^{II}, Co^{II}, or Mn^{II}.

Several new zinc metalloenzymes have been recognized during the past year. Pig renal amino-acylase[188] contains two tightly bound Zn atoms per mole of protein. This enzyme is dimeric, and hence it is presumed that the metal stoicheiometry is 1 gram atom of Zn^{II} per subunit. The enzyme is completely inactivated by chelating agents such as 1,10-phenanthroline and EDTA. Dialysis against 1,10-phenanthroline results in an inactive enzyme, which can be re-activated by the addition of Zn^{II}. The Zn-dissociation constant at pH 7.8 was estimated to be 10^{-10} mol l^{-1}.

Yeast nucleotide pyrophosphatase is reported to be a zinc metalloenzyme[189] containing 1 gram atom of Zn^{II} per mole of protein (mol. wt. 65000). Equine angiotensin-converting enzyme has also been shown[190] to be a zinc metalloenzyme (as is the enzyme from rabbit lung,[191] containing 1 gram atom of Zn per mole of protein). Both enzymes are completely inhibited by chelating agents. Monkey brain arylamidase is also reported[192] to be inhibited by EDTA; it can subsequently be re-activated by Zn, Co, or Mn, suggesting that this enzyme could be a zinc metalloenzyme, but more careful work is needed to ascertain this.

Rat liver fructose-1,6-biphosphatase has been shown[193] to contain twelve Zn^{II}-binding sites (since this enzyme is tetrameric, this is equivalent to three sites per subunit). The third set of sites, having the lowest affinity, appears to be identical to the binding sites for the activating cation, *i.e.* Mg^{II}, required for this enzyme. The binding of Zn^{II} to the first set of sites gives an enzyme of intermediate activity, while binding to the second set of sites results in almost complete inhibition. It has been suggested[193] on the basis of these results that Zn^{II} functions as an activator and as a negative allosteric regulator of fructose-1,6-biphosphatase.

[183] P. A. Meredith, M. R. Moore, and A. Goldberg, *Enzyme*, 1977, **22**, 22.
[184] J. Thompson, D. D. Jones, and W. H. Beasley, *Brit. J. Ind. Med.*, 1977, **34**, 32.
[185] N. Despaux, C. Bohvon, E. Comoy, and C. Boudene, *Biomedicine*, 1977, **27**, 358.
[186] E. Paus, *European J. Biochem.*, 1977, **73**, 155.
[187] S. M. Snaith, *Biochem. J.*, 1977, **163**, 557.
[188] W. Kordel and F. Schneider, *Z. Naturforsch.*, 1977, **32c**, 342.
[189] J. S. Twu, R. K. Haroz, and R. K. Bretthauer, *Arch. Biochem. Biophys.*, 1977, **184**, 249.
[190] M. Das and R. L. Soffer, *J. Biol. Chem.*, 1975, **250**, 6762.
[191] R. T. Fernley, *Clin. Exp. Pharmacol. Physiol.*, 1977, **4**, 267.
[192] H. Motoharu and O. Kiyoshi, *J. Biochem. (Japan)*, 1977, **81**, 631.
[193] F. O. Pedrosa, S. Pontremol, and B. L. Horecker, *Proc. Nat. Acad. Sci. U.S.A.*, 1977, **74**, 2742.

Human serum α_2-macroglobulin has been found[194] to be the major Cd^{II}-binding protein *in vitro*. This protein is known[195] to be the major zinc-containing protein in human serum. It has been shown that the levels of Zn bound to α_2-macroglobulin in human serum are unaltered during acute disease, while the level of zinc bound to smaller proteins (possibly albumins) is dramatically reduced (Figure 9). Similarly, the administration of ACTH reduces the zinc content of the latter protein fraction, but it does not alter that bound to α_2-macroglobulin.[196] These results represent the first step in the biochemical understanding of the well-known[196, 197] effects of corticosteroids and of acute disease in depressing serum zinc levels (Figure 10).

2 Manganese-activated Enzymes

Glycolytic Enzymes.—*Phosphoglucomutase* (E.C. 2.7.5.1). This enzyme catalyses the interconversion of glucose-1-phosphate and glucose-6-phosphate *via* a

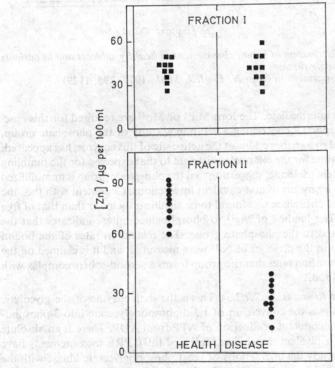

Figure 9 *The effect of acute disease on the α_2-macroglobulin fraction (I) and albumin fraction (II) of human serum*
(Reproduced by permission from *N. Engl. J. Med.*, 1977, **296**, 1129)

194 S. R. Watkins, R. M. Hodge, D. C. Cowman, and P. P. Wickham, *Biochem. Biophys. Res. Comm.*, 1977, **74**, 1403.
195 A. R. Parisi and B. L. Vallee, *Biochemistry*, 1970, **9**, 2421.
196 K. E. Falchuk, *N. Engl. J. Med.*, 1977, **296**, 1129.
197 G. Michaelsson, A. Vahlquist, and L. Juhlin, *Br. J. Dermatol.*, 1977, **96**, 283.

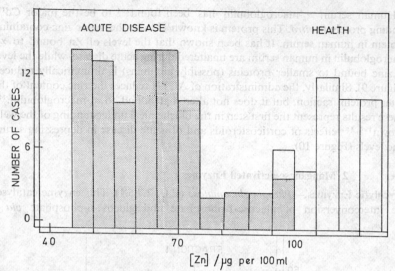

Figure 10 *The distribution of serum zinc content in healthy subjects and in patients with acute diseases*
(Reproduced by permission from *N. Engl. J. Med.*, 1977, **296**, 1129)

phosphoenzyme intermediate. The ions Mg^{II} or Mn^{II} are required for this reaction. A recent report[198] concerning a ^{31}P n.m.r. study of the phosphate group, covalently bound to a serine residue at the active site of this enzyme, has appeared. The correlation time for the ^{31}P nucleus is close to that expected for the tumbling of the entire protein molecule, suggesting that the phosphate group is immobilized at the active site by means of non-covalent interactions. Consistent with this, the pK_a of the bound phosphate was found to be significantly lower than that of free phosphoserine. The binding of Mg^{II} to phosphoglucomutase indicates that the Mg^{II} is not bound to the phosphate group. The relaxation rates of the bound phosphate group in the presence of Ni^{II} were measured, and it is claimed on the basis of these relaxation rates that the group forms a second-sphere complex with the bound metal ion.

Phosphoglycerate Kinase (E.C. 2.7.2.3). This is the sixth enzyme of the glycolytic pathway. It catalyses the conversion of 1,3-diphosphoglycerate into 3-phosphoglycerate with concomitant production of ATP from ADP. There is an absolute requirement for a bivalent metal ion (Mg^{II} or Mn^{II}). PRE measurements have been used[199] to study the interactions of yeast phosphoglycerate kinase with the Mn^{II} complexes of a number of nucleotides (ADP, GDP, IDP). Substantial synergism was found in the binding of the metal ion and nucleotide to the enzyme in the ternary complex. The metal nucleotide binds to the enzyme approximately two orders of magnitude more tightly than the free nucleotide.

[198] W. J. Ray, A. S. Mildvan, and J. B. Grutzner, *Arch. Biochem. Biophys.*, 1977, **184**, 453.
[199] B. E. Chapman, W. J. O'Sullivan, R. K. Scopes, and G. H. Reed, *Biochemistry*, 1977, **16**, 1005.

This synergic behaviour is not typical of other related kinases, such as creatine kinase and arginine kinase. The e.p.r. spectrum for the bound Mn^{II} in the enzyme–Mn–ADP complex was also found to differ substantially from those obtained for other kinases. These observations are consistent with previous suggestions that the structure of the Mn^{II}–nucleotide–enzyme ternary complex for phosphoglycerate kinase differs from that of other kinases.[199] It has been postulated that in this ternary complex the metal ion is bound to the phosphate groups of the nucleotide and also to a donor group on the enzyme. The PRE measurements on the Mn–ADP–enzyme complex in the presence of the product 3-phosphoglycerate indicate that a quaternary complex is formed between these species, despite the fact that previous X-ray diffraction studies had failed to detect such a quaternary complex.[200] It has been suggested that this discrepancy is due to the presence of high concentrations of $SO_4{}^{2-}$ in the solutions used for the X-ray work.

Pyruvate Kinase (E.C. 2.7.1.40). This is the last glycolytic enzyme, and it catalyses the reaction

$$\text{phosphoenolpyruvate} + \text{ADP} \rightleftharpoons \text{pyruvate} + \text{ATP}$$

Two metal ions per active site are required during this reaction; one of these is bound directly by the enzyme, while the other is co-ordinated to the phosphoryl groups of ATP or ADP. The distance between Mn^{II} and Cr^{III} in the Mn^{II}–enzyme–Cr^{III}–ATP complex has been measured,[201] by using a novel PRE method, based on the cross-relaxation effects of the two metal ions. The distance between Mn^{II} and Cr^{III} was estimated to be 5.2 ± 0.9 Å; this indicates the existence of a van der Waals contact between the hydration spheres of the enzyme- and nucleotide-bound metal ions.

Enzymes of the Tricarboxylic Acid Cycle.—NADP+-dependent isocitrate dehydrogenase (E.C. 1.1.1.42) is found in both the mitochondria and the cytoplasm of mammalian tissues. It is[202] monomeric, with mol. wt. of approximately 60000. The enzyme catalyses the oxidative decarboxylation of isocitric acid to α-ketoglutaric acid in the presence of a variety of bivalent ions. Of the catalytically active metals, Zn^{II} and Mn^{II} give the highest and Ni^{II} the lowest activity.[203] The metal ion is believed to stabilize the transiently formed enolate of α-ketoglutarate during catalysis. In an effort to understand better the role of the metal in this enzyme, magnetic resonance techniques have been employed[204, 205] in the study of the interaction of metals, nucleotide cofactors (NADP+/NADPH), and substrates with the enzyme. The authors[204] claim that the e.p.r. spectrum of the Mn^{II} enzyme shows six hyperfine lines, with no zero-field splitting (ZFS), indicating that the electronic environment of the metal ion is quite isotropic. The addition of α-ketoglutarate causes little alteration of the spectrum of the Mn^{II} enzyme, but the addition of isocitrate results in an extremely anisotropic spectrum. Under these

200 C. C. F. Blake and P. R. Evans, *J. Mol. Biol.*, 1974, **84**, 585.
201 R. K. Gupta, *J. Biol. Chem.*, 1977, **252**, 5183.
202 G. W. Plaut, G. Siebert, and M. Carsiotis, *J. Biol. Chem.*, 1957, **226**, 977.
203 D. B. Northrop and W. W. Cleland, *Fed. Proc. Fed. Am. Soc. Exp. Biol.*, 1970, **29**, 408 (Abstr.).
204 R. S. Levy and J. J. Villafranca, *Biochemistry*, 1977, **16**, 3293.
205 R. S. Levy and J. J. Villafranca, *Biochemistry*, 1977, **16**, 3301

conditions, pronounced ZFS and several broad, low-field fine-structure transitions are observed. On the basis of these results it was suggested[204] that the substrate binds directly to, or extremely near to, the Mn^{II} ion. This would be consistent with previous PRE data.[206] In addition, these e.p.r. studies[204] also demonstrated the mutually exclusive binding relationships between $NADP^+$ and NADPH, between Mn^{II} isocitrate and NADPH, and between HCO_3^- and formate or thiocyanate. Upper limits for the dissociation constants of the Co^{II} and Mg^{II} forms of the enzyme were obtained from this work by metal competition studies, involving the measurement of free Mn^{II} by e.p.r. spectroscopy. PRE studies[205] suggest that the nucleotide cofactors $NADP^+$/NADPH and the substrate analogue oxalylglycine bind near to, but not directly to, the metal ion. The PRE data also indicate that NADPH has a synergistic effect on the binding of substrate. Detailed studies[207] on the isotope effect in the mechanism of isocitrate dehydrogenase have been carried out. The carbon isotope effect for the carboxyl carbon undergoing decarboxylation is $k^{12}/k^{13} = 0.9989 \pm 0.0004$ for the Mg^{II}-activated enzyme and 1.0051 ± 0.0012 for the Ni^{II} form of the enzyme. The hydrogen isotope effect too is near unity for these two metallo-forms. This lack of a significant isotope effect indicates that oxidation and decarboxylation are much faster than release of product, which must therefore be the rate-limiting step.[207] Hence, the large rate decrease caused by substitution of Ni^{II} for Mg^{II} must result from the effects of metal on binding of substrate and dissociation of product, rather than effects of metal on catalysis. However, the authors[207] conclude that the different carbon isotope effects in the presence of Mg^{II} and Ni^{II} suggest that there is a large metal effect on the rate of the decarboxylation step, even though this step never becomes rate-limiting. This is consistent with the view that the carbonyl oxygen of the oxalosuccinate intermediate is co-ordinated to the metal during decarboxylation.

Isocitrate dehydrogenase has been reported[208] to be completely inactivated by methyl methanethiosulphonate, with a concomitant modification of one cysteine group. Manganese(II) or $NADP^+$ completely protect the enzyme against inactivation. Hence a cysteine group is implicated in the catalytic activity of the enzyme.

Succinyl-CoA Synthetase (E.C. 6.2.1.4). This enzyme catalyses the reversible hydrolysis of succinyl-CoA to succinate and CoA, which is coupled with phosphorylation of a nucleotide diphosphate to a nucleotide triphosphate.[209] The enzyme from *E. coli* is a tetramer with a molecular weight of approximately 140 000 and consists of two subunit types (α and β), of differing size. The obligatory requirement of succinyl-CoA synthetase for a bivalent cation can be satisfied by Mg^{II} or Mn^{II}. The reaction proceeds *via* a phosphoenzyme intermediate; a histidyl residue of the α-subunit has been identified as the site of phosphorylation. A magnetic resonance study of this enzyme has been reported.[210] The authors report that both the phosphoenzyme and the dephosphoenzyme bind two Mn^{II} ions per ($\alpha + \beta$) subunit with dissociation constants of 0.74 mmol

206   J. J. Villafranca and R. F. Coleman, *Biochemistry*, 1974, **13**, 1152.
207   M. H. O'Leary and J. A. Limburg, *Biochemistry*, 1977, **16**, 1129.
208   J. J. Villafranca, R. S. Levy, J. Kernich, and T. Vickroy, *Biochem. Biophys. Res. Comm.*, 1977, **77**, 457.
209   W. A. Bridger, in 'The Enzymes', ed. P. D. Boyer, Academic Press, New York, 1974, Vol. 10, p. 581.
210   D. H. Buttlaire, M. Cohn, and W. A. Bridger, *J. Biol. Chem.*, 1977, **252**, 1957.

l^{-1} and 1.4 mmol l^{-1} respectively. The enzyme binds Mn-ADP approximately 130 times more tightly than it binds free ADP, while Mn-ATP is bound very weakly, or not at all. The upper limits of the dissociation constant for CoA from the quaternary enzyme–Mn^{II}-ADP–succinate–CoA complex and for succinyl-CoA from the enzyme–Mn^{II}-ADP–succinyl-CoA complex are reported to be 390 μmol l^{-1} and 560 μmol l^{-1} respectively.[210] The successive occupation of substrate-binding sites resulted in a progressive decrease in PRE, indicative of alterations in the molecular dynamics and/or in the conformation of the active site. This is in agreement with the previously observed catalytic properties of the enzyme system, which had shown that the catalytic function of succinyl-CoA synthetase is potentiated by substrate binding.

Nucleotide Cyclases.—It is now realised that adenylate cyclase (E.C. 4.6.1.1) and guanylate cyclase (E.C. 4.6.1.2), and enzymes responsible for the formation of c-AMP and c-GMP respectively, have a central role in mediating cellular responses to a variety of stimuli. As yet, however, these enzymes are very poorly characterized, largely because of the low levels in which they occur in most tissues. It is known that these nucleotide cyclases are activated by bivalent metal ions, and that Mn^{II} is the most effective metal ion in this respect. Several papers[211-215] relating to the isolation of Mn^{II}-activated nucleotide cyclases have appeared during the past year.

Adenylate cyclase isolated from turkey erythrocyte membranes is reported[213] to have a mol. wt. of 316000. It is reported[213] that this enzyme is inhibited by Ca^{II} in a co-operative fashion, with a Hill coefficient of $n_H = 2.0$.

It has been known for a long time[216] that adenylate cyclase systems are activated by bivalent cations at concentrations in excess of those required to form the productive substrate, M^{II}-ATP. It has been proposed that such activation occurs *via* interaction with a metal-ion-binding site which is independent of the active site. However, it has been suggested that ATP is a potent inhibitor of adenylate cyclase, and that the activation by metal ions is due to a reduction in the free (uncomplexed) ATP concentration.[217] However, recent work[215] casts doubts on this interpretation. It is reported[215] that activation by Mg^{II} requires concentrations 50—100-fold greater than are required with Mn^{II}, a difference far greater than predicted by the relative stability constants for MnATP^{2-} and MgATP^{2-} if the sole action of these cations was to decrease the concentration of ATP^{4-}. Furthermore, activation is independent of changes in the ratio of ATP to metal-ATP^{2-}, and the activity with combinations of cations at their respective maximally effective concentrations is not additive.

The increase in activity of adenylate cyclase in the presence of appropriate hormones has also been attributed to reversal of ATP inhibition; it has been suggested that hormones activate by reducing the affinity of ATP for the enzyme.

[211] T. Braun, H. Frank, R. Dods, and S. Sepsenwool, *Biochim. Biophys. Acta*, 1977, **481**, 227.
[212] D. F. Malamud, C. C. Dirusso, and J. R. Aprille, *Biochim. Biophys. Acta*, 1977, **485**, 243
[213] E. Hanski, N. Sevilla, and A. Levitzki, *European J. Biochem.*, 1977, **76**, 513.
[214] W. H. Frey, B. M. Boman, D. Newman, and N. D. Goldberg, *J. Biol. Chem.*, 1977, **252**, 4298.
[215] C. Londos and M. S. Preston, *J. Biol. Chem.*, 1977, **252**, 5957.
[216] E. W. Sutherland and T. W. Rall, *Pharmacol. Rev.*, 1960, **23**, 265.
[217] C. DeHaen, *J. Biol. Chem.*, 1974, **249**, 2756.

However, since the above work[215] suggests that ATP does *not* inhibit adenylate cyclase, this interpretation is not borne out. These authors[215] further report that GTP alone, or in combination with glucogon, causes an increase in the apparent affinity for Mn^{II}, suggesting that the hormone, nucleotide, and metal-ion sites are linked heterotopically.

Biosynthetic Enzymes.—*Glycogen Synthase* (E.C. 2.4.1.11). This catalyses the formation of glycogen according to the reaction

$$\text{UDP-glucose} + (\text{glucose})_n \rightleftharpoons (\text{glucose})_{n+1}.$$

One form of this enzyme (called the D-form) is activated by glucose-6-phosphate. It has been reported that some bivalent cations, such as Mg^{II}, Mn^{II}, and Ca^{II}, which alone do not activate the enzyme, augment the activation by glucose-6-phosphate.[218, 219] In contrast, PO_4^{3-} and SO_4^{2-} reduce the activation by glucose-6-phosphate.[218] This decrease in the activation does not occur, however, if one of the metal ions is also present.[219] Although these metal ions, and the anions, separately have no effect on the activity of the enzyme in the absence of glucose-6-phosphate, Mn^{II} and SO_4^{2-} form a unique combination which greatly increases the enzyme activity. This activation by $MnSO_4$ has been further investigated[220] in detail for the D-form of glycogen synthetase isolated from human placenta. This study[220] confirms that $MnSO_4$ can activate this enzyme in the absence of glucose-6-phosphate, though Mn^{II} and SO_4^{2-} separately have no effect under these conditions. In the presence of glucose-6-phosphate, Mn^{II} activates the enzyme, SO_4^{2-} inhibits it, but $MnSO_4$ synergistically increases the enzyme activity. $MnSO_4$, like glucose-6-phosphate, increases the V_{max} of the enzyme in the presence of its substrate, UDP-glucose; $MnSO_4$ also slightly increases the K_m for UDP-glucose. At physiological concentrations of UDP-glucose and glucose-6-phosphate, either Mn^{II} or $MnSO_4$, at concentrations less than 1 mmol l^{-1}, greatly activate the enzyme. Manganese(II) can also relieve the inhibition of this enzyme by ATP. The D-form of glycogen synthase is considered to be physiologically inactive; this conclusion is based on the observation of the low activity of the enzyme at physiological concentrations of glucose-6-phosphate and its sensitivity to inhibition by nucleotides. However, this view may have to be modified in the light of the high activity of this enzyme in the presence of Mn^{II} that was reported above.[220]

Glutamine Synthetase (E.C. 6.3.1.2). This is a regulatory enzyme of great complexity in both structure and function. It catalyses the reaction

$$\text{NH}_3 + \text{glutamic acid} + \text{ATP} \rightarrow \text{glutamine} + \text{ADP} + \text{P}_i$$

E. coli glutamine synthetase contains three types of metal-ion-binding sites, with different affinities for Mn^{II} per dodecamer. To obtain a catalytically active protein molecule, at least two metal ions must be bound per subunit.[221] Conformational changes (both inter- and intra-subunit) accompany the binding of

[218] D. C. Lin, H. L. Segal, and E. J. Massaro, *Biochemistry*, 1972, **11**, 4466.
[219] K. P. Huang and J. C. Robinson, *Arch. Biochem. Biophys.*, 1976, **173**, 583.
[220] K. P. Huang and J. C. Robinson, *J. Biol. Chem.*, 1977, **252**, 3240.
[221] E. R. Stadtman and A. Ginsburg, in 'The Enzymes', ed. P. D. Boyer, Academic Press, New York, 1974, Vol. 10, p. 755.

metal ions and substrates to glutamine synthetase from *E. coli*.[222, 223] It has been reported[224] that Mn^{II} binds randomly to the two subunit metal ion sites and that an active enzyme can be formed by sequential binding of Mn^{II} and ADP to the second metal ion site or the direct binding of the Mn–ADP complex. The isolation and characterization of glutamine synthetase from *Salmonella typhimurium* has recently been reported.[225] Like the *E. coli* enzyme, *S. typhimurium* glutamine synthetase is a dodecamer; the enzyme molecule has been shown from electron micrographs to be a symmetrical aggregate of 12 subunits arranged in two hexagonal layers. The subunit molecular weight is approximately 50000. Two classes of metal-ion sites per subunit were found.[225] The apparent k_D values at pH 7.1 were 3.7×10^{-6} mol l^{-1} and 1.7×10^{-4} mol l^{-1}; these apparent k_D values drop with decreasing pH. PRE studies on the water protons at pH 7.1 indicate that both metal sites interact with the solvent. However, the interaction of the metal at the tight site is probably due to second-sphere interaction, while the metal at the second site probably has two or three rapidly exchanging water molecules in its co-ordination sphere. The e.p.r. spectrum of enzyme-bound Mn^{II} at the tight site is isotropic, and is dramatically sharpened by adding the substrate analogue methionine sulphoximine. Subsequent addition of ATP or the ATP analogue AMP-PCP (adenyl-methylene-diphosphate) produced anisotropic spectra that were similar, suggesting that both ATP and AMP-PCP bind similarly on the enzyme surface. However, a marked change in the Mn^{II} environment from anisotropic to nearly cubic results from the addition of ADP to the quaternary enzyme–Mn^{II}–sulphoximine–(AMP-PCP) complex, indicating that ADP displaces AMP-PCP. No change in the anisotropic spectrum due to the enzyme–Mn^{II}–sulphoximine–ATP complex is caused by the addition of ADP. This observation is consistent with previous findings[226] that ATP phosphorylates methionine sulphoximine, thereby producing an inactive enzyme. The allosteric effectors AMP and tryptophan have little effect on the e.p.r. spectrum of the Mn^{II} enzyme in the presence of sulphoximine and ADP,[225] suggesting the absence of direct co-ordination of AMP or tryptophan to the bound Mn^{II}.

Methionyl-tRNA Synthetase (E.C. 6.1.1.10). This enzyme catalyses the reaction

$$\text{enzyme} + \text{methionine} + \text{ATP} \rightleftharpoons \text{E—Met—AMP} + \text{PP}_i$$

A metal ion, usually Mg^{II}, is required for this reaction. Mechanistic studies have shown that the Mg^{II} is only involved as ATP–Mg^{II} and PP$_i$–Mg^{II} chelates. It has been reported[227] that, in the reaction catalysed by methionyl-tRNA synthetase isolated from *E. coli*, Mg^{II} can be efficiently replaced by Ni^{II}, Co^{II}, and Mn^{II}. These authors further report[227] that, while the standard ΔG value for the reaction (1) is close to zero in the case of Mg^{II}, Mn^{II} slows down the back reaction and

$$\text{E—Met—ATP—M}^{II} \rightleftharpoons \text{E—Met—AMP—PP}_i\text{—M}^{II} \tag{1}$$

222 A. Ginsburg, *Adv. Protein Chem.*, 1972, **11**, 1.
223 S. G. Rhee and P. B. Chock, *Proc. Nat. Acad. Sci. U.S.A.*, 1976, **59**, 476.
224 J. B. Hunt, P. Z. Smyrniotis, A. Ginsburg, and E. R. Stadtman, *Arch. Biochem. Biophys.*, 1975, **166**, 102.
225 M. S. Balakrishnan, J. J. Villafranca, and J. E. Brenchley, *Arch. Biochem. Biophys.*, 1977, **181**, 603.
226 A. R. Ronzio and A. Meister, *Proc. Nat. Acad. Sci. U.S.A.*, 1968, **59**, 164.
227 F. Hyafil and S. Blanquet, *European J. Biochem.*, 1977, **74**, 48.

hence shifts the reaction towards the AMP complex. The ΔG for the formation of the E–Met–AMP complex does not depend on the metal used, suggesting that the bivalent ion does not participate in the structure of this complex. The substitution of Mn^{II} for Mg^{II} also results in a ten-fold decrease of the dissociation constant of PP_i–M^{II} from the E–Met–AMP–PP_i complex and from the abortive analogue E–Met–adenosine–PP_i–M^{II}. Similarly, the dissociation constant of ATP–M^{II} from another dead-end analogue E–methioninol–ATP–M^{II} is decreased by Mn^{II}. Involvement of the N-7 atom of purine in the binding of the metal ion to the active site of methionyl-tRNA synthetase was ruled out[227] in this study by the use of 7-deaza-adenosine.

ATP Phosphoribosyltransferase (E.C. 2.2.2.17). It is the first enzyme of the histidine biosynthetic pathway, and is allosterically inhibited by the end product, histidine.[228] The reaction catalysed by this enzyme is

$$ATP + phosphoribosyl\ pyrophosphate \rightleftharpoons phosphoribosyl—ATP + PP_i$$

This reaction requires Mg^{II} or Mn^{II}. An e.p.r. study on the enzyme from *E. coli* has been carried out.[229] The dissociation constants for the dissociation of Mn^{II} from the substrates and products of the reaction were estimated from this study to be 10—15 times lower than those for the corresponding Mg^{II} complexes. These studies[229] do not show any significant direct interaction of either Mn-ADP or Mn^{II} with the enzyme, and hence it is concluded that the metal does not interact with the protein, and that in the ternary complex (enzyme–nucleotide–metal) the nucleotide acts as a bridge between the enzyme and the metal ion.

Prenyltransferase (E.C. 2.5.1.1). The enzyme that is involved in isoprenoid biosynthesis, responsible for the condensation of isopentenyl pyrophosphate with dimethylallyl pyrophosphate to give geranyl pyrophosphate, and also for the condensation of geranyl pyrophosphate with isopentenyl pyrophosphate to give farnesyl pyrophosphate, is prenyltransferase (Scheme 1). Either Mn^{II} or Mg^{II} is required for this reaction. It has been reported[230] that the enzyme is capable of binding either of its substrates in the absence of a bivalent cation, with a reduced affinity compared to that observed in the presence of a metal ion. In the absence of the bivalent cation, the binding of the substrates is non-competitive, but the binding of the product (farnesyl pyrophosphate) and isopentenyl pyrophosphate is competitive. In the absence of substrate, Mn^{II} is not bound to the enzyme, but when the product or either substrate is present, 2 gram atom of metal are bound per enzyme subunit. These results[230] suggest that the pyrophosphate groups of both substrates are linked *via* metal-ion bridges.

New Manganese(II)-requiring Enzymes.—A Mn^{II}-activated phosphohistone phosphatase has been isolated from canine heart. The molecular weight of the enzyme was calculated to be 61 000.[231] The enzyme is inactive in the absence of bivalent cations. The order of activation by bivalent metal ions is $Mn^{II} > Co^{II} > Mg^{II}$. Calcium(II) is ineffective in restoring activity, while Zn^{II}, Fe^{II}, and Cu^{II} are

[228] B. N. Ames, R. G. Martin, and B. J. Garry, *J. Biol. Chem.*, 1961, **226**, 2019.
[229] A. R. Tebar, A. Ballesteros, and J. Soria, *Experientia*, 1977, **33**, 1292.
[230] H. J. King and H. C. Rilling, *Biochemistry*, 1977, **16**, 3815.
[231] H.-C. Li and K.-J. Hsiao, *Arch. Biochem. Biophys.*, 1977, **179**, 147.

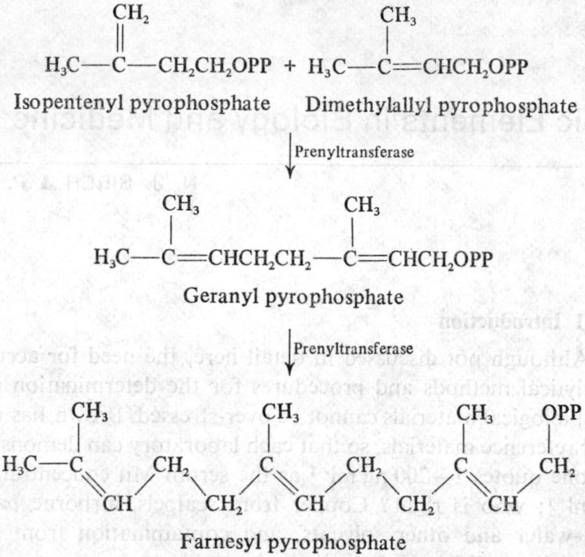

$$CH_2 \qquad\qquad CH_3$$
$$\|\qquad\qquad\qquad\quad \|$$
$$H_3C-C-CH_2CH_2OPP + H_3C-C{=}CHCH_2OPP$$

Isopentenyl pyrophosphate   Dimethylallyl pyrophosphate

↓ Prenyltransferase

$$CH_3 \qquad\qquad CH_3$$
$$| \qquad\qquad\qquad | $$
$$H_3C-C{=}CHCH_2CH_2-C{=}CHCH_2OPP$$

Geranyl pyrophosphate

↓ Prenyltransferase

$$CH_3 \qquad\quad CH_3 \qquad\quad CH_3 \quad OPP$$
$$| \qquad\qquad\quad | \qquad\qquad\quad | \qquad |$$
$$H_3C-C \qquad CH_2 \qquad C \qquad CH_2 \qquad C \qquad CH_2$$
$$\quad\ \ CH \qquad CH_2 \qquad CH \qquad CH_2 \qquad CH$$

Farnesyl pyrophosphate

The reactions catalysed by prenyltransferase

Scheme 1

inhibitory. The enzyme is reported to be maximally active between pH 7.0 and 7.5 and to have an apparent K_m for the substrate phosphohistone and for Mn^{II} of approximately 17 μmol l^{-1} and 500 μmol l^{-1} respectively. It is inhibited by a variety of nucleotide phosphates, and by inorganic phosphate and pyrophosphate, but is not affected by c-AMP or c-GMP. KCl and other salts greatly affect the rate of dephosphorylation of phosphohistone and the K_m for either Mn^{II} or phosphohistone.

Guanidinacetate Hydrolase (E.C. 3.5.3.2). As isolated from *Pseudomonas* sp. ATCC 14676, this enzyme requires Mn^{II} for activity.[232] It catalyses the hydrolytic cleavage of guanidinacetate to glycine and urea. The enzyme is said to be a tetramer, with subunit mol. wt. of 38000. The enzyme is completely inactive in the absence of metal ions. Maximal activity is restored by Mn^{II}, while Fe^{II}, Co^{II}, or Ni^{II} restore about 10% of the activity. The apparent K_m values for guanidinacetate and Mn^{II} were estimated to be 9.1 mmol l^{-1} and 1.3 μmol l^{-1} respectively. ATP inhibits the enzyme, and the inhibition is competitive with Mn^{II}.

An acid phosphatase isolated from soy bean[233] is said to be a Mn^{II} metalloenzyme containing 0.04% (by weight) of tightly bound Mn^{II}.

[232] T. Yorifuji, H. Tamai, and H. Usami, *Agric. and Biol. Chem. (Japan)*, 1977, **41**, 959.
[233] F. Sadaki, M. Tsutomu, and O. Akira, *Agric. and Biol. Chem. (Japan)*, 1977, **41**, 599.

Inorganic Elements in Biology and Medicine

N. J. BIRCH & P. J. SADLER

1 Introduction

Analysis.—Although not discussed in detail here, the need for accurate, reproducible analytical methods and procedures for the determination of inorganic elements in biological materials cannot be over-stressed. Brown has emphasized[1] the need for reference materials, so that each laboratory can demonstrate its own reliability: one quotes 1—200 ng ml^{-1} as the serum Mn concentration, another 20—30 ng ml^{-1}; who is right? Copper from scalpels, airborne particles, particulates in water and other solvents, and contamination from hands (even washed ones!) have all been known to cause problems.

On the basis of analysis alone, it would appear that the role of some elements in biological materials is grossly neglected at present. We see from Figure 1[2,3] that Br and Rb levels in whole blood are as high as those of Zn, and could it be that Sr, Y, Zr, and Sb have roles just as important as that of Mo, or Ba as important as I? Most of our discussion will be restricted to animals, and in the space available we can only highlight major talking points at recent conferences and in recent publications.

Hair and nails offer significant advantages as monitors of the body burden or of exposure to inorganic elements. Significant levels of Na, Mg, Al, Cl, K, Ca, Sc, V, Cr, Mn, Fe, Co, Cu, Zn, Se, and Sb have all been detected in toe-nail samples[4] (collected over a two-year period, 15 minutes after taking a bath!). In Iraq, Hg levels in hair were used to differentiate those who had eaten Hg-contaminated bread from those who had not,[5] as shown in Figure 2. The outbreak of organomercury poisoning during the winter of 1971—2 followed the distribution of alkyl-Hg-treated wheat and barley. This was subsequently made into bread or fed to livestock, and up to 1 g of Hg was ingested per person in three months. The rate of poisoning was 271 per 1000 population, with a mortality rate of 59 per 1000. For females, a time profile going back almost 18 months

[1] S. S. Brown, in 'Clinical Chemistry and Toxicology of Metals', ed. S. S. Brown, Elsevier/North Holland, Amsterdam, 1977, p. 381.

[2] W. Gooddy, E. I. Hamilton, and T. R. Williams, *Brain*, 1975, **98**, 65.

[3] J. D. Birchall, in 'New Trends in Bioinorganic Chemistry', ed. R. J. P. Williams and J. J. R. Frausto Da Silva, Academic Press, London, 1978.

[4] J. S. Hislop, A. G. Morton, and J. W. Haynes, in ref. 1, p. 323.

[5] (a) A. W. Al-Mufti, J. F. Copplestone, G. Kazantzis, R. M. Mahmond, and M. A. Majid, *Bull. W.H.O.*, 1976, **53**, Supplement, p. 23; (b) G. Kazantzis, A. W. Al-Mufti, A. Al-Jawad, Y. Al-Shahwani, M. A. Majid, R. M. Mahmond, M. Soufi, K. Tawfig, M. A. Ibrahim, and H. Dabagh, *ibid.*, p. 37; (c) G. Kazantzis, A. W. Al-Mufti, J. F. Copplestone, M. A. Majid, and R. M. Mahmond, *ibid.*, p. 49.

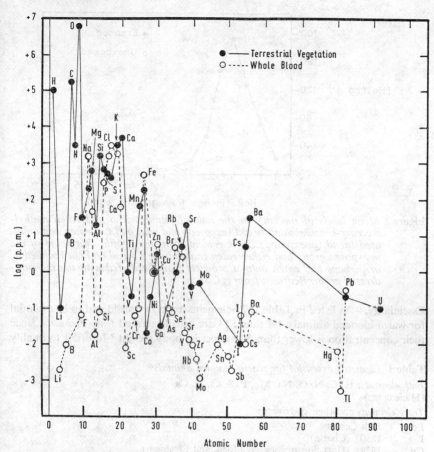

Figure 1 *The elemental composition of whole (U.K.) blood and terrestrial vegetation. Data are taken from refs. 2 and 3. Levels of H, C, N and O in blood will also be high, but are not shown, and lines joining points are for clarity only; some of the missing elements may have zero levels, whilst others were not listed in these particular compilations (blood level of Co is ca. 0.04 p.p.m. and of Ni ca. 0.005 p.p.m.)*

was constructed (hair growth was 1.13 cm per month), and hair levels were a more reliable guide to exposure than blood Hg levels. The mean half-life of methylmercury clearance from the blood is *ca.* 70 days.

Multi-element analysis of the levels of inorganic elements in food is in its infancy, and the U.S. Food and Drug Administration have concluded from studies of Ca, Cu, Fe, I, Mg, Mn, P, Se, and Zn in the average diet[6] that adults may be at some risk in their intakes of Cu, Fe, Mg, Mn, and Zn. Iodine occurs in appreciable excess in adult, infant, and toddler diets.

[6] B. F. Harland, L. Prosky, and J. E. Vanderveen, in 'Trace Element Metabolism in Man and Animals', ed. M. Kirchgessner, Arbeitskreis für Tierernährungs Forschung in Institut für Ernährungsphysiologie, Freising-Weihenstephan 1978 Vol. 3 311

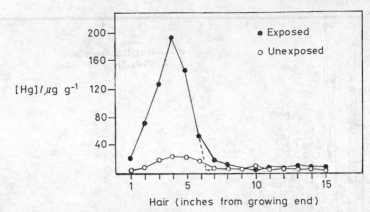

Figure 2 *Mean levels of mercury in the hair of females who had eaten methyl-mercury-contaminated bread (exposed) and who had not eaten contaminated bread (unexposed). It is evident from the profile that some of the unexposed group had either eaten contaminated bread and not realized it, or perhaps had eaten animal products from herds fed with the mercury-dressed grain. Redrawn from ref. 5b*

Essentiality.—As listed in Table 1, 25 elements are now considered to be essential for warm-blooded animals. Of these, 14 are usually termed *trace elements*, since their concentration is lower than 50 mg per kg body weight.[7] In order to qualify

Table 1 *Elements essential for warm-blooded animals[a]*

Bulk elements: H, C, N, O, Na, Mg, P, S, Cl, K, Ca
(11 elements)

Trace elements and their discovery:

Fe	17th Century
I	1850 (Chatin)
Cu	1928 (Hart, Steenbrock, Waddell, and Elvehjem)
Mn	1931 (Kemmerer and Todd)
Zn	1934 (Todd, Elvehjem, and Hart)
Co	1935 (Underwood and Filmer; Marston; Lines)
Mo	1953 (deRenzo, Kaleita, Heytler, Oleson, Hutchings, and Williams; Richert and Westerfeld)
Se	1957 (Schwarz and Foltz)
Cr	1959 (Schwarz and Mertz)
Sn	1970 (Schwarz, Milne, and Vinyard)
V	1971 (Schwarz and Milne; Hopkins and Mohr)
F	1971 (Schwarz and Milne)
Si	1972 (Schwarz and Milne; Carlisle)
Ni	1974 (Nielsen and Ollerich; Anke, Grün, Dittrich, Groppel, and Henning; Kirchgessner and Schnegg)

(14 elements)

Potential essentiality: As, Cd, Pb (growth effects demonstrated) (3 elements).

(a) Taken from 'Clinical Chemistry and Toxicology of Metals', ed. S. S. Brown, Elsevier/North-Holland Biomedical Press, Amsterdam, 1977, p. 3.

[7] M. Kirchgessner, E. Grassmann, H. P. Roth, R. Spoerl, and A. Schnegg, 'Nuclear Techniques in Animal Production and Health', International Atomic Energy Agency, Vienna, 1976, p. 61.

as an essential element, too low a concentration in blood must produce deficiency symptoms that can be prevented or cured by dietary addition of this element.[7] It must be remembered that trace elements which are toxic at relatively small dose levels may be highly essential and indispensible for normal body function at even lower, physiological amounts,[8] and toxicity itself is no argument against essentiality: in acute oral toxicity tests, simple Cu^{II} salts are as toxic as those of Cd^{II}. Most elements have a therapeutic index, defined as the daily effective dose divided by the minimum toxic dose, of about 1/100.[8] Thresholds for different elements, however, may vary by many orders of magnitude (Se is compared with Fe and K in Figure 3), and may vary greatly for different compounds of the same element (As^{III} is more toxic than As^{V}). Such chemical specificity must be further emphasized in future work.

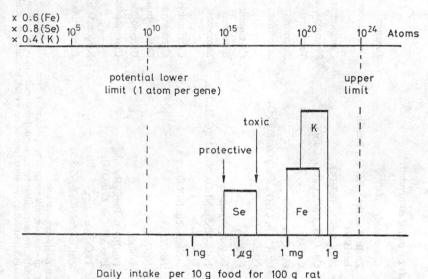

Figure 3 '*Concentration windows*' *where life is possible. The brackets span the minimum and maximum tolerated amounts of* Se (*as sodium selenite*), Fe (*as ferrous sulphate*), *and* K (*as various salts*) *in the diet. Redrawn from ref.* 8

Carcinogenicity and Mutagenicity.—Compounds of As, Cd, Cr, and Ni have been implicated in human carcinogenesis on the basis of epidemiological studies, and compounds of the nine metals Be, Cd, Co, Cr, Fe, Ni, Pb, Ti, and Zn have been reported to induce cancers in experimental animals.[9] Surprisingly, arsenic compounds have *not* been found to be carcinogenic in experimental animals, despite numerous investigations. Sunderman's summary of animal results is given in Table 2. Nickel subsulphide is probably one of the most potent carcinogens; iron–carbohydrate complexes (*e.g.* iron dextrans) are the only Fe compounds reported to have induced cancers, and Zn compounds are not thought to be carcinogenic except when given by intratesticular routes.

[8] K. Schwarz, in ref. 1, p. 3.
[9] F. W. Sunderman, jun. *Fed. Proc.*, 1978, **37**, 40.

Table 2 *Cancers induced in experimental animals by metal compounds*[a]

Metals	Species[b]	Routes[c]	Compounds	Tumours	Locations
Be	m, rb	iv	$ZnBeSiO_3$, BeO, $BeCO_3$, $BeSiO_3$, $BeHPO_4$	Osteosarcomas	Bone
Be	mk, r	inh	BeO, $BeSO_4$, $BeHPO_4$	Carcinomas	Lung
Cd	m, r	ih, im	Cd, CdS, CdO, $CdCl_2$, $CdSO_4$	Sarcomas, leydigiomas	Injection sites, testis
Cd	ch	its	$CdCl_2$	Teratomas	Testis
Co	r, rb	im, ios, sc	Co, CoS, CoO	Sarcomas	Injection sites
Cr	r, rb	im, ios, ip, ipl, sc	Cr, CrO_3, Cr_2O_3, $CaCrO_4$, $PbCrO_4$, $Na_2Cr_2O_7$	Sarcomas	Injection sites
Cr	m, r	ibr, inh	$CaCrO_4$	Carcinomas	Lung
Fe	h, m, r, rb	im, ip, sc	Fe-carbohydrates	Sarcomas	Injection sites
Ni	c, h, m, r, rb	icr, im, ipl, its, se	Ni, α-Ni_3S_2, Ni_2O_3, NiO, $Ni(C_2H_3O_2)_2$, $Ni(C_5H_5)_2$	Sarcomas	Injection sites
Ni	gp, r	inh	α-Ni_3S_2, Ni, $Ni(CO)_4$	Carcinomas	Lung
Ni	r	iv	$Ni(CO)_4$	Carcinomas, sarcomas	Diverse sites
Ni	m	ip	$Ni(C_2H_3O_2)_2$	Carcinomas	Lung
Ni	r	ir	α-Ni_3S_2	Carcinomas	Kidney
Pb	m, r	ip, po, sc	$Pb_3(PO_4)_2$, $Pb(C_2H_3O_2)_2$, $Pb(C_2H_3O_2)_2 \cdot 2Pb(OH)_2$	Carcinomas	Kidney, testes
Ti	r	im	$Ti(C_5H_5)_2$	Sarcomas	Injection sites
Zn	ch, h, q, r	its	$ZnCl_2$, $Zn(NO_3)_2$, $ZnSO_4$	Carcinomas, teratomas	Testis

(a) Taken from *Fed. Proc.*, 1978, **37**, 42, where original references are cited.

(b) c = cat, ch = chicken, gp = guinea pig, h = hamster, m = mouse, mk = monkey, q = quail, r = rat, rb = rabbit.

(c) ibr = intrabronchial, icr = intracerebral, ih = intrahepatic, il = intrapulmonary, im = intramuscular, ihn = inhalation, ios = intraosseous, ip = intraperitoneal, ipl = intrapleural, ir = intrarenal, itr = intratracheal, its = intratesticular, iv = intravenous, po = oral, sc = subcutaneous

An assay *in vitro*, based on the ability of metal compounds to affect the accuracy of DNA synthesis, is reported to identify carcinogenic compounds accurately.[10] Salts (most as the chloride or acetate) of Ag^I (the most potent), Cd^{II}, Cr^{II}, Co^{II}, Cr^{VI}, Mn^{II}, Ni^{II}, Pb^{II}, and Cu^{II} all increased the error frequency and decreased synthesis (compared to controls which contain only Mg^{II} and K^I). This assay may be superior to the Ames test,[11] in which compounds are evaluated by their ability to revert previously induced mutations in a microbe, since this cannot demonstrate the mutagenicity of nickel subsulphide.[9] An understanding of the molecular mechanism of these *in vitro* tests is required and, in particular, the nature of the DNA attack. This has been elusive even for the much-studied Pt drugs (see Section 4, p. 416). It is worth emphasizing that modification of the metal complex by other components of the assay medium (including the buffer!) can occur before DNA attack.

2 Essential Elements

Halogens.—Much of the present review is concerned with metals. However, one group of non-metallic elements, the halogens, provides a curious range of biological activities. Chloride is a universal counter-ion. It is freely permeable through a variety of membranes and is therefore found as an electro-osmotic balancing device both within cells and tissues. For instance, it re-equilibrates during uptake of oxygen in the red cell and during the secretion of cerebrospinal fluid. It is used in maintaining the external osmotic relations of the body, *e.g.* in the kidney, the salt gland of sea birds, and the gills of fishes.

Bromide is not clearly required by the body but has marked pharmacological effects on the nervous system, causing sedation. This has been ascribed to its direct effects on nerve membranes.

Iodine is an essential element, and plays its role largely in the function of the thyroid gland, whose hormones have a permissive, stimulatory function on overall metabolism together with specific actions on various tissues. Fluoride has recently emerged as an essential element, though the main interest must remain its effect on dental caries.

Fluorine.—*Fluoride and Dental Caries.* Fluoride metabolism has been investigated in depth because of the widespread use of fluoridated drinking water in the prevention of dental caries,[12] because of attempts to treat some metabolic bone diseases by administration of fluoride, and because of the potential toxicological hazards associated with exposure to high doses of fluoride in certain industries.[13]

Fluoride is incorporated into the mineralized tissues. Although some is subsequently released, a substantial proportion of an absorbed dose is retained (30—50%), and the concentration of the ion in the skeleton thus increases throughout life. The uptake of fluoride by the mineralized tissues is generally assumed to be associated with the mineral portion of the mineralized tissue, though it might be argued that the organic matrix must have some role in the ionic transactions which occur in its proximity. The mineral consists of calcium

[10] M. A. Sirover and L. A. Loeb, *Science*, 1976, **94**, 1434.
[11] J. McCann and B. N. Ames, *Proc. Nat. Acad. Sci.*, 1976, **73**, 950.
[12] *Caries Res.*, 1977, **11**, Supplement 1, ed. W. E. Brown and K. G. Konig, pp. 1–327.
[13] J. P. Hughes, *J. Occup. Med.*, 1977, **19**, (i), 11.

hydroxyapatite, $Ca_{10}(PO_4)_6(OH)_2$, and is capable of incorporating fluoride into the sites otherwise occupied by hydroxyl ions; this can occur either during crystallite formation or, later, at the crystal surface, by hetero-ionic exchange.

Dental caries is probably the most common disease in man. It results in progressive damage to the enamel and can ultimately lead to the total loss of the tooth. An excellent histological description of dental caries has been given by Scott *et al.*[14]

The initiation of caries is associated with the formation of a layer of bacterial material and cellular debris, plaque, which adheres to the tooth surface. Acids, produced by these bacteria, chiefly lactic acid, tend to dissolve the underlying dental enamel. For reasons which remain somewhat controversial, the initial attack involves the sub-surface regions of the tissue, and not until a relatively advanced state of the destructive process is the outer surface (approximately 50 μm thick) of the enamel mechanically disrupted. The carious attack is not uniform, and the development of a lesion may lead to comparatively little surface damage until a considerable quantity of mineral (up to 50%) has been removed from below the surface.[14] Until this occurs, the development of the lesion may be arrested. Indeed, many consider that the principal mechanism by which fluoride reduces caries is by encouraging the regression of the lesion at this stage of carious attack. It is believed that the action of fluoride is attributable to its capacity to facilitate the formation of apatite, leading to remineralization of the damaged tissue.

The chemical composition of enamel has been described by Weatherell *et al.*[15, 16] and the crystal chemistry and the varieties of atomic substitution have been discussed by Young.[17] A discussion of the physico-chemical basis of caries has been given by Brown[18] and by Moreno *et al.*[19]

Epidemiological Studies and Prophylactic Administration of Fluoride. The cariostatic action of fluoride is well documented and has been discussed in great detail at a recent symposium.[12] Ericsson has reviewed the clinical aspects.[20] Fluoride exerts both a systemic (*i.e.* as fluoride circulating in the blood and in the saliva) and a topical action on the dental surfaces. It can affect the microbiological flora of the mouth[21-24] and acts on damaged dental surface to promote remineralization of incipient lesions.

Caries can occur at various sites in the dentition. The most susceptible site is that of the fissure, the deep and narrow surface indentation seen, for instance, on the molars, but sites of stagnation between teeth and at the gingival (gum) margin are also at risk. The effectiveness of fluoride varies with site, and since

[14] D. B. Scott, J. W. Simmelink, and V. Nygaard, *J. Dent. Res.*, 1974, **53**, (Suppl. 2), 165.
[15] J. A. Weatherell, *Brit. Med. Bull.*, 1975, **31**, 115.
[16] J. A. Weatherell, C. Robinson, and A. S. Hallsworth, *J. Dent. Res.*, 1974, **53**, 180.
[17] R. A. Young, *J. Dent. Res.*, 1974, **53**, (Suppl. 2), 193.
[18] W. E. Brown, *J. Dent. Res.*, 1974, **53**, (Suppl. 2), 204.
[19] E. C. Moreno, M. Kresak, and R. T. Zahradnik, *Caries Res.*, 1977, **11**, (Suppl. 1), 142.
[20] S. Y. Ericsson, *Caries Res.*, 1977, **11** (Suppl. 1), 2.
[21] G. Rolla, *Caries Res.*, 1977, **11**, (Suppl. 1), 243.
[22] J. M. Hardie and S. H. Bowden, *Brit. Med. Bull.*, 1975, **31**, 131.
[23] I. R. Hamilton, *Caries Res.*, 1977, **11**, (Suppl. 1), 262.
[24] I. Kleinberg, R. Chatterjee, J. Reddy, and D. Craw, *Caries Res.*, 1977, **11**, (Suppl. 1), 292.

it is least effective against fissure caries it would seem that this difference may be the result of variations in the penetration of fluoride at different sites.

The optimal concentration of fluoride ion in drinking fluid for the prevention of caries is about 1 mg l^{-1} in temperate climates, although this will clearly depend on the access to fluoride from other sources and the local dental practice for fluoride supplementation or topical treatment. Concentrations of fluoride above 1.6 mg l^{-1} are considered to be potentially mildly fluorotic, producing slight mottling of the enamel. This, however, is not disfiguring, and the teeth are resistant to caries. In more severe fluorosis, with more than 2 mg of fluoride per litre, the changes are more important for cosmetic reasons. Brown spots may appear, and the enamel loses its smoothness and may become severely pitted.[25] Very high exposure to fluoride, such as in industrial contamination or at an intake above 8 mg of fluoride per litre in drinking water, may affect skeletal tissue, producing bony lesions and upsetting the normal balance between bone formation and resorption.

In the absence of water fluoridation, the intake of fluoride may be supplemented by daily administration of sodium fluoride tablets, though the exact dose is a matter of some dispute. Earlier suggestions have been that for children between 3 and 12 years of age, with drinking water fluoride concentrations of less than 0.4 mg l^{-1}, the dose should be one tablet (containing 1 mg of fluoride) per day.[26]

The fluoride requirement in infancy also is not clear. *In utero*, the placenta is a barrier to fluoride, and the newborn can be relatively deficient. Similarly, maternal milk is a poor source of fluoride, and it has been estimated[20] that bottle-fed infants in areas of water fluoridation may receive up to 50 times the fluoride intake of their breast-fed contemporaries. This does not noticeably affect the permanent dentition, but the deciduous teeth in bottle-fed babies may have fluoride concentrations some two to three times greater than those of breast-fed babies.[20, 27] Adair and Wei[28] have recently analysed a range of 'formula feeds', and suggest that it is unnecessary to supplement fluoride intake in any child until the age of six months. These authors also suggest that the optimal daily supplement recommendation should be somewhat lower than hitherto accepted, and this is supported by recent work of Aasenden and Peebles.[29]

The early epidemiological studies which established the effect of fluoride on dental caries have been reviewed recently.[20, 25, 30] These early studies showed a reduction in caries of well over 50% in the permanent dentition of school children who had received either naturally occurring or artificially added fluoride in the drinking water at a concentration of about 1 mg l^{-1}. A number of other vehicles for administration of fluoride have been investigated. An early review of these[25] lists salt, flour, and milk, and it can be administered as tablets containing sodium fluoride.

[25] P. Adler, 'Fluorides and Human Health', *W.H.O. Monograph Series*, 1970, No. 59.
[26] M. E. Bell, E. J. Largent, T. G. Ludwig, J. C. Muhler, and G. K. Stookey, in ref. 25, p. 17.
[27] B. Forsman, *Scand. J. Dent. Res.*, 1977, **85**, 22.
[28] S. M. Adair and S. H. Y. Wei, *Caries Res.*, 1978, **12**, 76.
[29] R. Aasenden and T. C. Peebles, *J. Dent. Res.*, 1977, **56**, (Suppl. B), B103.
[30] G. N. Jenkins, *Brit. Med. Bull.*, 1975, **31**, 142.

Topical Administration. Not only is fluoride effective systemically, but it has a very strong effect when administered topically, and this is the basis of the use of fluoride-containing dentifrices and topical painting with fluoride solutions. It is also possible that systemic fluoride may act locally, after secretion in the saliva,[31,32] since it appears that fluoride is not bound to plasma proteins,[33] and its appearance in the saliva is rapid after oral administration.

Jenkins[30] has reviewed the alternative modes of presentation, while Grøn[34] has described the chemistry of topical fluorides. The reagents used as vehicles for fluoridation by the topical route have the function of enhancing the penetration of fluoride into the enamel. The usual view is that this results in a stabilization of the apatitic enamel mineral. Some clinicians etch the teeth prior to the administration of fluoride, using the acid phosphate fluoride (APF) solutions (0.64 molar fluoride with 0.1 molar orthophosphate at pH 3) in order to increase fluoride penetration and uptake. The use of stannous fluoride (1.31 molar fluoride) is common, and it is suggested that the tin reacts with orthophosphate to form insoluble phosphates which have additional protective effects, though crystallographic and i.r. spectroscopic investigations have failed to identify the reaction products resulting from short-term application.[34]

Tin(II) hexafluorozirconate has been used, though the early clinical data were contradictory.[34] 'Amine fluorides' have been introduced in an attempt to combine the fluoride effect with the physico-chemical protection given by long-chain aliphatic amines against enamel decalcification. A recent report suggests that these amine fluorides provide greater protection against caries than does sodium fluoride when used as a daily mouth rinse.[35] Alexidine dihydrofluoride and a 1 : 1 mixture of cetylamine hydrofluoride and oleylamine hydrofluoride gave a significant reduction in caries in an experimental rat model. However, alexidine dihydrochloride also caused significant reduction, and it is possible that the effects can be related to bacteriostatic actions on micro-organisms in the mouth. Stannous fluoride also has actions on the formation of dental plaque,[36] though the significance of this is in dispute.[37]

Increased fluoride uptake at the surface of the enamel has also been demonstrated following the application of APF gel to the teeth of children and subsequent analysis of deciduous teeth.[38] Increased surface hardness has also been shown to be associated with treatment with stannous fluoride.[39] It has been shown that topical application of fluoride APF strongly promotes osteogenesis in bone of rats,[40] and for this reason fluoride has been investigated as a possible treatment for bone mineralization disorders.[41]

[31] L. D. Turtola, *Scand. J. Dent. Res.*, 1977, **85**, 535.
[32] J. Ekstrand, *Scand J. Dent. Res.*, 1977, **85**, 16.
[33] J. Ekstrand, Y. Ericsson, and S. Rosell, *Arch. Oral Biol.*, 1977, **22**, 229.
[34] P. Grøn, *Caries Res.*, 1977, **11**, (Suppl. 1), 172.
[35] R. J. Shern, S. M. Amsbaugh, and G. R. Reynolds, *J. Dent. Res.*, 1977, **56**, 1063.
[36] S. Hoffman, H. D. Tow, and J. S. Cole, *J. Dent. Res.*, 1977, **56**, 709.
[37] C. A. Ostrom, T. Koulourides, F. Hickman, and P. Phantumvanit, *J. Dent. Res.*, 1977, **56**, 212.
[38] J. R. Mellberg, C. R. Nicholson, G. J. Franchi, H. R. Englander, and G. W. Mosley, *J. Dent. Res.*, 1977, **56**, 716.
[39] D. J. Purdell-Lewis, J. Arends, and A. Groeneveld, *Caries Res.*, 1978, **12**, 43.
[40] T. Biller, Z. Yosipovitch, and I. Gedalia, *J. Dent. Res.*, 1977, **56**, 53.
[41] J. B. Rosenquist, *Clin. Orthop.*, 1977, **125**, 200.

Fluoride and the Structure of Enamel. Weatherell *et al.*[15, 16] have discussed structural aspects and have investigated the assimilation of fluoride by enamel.[42] The concentration of fluoride in enamel varies throughout the life of the teeth. There is a somewhat 'labile' peak of fluoride concentration which occurs during the mineralization period before the tooth has erupted, and subsequent accumulation of fluoride in the outer layers of enamel occurs during the later period of enamel maturation. This later stage of accumulation leads to the relatively high concentration [perhaps 1000 mg (kg tissue)$^{-1}$] of fluoride in the outer 25 μm of the enamel; a feature regarded by some workers as a protective barrier against carious attack. The physico-chemical basis of the action of fluorides on enamel material has been reviewed[19] and a study of enamel surface by scanning electron microscopy has been carried out by Boyde.[43]

The effects of fluoride on stabilization of hydroxyapatite may be by substitution for hydroxyl ions. Since fluoride is highly electronegative, the fluoride ion becomes hydrogen-bonded to the adjacent hydroxyl ions. Fluoride ions bound in this way restrict the mobility of hydroxyl ions, and it may be that the advance of carious attack is thereby delayed.[17] Alternatively, the same chemical mechanism will tend to favour remineralization. The roles of phosphate, carbonate, and water in the enamel structure are incompletely understood.

The effects of fluoride on the solubility characteristics of enamel have been discussed.[19, 44, 45] Brown *et al.*[44] concluded that under most circumstances fluoroapatite is more insoluble than hydroxyapatite, but the difference is so small that this is not likely to be a factor in cariostasis. Fluoride also promotes the precipitation rate of calcium phosphate, which increases the rate of apatitic growth,[19] and fluoride may cause acceleration of remineralization under all conditions.

Fluorosis. Fluorosis is the result of excessive fluoride absorption, and may be due to occupational exposure,[46] though moderate fluorosis may occur as a result of naturally occurring high concentrations of fluoride in drinking fluids. The range of naturally occurring fluoride in drinking water in England is reported to be between 0 and 5.8 mg l^{-1}, for the United States between 0 and 16.0, for Australia 0 and 13.5, and for U.S.S.R. between 0 and 7.0. Localized high concentrations of fluoride have been recorded in Czechoslovakia (28.0), Portugal (22.8), South Africa (53.0), Tanzania (95.0), and Kenya (2800). It is reported that small indigenous populations exist who are dependent on some of these waters that contain very high fluoride levels.[47] Clinical descriptions suggest that, if the concentration of fluoride is above 8 mg l^{-1} in the drinking water, bony lesions can occur, and with an intake of between 20 and 80 mg of fluoride per day, crippling fluorosis develops. With a concentration of 50 mg l^{-1} in drinking fluid, thyroid changes are seen, and at 100 mg l^{-1} growth retardation occurs,

[42] J. A. Weatherell, D. Deutsch, C. Robinson, and A. S. Hallsworth, *Caries Res.*, 1977, **11**, (Suppl. 1), 85.
[43] A. Boyde, *Brit. Med. Bull.*, 1975, **31**, 102.
[44] W. E. Brown, T. M. Gregort, and L. C. Chow, *Caries Res.*, 1977, **11**, (Suppl. 1), 118.
[45] S. Duke and G. C. Forward, *Caries Res.*, 1978, **12**, 12.
[46] H. C. Hodge, *J. Occup. Med.*, 1977, **19**, 12.
[47] B. R. Bhussry, V. Demole, H. C. Hodge, S. S. Jolly, Z. Singh, and D. R. Taves, *W.H.O. Monograph Series*, 1970, No. 59, p. 225.

while kidney damage is seen if the concentration is more than 125 mg l^{-1}. An acute dose of between 2.5 and 5 g of sodium fluoride, which is retained by the body, leads to death.

One effect of fluoride on bone when absorbed in high concentrations over a long period is to increase the mineralization. The mechanism may involve an increased precipitation of mineral or an effect on the bone-forming cells. Whichever it is, it results in heavy, deformed, and sometimes fused bones, with calcification of ligaments and (in cases of severe exposure) marked deformity.[47]

The absorption of fluoride by inhalation is the major source of occupational exposure to fluoride. It has been suggested that individuals living in central London may inhale 0.001—0.004 mg of fluoride daily, and this may be increased 10-fold on foggy days.[46] Large quantities of fluoride are emitted into the atmosphere by the production of superphosphate fertilizers, aluminium, and steel (both by open-hearth and electric furnaces) and in the manufacture of bricks. Smoke from coal fires is also a significant source of fluoride. The use of fluoride-containing fluxes in welding may lead to excessive local exposure. A study of urinary excretion of fluoride by workers in a fertilizer-manufacturing company showed a mean excretion of fluoride of 2.2 mg l^{-1}, with a maximum in those workers in the manufacturing department (1.5—7.5 mg l^{-1}). On radiography of the three workers with highest excretion, no evidence of fluorosis was found.[48]

Water Fluoridation. Newbrun[49] has considered the various arguments against water fluoridation and has concluded that there is no evidence that water containing optimal concentrations of fluoride impairs general health. The ill effects of fluoride that have been suggested are an association between fluoridation and Downs syndrome (mongolism), cardiovascular disease, cancer, allergies, and growth and development.

Various epidemiological studies have shown[50] and refuted[51] an effect on the rate of carcinogenesis in cities in the U.S.A. Another study, including the same cities, has shown a reduction in standardized mortality due to coronary heart disease in those cities where the water was fluoridated.[52] It is clear from these epidemiological studies that, whatever the effect of fluoride on populations subjected to fluoridation, the differences are small, except for a definite reduction in dental caries.

Iodine.—*Physiological Significance of Iodine.* Iodine is an essential element[53] whose main role is in the metabolism of the thyroid gland and its hormones. The thyroid gland, situated just below the larynx, and comprising two main lobes, regulates (by means of its hormones) the level of metabolism in tissues, and thus, indirectly, the metabolism of protein, lipid, and carbohydrate. The thyroid gland is not essential for life, but if absent there is poor resistance to cold; mental and physical slowing also occur. Excess thyroid hormone, *e.g.* as

---

[48] B. R. Gollop, *N. Z. Med. J.*, 1977, **85**, 143.
[49] E. Newbrun, *J. Am. Dent. Assoc.*, 1977, **94**, 301.
[50] J. Yiamouyiannis, *Lancet*, 1977, **ii**, 296.
[51] R. Doll and L. Kinlen, *Lancet*, 1977, **i**, 1300.
[52] D. R. Taves, *Nature*, 1978, **272**, 361.
[53] E. J. Underwood, 'Trace Elements in Human and Animal Nutrition', Academic Press, New York, 4th edn., 1977

caused by thyroid tumours, leads to excessive utilization of metabolic reserves, hyperactivity, and tremors. The most comprehensive treatise on thyroid function was published in 1974,[54] and a recent symposium has reviewed the metabolism of thyroid hormones.[55]

The iodide ion is accumulated by the thyroid gland (75 μg per day) from the very low concentrations circulating in the blood (3 μg l^{-1}) and that derived from dietary intake (minimum requirement is about 20 μg per day). Although it is clearly the most avid accumulator of iodide, the thyroid is not unique, since iodide-concentration mechanisms have been demonstrated in gastric mucosa, salivary gland, mammary gland, choroid plexus, ciliary body, small intestine, skin, hair, and placenta.[56] Salivary glands are commonly used for investigating iodide uptake, though species differences do occur. In human subjects whose thyroid has been totally removed surgically, 99.8% of administered radio-active ^{131}I was excreted immediately and the remaining 0.2% of iodide had a half life of excretion of 15 days.[57] This latter fraction has a distribution volume of 34 l, which suggests that its sequestration is extravascular. It is unaffected by subsequent administration of potassium iodide. It has been suggested that iodine probably has a major role only in thyroid function.

Iodide transport may be carrier-mediated, and is an active process working against an electrochemical gradient, is saturated by an excess of iodide, and is inhibited completely by related anions. In the thyroid, transport may be inhibited by interference with the 'iodide pump' by metabolic poisons, though inhibition may also occur as a result of interference with the binding of iodide by the tyrosine residues of thyroglobulins within the cell. Thionamide drugs, *e.g.* propylthiouracil (1) (PTU) or methimazole (2), which block the organification of iodide, are used in the treatment of hyperthyroidism.

(1)   (2)

An outline of the synthesis of thyroid hormones is seen in Scheme 1. The iodination steps of tyrosine residues and their condensation reactions to form the three hormones thyroxine (tetraiodothyronine, T_4), 3,5,3'-Tri-iodothyronine (T_3), and 3,3',5'-Tri-iodothyronine (reverse T_3) all occur whilst the residues are attached to the 'thyroglobulin' molecule. About 80 μg of T_4 and 50 μg of T_3 are secreted each day in normal man. Reverse T_3, traces of which are secreted by the thyroid,[58]

[54] 'Handbook of Physiology, Section 7, Endocrinology; Vol, III, Thyroid,' ed. M.A. Greer and D. H. Solomon, American Physiological Society, Washington, 1974, 491 pp.
[55] W. A. Harland and J. S. Orr, 'Thyroid Hormone Metabolism', Academic Press, London, 1975, 439 pp.
[56] C. H. Bastomsky, in ref. 54, p. 81.
[57] T. Smith and C. J. Edmonds, *Clin. Sci. Mol. Med.*, 1977, **53**, 81.
[58] V. Westgren, A. Melander, S. Ingemansson, A. Burger, S. Tibblin, and E. Wahlin, *Acta Endocrinol. (Copenhagen)*, 1977, **87**, 281.

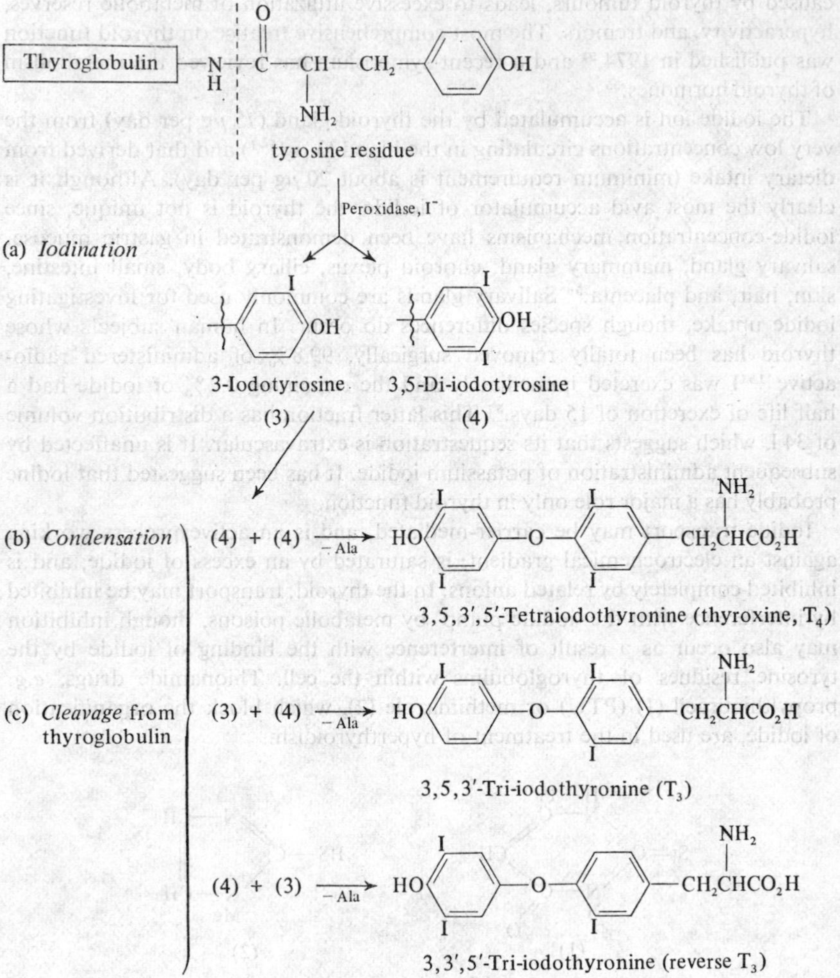

Thyroglobulin —N—C—CH—CH$_2$—⟨ ⟩OH
               H   ‖ |
                   O   NH$_2$

tyrosine residue

Peroxidase, I$^-$

(a) *Iodination*

3-Iodotyrosine   3,5-Di-iodotyrosine
      (3)   (4)

(b) *Condensation*   (4) + (4) $\xrightarrow{- \text{Ala}}$ HO⟨ ⟩—O—⟨ ⟩—CH$_2$CHCO$_2$H
                                                                    |
                                                                   NH$_2$

3,5,3′,5′-Tetraiodothyronine (thyroxine, T$_4$)

(c) *Cleavage* from   (3) + (4) $\xrightarrow{- \text{Ala}}$ HO⟨ ⟩—O—⟨ ⟩—CH$_2$CHCO$_2$H
    thyroglobulin |
                                                                   NH$_2$

3,5,3′-Tri-iodothyronine (T$_3$)

                      (4) + (3) $\xrightarrow{- \text{Ala}}$ HO⟨ ⟩—O—⟨ ⟩—CH$_2$CHCO$_2$H
                                                                    |
                                                                   NH$_2$

3,3′,5′-Tri-iodothyronine (reverse T$_3$)

Biosynthesis of hormones in the thyroid cell
Scheme 1

has thyroxine antagonistic activity in some systems, though the significance of this is not known.[59]

It is likely that a range of iodo-proteins occurs in the thyroid, but that a protein of sedimentation coefficient 19S (which is insoluble in 38% ammonium sulphate) is the major component of the functional 'thyroglobulin', though a further thyroid iodo-protein with a sedimentation coefficient of 27S has been shown which may be a dimer.[60] Extra-thyroidal iodo-proteins have also been reported, though these may be residual traces of the thyroglobulin contained in residues

[59] C. S. Pittman and J. A. Pittman, in ref. 54, p. 233.
[60] N. Ui, in ref. 54, p. 55.

which have been discharged during secretion of thyroid hormones. In normal human blood, iodo-proteins may comprise 15% of total iodine, and most of this is due to iodination of serum albumin. Iodo-proteins are also formed in the mammary gland, and are secreted into milk, though thyroxine synthesis has not been seen.[60]

Other anions are accumulated by the thyroid gland, particularly bromide and astatide as well as peroxy-anions of Group VIIA, perrhenate (ReO_4^-) and pertechnetate (TcO_4^-), and also perchlorate (ClO_4^-) and fluoroborate (BF_4^-). Thiocyanate (SCN^-) and selenocyanate ($SeCN^-$) competitively inhibit iodide transport but are not themselves transported.[56]

Biosynthesis, Action, and Control of Thyroid Hormones. The synthesis of the three active hormones (T_3, reverse T_3, and T_4) occurs by iodination of tyrosine residues in the presence of peroxidase, and it has been suggested that an enzyme-bound iodinium intermediate is formed which may either be converted into iodinated tyrosine residue (in the presence of thyroglobulins) or (in the presence of thiouracil and other anti-thyroid agents) oxidized. The evidence for the various mechanisms proposed for the iodination step has been reviewed.[61]

Both the iodination of tyrosine and the synthesis of thyroxine on the thyroglobulin are catalysed by thyroid peroxidase. Kinetic studies have suggested that a step-wise reaction does not occur in the iodination and coupling of iodotyrosines, since a considerable time-lag occurs between the appearance of di-iodotyrosine residues and T_4 following the labelling of precursors.[62] It has also been shown that hydrogen peroxide is required for the coupling to occur and that the concentration of iodide ion is critical. This suggests that iodide itself may have effects on the control of thyroid peroxidase activity.[62] Structural studies have shown that the native configuration of T_4 contributes to the efficiency of the coupling reaction, though the specificity of the peroxidase is not critical.[63]

Extra-thyroidal peroxidase and iodinase are affected by circulating thyroid hormones. In rat submaxillary gland, thyroidectomy (removal of the thyroid) leads to a compensatory increase in both peroxidase and iodinase activity, and this increase is prevented by administration of thyroxine. Inhibition of protein synthesis by puromycin, cycloheximide, or actinomycin D or by the administration of thiouracil partially abolished the increase in activity, which suggests that thyroxine may regulate the enzyme synthesis.[64]

At cellular sites in the target organs, thyroid hormones have effects on a number of metabolic systems. Oxygen consumption and heat production are stimulated, and parallel changes occur in the metabolic regulation of carbohydrates, lipids, and proteins. The all-pervasive action on cellular metabolism is consistent with a homeostatic role of thyroid hormones in the control of body temperature in the adult animal. In the immature animal, thyroid hormones have the function of regulating aspects of growth and development. Metabolic effects of thyroid hormones may be divided into those which occur rather rapidly, usually resulting in changes in respiration or heat production, and those which

[61] A. Taurog, in ref. 54, p. 101.
[62] D. Deme, J. Pommier, and J. Nunez, *European J. Biochem.*, 1976, **70**, 435.
[63] L. Lamas and A. Taurog, *Endocrinology*, 1977, **100**, 1129.
[64] T. Chandra and R. Das, *European J. Biochem.*, 1977, **72**, 259.

13

occur through the induction of enzyme synthesis, and whose onset of action is much slower.[65]

The regulation of the synthesis and release of thyroid hormones is under the control of the pituitary gland, which releases thyroid-stimulating hormone (TSH) into the peripheral blood circulation, where it acts on the thyroid gland. TSH is in turn released as a response by the pituitary gland to thyrotropin-releasing factor (TRF), a local hormone secreted by the hypothalamus as a result of an integrative processes involving neuronal regulatory centres within the mid-brain.

Iodine Deficiency. In iodine deficiency, for any reason, a wide spectrum of effects is seen at all levels of organization within the body. Lack of iodinated thyroid hormones leads to feed-back stimulation of the control systems, leading ultimately to enlargement of the thyroid gland in an attempt to compensate (goitre). Lack of the hormones at the cellular level leads to dysregulation of carbohydrate, lipid, and protein metabolism and to poor response to changes in environmental temperatures, together with lack of readily available energy supply for the voluntary aspects of the bodily function. In the adult human, therefore, hypothyroidism (myxoedema) presents as a patient who complains of feeling cold, has dry, yellowish skin, lack of energy, and retarded thought and action. These symptoms are readily reversed by administration of thyroid hormones or, in the case of primary deficiency, of iodine itself.

Undiagnosed iodine deficiency or defects in thyroid metabolism in children have serious effects both on physical and mental development. Cretins (children who are hypothyroid from birth) are most commonly associated with mothers who have been iodine-deficient during pregnancy. If cretinism is detected soon after birth, considerable improvement may be obtained, and the physical and mental development may approach that of a normal child. However, once the typical clinical picture appears, permanent mental retardation is likely to have occurred, though bodily growth may still be stimulated by appropriate therapy.

The effects on growth and development of iodine and thyroid hormones are not restricted to mammalian species. The metamorphosis of amphibian species is initiated by thyroid hormone, though examples of evolutionary adaptation to low iodine intake occur. The classical case is the paedogenetic development of the Mexican axolotl, which is sexually mature and able to reproduce though it is morphologically immature, staying in a late larval stage as a result of iodine deficiency in the lakes that it inhabits. If small quantities of iodine, or thyroid hormones, are added to the aqueous environment, the axolotl loses its external gills, develops lungs, and becomes terrestrial. It undergoes metamorphosis to an adult form which is similar to the common North American tiger salamander (*Amblystoma tigrinum*).

Goitre. Goitrous enlargement of the thyroid gland in the neck may be due to iodine deficiency or to goitrogens, either naturally occurring or resulting from unwanted actions of administered drugs. Thiocyanates and other naturally occurring goitrogens may be ingested in food. Progoitrin occurs in vegetables of the brassica family, particularly cabbage and turnips, and this substance is converted into an active antithyroid agent, goitrin (5). On a normal diet, this

[65] F. L. Hoch, in ref. 54, p. 391.

$$H_2C-\overset{\overset{\displaystyle H}{|}}{N}$$
$$H_2C=\overset{\overset{\displaystyle H}{|}}{C}-\overset{\overset{\displaystyle |}{C}}{\underset{\underset{\displaystyle H}{|}}{}}-O$$
$$C=S$$

(5)

goitrogenic activity is of little consequence. However, in vegetarians, 'cabbage goitres' may occasionally be seen.

Endemic goitre has been recognized since ancient times, and it is widespread public health practice to add iodine to table salt as a prophylactic measure against iodine-deficiency goitre. Endemic goitres may also appear following pollution of water supplies by goitrogens. The public health aspects and natural history of goitre have been discussed in a recent monograph.[66] A simple goitre often occurs in adolescent girls with apparently normal thyroid function. Spontaneous remission of symptoms usually occurs.

The use of lithium in psychiatric practice, which is discussed in another section, is associated with development of hypothyroidism in about 10—15% of cases. The usual treatment is administration of thyroxine; however, a recent report describes the investigation of the simultaneous administration of iodine and lithium.[67] It was concluded that this apparently simple solution produced a wide variety of effects on thyroid function in different individuals, and this would preclude its use as a routine practice.

Hyperthyroidism. Hyperthyroidism, or thyrotoxicosis, presents symptoms of nervousness, weight loss, tremors, sweating, and heat intolerance, together with cardiovascular symptoms such as palpitations. A study of twelve patients with toxic diffuse goitre showed increased iodination of thyroglobulins due to increased peroxidase activity in the particular fraction of a sample of thyroid tissue taken at thyroidectomy. The iodination of thyroglobulins *in vitro* was increased in toxic diffuse goitre even when the patients had apparently normal thyroid function, after treatment with thionamide drugs.[68] A study of long-term effects of the treatment of hyperthyroid patients with carbimazole showed that, after about eighteen months therapy, no significant difference was seen in iodine excretion between normal and thyrotoxic patients. However, a high relapse rate was noted in the subsequent drug-free period following long-term administration of antithyroid drugs in areas with low iodine intake.[69]

An alternative treatment for hyperthyroidism is the administration of radio-isotopes of iodine. These are readily taken up by thyroid cells, which in turn are damaged by the radiation emitted by the isotope. Partial thyroidectomy by means of radioisotope administration has advantages in older patients since surgery is thus avoided. It has been suggested that ^{131}I is specifically accumulated in thyroid carcinoma cells ('hot' nodules) and that this makes it particularly

[66] F. W. Clements, *W.H.O. Monograph Series*, No. 62, p. 83, 1976.
[67] S. W. Spaulding, G. N. Burrow, J. N. Ramey, and R. K. Donebedian, *Acta Endocrinol. (Copenhagen)*, 1977, **84**, 290.
[68] S. Nagataki, H. Uchimura, H. Ikeda, N. Kuzuya, Y. Masuyama, L. F. Kumagai, and I. K. Izo, *J. Clin. Invest.*, 1977, **59**, 723.
[69] I. B. Lumholtz, *Acta Endocrinol. (Copenhagen)*, 1977, **84**, 538.

suitable for treatment of these diseases. A recent study, comparing the auto-radiographic scans of the thyroid gland by both ^{131}I and [99m]Tc (in the form of pertechnetate), showed discrepancies, though kinetic calculations indicated that there was no difference between the uptake of the iodine or the technetium. This suggests that the major discrepancy in handling of iodine seen after pertech-netate and ^{131}I screening was possibly a result of deficient organification of iodine, and was not due to 'hot' nodules.[70]

The uptake of small doses of radioactive iodine is used as a diagnostic test for the ability of the thyroid gland to trap iodine. Recent studies suggest that this test is a poor one for hyperthyroidism, particularly toxic nodular goitre, though it is useful in calculating the dose of ^{131}I required for isotope therapy.[71]

Iodine-131 is also used as a adjunct to surgical removal of the thyroid, and the survival rate in cases of thyroid carcinoma is claimed to be significantly improved.[72] The short- and long-term effects of radio-iodine and antithyroid drugs have been reported.[73] After treatment with either drugs or a radioisotope, the levels of both thyroxin-binding plasma globulin and thyroxin-binding plasma albumin were the same as those of normal controls one year after treatment, as compared with the pre-treatment finding that there was less thyroxin-binding plasma albumin in the hyperthyroid patients while the globulin fraction showed no difference. The use of radionuclides in the diagnosis and treatment of thyroid disease has been reviewed,[74] while fears that therapy with ^{131}I might cause damage to the bone marrow have been allayed by a recent study[75] which showed that the absorbed dose of ^{131}I in the bone marrow was equivalent to about 0.59 rad per millicurie of isotope dose administered during treatment of hyperthyroidism, compared with the level in normal subjects of 0.36 rad per millicurie administered. Even allowing for the high dose of isotope administered in treatment, this exposure would not seem to be sufficient to cause leukaemia.

Calcium.—Calcium has been discussed widely in the past few years, since the recognition of its large number of roles in initiation of energetic events such as muscular contraction[76-78] and release of neurotransmitters and hormones[79] and in membrane phenomena in mitochondria.[78] The traditional role of calcium in the formation of hard tissues has been extensively studied.[80] A rapidly expanding field of interest is derived from the early work of Wasserman, who identified the calcium-binding proteins, used in calcium transport at membranes. A recent

[70] M. K. O'Connor, M. J. Cullen, and J. F. Malone, *J. Nucl. Med.*, 1977, **18**, 796.
[71] P. L. Hooper and R. H. Caplan, *J. Am. Med. Assoc.* 1977, **238**, 411.
[72] U. Y. Ryo, *Am. J. Med.*, 1977, **63**, 167.
[73] C. Jaffiol, L. Baldet, M. Robin, C. Parachristou, H. Lapinski, and J. Mirouze, *Horm. Metab. Res.*, 1977, **9**, 73.
[74] G. Riccabona, *Wien Med. Wochenschr.*, 1977, **127**, 97.
[75] A. C. McEwan, *Br. J. Radiol.*, 1977, **50**, 329.
[76] C. C. Ashley and P. C. Caldwell, in 'Calcium and Cell Regulation' ed. R. M. S. Smellie, Biochemical Society, London, 1974, p. 29.
[77] S. V. Perry, in ref. 76, p. 115.
[78] 'Calcium Transport in Contraction and Secretion' ed. E. Carafoli, F. Clementi, W. Drabi-kowski, and A. Margreth, North-Holland/Elsevier, Amsterdam, 1975, 588 pp.
[79] W. W. Douglas, in ref. 76, p. 1.
[80] 'Calcium, Phosphate and Magnesium Metabolism', ed. B. E. C. Nordin, Livingstone, Edinburgh, 1976.

colloquium reports the present state of knowledge.[81] Many aspects of calcium have been discussed in an earlier chapter of these Reports.

Clinical signs of calcium deficiency have been recognized for many years. Acute deficiency may arise as a result of chronic low calcium intake, when a sudden increase in calcium loss precipitates tetany, a neurological syndrome. Chronic deficiency occurs when the dietary intake is low compared with excretory loss over a long period, and it may be due either to low dietary content or availability of calcium or to pathologically increased excretion. Malabsorption of calcium from the gut occurs in a number of diseases, *e.g.* steatorrhoea, and this may lead to relative deficiency. Chronic deficiency is usually manifest in bone diseases.[80, 82, 83] Wilkinson[84] has reviewed the clinical physiology of calcium. A recent monograph has discussed aspects of nutritional calcium requirements, and the W.H.O. recommendations are (mg per day): infants less than 1 yr, 500–600; children 1—9 yr, 400—500; adolescents 10—15 yr, 600—700; adolescents 16—19 yr, 500—600; adults, 400—500. In pregnancy and lactation, the calcium intake should be increased to 1000—1200 mg per day.[85] Other aspects of nutritional deficiencies affecting calcium metabolism have been reviewed by Paunier.[86]

The major hormones controlling calcium metabolism are parathyroid hormone (PTH), secreted by the parathyroid gland, and calcitonin (CT), secreted by the 'C' cells of the thyroid gland in mammals and by the ultimobranchial bodies of birds, reptiles, and fishes. The prohormone calciferol (vitamin D) gives rise to at least one (and possibly more) hormones, and a review of the current evidence of its metabolism and its effects on calcium metabolism has been presented by de Luca (see Figure 4).[87] The proceedings of a comprehensive workshop on vitamin D metabolism have also appeared[88] and a recent European symposium contained extended discussion of vitamin D and its metabolism.[89]

Control of Calcium Metabolism by Hormones. The function of the three hormones controlling calcium metabolism is intimately intertwined. Parathyroid hormone is a polypeptide whose functions include actions directly on bone *via* the osteoclast cells, to promote dissolution of the matrix and release of bone mineral.[90] PTH also acts at the kidney to diminish tubular re-absorption of phosphate and

81 'Calcium Binding Proteins and Calcium Function', ed. R. H. Wasserman, R. A. Caradino, E. Carafoli, R. H. Kretsinger, D. H. MacLennan, and F. L. Siegel, Elsevier/North-Holland, New York, 1977, 514 pp.
82 J. Vaughan, 'The Physiology of Bone', Clarendon Press, Oxford, 2nd edn., 1975.
83 'The Biochemistry and Physiology of Bone', ed. G. H. Bourne, Academic Press, New York, Vols. 1—3 1972, Vol. 4 1977.
84 R. Wilkinson, in ref. 80, p. 36.
85 R. Passmore and B. M. Nicol, *W.H.O. Monograph Series*, 1974, No. 61, p. 49.
86 L. Paunier, *W.H.O. Monograph Series*, 1976, No. 62, p. 111.
87 H. F. de Luca, in 'Inborn Errors of Calcium and Bone Metabolism', ed. H. Bickel and J. Stern, M.T.P. Press, Lancaster, 1976, p. 1.
88 'Vitamin D; Biochemical, Chemical and Clinical Aspects Related to Calcium Metabolism' ed. A. W. Norman, K. Schaefer, J. W. Coburn, H. F. de Luca, D. Frazer, H. G. Grigdleit, and D. V. Herrath, deGruyter, Berlin, 1977, 973 pp.
89 *Calcif. Tissue Res.*, Supplement to Vol. 22, ed. W. G. Robertson, B. E. C. Nordin, and F. G. E. Pautard, Springer-Verlag, Heidelberg, 1977.
90 R. V. Talmage and R. A. Meyer, in 'Handbook of Physiology, Section 7, Endocrinology, Vol. VII, The Parathyroid Gland' American Physiological Society, Washington, 1976, p. 343.

to increase re-absorption of calcium, with a consequent rise in plasma calcium.[90] Calcitonin, also a polypeptide, has opposite effects;[91] it diminishes bone resorption and lowers the plasma calcium level. Recent work suggests that vitamin D plays a mutually facilitatory role in the metabolic expression of PTH action[92] in addition to its own direct effects, which are to promote absorption of calcium from the gut and to increase mineralization of bone, paradoxically, by increasing the mobilization of calcium from pre-formed bone.[92] The role of the kidney in maintaining homeostasis of plasma calcium under the influence of these various hormones has been discussed by Nordin.[93] An elegant theoretical treatment of the regulatory system for calcium homeostasis has been presented by Pheng and Weiss.[94]

Vitamin D, or more correctly vitamin D_3 (cholecalciferol), is derived either by absorption from dietary sources in the gut or from conversion of 7-dehydrocholesterol,[95] present in the skin, under the influence of sunlight. Historically, rickets, a deficiency disease in children, was associated with life in crowded, shadowed streets in industrial cities with high levels of airborne pollution. The disease was reversed either by the administration of cod-liver oil or by exposure to sunlight, and it was therefore a source of some dispute as to whether the disease was environmental or resulted from a deficiency. It is now known that two alternative sources of vitamin D are operative. The synthetic pathway is seen in Scheme 2.

Vitamin D is in fact the precursor of a series of *hormones* which result from metabolic conversion in the liver and kidney into mono- and di-hydroxycholecalciferols. Vitamin D itself does not have significant direct actions of its own.[87] Vitamin D_3 is converted in the liver into the 25-hydroxy-D_3 (25-hydroxycholecalciferol, 25-HCC). The fate of this metabolite, which itself has some activity, is decided by the concentration of serum calcium. At abnormal serum calcium levels, both $1,25(OH)_2D_3$ (1,25-dihydroxycholecalciferol, 1,25-DHCC) and $24,25(OH)_2D_3$ (24,25-dihydroxycholecalciferol, 24,25-DHCC) are synthesized in the kidney. Under conditions of even slight hypocalcaemia, the synthesis of $1,25(OH)_2D_3$ is stimulated, whilst it is relatively inactive at normal and high serum calcium levels. $1,25(OH)_2D_3$ is therefore the hormone that mobilizes calcium from both bone and intestine, and has its synthesis regulated in turn by serum calcium. This feedback inhibition is very characteristic of endocrine glands. The function of the $24,25(OH)_2D_3$ is unknown, but its synthesis is stimulated whenever $1,25(OH)_2D_3$ synthesis is inhibited.[87]

de Luca has presented evidence that parathyroid hormone itself stimulates the synthesis of $1,25(OH)_2D_3$, and it has been suggested that the action of the dihydroxy-D_3 in intestine does not require PTH, whilst its action on bone requires the facilitation of PTH. The further suggestion is therefore that the

[91] P. L. Munson, in ref. 90, p. 443.
[92] H. F. de Luca, in ref. 90, p. 265.
[93] B. E. C. Nordin, in 'Calcium, Phosphate and Magnesium Metabolism', ed. B. E. C. Nordin, Livingstone, Edinburgh, 1976, p. 186.
[94] J. M. Phang and I. W. Weiss, in ref. 90, p. 157.
[95] M. F. Holick, J. Frommer, S. McNeill, N. Richtand, J. Henley, and J. T. Potts, in ref. 88, p. 135.

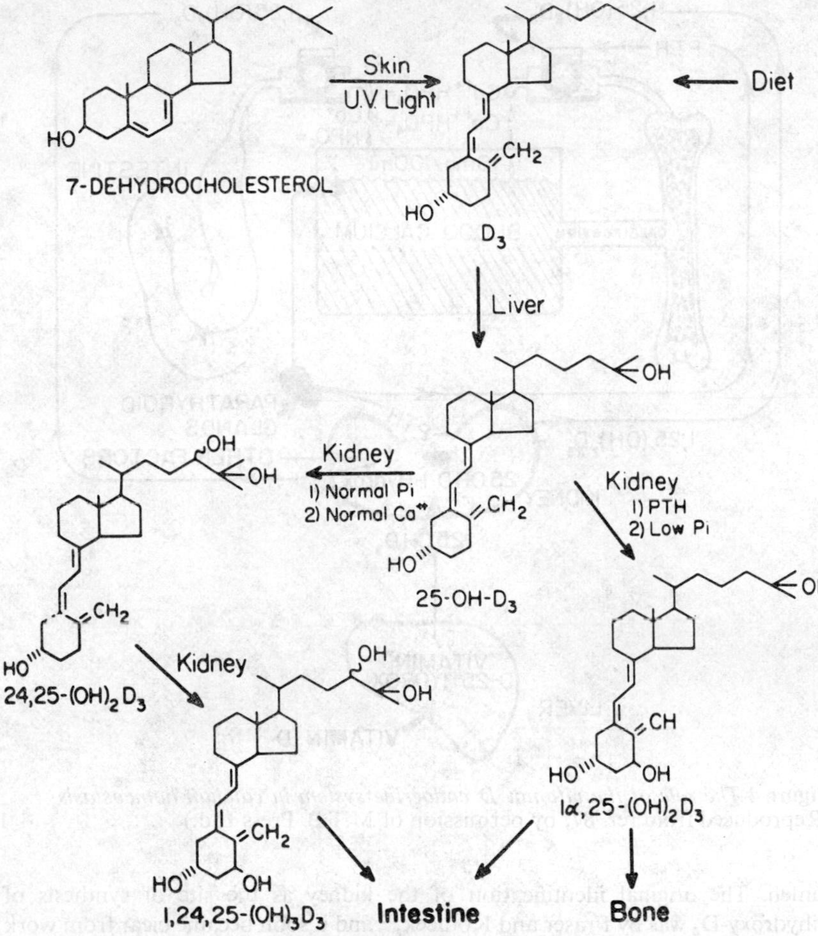

Synthesis and metabolism of vitamin D
(Reproduced from ref. 87 by permission of the copyright owners)

Scheme 2

action of PTH in stimulating intestinal calcium transport is mediated by its stimulation of the synthesis of $1,25(OH)_2D_3$. This is summarized in Figure 4. A cellular mechanism for the action of $1,25(OH)_2D_3$ on calcium transport in intestine has been proposed by Haussler and his colleagues.[90] This suggests that the hormone is bound to cytoplasmic receptor proteins which allow it to migrate to the nuclear chromatin, where it initiates the transcription of mRNAs. mRNA is translated into functional calcium-binding proteins (CaBP), which are then secreted into the cell, to be employed in the uptake of calcium from the intestinal

[96] M. R. Haussler, M. R. Hughes, T. A. McCain, P. F. Terwekh, J. W. Brumbaugh, and R. H. Wasserman, *Calcif. Tissue Res.*, 1977, **22**, (Suppl.), 1.

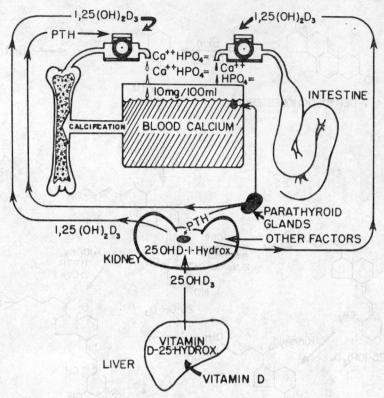

Figure 4 *The role of the vitamin D endocrine system in calcium homeostasis* (Reproduced from ref. 87, by permission of M.T.P. Press Ltd.)

lumen. The original identification of the kidney as the site of synthesis of dihydroxy-D_3 was by Fraser and Kodicek,[97] and it soon became clear from work in de Luca's lab that the kidney was the sole site of the synthesis of $1,25(OH)_2D_3$. The enzyme $25(OH)D_3$ 1α-hydroxylase is exclusively a mitochondrial enzyme, depending entirely on cytochrome P-450.

The synthesis and conformational analysis of vitamin D metabolites has been discussed at length by Okamura *et al.*,[98] who concluded that the 1α-hydroxy-group appears to be particularly critical for biological activity. Salmond[99] has discussed the synthesis of vitamin D analogues. The therapeutic use of vitamin D and its metabolites in the treatment of osteomalacia, rickets, hypoparathyroidism, osteoporosis, renal failure, and renal osteodystrophy has been discussed in symposia previously cited.[88, 89, 100]

[97] D. R. Fraser and E. Kodice, *Nature*, 1970, **228**, 764.
[98] W. H. Okamura, M. L. Hammond, P. Condran, and A. Mourino, in ref. 88, p. 33.
[99] W. G. Salmond, in ref. 88, p. 61.
[100] *Calcif. Tissue Res.*, Supplement to Vol. 21, ed. S. Pors Nielsen and E. Hjørting-Hansen, FADL, Copenhagen, 1976, 483 pp.

Parathyroid hormone and calcitonin. Interest in the other hormones controlling calcium metabolism, *i.e.* PTH and CT, has been somewhat overshadowed by the rapid advances made in the area of vitamin D metabolism. Studies of the role of cyclic nucleotides in the secretion and action of PTH have allowed a number of fundamental conclusions to be drawn. PTH is secreted from parathyroid cells in response to both calcium ion and catecholamines, and it is thought that this is a result of the stimulation of cyclic adenosine 3' : 5'-monophosphate (cyclic AMP). The most effective stimulus to PTH release is isoprenaline. The next most effective is adrenaline, and then noradrenaline. The order of responses to adrenergic agonists therefore is characteristic of β_2-type catecholamine receptors, and cell membrane prepared from parathyroid gland cells contains adenylate cyclase which is activated by catecholamines with the same order potency.[101] The presence of the catecholamine receptor has been confirmed by the use of stereospecific β adrenergic blockers.[101]

It is therefore possible to conclude that there are specific β adrenergic receptors on the parathyroid gland cells and that adrenergic catecholamines interact with these receptors to stimulate adenylcyclase in the membrane, causing the release of cyclic AMP. Cyclic AMP in turn releases PTH. This mechanism is distinct from that whereby calcium itself may cause the release of PTH.[101]

Cyclic AMP has also been implicated in the action of PTH at its target site in the kidney. Receptors for PTH lie in different segments of the nephron, and differ from those receptors which are sensitive to calcitonin, vasopressin, or catecholamines.[101, 102] The action of PTH at the kidney may therefore be independent of the simultaneous action of these other hormones.

Calcitonin is secreted by cells of the thyroid gland in response to increased plasma calcium, and it has an inhibitory effect on bone resorption, though the physiological significance of this is not clear.[103] However, disorders of secretion and metabolism of calcitonin have recently been reported in renal disease.[104] Twenty-five patients with chronic renal failure showed elevated basal concentrations of calcitonin and a significant increase in plasma calcitonin following renal dialysis, when the plasma calcium concentration also rose. It was suggested that a decrease in the catabolic and excretory function of the damaged kidney could lead to the failure of one of the mechanisms for removal of the hormones and thus for maintaining normal circulating concentrations of the hormone. This in turn would lead to hypocalcaemia. In consequence, after dialysis against normalized fluids, an increase in secretion of endogenous calcitonin occurs, in response to plasma calcium increase, exacerbating the hypercalcitoninaemia.

Interrelationships of calcium-regulating hormones. The relationship between calcium intake and excretion, intermediary metabolism, and the calcium-regulating hormones has been summarized succinctly by Aurbach in the introduction to a treatise on parathyroid hormone.[105] Despite the limitations of its title, this

---

[101] G. D. Aurbach, E. M. Brown, and S. J. Marx, *Calcif. Tissue Res.*, 1977, **22**, (Suppl.), 117.
[102] G. D. Aurbach and L. R. Chase, in 'Handbook of Physiology, Section 7, Endocrinology, Vol. VII, The Parathyroid Gland' American Physiological Society, Washington, 1976, p. 353.
[103] G. Coen and B. Palagi, *Calcif. Tissue Res.*, 1976, **21**, (Suppl.), 294.
[104] J. C. Lee, J. G. Parthemore, and L. J. Deftos, *Calcif. Tissue Res.*, 1977, **22**, (Suppl.), 154.
[105] G. D. Aurbach, in ref. 102, p. 1.

volume contains a most authoritive compendium of knowledge of the control of calcium and magnesium metabolism and of their roles in bone and cellular metabolism,[90-92, 102, 106] and a monumental review of the composition, structure, and organization of bone.[107] In addition, the chemistry, biosynthesis, secretion, and metabolism of parathyroid hormones[108] and calcitonin[109] are reviewed in depth.

Disorders of Calcium Metabolism. The relationships between the various controlling factors and diseases involving abnormal calcium metabolism have been summarized by Norman (see Figure 5).[110] The range of diseases identified exemplifies the number of loci of action of the various hormones and their control systems.

The expression of disordered calcium metabolism is usually in one of three ways. Neurological signs usually manifest as tetany during hypocalcaemia, and increased calcium turnover may result in calcification of soft tissues, more

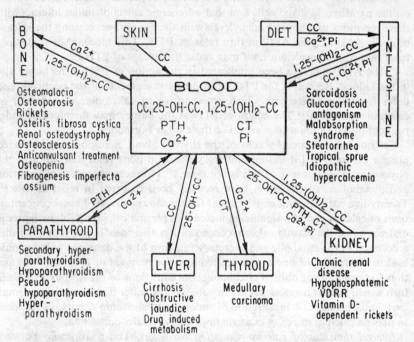

Figure 5 *Disease states seen in man which are related to metabolism of vitamin D* (Taken from ref. 110, by permission of Springer-Verlag, Berlin).

106 H. Rasmussen, D. B. P. Goodman, N. Friedmann, J. E. Allen, and K. Kurokawa, in ref. 102, p. 225.
107 M. J. Glimcher, in ref. 102, p. 25.
108 J. F. Habener and J. T. Potts, in ref. 102, p, 313.
109 J. T. Potts and G. D. Aurbach, in ref. 102, p. 423.
110 A. W. Norman, W. H. Okamura, M. N. Mitra, and R. L. Johnson, in 'Calcium Metabolism, Bone, and Metabolic Bone Diseases' ed. F. Kuhlencordt and H. P. Kruse, Springer-Verlag, Berlin, 1975, p. 60.

particularly the formation of renal stones and nephrocalcinosis, but also deposition of calcium in the conjunctivae of the eye and around limb joints. Finally, long-standing defects in calcium metabolism produce characteristic changes in bone, though their development may be relatively slow.

Metabolic bone disease is a complex subject, and beyond the remit of the present chapter, though the following broad simplification may help in the interpretation of literature, and references are given to recent studies. For a more detailed exposition the reader is referred to standard texts.[80-83, 111]

Osteomalacia and rickets[86, 112] are both due to lack of calcium absorption following vitamin D deficiency. The difference between the two diseases is merely in time of onset, rickets being a result of deficiency during bone development in childhood. Osteoporosis may be defined as atrophy of bone: the volume of bone remains the same though its density decreases as a result of loss of both bone matrix and mineral.[113] Osteoporosis may be associated with a number of disorders,[113] though the biochemical basis of the disease is unknown. Renal osteodystrophy[114] is associated with chronic renal failure and may be a result of the failure of the kidney to convert $25(OH)D_3$ into $1,25(OH)_2D_3$. Secondary hyperparathyroidism often occurs with renal osteodystrophy.

Hypocalcaemia is seen in parathyroid hormone deficiency, and results in tetany, whose most characteristic symptom is spontaneous muscle cramps; those of the face are usually most readily recognized, and may be elicited by tapping with the finger over the facial nerve in the cheek (Chvostek's sign). A similar response may be obtained in normal subjects following voluntary overbreathing for two or three minutes. The level of plasma ionized calcium is decreased in respiratory alkalosis (excessive loss of carbon dioxide from the blood) due to sequestration by plasma proteins.

Disorders of calcium metabolism may also be induced by immobilization,[115] weightlessness,[116] and the administration of drugs or hormones for the treatment of other disorders. Osteomalacia induced by anti-convulsants has been shown in rats to be a result of inhibition of absorption or metabolism of vitamin D,[117] and high doses of cortisol in rats have been shown to increase calcium absorption and turnover.[118]

Milk fever in cows occurs shortly after calving, due to hypocalcaemia that results from the high calcium demand of milk production for suckling and from

111 P. Fourman and P. Royer, 'Calcium Metabolism and the Bone', Blackwell, Oxford, 2nd edn., 1968.
112 C. E. Dent, in 'Inborn Errors of Calcium and Bone Metabolism', ed. H. Bickel and J. Stern, M.T.P. Press, Lancaster, 1976, p. 124.
113 F. Kuhlencordt, *Calcif. Tissue Res.*, 1976, **21**, (Suppl.), 405; R. P. Heaney, in ref. 88, p. 627; A. Horsman, in, Calcium, Phosphate, and Magnesium Metabolism', ed. B. E. C. Nordin, Livingstone, Edinburgh, 1976, p. 357; A. Horsman, B. E. C. Nordin, J. C. Gallagher, P. A. Kirby, R. M. Milnery, and M. Simpson, *Calcif. Tissue Res.*, 1977, **22**, (Suppl.), 217.
114 J. W. Coburn, A. S. Brickman, D. J. Sherrard, F. R. Singer, D. J. Baylink, E. G. C. Wong, S. G. Massry, and A. W. Norman, in ref. 88, p. 657.
115 J. U. Lindgren, *Calcif. Tissue Res.*, 1976, **22**, 41.
116 G. D. Whedon, L. Lutwak, P. Rambaut, M. Whittle, C. Leach, J. Reid, and M. Smith, *Calcif. Tissue Res.*, 1976, **21**, (Suppl.), 423; N.A.S.A. SP-377, ed. R. S. Johnson and L. F Dietlein, National Aeronautics and Space Administration, Washington, 1977, 491 pp.
117 M. Harris, D. J. F. Rowe, and A. J. Darby, *Calcif. Tissue Res.*, 1978, **25**, 13.
118 J. L. Ferretti, J. L. Bazan, D. Alloatti, and R. C. Puche, *Calcif. Tissue Res.*, 1978, **25**, 1.

the inability of the parathyroid gland to mobilize calcium from bone.[119] It has been suggested that two apparently divergent alternatives may contribute to the onset of the disease; (*a*) overfeeding with calcium-containing supplements before calving or (*b*) not providing any supplementary diet after calving. The mechanism proposed to rationalize these apparently contradictory aetiological factors is that adaptation to a high-calcium diet occurs, so that a high calcium intake over prolonged periods results in a decreased absorption rate. The body responds to a low-calcium diet by increasing the production of $1,25(OH)_2D_3$ and PTH, which together promote synthesis of CaBP. An increase in calcium in a diet leads to the 'switching off' of $1,25(OH)_2D_3$ synthesis in kidney, and thus absorption of CaBP and calcium is impaired. After calving, the calcium demand is high, but it cannot be satisfied, because of the reduced absorption rate.

In the case of low calcium intake prior to parturition, the demands of milk production immediately after birth are unable to be met by dietary absorption, and they therefore cause mobilization of bone calcium by PTH. Young cows may be able to cope with this sudden demand, but in older animals the sensitivity of bone to PTH is reduced and the mobilization is impaired, leading to rapidly falling plasma calcium and consequent tetany. A new feeding technique has been developed to match dietary intake with bodily requirements of calcium and phophorus in the ante- and post-natal periods, so that bone sensitivity to PTH is not the controlling factor, but this role is taken over by the gut, with its enhanced rate of calcium absorption that results from previous dietary restriction. Such a regimen has reduced the incidence of milk fever from 124 cases during the winter of 1973—4 to 10 during the winter of 1974—5 in the same group of 216 cows held at 24 different farms. Seven of the ten cases reported in 1974—5 were on farms which have previously had cases of hypomagnesaemia, and it has been suggested that hypomagnesaemia contributes to the reduced sensitivity to PTH.[119]

Magnesium.—*Photosynthesis*. One of the most fundamental advances of early evolution, the incorporation of magnesium into the chloroplast, the subcellular organelle which contains chlorophyll, permitted the direct conversion of solar radiation into chemical potential energy. Upon this single process has depended the subsequent evolution of higher plants, and in turn the development of multi-cellular animals. Since Nature is essentially conservative, this early evolutionary use of magnesium is reflected also in its requirement by a large number of enzymes, particularly those involving ATP. The common evolutionary heritage of magnesium-dependent enzymes is demonstrated by the range of non-specialized processes with which the ion is implicated. Thus both plants and animals show oxidative phosphorylation, synthesis of DNA and RNA, and the translation of the genetic code into protein synthesis, all of which are magnesium-dependent. The incorporation of solar energy into the synthesis of biological materials is the primary energy source for both plants and animals, but this energy must be capable of being stored, to maintain the organism during the absence of sunlight. The conversion of stored chemical energy from photosynthesis, in the form of

[119] D. W. Pickard, in 'Nutrition and the Climatic Environment', ed. W. Haresign, H. Swan, and D. Lewis, Butterworth, London, 1977, p. 113.

carbohydrate, to readily utilized ATP occurs, under the influence of magnesium, in the mitochondrion, in both plants and animals.

Chlorophyll *a* is synthesized from protoporphyrin-IX by the incorporation of magnesium and sequential reduction, esterification, and ring closure. Recent views of the structure, synthesis, and photochemistry of chlorophyll have been reviewed by Seely.[120] Other workers have studied the biosynthesis of magnesium–protoporphyrin complex, and concluded that the insertion of magnesium into chloroplasts, at least *in vitro*, is an enzyme-catalysed reaction.[121]

Not only is magnesium involved in the structure of chlorophyll, but it also has a role in the regulation of the interaction between the two electron-transport chains, photosystems I and II.[121] Photosystem II is associated with the oxidation of water and photosystem I with the reduction of $NADP^+$, and these two systems are linked by an electron-transport chain. Light energy is absorbed into photosystem II by a specialized form of chlorophyll, P-680, which acts as an electron donor whose electrons ultimately traverse photosystem II and enter the cytochrome chain that bridges the two photosystems. A model for photosystem I has been proposed in which the photosensitive electron donor, in this case P-700, lies on the inside of the thylakoid membrane of the chloroplast. P-700 is excited by light and releases electrons that are accepted ultimately by ferredoxin–NADP reductase outside the thylakoid.[122] A light-driven transfer occurs in the centre of the thylakoid space, and it has been suggested that Mg^{2+} is the main counter-ion for the proton pump, since there is no evidence of concomitant anion transport. The associated volume, pMg, and pH changes in the intra-thylakoid space are postulated to be major regulators of various stages of the photosynthetic reaction chain.[123] It has further been suggested that the con-centration of ionized Mg^{2+} has a role in the regulation of 'phosphate potential' in the coupling of photophosphorylation to electron transport.[124] The effect of Mg^{2+} on the quantum efficiency of photophosphorylation coupling has been reported.[125]

The role of magnesium in the action of ATPases and in subcellular metabolism has been discussed in an earlier chapter of these Reports. We will therefore consider magnesium now from the point of view of studies in whole animals.

Magnesium Metabolism in Animals. Until comparatively recently, the role of magnesium in biological processes was obscure. The advent of commercially available atomic absorption spectrometers in the early 1960's signalled the expansion of interest in magnesium and enabled, for the first time, the simple

[120] G. R. Seely, in 'Primary Processes of Photosynthesis', ed. J. Barber, Elsevier/North-Holland, Amsterdam, 1977, p. 1.
[121] B. B. Smith and C. A. Rebeiz, *Arch. Biochem. Biophys.*, 1977, **180**, 178; W. P. Williams, in ref. 120, p. 101; C. J. Arntzen, P. A. Armond, J. M. Briantais, J. J. Burke, and W. P. Novitzky, in 'Chlorophyll–Proteins, Reaction Centres and Photosynthetic Membranes', eds. J. M. Olsen and G. Hind, Brookhaven National Laboratory, Upton, N.Y., 1977, p. 316; E. L. Gross and D. J. Davis, *ibid.*, p. 363.
[122] J. R. Bolton, in ref. 120, p. 189.
[123] J. Barber, in 'The Intact Chloroplast', ed. J. Barber, Elsevier/North-Holland, Amsterdam, 1976 p. 89.
[124] D. O. Hall, in ref. 123. p. 135.
[125] H. Huzisige, in 'Photosynthetic Organelles', ed. S. Miyachi, S. Katoh, Y. Fujita, and K. Shibata, Special Issue No. 3 of *Plant Cell Physiol.*, Japanese Society of Plant Physiology Kyoto, Japan, 1977.

and accurate determination of the element in biological tissues. The availability of the isotope [28]Mg, despite the disadvantage of its short half-life (21.3 h), has also simplified the study of the metabolism of magnesium, though much remains to be discovered of its cellular and subcellular transactions. A number of reviews of the present state of knowledge of magnesium have appeared during the period of the present survey,[126-128] and background information may be obtained from exhaustive reviews of our knowledge of magnesium reported at two comprehensive symposia at the end of the last decade.[129]

The role of magnesium in the action of a large number of enzymes, hormones, and different systems of the body has previously been discussed, though there are a number of miscellaneous aspects which have received attention recently. Intracellular regulation of squid giant axons (nerve cells)[130, 131] and the relationships between free ionized calcium and free ionized magnesium have been considered in the light of their relative affinities for intracellular protein, amine, and nucleotide sites.[130] The relative excess of magnesium over calcium within the cell is stressed, and also the importance of the rapid flux of calcium as a switching device both at the nerve-cell membrane and at the mitochondrion. The role of magnesium at membranes has been comprehensively reviewed.[132] Baker has, however, warned that extrapolation from results for squid axons may not be justified, since the solutions used and tolerated by the tissue may be far from physiologic, and results may not represent true responses to the experimental alterations made.[131] This difficulty of interpretation should be constantly borne in mind. Chutkow has reviewed the role of magnesium in the mammalian central nervous system.[133]

It has been suggested that part of the function of magnesium in bone may be to stabilize, with ATP, the structure of amorphous calcium phosphate.[134] Magnesium and ATP have a synergistic effect *in vitro*, delaying the conversion of a slurry of amorphous calcium phosphate into crystalline hydroxyapatite. Magnesium decreases the rate of ATP hydrolysis, and the ATP prevents formation of hydroxyapatite from metastable solutions of calcium and phosphate. ATP stabilizes amorphous calcium phosphate by 'poisoning' heteronuclear growth sites or by 'poisoning' growth of embryonic hydroxyapatite nuclei (or both).

[126] D. M. Livingstone and W. E. C. Wacker, in 'Handbook of Physiology, Section 7, Endocrinology, Vol. VII, The Parathyroid Gland', 1976, American Physiological Society, Washington, 1976, p. 215.
[127] 'Nutritional Imbalances in Infant and Adult Disease', ed. M. S. Seelig, Monograph of A.C.N., Spectrum Press, New York, 1976.
[128] J. K. Aikawa, in 'Trace Elements in Human Health and Disease', ed. A. S. Prasad, Academic Press, New York, Vol II, p. 47.
[129] W. E. C. Wacker, *Ann New York Acad. Sci.*, 1969, **162**, Art. 2, 717; '1st International Symposium on Magnesium' (Vittel, France, 1971), ed. J. Durlach, 'I, Volume des Rapports', S.G.E.M.V., Vittel, France, 1971, 'II, Volume des Communications, S.G.E.M.V., Vittel, France, 1973.
[130] F. J. Brinley, *Fed. Proc.*, 1976, **35**, 2572.
[131] P. F. Baker, *Fed. Proc.*, 1976, **35**, 2589.
[132] M. Bara, *Année Biol.*, 1977, **16**, 290.
[133] J. G. Chutkow, *Neurology*, 1974, **24**, 780.
[134] N. C. Blumenthal, F. Betts, and A. S. Posner, *Calcif. Tissue Res.*, 1977, **23**, 245; A. S. Posner, F. Betts, and N. C. Blumenthal, *ibid.*, 1977, **22**, (Suppl.), 208.

Dietary requirements. The daily intake of magnesium should be between 0.15 and 0.175 mmol per kg per day to prevent net loss of the element,[126] though an early review has drawn attention to the wide variation and misinterpretation of data which has occurred in the determination of such requirements.[135] It is reported that the average American adult (weight 70 kg) ingests about 12.5 mmol per day, mainly in the form of green vegetables, suggesting that the intake is only slightly more than the requirement, and that the chronic avoidance of green vegetables might induce significant magnesium depletion. Furthermore, it has been reported that the magnesium intake in the average American diet has been declining steadily since about 1910. A recent factor in this decline is the widespread commercial use of chelating agents to preserve the green and 'wholesome' appearance of frozen vegetables.[136] The widespread excessive supplementation of milk and foodstuffs with vitamin D also leads to an increase in the magnesium requirement to offset the increased renal losses.

Magnesium is the fourth most abundant cation in the body, though 50% of total magnesium is in bone. In the soft tissues, magnesium is the second most abundant, intracellularly. The concentration of magnesium in plasma appears to be very closely regulated, though the homeostatic mechanism involved has not been identified.[126] Magnesium metabolism is, however, affected by parathyroid hormone, though it is not clear whether this is secondary to the hormone's effects on calcium metabolism.

The circadian variation in plasma magnesium was disturbed following thyroparathyroidectomy (TPTX) in rats, suggesting that both calcitonin and parathyroid hormones have a role in the control of magnesium metabolism.[137]

Absorption and excretion of magnesium. Absorption of magnesium occurs mainly in the small intestine, and the peak concentration of the isotope ^{28}Mg in human plasma is reached two to three hours after oral administration. Absorption does occur, however, in other parts of the gastrointestinal tract. A recent study suggests that the caecum plays some role in Mg absorption and that this is related to the degree of caecal distention.[138] Hypermagnesaemia may occur in humans following the use of rectal enemas containing magnesium.[126] In normal subjects, much of the dietary magnesium is lost in faeces. In 26 subjects given ^{28}Mg orally, 60—88% of the radioactivity was recovered in the faeces within 120 h. That this represents non-absorbed magnesium rather than absorption and re-secretion into the gut is indicated by an experiment in which ^{28}Mg was injected intravenously, when only 1.8% of the isotope was recovered from faeces.[128]

The major route of excretion of magnesium is therefore through the kidney, and there is evidence that renal mechanisms play the major part in regulation of plasma magnesium.[139] Renal tubular re-absorption of magnesium is stimulated by parathyroid hormone and also by aldosterone, though there is little evidence of homeostatic regulation of magnesium by the latter hormone.[139] Recent

[135] M. S. Seelig, *Am. J. Clin Nutr.*, 1964, **14**, 342.
[136] M. S. Seelig, in 'Proceedings of the International Symposium on Magnesium', ed. M. S. Seelig and M. Cantin, Spectrum Press, New York, 1979, in the press.
[137] Y. Rayssiguier, *Horm. Metab. Res.*, 1977, **9**, 161.
[138] Y. Rayssiguier and C. Remesy, *Ann. Rech. Vét.*, 1977, **8**, 105.
[139] B. E. C. Nordin, in 'Calcium, Phosphate and Magnesium Metabolism', Livingstone, Edinburgh, 1976, p. 186.

micropuncture studies in the dog and rat have suggested that magnesium is readily ultrafiltered in the glomerulus of the kidney; some is probably re-absorbed in the proximal tubule, though the majority of filtered Mg^{2+} (about 60%) is re-absorbed in the ascending limb of the loop of Henle. Little further magnesium re-absorption occurs in the distal tubule and collecting duct.[140]

Magnesium deficiency. Because of the dietary factors previously discussed, and because its symptoms are generalized and non-specific, magnesium deficiency remains unrecognized in a large number of cases. The most distinctive symptom is a tetany which is resistant to calcium administration and which is either overt or which may be precipitated by ischaemia.[141, 142] The latter has been called 'spasmophilia',[143] and was the subject of a number of papers at an international meeting.[129] Other symptoms which are seen are muscular tremor, choreic movements, ataxia, hallucinations, agitation, confusion, delirium, and muscular weakness. Decreased serum magnesium does not necessarily occur, and this is one of the particular diagnostic difficulties.[144] Magnesium deficiency can only be confirmed following higher than normal retention of an orally administered magnesium load. Because the concentration of magnesium is critical for enzyme action,[145-147] the maintenance of a steady plasma magnesium level appears to have precedence, and depletion of tissue stores of the ion may be severe before plasma magnesium levels fall. It has been suggested that plasma magnesium should be merely considered to be 'in transit' between bone stores and actively metabolizing tissues.[148]

Magnesium depletion may occur under a number of circumstances. Prolonged dietary deficiency may result from inadequate intake of green vegetables or from dependence on synthetic diets. Prolonged fasting, anorexia nervosa, kwashiorkor, or protein-calorie malnutrition of infants and young children all produce magnesium deficiency, which may be masked by other symptoms. The acute phase of the magnesium-deficiency syndrome appears usually during dietary repletion, and the syndrome may be obviated by administration of magnesium supplements.

Magnesium deficiency may also occur as a result of excessive losses from the gastrointestinal tract due to persistent vomiting or diarrhoea. Malabsorption syndromes result in deficient magnesium uptake.[149] Magnesium depletion occurs following enhanced renal losses due to glomerulonephritis, nephrosclerosis, and

[140] J. H. Dirks, in 'Proceedings of the 2nd International Symposium on Magnesium', ed. M. S. Seelig and M. Cantin, Spectrum Press, New York, 1979, in the press.
[141] J. Durlach, *Z. Ernährungswiss.*, 1975, **14**, 75.
[142] M. S. Seelig, A. R. Berger, and N. Spielholz, *Dis. Nerv. Syst.*, 1975, **36**, 461.
[143] J. Durlach, 'Spasmophilie et Déficit Magnésique', Masson, Paris, 1969, 141 pp.
[144] J. Durlach, in 'Proceedings of the 2nd International Symposium on Magnesium', ed. M. S. Seeling and M. Cantin, Spectrum Press, New York, 1979, in the press.
[145] F. W. Heaton, *Biochem. Soc. Trans.*, 1973, **1**, 67.
[146] H. Rubin, *Proc. Nat. Acad. Sci. U.S.A.*, 1975, **72**, 3551.
[147] T. Guenther, *J. Clin. Chem. Clin. Biochem.*, 1977, **158**, 433.
[148] M. Walser, in 'Proceedings of the 2nd International Symposium on Magnesium', ed. M. S. Seeling and M. Cantin, Spectrum Press, New York, 1979, in the press.
[149] M. S. Seelig and G. E. Bunce, in 'Magnesium in the Environment: Soils, Crops, Animals and Man', ed. J. B. Jones, M. C. Blunt, and S. R. Wilkinson, 1972, Fort Valley State College, Fort Valley, Georgia, U.S.A., p. 61.

renal tubular acidosis.[150] Chronic magnesium deficiency leads to nephrocalcinosis.[126]

A number of metabolic responses to experimental magnesium deficiency have been reported. Serum proteins are rapidly decreased in the rat, there is an immediate fall in albumin, and a longer-term fall in gamma globulins. This hypoproteinaemia has been attributed to effects on protein synthesis.[151]

A significant decrease in bone strength and magnesium content occurred in rats which were treated with chronic sub-optimal magnesium diet for 500 days but which did not develop overt magnesium deficiency.[152] Sub-optimal magnesium intake may lead, also, to decreased resistance to cold stress.[153]

The relationship between magnesium deficiency, calcium, parathyroid hormone, and bone has been reviewed by Shils,[154] who concludes that there are three possible scenarios for the progress of magnesium deficiency, depending on the initiation, perpetuation, or regression of concomitant hypocalcaemia which may result from reduction in sensitivity of the bone to PTH. Possibly there occurs also a reduction in the metabolic activity of bone cells resulting from the decreased cellular magnesium. Others have demonstrated that there is no impairment of the primary action of PTH[155] since a PTH-stimulated shift of calcium into bone occurred in magnesium-deficient rats both on normo- and hypo-calcaemic diets. However, though magnesium deficiency itself did not affect the basal concentration of cyclic AMP or of PTH-sensitive production of cyclic AMP in renal cortex, hypocalcaemic rats with magnesium deficiency excreted more cyclic AMP than did controls, while similar rats with hypocalcaemia alone excreted less. A similar relationship was seen in renal accumulation of cyclic AMP in renal cortex in response to PTH.[156] It has therefore been suggested that PTH-stimulated cyclic AMP metabolism is affected by magnesium deficiency due to its consequences on other aspects of metabolism, rather than by a direct effect.

Magnesium deficiency in chronic alcoholism. The acute magnesium-deficiency syndrome in man was first observed in alcoholics, who exhibit hypomagnesaemia and a spectrum of symptoms that are now known to resemble those seen in experimentally induced acute magnesium deficiency.[149] Administration of alcohol to normal subjects may increase the renal output of magnesium, but it is clear that in chronic alcoholism a number of additional factors are involved. Chronic alcoholics frequently are malnourished, have increased losses of intestinal contents through vomiting, and may exhibit polyuria (increased urinary volume), which is a result of inhibition by ethanol of the anti-diuretic hormone (ADH).[157, 158] In addition, patients suffering from cirrhosis of the liver as a result

150 R. E. Randall, *Ann. New York Acad. Sci.*, 1969, **162**, 831.
151 Y. Rayssiguier, P. Larvor, Y. Augusti, and J. Durlach, *Ann. Biol. Anim., Biochim. Biophys.*, 1977, **17**, 147.
152 O. Héroux, D. Peter, and A. Tawner, *Canad. J. Physiol. Pharmacol.*, 1975, **53**, 304.
153 O. Héroux, D. Peter, and A. Heggtvelt, *J. Nutr.*, 1977, **107**, 1640.
154 M. E. Shils, in 'Trace Elements in Human Health and Disease', ed. A. S. Prasad, Academic Press, New York, 1976, Vol. II, p. 23.
155 J. P. Ashby and F. W. Heaton, *Horm. Metab. Res.*, 1975, **5**, 259.
156 J. P. Ashby and F. W. Heaton, *J. Endocrinol.*, 1975, **67**, 103.
157 J. F. Sullivan, P. W. Wolpert, R. Williams, and J. D. Egan, *Ann. New York Acad. Sci.*, 1969, **162**, 947.
158 J. E. Jones, S. R. Shane, W. H. Jacobs, and E. B. Flink, *Ann. New York Acad. Sci.*, 1969, **162**, 934.

of alcoholism may often be treated with diuretics, which will exacerbate the loss of magnesium.

It has been suggested that even the lower ethanol intake associated with social drinking may increase magnesium requirement, and this should be considered in subjects whose magnesium intake is not optimal. In this case incipient magnesium deficiency may occur with no evidence of hypomagnesaemia.

Recent studies on chronic alcoholic subjects and subjects suffering from malabsorption, both exhibiting hypomagnesaemia, have shown that, compared with controls, there is a decrease in isometric voluntary muscle contraction force, and this is significantly correlated with a decrease in the muscle content of magnesium as determined by muscle biopsy.[159] The ATP content of both hypomagnesaemic and control muscles was normal, but there was a decrease in ADP and creatine phosphate in the hypomagnesaemic group, and it has been suggested that these changes may result in a decreased ability of the muscle to perform sustained effort.

Magnesium, lipid metabolism, and heart disease. The inverse correlation between death rate from ischaemic heart disease and the hardness of the local water supply is well known.[160] Recent studies have suggested that part of the protective effect of high mineral content in drinking fluid may be due to the water-borne magnesium intake.[161, 162] A comparison of accident victims in two areas in Ontario at autopsy showed a significant difference in myocardial muscle magnesium concentration, residents of the soft-water areas having 7% less magnesium than those from hard-water areas. It has been suggested that some residents of soft-water areas may be in a state of subclinical magnesium deficiency, and that these individuals have an increased risk of developing fatal cardiac arrhythmia should they suffer a myocardial infarction.[161] In other studies, heart muscle from patients who died of myocardial infarction had significantly lower magnesium content than that of subjects who died from other causes.[162] This difference was further exaggerated when the magnesium content of the infarcted segment of heart muscle was compared with controls. The role of Mg in the pathology of myocardial infarction has been reviewed.[163, 164]

In laboratory models of cardiac hypoxia, mainly in dogs, very severe myocardial magnesium depletion was shown. Intravenous magnesium salts are protective against anoxia, and ECG changes following experimental coronary ligation in dogs[162] and the clinical use of intravenous magnesium in man to treat a resistant arrhythmia in a patient suffering from lithium toxicity[165] may support the evidence for this cardio-protective action of magnesium. The use of intravenous magnesium sulphate as supportive therapy for acute myocardial infarction has been reviewed.[162]

[159] G. Stendig-Lindberg, J. Bergström,, and G. Hultman, *Acta Med. Scand.*, 1977, **201**, 272.
[160] D. Harman, in 'Nutritional Imbalances in Infant and Adult Diseases', ed. M. S. Seelig, Spectrum Press, New York, 1977, p. 1.
[161] T. J. Anderson, Abstracts of the 2nd International Symposium on Magnesium, American College of Nutrition, 1976, p. 27.
[162] M. S. Seelig and H. A. Heggtveit, *Am. J. Clin. Nutr.*, 1974, **27**, 59.
[163] D. Lehr and I. S. Chaur, *Recent Adv. Stud. Card. Struct. Metab.*, 1975, **6**, 85.
[164] D. Lehr, *Comp. Ther.*, 1975, **1**, 47.
[165] L. I. Worthley, *Anaesthesia and Intensive Care*, 1974, **2**, 357.

It has also been suggested that magnesium deficiency may increase the risk of cardiovascular disease, owing to its effects on lipid metabolism.[166, 167] Hyperlipoproteinaemia was induced in rabbits by a high-lipid diet and administration of ethanol, and this led to the distortion of the Golgi apparatus and increased levels of lipid in the smooth endoplasmic reticulum (SER) of liver cells. When the diet was supplemented with magnesium, plasma cholesterol decreased and the lipid desposit in the SER was decreased. The authors postulated that lysosomes are involved in the SER effects.[167]

In a study of seventy-three 60-year-old Danish subjects undergoing a 12 hour fast, the serum magnesium concentration was inversely correlated with the systolic blood pressure. Furthermore, the concentration of magnesium in the red blood cells correlated with that of serum cholesterol.[168] It was suggested that alterations in magnesium metabolism might play a role in ischaemic heart disease.[168] However, results from the same laboratory conflict with this, since a daily supplement of 3 g of magnesium oxide for six weeks had no effect on fasting serum cholesterol and serum triglyceride levels in 17 patients with hypercholesterolaemia and hypertriglycerideaemia.[169] Intravenous infusion of adrenaline in livers resulted in hypomagnesaemia, and this is enhanced by phentolamine and inhibited by propranolol, suggesting that β-adrenergic receptors are involved. Hypomagnesaemia is also related to initiation of lipolysis.[170]

A recent study of guinea-pig heart muscle has indicated that the stimulation of adenyl cyclase by histamine and adrenaline occurs by decreasing the requirement of magnesium as activator for this enzyme. The authors suggest that the mechanism may be either an increased affinity for magnesium of the enzyme binding site or a decrease in the sensitivity of adenyl cyclase to inhibition by ATP.[171] It is also suggested that magnesium interacts with the GTP regulatory site. It is clear therefore that magnesium has effects in the cardiovascular system, though presently these are not well understood.

Grass staggers and milk tetany. The hypomagnesaemic origin of these two disorders in sheep and cattle was identified in the 1930's. Milk tetany is seen in cows reared for longer than usual on an exclusively milk diet and reflects the relatively low magnesium content of milk. Recent studies have shown that, during induced magnesium deficiency in calves, the parathyroid gland failed to respond satisfactorily to increased plasma calcium, though calcitonin secretion was unaffected.[172]

Grass staggers or grass tetany occurs in cows stall-fed during winter and put out to graze in early spring. Neuromuscular symptoms, motor inco-ordination, tetany, and convulsions predominate, as suggested by the name of the syndrome, and a characteristic fall in plasma magnesium is seen. If the disease has a fatal

[166] G. E. Burch and T. D. Giles, *Am. Heart J.*, 1977, **94**, 600.
[167] D. A. Werderitsh, K. Hess, D. Sandford, K. R. Safranski, and D. J. Moore, *Proc. Indiana Acad. Sci.*, 1976, **85**, 113.
[168] B. Petersen, M. Schroll, C. Christiansen, and I. Transbøl, *Acta Med. Scand.*, 1977, **201**, 31.
[169] B. Petersen, C. Christiansen, and P. I. Hansen, *Acta Med. Scand.*, 1976, **200**, 59.
[170] Y. Rayssiguier, *Horm. Metab. Res.*, 1977, **9**, 309.
[171] R. Alvarez and J. J. Bruno, *Proc. Nat. Acad. Sci. U.S.A.*, 1977, **74**, 92.
[172] Y. Rayssiguier, J. M. Garel, and M.-J. Davicco, *Horm. Metab. Res.*, 1977, **9**, 438.

outcome, cardiovascular lesions similar to those of magnesium deficiency are often reported.[149] The incidence in Europe is fairly low (1—2%), but in individual herds the mortality rate may be as high as 30%,[173, 174] and specific breeds, *e.g.* Ayrshires, seem to be particularly susceptible. It has been suggested that marginal magnesium deficit is induced by the artificial winter feeds, and that the change to fresh grass, particularly that which may have been previously dressed with high phosphate or nitrogen fertilizers, may induce changes in magnesium absorption rate in the gut and lead to the precipitous fall in serum magnesium. It has also been suggested that the stress induced by exposure to cold, wet, and windy conditions may induce a fall in magnesium intake. Relief of symptoms is obtained by subcutaneous injection of magnesium, provided that it is carried out at an early stage. Prophylactic treatment of animals by magnesium supplements in artificial feeds or by top-dressing pastures with magnesium-containing fertilizer is successful.

The similarity between the cardiovascular symptoms of grass tetany in cattle and those of pre-eclampsia (toxaemia) during pregnancy in the human female has been stressed by Seelig and Bunce.[149] Intravenous magnesium has been used empirically for many years in the treatment of this condition. The action of magnesium has been attributed to its sedative and anti-hypertensive pharmacological effects, though it may be that its real effect is in repletion of body stores of the element. Magnesium depletion may also occur following surgery[175-177] and in protein-calorie malnutrition.[178]

Magnesium in Biology. It is clear from the range of biological activity reported that magnesium is of very wide significance. Indeed, recent work has shown that, for instance, it may be involved in the recognition of potential host snails by parasitic flatworm miracidiae, which are the free-swimming stages of the life cycle that penetrate the snail, where they ultimately develop into larvae, which in turn affect man. Schistosomiasis, a debilitating and intractable disease, results. It is very widespread in tropical areas. The adult flatworm infests the cardiovascular system of the human host after the infection has been acquired from water containing infected snails, the secondary host.[179]

Changes in magnesium concentration in blood and tissues have been reported during experimental carcinogenesis in rats, and it has been suggested that the altered permeability to magnesium may be one of the triggers for the development of the neoplasia.[180, 181]

[173] D. C. Church, 'Digestive Physiology & Nutrition of Animals', Vol. 2, 'Nutrition', publ. D. C. Church, Corvallis, Oregon, 1972, p. 704.
[174] P. McDonald, R. A. Edwards, and J. F. G. Greenhalgh, 'Animal Nutrition', Oliver and Boyd, Edinburgh, 2nd edn., 1973, p. 98.
[175] F. Annoni, F. Longoni, and I. Brattini, *Rev. Franc. Endocrinol. Clin., Nutr. Metab.*, 1974, **15**, 510.
[176] F. Annoni, F. Longoni, I. Brattini, F. Lavorato, M. Belloni, and R. Sacos, *Rev. Franc. Endocrinol. Clin., Nutr. Metab.*, 1976, **17**, 241.
[177] G. S. Fell and R. R. Burns, *Proc Roy. Soc. Med.*, 1976, **69**, 474.
[178] J. L. Caddell, *Ann. New York Acad. Sci.*, 1969, **162**, Art. 2.
[179] H. H. Stibbs, E. Chernin, S. Ward, and M. L. Karnovsky, *Nature*, 1976, **260**, 702.
[180] L. J. Anghileri and M. Heidbreder, *Eur. J. Cancer*, 1977, **13**, 291.
[181] G. A. Young and F. M. Parsons, *Eur. J. Cancer*, 1977, **13**, 103.

These and a number of other apparently unrelated phenomena are presently unexplained, but it is perhaps significant that magnesium is involved in primitive processes, primitive organisms, and unspecialized cells in the highest organisms.

Zinc.—Vallee has remarked that the notable abundance of zinc in biological matter is appreciated by few, even though it has been known for 30 years that there are 1.4—2.3 g in an average 70 kg adult.[182-184] Zinc is essential to the normal growth and development of all living matter, and its deficiency results in major abnormalities (see Table 3). For zinc, the problems of contamination control and analysis (microwave excitation spectroscopy can detect 10^{-14} g atom of zinc in 1—2 μg of enzyme[185]) have largely been overcome, leading to a wealth of information about zinc enzymes, and this has directed critical experiments in biochemistry, physiology, pathology, nutrition, and medicine.[186, 187]

Acrodermatitis Enteropathica: Zinc Malabsorption. This is a rare autosomal recessive condition which was invariably fatal in infancy until 5,7-di-iodo-8-hydroxyquinoline (6) was found to induce a complete clinical remission.[188] For

(6)

20 years the drug was used purely empirically, without any knowledge of its interactions with zinc. Oral zinc supplements alone were then found to induce complete remission [189] The Adema disease in cattle is a closely related condition, and again the quinoline enhances transport of zinc through the intestinal barrier.[190] Typical Zn levels before and after therapy are given in Table 4. Normally the body strives to absorb zinc in amounts above its immediate needs and, in situations of deficient or sub-optimal supply, is able to increase the efficiency of Zn utilization homeostatically by fully absorbing the injected Zn and reducing the endogenous Zn excretion.[191, 192] As dietary Zn supply increases, absorbility continually declines, and becomes the major factor controlling an excessive inflow of zinc into metabolism. An important finding by Hurley[193, 194]

182 E. J. Underwood, 'Trace Elements in Human and Animal Nutrition', Academic Press, New York, 4th edn., 1977, p. 196.
183 B. L. Vallee, *Trends Biochem. Sci.*, 1976 (April), 88.
184 B. L. Vallee, in 'Biological Aspects of Inorganic Chemistry', ed. D. Dolphin, Wiley, New York, 1977, p. 38.
185 H. Kawaguchi and B. L. Vallee, *Analyt. Chem.*, 1975, **47**, 1029.
186 J. F. Riordan, *Med. Clin. North Am.*, 1976, **60**, 661.
187 J. F. Riordan and B. L. Vallee, in 'Trace Elements in Human Health and Disease', ed. A. S. Prasad, Academic Press, New York, 1976, Vol. 1, p. 227.
188 H. T. Delves, J. T. Harries, M. S. Lawson, and J. D. Mitchell, *Lancet*, 1975, **ii**, 929.
189 P. M. Barnes and E. J. Moynahan, *Proc. Roy. Soc. Med.*, 1973, **66**, 327.
190 T. Flagstad, in ref. 6, p. 423.
191 E. Weigand and M. Kirchgessner, *Nutr. Metab.*, 1978, **22**, 101; and in ref. 6, p. 106.
192 F J. Schwarz and M. Kirchgessner, in ref. 6, p. 100.
193 L. S. Hurley, J. R. Duncan, C. D. Eckhert, and M. S. Sloan, in ref. 6, p. 449.
194 L. S. Hurley, *Proc. Nat. Acad. Sci. U.S.A.*, 1977, **74**, 3547.

Table 3 *Zinc deficiency in different phyla[a]*

	Growth	Development	Changes in chemical composition		Decreased enzymatic activities
			Decrease	Increase	
Micro-organisms	Retardation	Increase in cellular size	Protein RNA (ribosomal) Pyridine nucleotides	DNA Amino-acids Polyphosphate Phospholipids ATP Organic acids	Alkaline phosphatase Alcohol and D-lactate dehydrogenase Tryptophan desmolase
Plants	Retardation	Small, abnormal leaves Chlorotic mottling Decreased fruit production	Protein Auxin Ethanolamine	Amino-acids	Tryptophan desmolase Carbonic anhydrase Aldolase Pyruvic carboxylase
Vertebrates	Retardation	Testicular atrophy Parakeratosis Dermatitis Coarse, sparse hair	Red blood cells Serum proteins	Uric acid	Alkaline phosphatase Pancreatic proteases Malate, lactic alcohol dehydrogenase NADH diaphorase

(a) Taken from 'Biological Aspects of Inorganic Chemistry', ed. D. Dolphin, Wiley, New York, 1977, p. 38.

Table 4 *Some zinc levels in acrodermatitis enteropathica[a] (data for 4 patients)*

	Normal	Before	After oral zinc
Plasma (μg per 100 ml)	68—110	15—39	96—153
Urine (μg per 24 h)	100—700	37—47	286—1168
Erythrocyte (μg per ml)	10.1—13.4	5.7—10.1	6.4—12.3
Hair (μg per gram)	105	66—154	75—167
Skin (μg per gram)	15—25	6.7	18.3

(a) Taken from 'Zinc and Copper in Clinical Medicine', ed. K. M. Hambidge and B. L. Nichols, Jr., SP Medical and Scientific Books, New York, 1978, Ch. 7, by K. M. Hambidge.

is that human milk contains a small zinc-binding ligand (mol. wt. 8700) which may be required for intestinal absorption of Zn by the neonate before its own mechanisms for Zn absorption are mature. The absence of this ligand in some species of cattle and sheep is unexplained at present. Analysis of the trace-metal content of human milk at various stages of lactation shows that a large decline in Ca and Zn levels occurs during the first six months of lactation.[195] Zinc deficiency during gestation has been induced experimentally in pigs,[196] and it consistently causes abnormal ossification of the skeleton. The recessive mutation *lethal milk* (lm) in mice has been correlated with a reduction in Zn levels of both milk and pup carcass,[197] and may provide a useful animal model for acrodermatitis.

Syndrome of Acute Zinc Deficiency. Zinc deficiency was once thought to be unlikely in man, but is now known to occur in some Middle Eastern countries, probably because of the high content of phytic acid (7) (myo-inositol hexaphosphate) of

(7)

cereals. Animal studies have shown a decreased absorption of Zn from diets with a high phytic acid content (*e.g.* unleavened bread), and phytic acid itself can induce Zn deficiency.[198] Changes in the serum Zn level do not seem to be a useful measure of the degree of absorption. Nearly all the circulating zinc in human plasma is bound to two serum proteins. Two-thirds is 'loosely bound' to albumin and one third 'tightly bound' to α_2-macroglobulin. The α_2-macroglobulin affinity constants for Zn^{199} are K_1 3×10^7 and K_2 1×10^5, a Zn content of 150—180 μg (g protein)$^{-1}$. Chronic deficiency in this group is characterized by a hypogonadal type of dwarfism, and both growth and sexual development improve rapidly with Zn supplementation.[200] Administration of α-mercapto-β-(2-furyl)acrylic acid and related compounds to rats causes a long-lasting and dose-related increase in the serum concentration of Zn without affecting serum Cu and Fe concentrations.[201]

A syndrome of acute Zn deficiency has now been described for four patients totally supported by intravenous alimentation.[202] It resembles the parakeratotic syndrome of swine and is very similar to acrodermatitis enteropathica (alopecia, diarrhoea, and pustular eruption around the orifices, plus bullous or verrucous

[195] L. A. Vaughan, C. W. Weber, and S. R. Keinberling, in ref. 6, p. 452.
[196] I. Wegger and B. Palludan, in ref. 6, p. 428.
[197] J. E. Piletz and R. E. Ganschow, *Science*, 1978, **199**, 181.
[198] N. T. Davies and R. Nightingale, *Br. J. Nutr.*, 1975, **34**, 243.
[199] N. F. Adham. M. K. Song, and H. Rinderknecht, *Biochim. Biophys. Acta*, 1977, **459**, 212.
[200] J. A. Halstead, H. A. Ronaghy, and P. Abadi, *Am. J. Med.*, 1972, **53**, 277.
[201] E. Girous, N. J. Prakash, P. J. Schechter, and J. Wagner, in ref. 6, p. 68.
[202] R. G. Kay, C. T. Jones, J. Pybus, R. Whiting, and M. Black, *Ann. Surg.*, 1976, **183**, 331.

eruption on the extremities). The response to oral or intravenous therapy is striking, except for hair growth, which is delayed but eventually complete.

Zinc Therapy for Arthritis? Simkin [203] has described the effective use of oral zinc sulphate (220 mg, three times daily) to alleviate the symptoms of active rheumatoid arthritis, and suggested that D-penicillamine (which is also effective) may act by promoting Zn absorption. Whitehouse *et al.* carried this argument one stage further, in the hope of discovering an improved drug:[204] their complex $Zn(D\text{-pen})_2$, however, was *lethal* (75% mortality in adult male rats at 140 μmol kg^{-1}, intraperitoneally). Similarly, neutralized aqueous mixtures of $ZnCl_2 \cdot 2D\text{-pen}$ were also toxic.

Penicillamine may also be effective for Zn excretion, since Zn deficiency can occur in patients receiving penicillamine therapy for Wilson's disease.[205] The manifestations consisted of parakeratosis, dead hair and alopecia, keratitis, and centrocecal scotoma, which disappeared on Zn supplementation.

Wound Healing, Taste, and Night Blindness. Zinc oxide skin ointments have been faithfully used for many years, but although recent studies indicate that wound healing is impaired in patients with Zn deficiency,[206, 207] there is little evidence that extra dietary Zn assists those with normal serum Zn concentrations. Perhaps serum Zn is a poor indicator of deficiency in this instance.

It was noted in 1967 that 30% of a group of patients receiving D-penicillamine therapy for rheumatoid arthritis, scleroderma, cystinuria, or pulmonary fibrosis developed diminished taste sensation (hypogeusia),[208] which could be corrected by oral administration of Ni^{II}, Zn^{II}, or Cu^{II}.[209] The fluids that bathe the taste buds have been found to contain the zinc protein gustin (mol. wt. 37 000; 2Zn), and there seems to be a relationship between gustin and nerve-growth factors at the taste-bud membrane. However, the role of transition metals in both the preneural and neural events associated with taste is largely unexplored territory.

Perception in dim light by the human eye depends on rhodopsin, the photosensitive chromophore in rods. 11-*cis*-Retinal, the prosthetic group of rhodopsin, is formed from vitamin A in three steps, the first of these being the oxidation of the alcohol group to an aldehyde by the Zn enzyme retinol dehydrogenase. Massive doses of vitamin A given to Zn-deficient animals fail to restore plasma vitamin A levels to normal.[210] Some patients with alcoholic liver disease are resistant to vitamin A therapy, presumably due to altered Zn metabolism.

Copper.—Menkes' syndrome was first described by John Menkes in 1962[211] and is probably caused by a genetic defect in Cu transport.[212] It appears to be

203 P. A. Simkin, *Lancet*, 1976, **ii**, 539.
204 M. W. Whitehouse, W. S. Hanley, and L. Field, *Arthritis Rheum.*, 1977, **20**, 1035.
205 W. G. Klingberg, A. S. Prasad, and D. Oberleas, in 'Trace Elements in Human Disease and Health', ed. A. S. Prasad, Academic Press, New York, 1976, Vol. I, p. 51.
206 W. E. C. Wacker, in 'Zinc and Copper in Clinical Medicine', ed. K. M. Hambidge and B. L. Nichols, Jr., SP Medical and Scientific Books, New York, 1978, p. 15.
207 F. W. Sunderman, Jr., *Ann. Clin. Lab. Sci.*, 1975, **5**, 132.
208 R. I. Henkin, H. R. Keiser, I. A. Jaffe, I. Sternlieb, and I. H. Scheinberg, *Lancet*, 1967, **ii**, 1268.
209 R. I. Henkin and D. F. Bradley, *Life Sci.*, 1970, **II**, 701.
210 W. A. Cassidy, E. D. Brown, and J. C. Smith, in ref. 206, p. 59.
211 J. H. Menkes, M. Alter, G. K. Steigleder, D. R. Weakley, and J. H. Sung, *Pediatrics*, 1962, **29**, 764.
212 D. M. Danks, *Inorganic Perspectives in Biology and Medicine*, 1977, **1**, 73.

inherited as a recessive X-linked trait, and only males are affected. Symptoms start between birth and three months: abnormally spirally twisted hair (pili torti), signs of progressive cerebral degeneration (with frequent seizures), temperature instability (hypothermia), arterial abnormalities, low plasma Cu and caeruloplasmin, and often death at an early age.[213]

Low serum Cu, caeruloplasmin, and liver Cu levels can be restored to normality by administration of intramuscular Cu^{II}(EDTA) (or intravenous Cu acetate, sulphate, albumin, or histidine complexes), but the disease is not cured, and no clinical benefits have been recorded where treatment has started after age 2 months. There is need for a Cu complex that will deliver Cu to intracellular sites. Since increased Cu levels are found in the gut mucosa, the defect must affect transit through the gut cell rather than uptake by these cells.[214]

Mottled mice appear to contain an homologous mutation, and may be a valuable experimental model of Menkes' disease.[212] Abnormal pigmentation is due to altered tyrosinase levels. This is a Cu enzyme which converts dopamine into melanine.

It has been postulated that an intracellular Cu-binding protein has an increased affinity for Cu in Menkes' disease and is absent from liver.[212] Metallothionein can constitute a major Cu-binding protein in liver, and its synthesis can be induced by Cu. (Cu,Zn)-Thioneins have been isolated and characterized from the livers of Cu-injected rats,[215] pigs,[216] and foetal calves.[217] They are best handled at low temperature under anaerobic conditions, and Cu may be present as copper(I). The general need for more monitoring of tissue Cu levels and Cu enzymes at all stages of development has been emphasized.[212]

Overall Cu homeostasis depends on the balance between intestinal absorption and biliary excretion, and whereas in Menkes' syndrome and mottled mice there is an absorption fault, in Wilson's disease there is a biliary excretion defect, leading to severe Cu overload.[218] Copper normally enters the bile *via* lysosomes of the hepatic cells, and renal excretion accounts for less than 1 % of the daily Cu loss.[219] D-Penicillamine is usually administered as therapy for Wilson's disease, and attempts are being made to determine whether Cu is excreted as the violet mixed-valence complex $[Cu^{II}_6Cu^{I}_8pen_{12}Cl]^{5-}$.[220] The binding site on albumin, which is responsible for much Cu transport, has been defined.[221]

Persistent neutropenia, which occurs prior to a fall in serum caeruloplasmin and Cu levels, is the earliest detectable evidence of Cu deficiency,[222] and in the presence of Cu deficiency, anaemia cannot be overcome unless oral Fe and Cu

[213] A. Cordano, in ref. 206. Ch. 10.
[214] D. M. Danks, E. Cartwright, B. J. Stevens, and R. R. W. Townley, *Science*, 1973, **179**, 1140.
[215] I. Bremner and B. W. Young, *Biochem. J.*, 1976, **157**, 517.
[216] I. Bremner and B. W. Young, *Biochem. J.*, 1976, **155**, 631.
[217] H. J. Hartman and U. Weser, *Biochim. Biophys. Acta*, 1977, **491**, 211.
[218] I. H. Scheinberg and I. Sternlieb, in 'Trace Elements in Human Health and Disease', ed. A. S. Prasad, Academic Press, New York, 1976, Vol. I, p. 415.
[219] G. G. Cartwright and M. M. Wintrobe, *Am. J. Clin. Nutr.*, 1974, **14**, 224.
[220] P. J. Birker, H. C. Freeman, and J. A. Ramshaw, in ref. 6, p. 409.
[221] S. J. Lau and B. Sarkar, *Canad. J. Chem.*, 1975, **53**, 710.
[222] A. Cordano, in 'Zinc and Copper in Clinical Medicine', ed. K. M. Hambidge and B. L. Nichols, Jr., SP Medical and Scientific Books, New York, 1978, p. 119.

are given simultaneously (role of caeruloplasmin as a ferroxidase). Despite these observations, little attention has been given to the clinical administration of Cu as an essential trace element. Cordano has suggested[223] that it should be included as a Recommended Dietary Allowance. The high serum Cu level in acute lymphoblastic leukaemia (> 150 μg per 100 ml), which is depressed on the first emission (H. T. Delves, personal communication), is of current interest, and there is speculation that leukaemic white cells secrete a material of low molecular weight which can cause a redistribution of zinc and copper.[224]

Chromium.—Chromium has been recognized as an essential nutrient for animals since 1959,[225] following the observation that rats fed with *Torula* yeast develop impaired glucose tolerance.[226] The active ingredient in other yeasts, such as Brewers yeast, is a chromium complex termed 'glucose tolerance factor' (GTF).[227-229] The National Bureau of Standards (U.S.A.) reference Brewers yeast contains 2.12 μg of Cr per gram of yeast, although results from standard determinations in seventeen laboratories range from 0.35 to 5.4 μg per gram,[230] emphasizing the considerable difficulties in analysing for Cr in biological material and the need for careful sample preparation.

Mertz has extracted GTF from yeast and tested its biological activity in an assay which involves potentiation of the action of insulin on the oxidation of glucose by Cr-deficient rat adipose tissue *in vitro*.[231] No system has yet been found where Cr supplementation has any effects in the absence of insulin. After extraction of GTF with 50% EtOH, adsorption on to charcoal, elution with NH_4OH–ether, hydrolysis with 5M-HCl, and ion-exchange chromatography, Cr fractions with distinct biological activity are found to contain predominantly Gly, Glu, and Cys (the elements of glutathione) and nicotinic acid. Vigorous refluxing of mixtures of Cr^{III} acetate, 2 moles of nicotinic acid, 2 Gly, 1 Glu, and 1 Cys can produce a synthetic complex of composition (8), with very similar electronic absorption, i.r., and chromatographic behaviour to GTF. The aquo–nicotinic acid–chromium(III) complex (9) is also biologically active, but hydrolytically unstable. The kinetic inertness of Cr^{III} favours the isolation of an intact Cr

$Cr^{III}(nic)_2(Gly)_2(Glu)(Cys)$

(8)

(9)

[223] G. G. Graham and A. Cordano, in ref. 218, p. 363.
[224] W. R. Beisel, R. S. Pekarek, and R. W. Wannenacher, in ref. 218, p. 97.
[225] K. Schwarz and W. Mertz, *Arch. Biochem. Biophys.*, 1959, **85**, 292.
[226] W. Mertz and K. Schwarz, *Arch. Biochem. Biophys.*, 1955, **58**, 504.
[227] W. Mertz, *Physiol. Rev.*, 1969, **49**, 163.
[228] W. Mertz, *Nutr. Rev.*, 1975, **33**, 129.
[229] W. Mertz, *Med. Clin. North Am.* 1976, **60**, 739.
[230] W. Mertz, R. A. Anderson, W. R. Wolf, and E. E. Roginski, in 'Trace Element Metabolism in Man and Animals, ed. M. Kirchgessner, Arbeitskreis für Tierernährungs Forschung in Institut für Ernahrungsphysiologie, Freising-Weihenstephan, 1978, p. 272.
[231] E. W. Toepfer, W. Mertz, M. M. Polansky, E. E. Roginski, and W. R. Wolf, *J. Agric. Food Chem.*, 1977, **25**, 162.

complex from yeast but makes synthetic work difficult. Surprisingly little co-ordination chemistry of Cr^{III} complexes of this type seems to have been reported. Chromium(IV) is apparently inactive in restoring glucose metabolism,[228] but can be incorporated into yeast at similar rates as Cr^{III}.[232]

Dietary supplements of 150—250 μg of Cr, administered as $CrCl_3$, have been demonstrated to improve glucose tolerance in adults and in elderly[233, 234] and malnourished children who are Cr-deficient.[235, 236] Present estimates are that the absolute requirement of man for absorbable Cr is > 8 μg per day (probably about 15), and dietary requirement is 50—200 μg per day.[228, 229] The normal urinary excretion rate is about 10 μg per day and is a promising measure of the chromium status in man.

GTF, when compared to aquo, chloro, acetato, or organic acid Cr^{III} complexes, is absorbed faster, is transported across the placenta (the foetus acquires high Cr concentrations during development), has access to the Cr pool that is the source of acute plasma increment in response to insulin, and has a different tissue distribution[229] (the average man contains about 6 mg of Cr).

A case of pronounced Cr deficiency has been reported.[237] A female patient had been receiving intravenous nutrition which furnished only 8 μg of Cr per day, and after $3\frac{1}{2}$ years on this regime developed a 15% weight loss together with peripheral neuropathy. The Cr balance was negative: 0.55 ng ml^{-1} (normal is 4.9—9.5) in blood and 154—175 ng g^{-1} (normal is > 500) in hair. Intravenous administration of chromium (as $CrCl_3$) at 250 μg per day for two weeks corrected all abnormalities, and the patient has since had a daily level of 20 μg of Cr per day added to the infusate. It is notable that the administration of insulin had failed to correct this case of impaired glucose tolerance.

Members of the Pima Indian tribe (near Phoenix, Arizona) have the highest incidence of diabetes of any reported population group (reaching 50%), but Eatough *et al.*[238] could not find any significant differences between Cr tissue levels in diabetic and non-diabetic autopsy samples. Hair levels were lower (254 ng g^{-1} compared to 518 ng g^{-1} for non-diabetic), but the average age of the diabetic group was 21 years more, and Cr concentrations in hair are thought to decrease with age.

Does chromium potentiate the action of insulin by forming a ternary complex with it at the membrane receptor? It is certainly a possibility, since porcine insulin binds a $Cr(nic)_2(glutathione)$ complex very tightly,[232, 239] and the Cr–insulin complex is only partially dissociated by trichloroacetic acid, $HClO_4$, or ammonium sulphate precipitation.

232 R. A. Anderson, M. M. Polansky, J. H. Brantner, and E. E. Roginski, in ref. 6, p. 269.
233 W. H. Glinsmann and W. Mertz, *Metabolism*, 1966, **15**, 510.
234 R. A. Levine, D. H. P. Streeten, and R. J. Doisy, *Metabolism*, 1968, **17**, 114.
235 C. T. Gürson and G. Saner, *Am. J. Clin. Nutr.*, 1971, **24**, 1313.
236 J. P. Carter, A. Kattab, K. Abd-El-Hadi, J. T. Davis, A. El Cholmi, and V. N. Patwardhan, *Am. J. Clin. Nutr.*, 1968, **21**, 195.
237 N. Jeejeebhoy, R. C. Chu, E. B. Marliss, G. R. Greenberg, and A. Bruce-Robertson, *Am. J. Clin. Nutr.*, 1977, **30**, 531.
238 D. J. Eatough, L. O. Hansen, S. E. Starr, M. S. Astin, S. B. Larsen, R. M. Izatt, J. J. Christensen, and R. F. Hamman, in ref. 6, p. 259.
239 R. A. Anderson and J. H. Brantner, *Fed. Proc.*, 1977, **36**, 1123.

Molybdenum.—Mertz has stressed that although acute and chronic toxicity data exist for Mo, no direct information is available on which a minimal or optimal human requirement can be based. A safe dietary level may be about 140 ng of Mo per gram, but the consequences of Mo deficiencies in man are unknown.

Payne[240] has detected Mo-responsive diseases in chickens, and notes that feed compounders are now routinely adding 1 p.p.m. of Mo as Na_2MoO_4 (sodium molybdate) to breeder and broiler feeds.

Tungsten.—Tungsten usually acts antagonistically against molybdenum. Feeding W to rats can produce Mo deficiency, and their tissues contain inactive demolybdo-forms of the enzymes sulphite oxidase and xanthine oxidase.[241] Demolybdo sulphite oxidase containing tungsten at the active site can be reconstituted by inorganic molybdate, but the metal-free form can be reconstituted only by a molybdenum cofactor which can be isolated from the mitochondrial outer membrane of rat liver.[242] The cofactor is the source of a labile pool of Mo in this tissue and has been identified in many other tissues (including human) and in *E. coli* and *N. crassa*.

Surprisingly, when tungstate is substituted for molybdate, a substantial increase in the formate dehydrogenase activity of *Clostridium thermoaceticum* cells is observed,[243] especially when administered together with selenite. Similar results have been obtained with *C. formicoaceticum*, and it has been shown that tungsten is the preferred metal for the formate dehydrogenases of these two micro-organisms.

Balaeva[244] considers that tungsten may have a physiological role in animals (typical bovine heart level 0.005 p.p.m.), and has found that administration of an aqueous solution of Na_2WO_4 to guinea-pigs at a dose of 13.4 μg per kg live weight significantly increases their oxygen absorption and weight (12%), whereas at higher doses either no change (0.5 mg per kg) or a 20% decrease (21 mg per kg) was observed.

Nickel.—Following the suggestions of Nielson *et al.*,[245] Sunderman *et al.*,[246] and Anke *et al.*[247] from 1970—74 that Ni might be an essential trace element, Schnegg and Kirchgessner have now demonstrated this unequivocally.[248] Nickel deficiency in rats is associated with retarded growth and reduction of blood haemoglobin, haemocrit values, and erythrocyte counts. Nickel deprivation profoundly impairs intestinal absorption of iron and thereby causes anaemia.[249, 250] Typically a Ni-deficient diet (containing only 15 ng of Ni per gram) reduces the live weight of 30-day-old rats by as much as 30%, and, in order to satisfy biochemical

[240] C. G. Payne, in ref. 6, p. 515.
[241] J. L. Johnson, H. P. Jones, and K. V. Rajagopalan, *J. Biol. Chem.*, 1977, **252**, 4994.
[242] H. P. Jones, J. L. Johnson, and K. V. Rajagopalan, *J. Biol. Chem.*, 1977, **252**, 4988.
[243] L. G. Ljungdahl, *Trends Biochem. Sci.*, 1976 (March), 63.
[244] M. S. Balaeva, in ref. 6, p. 617.
[245] F. H. Nielsen and H. E. Sauberlich, *Proc. Soc. Exp. Biol. Med.*, 1970, **134**, 845.
[246] F. H. Nielsen and D. A. Ollerich, *Fed. Proc.*, 1974, **33**, 1767.
[247] F. W. Sunderman, jun., S. Nomoto, R. Morang, N. W. Necay, C. N. Burke, and S. W. Nielsen, in 'Trace Element Metabolism in Man and Animals', ed. W. G. Hoekstra, University Park Press, Baltimore, 1974, Vol. 2, p. 715.
[248] A. Schnegg and M. Kirchgessner, *Z. Tierphysiol., Tierernaehr. Futtermittelkd.*, 1975, **36**, 63.
[249] A. Schnegg and M. Kirchgessner, *Int. J. Vitam. Nutr. Res.*, 1976, **46**, 96.
[250] A. Schnegg and M. Kirchgessner, in ref. 6, p. 236.

criteria on the essentiality of Ni, enzyme activities in the liver have been measured:[251, 252]

malate dehydrogenase	-29%
glucose-6-phosphate dehydrogenase	-82%
isocitrate dehydrogenase	-73%
lactate dehydrogenase	-66%
glutamate dehydrogenase	-45%
glutamate–pyruvate transaminase	-45%
alkaline phosphatase	same as control
creatine kinase	$+27\%$

Remarkably, both glucose and glycogen concentrations in the liver and serum are reduced by 90% on Ni depletion. Prompt increases of plasma glucagon and glucose following Ni administration ($NiCl_2$ at about 68 μmol per kg) are observed with fasting rats.[253]

Little is known about nickel-binding sites *in vivo*, and only one Ni protein has been characterized – jack-bean urease[254] (the Ni in this lay undetected for over 30 years!). A protein which is rich in Ni but not other trace metals can be fractionated from human serum (an α_1-macroglobulin),[255] and a Ni-binding 9.5S α_1-glycoprotein can also be isolated. Albumin is the principal Ni^{II}-binding protein in human, bovine, rabbit, and rat serums, and Ni appears to compete with Cu^{II} for the site formed from the NH_2 terminus, the adjacent two peptide nitrogens, and the histidine in the third position from this terminus[256] ($K_1 = 6 \times 10^5$ mol^{-1} in man). Nickel analyses of urine and serum can serve as laboratory indices of environmental exposure, as demonstrated by the comparison of Ni levels in adults living in areas of low and high environmental exposure:[256, 257]

	Hartford, U.S.A. (low)	Sudbury, Canada (high)
Serum	2.6±0.9 (0.8—5.2)	4.6±1.4 (2.0 7.3) μg per litre
Urinary excretion	2.6±1.4 (0.5—6.4)	7.9±3.7 (2.3—15.7) μg per day

Our daily intake of Ni is probably about 300—600 μg per day, most of which is excreted in the faeces (260 μg per day).

An increase in Ni concentration has been observed in serum from patients after acute myocardial infaction (heart attack),[256] and this is probably a secondary manifestation of leukocytosis and leukocytolysis. These data have been confirmed independently in other laboratories. Hypernickelaemia is also observed in patients with acute stroke or extensive thermal burns.

Poisoning with $Ni(CO)_4$ is an industrial problem.[257] Type I alveolar cells are the primary target (severe pulmonary symptoms). The antidotal efficacy of

[251] M. Kirchgessner and A. Schnegg, *Bioinorg. Chem.*, 1976, **6**, 151.
[252] A. Schnegg and M. Kirchgessner, in ref. 6, p. 236.
[253] E. Horak and F. W. Sunderman, jun., *Toxicol. Appl. Pharmacol.*, 1975, **33**, 388.
[254] N. E. Dixon, C. Gazzola, R. L. Blakeley, and B. Zerner, *J. Amer. Chem. Soc.*, 1975, **97**, 4131.
[255] M. I. Decsy and F. W. Sunderman, jun., *Bioinorg. Chem.*, 1974, **3**, 95.
[256] F. W. Sunderman, jun., in ref. 1, p. 231.
[257] F. W. Sunderman, jun., *Ann. Clin. Lab. Sci.*, 1977, **7**, 377.

sodium diethyldithiocarbamate is greater than that of D-pen or that of trien.[258, 259] $Ni(CO)_4$ can cross alveolar membranes in either direction without metabolic alteration. The lung is the major excretory organ, and Ni^{II} is also excreted in the urine, accompanied by CO in the breath, following oxidation of $Ni(CO)_4$ in erythrocytes and other cells. An initial, transient hyperglycaemia is also observed (as with Ni^{II} injections alone).

Nickel subsulphide (α-Ni_3S_2) is one of the most potent carcinogenic metal compounds that has been studied.[9] A single intramuscular injection of only 1.2 mg gives a 77% incidence of comas in Fischer rats in 20 months (suppressed by simultaneous injection of Mn dust).

There is considerable interest and concern about allergic hazards following exposure to Ni, and Wahlberg has predicted that it may be feasible to perform passive transfer of Ni allergy in experimental animals.[260] Lymphocyte blast transformation is a practical test for Ni sensitivity *in vitro*.[261]

Silicon.—Birchall, in his timely review of silicon in the biosphere,[3] has concluded that there is an association between silicon (as silicic acid, $[Si(OH)_4]_n$) and carbohydrates (Si—O—C bonds) and perhaps calcium, resulting in subtle effects on organization and structure, mineral binding, cohesion, and adhesion.

Schwarz began experiments on the essentiality of Si in his all-plastic trace-element-controlled isolator system (no glass!) in 1965,[262] and by 1971 was recording 25—34% increases in growth rates on supplementation of rat diets with metasilicate ($Na_2SiO_3 \cdot 9H_2O$) at 50 mg per 100 g of diet. Silicon-deficient rats developed disturbances in bone structure development,[263] and Si is particularly associated with the polysaccharides of connective tissue:[264] cartilage, skin, and bone. Hyaluronic acid of the human umbilical chord is very rich in Si, for example, containing a total of 1.89 mg per gram (1.53 mg is free).[265] In bone, Si is localized in regions where there is active calcification in progress, especially in the osteoblast, the active bone-forming cell.[266] The normal human plasma level is low (*ca.* 1 p.p.m.) and the daily urinary output is about 9 mg.[*3] Analytical electron microscopy is a promising technique for mapping out the distribution of silicon in biological material.[†267, 268]

Very little is known about the incorporation of Si into the biopolymers of connective tissue or about the absorption of dietary Si, but Si levels decrease

---

* Bone and ligament levels are about 100 p.p.m., and an adult contains about 7 g of Si.
† In Alzheimer's disease there is a curiously high level of Si in plaques; 1600—14 600 μg per g, compared to the normal cerebrum level of 7 μg per g.[269]

258 E. Horak, F. W. Sunderman, jun., and B. Sarkar, *Res. Commun. Chem. Pathol. Pharmacol.*, 1976, **14**, 153.
259 R. C. Baselt, F. W. Sunderman, Jr., J. Mitchell, and E. Horak, *Res. Commun. Chem. Pathol. Pharmacol.*, 1977, **18**, 677.
260 J. E. Wahlberg, *Dermatologica*, 1976, **152**, 321.
261 I. E. Millikan, F. Conway, and J. E. Foote, *J. Invest. Dermatol.*, 1973, **60**, 88.
262 K. Schwarz and D. B. Milne, *Nature*, 1972, **239**, 333.
263 K. Schwarz and S. C. Chen, *Fed. Proc.*, 1974, **33**, 3.
264 K. Schwarz, *Fed. Proc.*, 1974, **33**, 1748.
265 K. Schwarz, *Proc. Nat. Acad. Sci. U.S.A.*, 1973, **70**, 1608.
266 E. Carlisle, *Fed. Proc.*, 1975, **34**, 927.
267 C. Appleton and P. F. Newell, *Nature*, 1977, **266**, 854.
268 C. W. Mehard and B. E. Volcani, *Cell Tissue Res.*, 1976, **166**, 255.
269 J. H. Austin, *Prog. Brain Res.*, 1973, **40**, 485.

with age, and there is a suspicion that the lack of biologically available Si plays a part in the aetiology of atherosclerosis.[270] A study of males in Finland with a high risk of coronary disease discovered significantly less Si in their water supply than a low-risk group.[271]

Thus, silicon can no longer be neglected because of its insolubility at neutral pH (100 p.p.m. for silicic acid); indeed the biospheric incorporation of Si per year greatly exceeds that of carbon! There remains much work to be done on the inorganic biochemistry of Si, but the rewards may be considerable.[3]

Selenium.—The detection of the essentiality of selenium resulted from work on necrotic liver degeneration in rats,[272] which is a fatal disease induced by diets low in sulphur-containing amino-acids, lacking vitamin E (10) and Factor 3, a substance which prevents this disease independently of the other two agents. Factor 3 preparations concentrated 10^4 times from kidney powder were found to contain organically bound Se (degradation produced garlic-like odours characteristic of Se compounds, well known from persons exposed to Se and from cattle grazing on seleniferous pastures, and mainly due to the excretion of Me_2Se). Factor 3 preparations afford 50% protection against liver necrosis, at a dietary level supplying only 0.007 p.p.m. of Se! Sodium selenite, Na_2SeO_3, is about one-third as active, and complete protection is afforded by 0.05—0.01 p.p.m. of Se.[273]

$$HO \underset{Me}{\overset{Me}{\bigcirc}} \underset{Me}{\overset{}{O}} \underset{Me}{\overset{}{}} CH_2 \left(CH_2CH_2CHCH_2 \right)_3 H$$

(10)

There is significant interest now in the inter-relationship of selenium toxicity with other elements, notably As,[274] Hg,[275] Cd,[275] and Ag,[276] in vitamin-E-deficient rats and chicks. Silver is thought to cause Se deficiency in the absence of vitamin E. Schwarz[8] has used Se to emphasize the narrow concentration limits in which life is possible: in the rat, 0.1 p.p.m. in the diet is protective, 4 p.p.m. a minimum chronic toxic dose, and 40 p.p.m. an acute toxic dose. A normal liver cell contains only a million atoms of Se. Schwarz and Fredga have found that monoselenodicarboxylic acids $HO_2C(CH_2)_nSe(CH_2)_nCO_2H$ are of similar or better potency than selenite in the prevention of dietary liver necrosis, but only when n is odd (5, 7, 9, or 11).[277]

Despite good evidence that selenium deficiency diseases occur in farm animals, with important consequences in the livestock industry, there is no firm evidence

[270] K. Schwarz, *Lancet*, 1977, **i**, 454.
[271] K. Schwarz, B. A. Ricci, S. Punsar, and M. J. Karvoneh, *Lancet*, 1977, **i**, 538.
[272] K. Schwarz and C. M. Foltz, *J. Amer. Chem. Soc.*, 1957, **79**, 3292.
[273] K. Schwarz, *Med. Clin. North Am.*, 1976, **60**, 745.
[274] H. S. Hsieh and H. E. Ganther, *Biochim. Biophys. Acta*, 1977, **497**, 205.
[275] J. Parizek, J. Kalonskova, A. Babicky, J. Benes, and L. Pavlik, in 'Trace Element Metabolism in Animals', ed. W. G. Hoekstra, University Park Press, Baltimore, 1974, p. 119.
[276] A. T. Diplock, J. Green, J. Bunyan, D. McHale, and I. R. Muthy, *Br. J. Nutr.*, 1967, **21**, 115.
[277] K. Schwarz and A. Fredga, *Bioinorg. Chem.*, 1975, **4**, 235.

that the same is true in man. People living in Se-deficient areas and having low Se blood counts (*e.g.* New Zealand) appear to be none the worse off. It may be that Se and vitamin E are the 'belt and braces' (to quote a conversation with Professor A. Diplock) of the same problem: we do not need them both to remain healthy!

Difficulties in Analysis for Selenium. Selenium is a particularly difficult element to estimate because of its volatility, which can be overcome, *e.g.* by the addition of $NiSO_4$.[278] The normal blood Se levels (about 0.3 p.p.m.) are beyond the range of conventional atomic absorption spectroscopy, although graphite-rod procedures can be used. A neutron activation analysis method has been developed[279] with an absolute sensitivity of 4 ng per gram and a mean detection limit of 50 ng per gram, utilizing cyclic activation of the short-lived isotope ^{77m}Se ($T_{\frac{1}{2}} = 17$ s). N.B.S. bovine liver (1.045 p.p.m. of Se, dry weight) was checked as a standard.

Clinton has reported[280] a novel determination of Se in blood and plant material by SeH_4 generation (treat with HNO_3, $HClO_4$, then BH_4^-) coupled with atomic absorption spectroscopy. Selenium(II) can be differentiated from Se^{IV} at a level of 0.01—0.5 p.p.m. Fluorescence spectrometry of the 2,4-diaminonaphthalene complex constitutes another sensitive analytical method for the determination of Se, and has the advantage that several valence states can be selectively measured.[281]

The toxic effects of Se compounds have been recognized much longer than the nutritional ones, but are difficult to generalize on account of the modifying effects of the chemical form (elemental Se is insoluble in H_2O and harmless when ingested, whereas H_2Se is one of the most toxic substances known: exposure limit 0.2 mg of Se per m^3 of air[282, 283]), quantity consumed, and nature of the diet (including such considerations as As, Hg, Cd, Ag, and sulphur amino-acid content[283]).

Selenium can have teratogenic effects in chickens, and possibly in humans.[284] Of ten women using Se in an industrial situation, five became pregnant during the study period and four aborted; a full-term infant with a club foot was born to another. Although hotly debated,[283] there seems to be no good evidence that Se is carcinogenic, and in fact some compounds are claimed to have anti-carcinogenic properties.[285, 286]

Three categories of chronic toxicity have been described:[283]

(1) 'Blind staggers', caused by organic Se compounds (with or without additional selenate) that are readily extracted by water from native seleniferous plants. It occurs in both cattle and sheep grazing pastures containing plants of the genera *Astragalus*, *Xylorhiza*, *Stanleya*, and *Oonopsis*, and the symptoms involve mainly

278 J. E. Cove, R. M. Del Rio, and T. C. Stadtman, *J. Biol. Chem.*, 1977, **252**, 5337.
279 A. Egan, S. Kerr, and M. J. Minski, in ref. 1, p. 353.
280 O. E. Clinton, *Analyst*, 1977, **102**, 187.
281 P. J. Peterson, C. A. Girling, D. W. Klumpp, and M. J. Minski, Proceedings of the International Symposium on Nuclear Activation Techniques in Life Sciences, 22 May 1978, Vienna.
282 E. A. Cerwenka and W. C. Cooper, *Arch. Environ. Health*, 1961, **3**, 189.
283 A. T. Diplock, *CRC Crit. Rev. Toxicol.*, February 1976, p. 271.
284 D. S. F. Robertson, *Lancet*, 1970, **i**, 518.
285 G. N. Schrauzer and D. Ismael, *Ann. Clin. Lab. Sci.*, 1974, **4**, 441.
286 G. N. Schrauzer, *Bioinorg. Chem.*, 1976, **5**, 275.

central nervous disorder with loss of appetite, paralysis, and general loss of muscular tone.

(2) 'Alkali disease', produced in farm livestock that has consumed plants or grain in which Se is bound to protein and relatively insoluble in water. The clinical picture in horses, cattle, sheep, and pigs involves growth retardation, emaciation, deformities and the shedding of hoofs (said to have been observed in about 1295 by Marco Polo, during his travels), loss of hair, and arthritic disorders of the joints.

(3) Chronic selenosis, produced in experimental animals by administration of SeO_3^{2-} or SeO_4^{2-}.

Of foodstuffs, fish and fish products contain the largest amount of Se, usually > 1 μg per gram, and a similar level is found in grains from seleniferous areas.[287] Animal meats contain about 0.2 μg of Se per gram, but food that is low in protein, such as fruit, contains very little Se. Typical Se levels are:

Dietary intake	58—108 mg per year
U.S.A. diet	151 μg per day
Whole human blood	0.09—0.29 μg ml^{-1}

(in New Zealand the level of 0.068 μg ml^{-1} is thought to be indicative of a Se-deficient country)

Schrauzer has concluded,[288] from studies of age-corrected mortalities at 17 major body sites, that a correlation exists with the apparent dietary Se intake estimated from food-consumption data in 27 countries. He has postulated that the cancer mortalities in Western industrialized nations would decline significantly if the dietary Se intakes were increased to approximately twice that of the current U.S. diet. This conclusion must be treated with great caution in view of the complex inter-relationships between Se utilization and levels of vitamin E and of metals which have been demonstrated with experimental animals. Shamberger *et al.* have concluded that high Se levels correlate with low cancer mortality in a similar epidemiological study.[289]

The primary target organs for Se following oral administration appear to be the kidney and liver, and when SeO_3^{2-} is added to whole human blood, 50—70% is present in the red cells 1—2 minutes after its addition, but this erythrocyte-bound Se is completely liberated into the plasma after 15—20 minutes.[290]

Administration of arsenite ($NaAsO_2$) has been shown to aid the excretion of Se *via* the gastrointestinal tract,[291] perhaps by biliary excretion of a selenol-arsenic adduct.

Selenium, usually as selenite, can be regarded as preventing many of the manifestations of Cd toxicity.[275] The protection mechanism is not understood. Selenium does not prevent Cd uptake by the testes, for example, but causes a marked diversion of Cd in the soluble fraction of proteins of mol. wt. 10—30 000

[287] R. F. Burk, in 'Trace Elements in Human Health and Disease', ed. A. S. Prasad, Academic Press, London, 1976, Vol. II, p. 105.
[288] G. N. Schrauzer, D. A. White, and C. J. Schneider, *Bioinorg. Chem.*, 1977, **7**, 23.
[289] R. J. Shamberger, S. A. Tytko, and C. E. Willis, *Arch. Environ. Health*, 1976, 231.
[290] M. Lee, A. Dong, and J. Yano, *Canad. J. Biochem.*, 1969, **47**, 791
[291] O. A. Levander, *Ann. New York Acad. Sci.*, 1972, **192**, 181.

to a protein of mol. wt. 110 000.[292] Similarly, the metabolism of chronically administered $HgCl_2$ is markedly affected by the Se status of the animal,[293] particularly in facilitating Hg accumulation by the kidney. Administration of methylmercury induces a mobilization of Se from the cytosol to the mitochondrial fraction in brain.[294]

At least four selenium-containing proteins or enzymes have now been characterized, and these are shown in Table 5.

Table 5 *Selenium proteins that have been characterized*

Protein	Mol. wt.	Number of selenium atoms
Formate dehydrogenase	high	?
Glycine reductase	12 000	1; selenocysteine in reduced form
Glutathione peroxidase	84 000	4; selenocysteine ?
	$(4 \times 21\ 000)$	Se—O bonds present after oxidation with H_2O_2
Muscle protein	10 000	1; contains a haem group

Selenium Enzymes. Glutathione peroxidase. In 1957 Mills discovered an enzyme in erythrocytes which catalysed the reduction of H_2O_2 to water, using glutathione (GSH) as the donor substrate:[295]

$$2\ GSH + H_2O_2 \longrightarrow GSSG + 2H_2O$$

It was only recognized as a seleno-enzyme in 1971 by Rotuck, Hoekstra, and co-workers, at the University of Wisconsin, who reasoned[296–298] that Se must play a role in preventing oxidative damage because the effects of Se deficiency in animals could be prevented by vitamin E or antioxidants as well as by Se. The key observation was that Se added to the diet of vitamin-E-deficient animals prevented the haemolysis of their red cells *in vitro*, but only if glucose was present in the incubation medium. The glucose-dependent pathway for the reduction of peroxides (responsible for rupture of the cell and release of haemoglobin) travels *via* glutathione and the selenium-containing enzyme glutathione peroxidase (GSH Px), as shown in Scheme 3. GSH Px, mol. wt. 84 000, contains

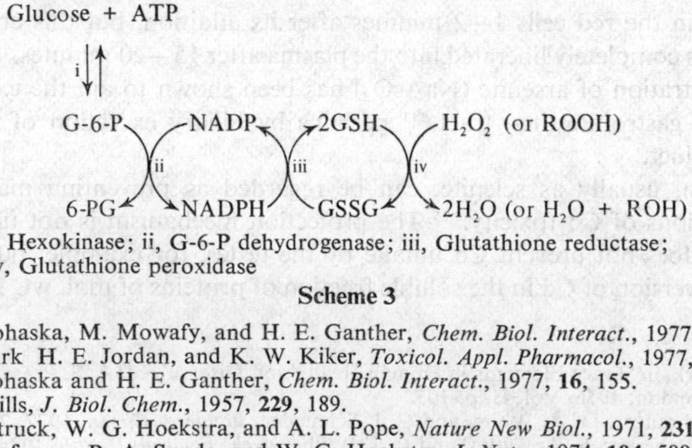

Glucose + ATP

Reagents: i, Hexokinase; ii, G-6-P dehydrogenase; iii, Glutathione reductase; iv, Glutathione peroxidase

Scheme 3

[292] J. R. Prohaska, M. Mowafy, and H. E. Ganther, *Chem. Biol. Interact.*, 1977, **18**, 253.
[293] R. F. Burk H. E. Jordan, and K. W. Kiker, *Toxicol. Appl. Pharmacol.*, 1977, **40**, 71.
[294] J. R. Prohaska and H. E. Ganther, *Chem. Biol. Interact.*, 1977, **16**, 155.
[295] G. C. Mills, *J. Biol. Chem.*, 1957, **229**, 189.
[296] J. T. Rotruck, W. G. Hoekstra, and A. L. Pope, *Nature New Biol.*, 1971, **231**, 223.
[297] D. G. Hafeman, R. A. Sunde, and W. G. Hoekstra, *J. Nutr.*, 1974, **104**, 580.
[298] J. T. Rotruck, A. L. Pope, H. E. Ganther, A. B. Swanson, D. G. Hafeman, and W. G. Hoekstra, *Science*, 1973, **179**, 588.

4 gram atom of Se^{299} and has subsequently been found in a variety of other tissues. Unlike catalase (which also employs H_2O_2 as a substrate) GSH Px is not inhibited by CN^- (at least in the reduced state) and can utilize a variety of lipoperoxides, including fatty acid hydroperoxides, as substrates.[300] Indeed, GSH Px may play a decisive role in the biosynthesis of prostaglandins. It converts the G series, which contain 15-hydroperoxy-groups, into prostaglandins E and F.

As originally noted by Mills, the enzyme becomes progressively more unstable during purification, even under mild conditions. Addition of GSH, 10% ethanol, or BSA stabilizes the enzyme.[301] Dilution of GSH Px in water results in the spontaneous release of 50% or more of Se, in a volatile form, in several days.

Selenium at the active site undergoes redox changes which determine whether or not it is sensitive to inhibitors. Purified GSH Px is inhibited by CN^-,[302] but enzyme that has been pre-incubated with GSH is not, and the addition of peroxide substrates to reduced enzyme restores CN^- sensitivity and also eliminates its sensitivity to iodoacetate. Treatment with CN^- releases Se from the enzyme, partly as $SeCN^-$. A mechanism for enzyme action has been proposed[303] which seems to be supported by X-ray photoelectron studies of the selenium 3d ($\frac{3}{2}$, $\frac{5}{2}$) levels in the reduced and oxidized forms,[304] as shown in Scheme 4. In contrast to the sulphydryl group of cysteine (pK_a 8.3), the selenocysteine group will be ionized at physiological pH since it is a considerably stronger acid (pK_a 5.2).

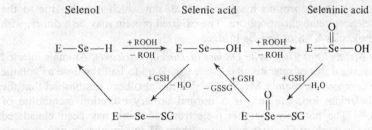

Scheme 4

Correlations have been noted between Se levels of whole blood and levels of glutathione peroxidase in whole blood.[305-307] The enzyme assay is faster and more simple than methods for direct determination of Se.

[299] S. H. Oh, H. E. Ganther, and W. G. Hoekstra, *Biochemistry*, 1974, **13**, 1825.

[300] C. Little and P. J. O'Brien, *Biochem. Biophys. Res. Comm.*, 1968, **31**, 145.

[301] H. E. Ganther, J. R. Prohaska, S. H. Oh, and W. G. Hoekstra, in ref. 6, p. 77.

[302] J. R. Prohaska, S. H. Oh, W. G. Hoekstra, and H. E. Ganther, *Biochem. Biophys. Res. Comm.*, 1977, **74**, 64.

[303] H. E. Ganther, *Chemica Scripta*, 1975, **8A**, 79.

[304] A. Wendel, W. Pilz, R. Ladenstein, G. Sawatzki, and U. Weser, *Biochim. Biophys. Acta*, 1975, **377**, 211.

[305] C. D. Thomson, H. M. Rea, V. M. Doenburg, and M. F. Robinson, *Br. J. Nutr.*, 1977, **37**, 457.

[306] G. Chauvaux, F. Lomba, I. Fumiere, and U. Bienfet, *Ann. Med. Vet.*, 1977, **121**, 111.

[307] G. Perona, R. Cellerino, G. C. Guidi, G. Moschini, B. M. Strerano, and C. Tregnaghi, *Scand. J. Haematol.*, 1977, **19**, 116

Muscle protein. 'White muscle' disease in lambs appears to be associated with the absence of a small selenoprotein (mol. wt. 10 000) containing one Se, with spectral properties very similar to those of cytochrome c and an amino-acid composition similar to that of cytochrome b_5.[308,309] The disease can be completely prevented by Se supplementation of the diet of ewes during pregnancy.

Microbial selenoproteins. In 1954 Pinsent demonstrated that *E. coli* has a natural requirement for Se in order to synthesize the enzyme formate dehydrogenase, which catalyses the oxidation of formic acid to CO_2.[310] She noted that a factor present in London tap-water that was absent in distilled water was involved in the formation of this enzyme. Maximum stimulation was achieved by adding selenite to the growth medium (10^{-8} mol l^{-1}), but selenate, tellurite, and tellurate were without effect. The enzyme is difficult to isolate and purify, but has since been shown to be a selenoprotein of high molecular weight[311] and is present in a variety of micro-organisms.

Similar observations were made by Stadtman and co-workers for the glycine reductase system of *Clostridium stricklandii*, a multi-enzyme system that catalyses the reductive deamination of glycine.[312] Prior to the recognition of Se as an essential constituent, cells were frequently devoid of glycine reductase activity because one of the components (protein A) of the reductase was missing.[313,314] Cells cultured in 1 μM-selenite exhibited high reductase activity and were rich in protein A. Purified protein A (traced by radioactive ^{75}Se) has a molecular weight of 12 000, is glycosylated, and contains one selenium as selenocysteine in the fully reduced form.[278] The (borohydride-)reduced protein is highly oxygen-sensitive, with an absorption maximum at 238 nm which may be due to the ionized selenocysteine chromophore. The oxidized protein may be a dimer, with an intermolecular S—S or Se—Se bridge.

Selenium Metabolism: The Garlic Odours of Dimethyl Selenide. Animals injected with sub-acute doses of selenite or selenate exhale up to half the dose as volatile selenium compounds, mainly Me_2Se.[315,316] It has also been established that the trimethylselenium ion, Me_3Se^+, is a normal urinary excretion metabolite of selenium.[283] The pathway for this 6-electron reduction has been elucidated largely by the work of Ganther and co-workers,[303] and involves a non-enzymic reaction between selenious acid and glutathione, producing a selenotrisulphide GSSeSG, which is enzymatically reduced by NADPH with glutathione reductase to glutathione selenopersulphide GSSeH and then, under anaerobic conditions, to H_2Se. The methylation process is enzyme-catalysed, with *S*-adenosyl-methionine as the methyl donor.[317] In the kidney, synthesis of Me_2Se occurs

308 P. D. Wanger, *Fed. Proc.*, 1972, **31**, 691.
309 P. D. Wanger, N. D. Pedersen, and P. H. Weswig, in ref. 247, p. 571.
310 J. Pinsent, *Biochem. J.*, 1954, **57**, 10.
311 H. G. Enoch and R. L. Lester, *J. Biol. Chem.*, 1975, **250**, 6693.
312 D. C. Turner and T. C. Stadtman, *Arch. Biochem. Biophys.*, 1973, **154**, 366.
313 T. C. Stadtman, *Arch. Biochem. Biophys.*, 1966, **113**, 9.
314 J. E. Cone, R. Martin del Rio, J. N. Davis, and T. C. Stadtman, *Proc. Nat. Acad. Sci. U.S.A.*, 1976, **73**, 2659.
315 H. S. Hsieh and H. E. Ganther, *J. Nutr.*, 1976, **11**, 1577.
316 H. S. Hsieh and H. E. Ganther, *Biochemistry*, 1975, **14** 1632.
317 J. R. Prohaska and H. E. Ganther, *Biochem. Biophys. Res. Comm.*, 1977, **76**, 437.

entirely in the soluble fraction, whereas in the liver the microsomal fraction is a requirement.[318]

The pathway proposed by Ganther[303] for the reductive methylation of selenite is shown in Scheme 5.

$$H_2SeO_3 + 4GSH \longrightarrow GSSeSG + GSSG + 3H_2O$$

$$(Se^{IV}) \qquad\qquad (Se^{II})$$

$$GSSeSG + NADPH + H^+ \xrightarrow[\text{reductase}]{GSH} GSSeH + GSH + NADP$$

$$(Se^{II}) \qquad\qquad (Se^0)$$

$$GSSeH + NADPH + H^+ \xrightarrow[\text{reductase}]{GSH \text{ or other}} H_2Se + GSH + NADP$$

$$(Se^0) \qquad\qquad (Se^{-II})$$

$$H_2Se \xrightarrow[\text{transferase}]{SAM \text{ methyl}} Me_2Se + Me_3Se^+$$

$$(Se^{-II})$$

Scheme 5

A curious feature of the pathway is the strong inhibition by arsenite, AsO_2^- (50% at 10^{-6} mol l^{-1} in the presence of a large excess of monothiols);[319] in fact, in early animal work it was realized that the chronic and acute selenoses produced by feeding seleniferous grains containing 15 p.p.m. Se could be alleviated (or prevented) by administering 5 p.p.m. of As as sodium arsenite in the drinking water. Arsenic increases the excretion of Se *via* the gastrointestinal tract.[320]

3 Potentially Essential Elements

Cadmium.—This is generally regarded as a serious occupational and general population hazard. There is a vast literature on occurrence, metabolism, and effects of cadmium, and this has recently been extensively reviewed.[321,322] Cd^{2+} is teratogenic in experimental animals, although this can be prevented by Zn^{2+}.[323] Once absorbed, most of the Cd^{2+} (over 50%) accumulates in the mammalian kidney and liver and is bound by *inducible* proteins (thioneins) that are synthesized in these organs in response to Cd^{2+} uptake.[324] One third of the amino-acid residues in thioneins of mol. wt. 10 000 are cysteinyl, about three of which are bound to each Cd^{2+} (or Zn^{2+}, and perhaps Cu^+). Synthesis of these heat-stable, cytoplasmic thioneins, when primed by one dose of Cd^{2+}, occurs (without lag) in response to a second dose.[325,326] Thionein binding accounts for the extremely long half-life [> 20 years; it accumulates in man at a rate of about

[318] H. S. Hsieh and H. E. Ganther, *Biochemistry*, 1975, **14**, 1632.
[319] H. S. Hsieh and H. E. Ganther, *Biochim. Biophys. Acta*, 1977, **497**, 205.
[320] H. E. Ganther and C. A. Baumann, *J. Nutr.*, 1962, **77**, 210.
[321] L. Friberg, M. Piscator, G. Nordberg, and T. Kjellström, 'Cadmium in the Environment', CRC Press, Cleveland, Ohio, 1974.
[322] W.H.O., Environmental Health Criteria 6, Cadmium, 1977.
[323] R. Semba, K. Ohta, and H. Yanamura, *Teratology*, 1974, **10**, 96.
[324] M. Webb, in ref. 1, p. 51.
[325] M. Cempel and M. Webb, *Biochem. Pharmacol.*, 1976, **25**, 2067.
[326] M. Webb and R. D. Verschoyle, *Biochem. Pharmacol.*, 1976, **25**, 673.

15 μg per gram of kidney cortex by the age of 50 (up to 30 μg per gram in heavy smokers) and to about five times this level in horses[327]. Concentrations of cadmium in kidney cortex exceeding 200 μg per gram net weight may give rise to tubular proteinuria,[327] *i.e.* renal failure. Hg^{2+}, but not methylmercury, induces thionein synthesis, although Hg^{2+}-thionein has a shorter half-life.[324] Some believe that thionein provides protection against Cd^{2+} toxicity but, paradoxically parenterally injected Cd-thionein is very toxic to renal tubular lining cells[328, 329] at Cd doses where $CdCl_2$ alone has no effect on the kidney, nor does a similar injection of Zn-thionein. Evidently extracellular Cd-metallothionein is a potent nephrotoxin. A protein of mol. wt. 30 000, with an even higher affinity for Cd^{2+} than thionein, has been found only in the testes (of rats) and is a possible target of Cd-induced testicular injury, to which the rat is particularly prone.[330]

New methods are urgently required for monitoring the body burden of cadmium. After intensive exposure, urine and blood levels may only reflect recent exposure (but should still be monitored), and the excretion of proteins of low molecular weight is usually regarded as the first sign of renal tubular dysfunction.[331] In experimental animals, 'interactions' between toxic metals and essential elements have been noted (see Table 6).

Table 6 *Interactions of toxic and essential elements*[a]

Toxic metal	Essential element	Effect on toxicity
Pb	Deficiencies of Ca and Fe	Enhances
Cd	Deficiencies of Zn, Ca, and Fe	Enhances
Hg	Excess of Se	Protects against methylmercury

(a) Taken from 'Clinical Chemistry and Toxicology of Metals', ed. S. S. Brown, Elsevier/North-Holland, Amsterdam, 1977, p. 102.

Although noteworthy for its toxicity, Schwarz has postulated a natural physiological role for cadmium.[8, 332] He noted small but consistent increases in the growth of weaning rats (improved weight, body length, and appearance) fed on diets containing < 4 p.p.b. of Cd supplemented by 0.05—0.5 p.p.m. of Cd. Numerous tests have been carried out in trace-element-controlled isolators over several years. Typically, Cd^{2+} sulphate at 0.2 p.p.m. causes a 13 % growth increase. Since these are the Cd levels *normally* found in foods, feeds, and animal and human tissue, he speculates that Cd may be an essential element.

Arsenic.—Although exposure to high concentrations of arsenic causes debilitating and often fatal illness, and chronic exposure appears to cause cancer of the skin and lung, arsenic deficiency diseases have been produced in goats and pigs, and in rats[333] fed diets containing < 50 p.p.b. of As over several generations. The pigs and goats had low birth rates and retardation of growth, and As-deficient

[327] C. G. Elinder and M. Piscator, in ref. 6, p. 569.
[328] R. A. Goyer and M. G. Cherian, in ref. 1, p. 89.
[329] G. F. Nordberg, R. A. Goyer, and M. Nordberg, *Arch. Pathol.*, 1975, **99**, 192.
[330] R. W. Chen and H. E. Ganther, *Environ. Physiol. Biochem.*, 1975, **5**, 378.
[331] M. Piscator and B. Pettersson, in ref. 1, p. 143.
[332] K. Schwarz and J. Spallholz, *Fed. Proc.*, 1976, **35**, No. 3.
[333] (a) M. Anke, *Arch. Tierernaehr.*, 1976, **26**, 742; (b) F. H. Nielsen, *Fed. Proc.*, 1975, **34**, 923.

goats frequently died, with myocardial damage, during lactation. Schwarz has confirmed these findings in rats.[8] An average growth increase of 22% was observed with 0.5 p.p.m. of As supplement (as sodium arsenite, $NaAsO_2$). Higher arsenic levels did not increase the response, and sodium arsenate, methanearsonic acid, and cacodylic acid were less effective than arsenite, As_2O_3, or arsanilic acid.

Arsenic is ubiquitous in the environment, and consequently in food, especially in seafood (the black bass contains up to 40 p.p.m.) and certain vegetables.[334]

The predominant concern about industrial exposure to arsenic centres on procedures which produce or use As_2O_3, *e.g.* fungicides and insecticides. Arsenic(III) is considered to be far more toxic than As^V, and AsH_3 and $AsCl_3$ are dangerous hameolytic agents, giving rise to anaemia.

Estimation. Arsenic is a difficult element to measure in biological materials. The most successful approaches involve the liberation of AsH_3 from the sample and its detection by atomic absorption or emission.[335] Spectrophotometric, electrochemical, gas chromatographic, and neutron activation approaches have also been successful.

Lead.—Again, the evidence for the potential essentiality of lead comes from Schwarz's laboratory, from rats maintained in his trace-element-isolator system on a basal diet of 53 individual components (21 amino-acids, 13 vitamins, sucrose, two fats, three salts, and 13 trace-element compounds), purified to a Pb level of 0.2 p.p.m. Before purification, typical 'natural' Pb levels were: arginine hydrochloride 0.2—14 p.p.m., lysine 5—10 p.p.m., and the labelling tape used for numbering the cages 56 000 p.p.m.! Initially, experiments were not reproducible, because of problems of contamination, but typical results obtained now are 17—30% weight increase at 1—2.5 p.p.m. dietary Pb levels (as Pb suboxide, oxide or nitrate).[8, 336]

In one study of hospitalized children[337] the airborne intake averaged 3.4 μg of Pb per day, *i.e.* 6.4% of the average dietary intake of 53 μg per day (6.3 μg per kg per day). Both adults and children can absorb a large fraction of their ingested Pb, and may be in negative balance for long periods.

4 Drugs

Lithium.—*Introduction.* Lithium has been used in medicine for about 120 years, being introduced for the treatment of 'gout and rheumatics' following the demonstration that lithium urate was the most soluble salt of uric acid. The element was later detected in spa waters and was advocated for various disorders. Lithium bromide was said to be the most effective of the bromides, though the role of the lithium ion was not considered once the specific action had been ascribed to the bromide anion.

[334] H. Kraybill, in 'Health Effects of Occupational Lead and Arsenic Exposure; A Symposium, Cincinnati, Ohio, U.S.A.' Dept. of Health Education and Welfare, HE 20. 7102:L4613, 1976, p. 272.
[335] J. Savory and F. A. Sedor, in ref. 1, p. 271.
[336] K. Schwarz, *Arh. Hig. Rada Toksikol.*, 1976, **26**, 13.
[337] D. Barltrop and C. D. Strehlow, in ref. 6, p. 332.

In common with many mineral and spa treatments, lithium salts were prescribed in small doses, with large quantities of water. However, in the late 1940's a salt substitute, 'Westral', was promoted in the United States which was in fact lithium chloride, and this was intended for use by patients who were on salt-restricted diets for cardiovascular diseases (a group now known to have a high risk of lithium intoxication). Unrestricted use of 'Westral' led to a number of deaths from lithium toxicity, and the drug was abandoned. At about the same time, in Australia, Cade had administered lithium carbonate in small doses to manic depressive patients and found a clear anti-manic effect. His original animal and human experiments and the subsequent history of lithium have been described recently.[338]

As a result of the doubt engendered by toxicity studies, the use of lithium in psychiatry was slow to become established. However, the current status of lithium in psychiatry, at least in the United Kingdom, may be measured by the data presented by Glen[339] that one in a thousand of the population of Edinburgh receives lithium prophylactically for manic depressive psychoses, and this is supported by the total annual sales of lithium preparations which suggest that approximately 1 in 2000 of the population of the United Kingdom is receiving lithium.[340] Lithium is not, therefore, only an oddity but has wide-ranging significance both socially and in the economy.

Manic-depressive Psychoses. Lithium is specific for the treatment of manic depressive psychoses, and these diseases are characterized by recurrent episodes of depression which may be interspersed with periods of mania. They occur more frequently in women than men, and predominantly after the menopause. Depression is characterized by lack of energy and drive, retarded thought and speech, and low self esteem, which may be accompanied by suicidal thoughts. In contrast, during mania, the patient is hyperactive, has rapid speech and thought content, is easily distracted, is interfering, aggressive, and cheerful. As far as the patient is concerned, therefore, the most traumatic part of the disease is depression; mania may be very enjoyable, and indeed in intelligent and well-motivated subjects it can be a productive period for creative work. The patient's relatives, on the other hand, may find the constant high level of activity and speech difficult to tolerate. Social family problems frequently arise as a result of the excessive spending, alcohol intake, and aggression of the manic phase.

There is a clear distinction between manic depressive psychoses and the schizophrenias. In the latter there is very great difficulty in obtaining any emotional contact with the patient (the so called 'glass wall') whilst with manic depressive patients, rapport is readily obtained, and indeed patients in the manic phase are often witty and amusing, though their stories are frequently improbable and grandiose.

Three main types of periodic affective disorder are recognized. The bipolar or manic depressive type has both depression and mania, unipolar or recurrent depressive type has only depression, while in schizoaffective disorder the primary

[338] J. F. J. Cade, in 'Lithium in Medical Practice' ed. F. N. Johnson and S. Johnson, M.T.P. Press, Lancaster, 1978, p. 5.
[339] A.I.M. Glen, in ref. 338, p. 183.
[340] R. P. Hullin, in ref. 338, p. 433.

defect is in mood, though schizoid symptoms may be seen. Aspects of periodicity in patients with short-cycle affective disorder have been reviewed.[341] Lithium is considered to be effective in the prevention of manic depressive cycles, though the prophylactic effect against recurrent depression or schizoaffective disorder is so far unproven.[342]

Clinical Use of Lithium. Lithium is administered in the form of lithium carbonate tablets, though the daily dose varies depending on age, body weight, and kidney efficiency.[343] The objective is to maintain a plasma lithium level in the range $0.6-1.2$ mmol l^{-1}.[344] If plasma lithium is allowed to rise above 2 mmol l^{-1}, toxic symptoms may be seen, and these usually take the form of tremor, confusion, and gastrointestinal disturbances. Recent studies have suggested that lower levels of lithium may be adequate for prophylactic treatment.[345] To maintain the recommended plasma lithium, most patients require between 1 and 2 grams of lithium carbonate per day, and this is usually given in divided doses to try and maintain a stable level of plasma lithium. 'Slow release' preparations have been marketed, though evidence suggests that, *in vivo*, these do no live up to the expectation of dissolution tests *in vitro*, and little difference may be seen between the rate of attainment or magnitude of the peak in plasma lithium following the dose except where incomplete absorption occurs.[346]

Side Effects and Toxicity. A survey of toxicological aspects has appeared recently.[347]

Minor side effects are common in the early stages of lithium treatment, though most of them disappear within a few weeks. The most common side effect is slight tremor of the hand, which, however, may be treated by β-adrenergic receptor blockers such as propranolol. A side effect in the longer term is the development of hypothyroidism, and in a few patients polydipsia–polyuria occurs, though it is possible that the latter is a sign of marginal toxicity.

The signs of impending intoxication are vomiting and diarrhoea, development of coarse tremor of the hand, sluggishness, drowsiness, vertigo, and slurred speech.[344] If intoxication is not treated, the effects are mainly on the central nervous system, leading to coma and death. Kidney damage occurs and a progressive cycle is set up in which plasma lithium rises, causing more kidney damage, causing a further rise in plasma lithium, having its ultimately fatal effects on the central nervous system. Haemodialysis is the treatment of choice in lithium toxicity, though it must be maintained for some considerable period, since the removal of lithium from tissues is slow.

Management of Lithium Therapy. Because of the danger of toxicity occurring, it is usual to institute therapy by gradually increasing the dose, with frequent determination of serum lithium, until a dose is found which will stabilize the

341 N. J. Birch, *Inorg. Persp. Biol. Med.*, 1978, **1**, 173.
342 M. Schou, in ref. 338, p. 21.
343 D. S. Hewick, in ref. 338, p. 355.
344 M. Schou, *Psychiatr., Neurol., Neurochir.*, 1973, **76**, 511.
345 T. C. Jerram and R. McDonald, in ref. 338, p. 407.
346 S. Tyrer, R. P. Hullin, N. J. Birch, and J. C. Goodwin, *Psychol. Med.*, 1976, **6**, 51; S. Tyrer, in ref. 338, p. 395.
347 G. M. Dempsey and H. L. Meltzer, in 'Neurotoxicology', ed. L. Roizin, H. Shiraki, and N. Grcevic, Raven, New York, 1977, p. 171.

subject within the normal limits. In practice, this means weekly blood samples for the first month, with a gradual lengthening of the interval between monitoring until the normal frequency is about eight weeks. In the case of elderly patients, those who might be unreliable in taking tablets, and in subjects with cardiovascular or renal disease the maximum period between lithium estimation should be reduced. It is important that good records be kept of plasma lithium, plasma urea, and creatinine and that occasional tests of thyroid function be made.

An alternative approach has been suggested in which the dose required may be predicted by a single test dose followed by the determination of the plasma lithium 24 hours later.[348] The required dose can be ascertained from a standard nomogram prepared by pharmacokinetic curve-fitting techniques.[348]

Biochemistry of Lithium. Aspects of the pharmacology and biochemistry of lithium have been extensively reviewed in a number of recent papers,[341, 349] and the early work on lithium was reviewed by Schou,[350] who has also published, at intervals, a cumulative bibliography of the biology of lithium.[351]

The breadth of the range of effects of lithium on biochemical processes has been discussed, although it must be stressed that many of the effects reported may have no significance in the pharmacology of lithium in the manic depressive psychoses because of the high concentrations of lithium used. Thus, many studies are carried out in which sodium in physiological solutions is replaced in equimolar concentrations by lithium, leading to solutions containing anything up to 150 mmol l^{-1} of lithium. The tissue concentration of lithium is similar to that of plasma (about 1 mmol l^{-1}), except in bone and thyroid, where concentrations of up to 5 or 6 mmol per kg wet weight may be seen, and in localized areas of the kidney, where somewhat higher concentrations occur. The very high concentrations used experimentally, therefore, are clearly not relevant to the pharmacological action of lithium in the periodic affective disorders. Indeed, it could be argued that the effects seen in these studies are due to the toxic action of lithium rather than to the lack of sodium, which is the commonly accepted dogma.[341]

Absorption and Distribution of Lithium. When lithium is administered orally, its absorption is rapid and occurs mainly in the upper small intestine. As soon as it appears in the bloodstream, it distributes rapidly in the extracellular fluids, though its transfer into the tissues varies, its uptake rate being rapid in kidney but slower in liver and muscle, and very slow in bone and nervous tissue.

Following long-term lithium treatment, the distribution between the tissues and even within one tissue may not be uniform, and indeed a number of studies have shown variations of the lithium concentration between the different areas

[348] T. B. Cooper, P. E. E. Bergner, and G. M. Simpson. *Am. J. Psychiatry*, 1973, **130**, 601; T. B. Cooper and G. M. Simpson, *ibid.*, 1976, **133**, 440; P. E. E. Bergner, K. Berniker, and T. B. Cooper, *Br. J. Pharmacol.*, 1973, **49**, 328.

[349] M. Schou, *Ann. Rev. Pharmacol.*, 1976, **16**, 231; W. E. Bunney and D. L. Murphy, in 'Neurobiology in Lithium' Neuroscience Research Program Bulletin, NSRPB, Boston, 1976, **14**, 111; N. J. Birch, in ref. 338, p. 173.

[350] M. Schou, *Pharmacol. Rev.*, 1957, **9**, 17.

[351] M. Schou, *Psychopharmacol. Bull.*, 1969, **5**, No. 4, p. 33; *ibid.*, 1972, **8**, No. 4, p. 36; *ibid.*, 1976, **12**, No. 1, p. 49; No. 2 p. 69; No. 3, p. 86; *Neuropsychobiol.*, 1976, **2**, 161.

of the brain.[352] A recent elegant study has localized lithium histologically in mouse brain. Following administration of the stable isotope ^{6}Li, a section was irradiated with fast neutrons and an autoradiograph was made.[353, 354] This is the first time that it has become possible directly to visualize the distribution of the metal. Though the quantitative aspects are presently imprecise, it appears that lithium accumulates in certain areas of the brain, notably the cerebral cortex, olfactory lobe, and hippocampus, and in localized areas of the corpus striatum. The thalamus has a particularly low lithium content. This agrees with tissue analysis studies,[352] though examination of the autoradiographs might suggest that the differences in distribution reflect merely the distribution of aqueous phase in the brain, and that areas of low lithium content are associated with major nerve tracts and hence cerebral lipids. However, accumulation does occur in the basal ganglia, and this may relate to the movement disorders Huntington's chorea and Tardive dyskinesia, both of which have been treated (with variable success) by lithium.[342] Again, the high hippocampal lithium content may reflect an effect on body weight through regulation of feeding behaviour, and also effects on expression of rage. Weight gain is a well-known side effect of lithium, and the element has also been suggested for the treatment of pathological aggression.[355]

A method has been reported recently for the determination of ^{6}Li in biological samples by atomic absorption spectroscopy, and this suggests a number of possible pharmacokinetic and membrane-transport studies.[356]

Effects of Lithium on Metabolic Processes. Lithium readily permeates cell membranes, and is potentially able, therefore, to affect a large number of processes at different levels of organization of the body. It is known to affect a number of enzymes, though again caution must be advised in interpretation due to the concentration factors discussed above.[341] The range of enzymes affected by lithium includes those of the glycolytic cycle, the synthetic enzymes of RNA and DNA, and enzymes in the synthesis of neurotransmitters. Furthermore, studies have shown that lithium might affect the action of ATPases, which are the enzymes responsible for the pumping of metal ions across membranes,[357] and in the action of cyclic AMP, which is the 'second messenger' for a number of hormone–cell interactions.

The actions of lithium on cyclic AMP may relate to effects seen on thyroid gland (thyroid adenyl cyclase is inhibited by lithium)[358] and also to renal effects of lithium, since the ion inhibits kidney adenyl cyclases that are sensitive to antidiuretic hormone and parathyroid hormone.[359] This integrative hypothesis

352 P. A. Bond, in 'Lithium in Medical Practice', ed. F. N. Johnson and S. Johnson, M.T.P. Press, Lancaster, 1978, p. 215.
353 M. Thellier, T. Stelz, and J.-C. Wissocq, *J. Microsc. Biol. Cell.*, 1976, **27**, 157.
354 M. Thellier, T. Stelz, and J. C. Wissocq, *Biochim. Biophys. Acta*, 1976, **437**, 604.
355 E. P. Worral, in ref. 352, p. 69.
356 N. J. Birch, R. P. Hullin, R. A. Inie, and D. Robinson, *Brit. J. Clin. Pharmacol.*, 1978, **5**, 351P.
357 J. E. Hesketh, J. B. Loudon, H. W. Reading, and A. I. M. Glen. *Brit. J. Clin. Pharmacol.*, 1978, **5**, 323.
358 S. C. Berens and J. Wolff, in 'Lithium Research and Therapy', ed. F. N. Johnson, Academic Press, London, 1975, p. 443.
359 J. Forn, in ref. 358, p. 485.

for the action of lithium at hormone–receptor interfaces has yet to be fully tested.

A further, more general, view has been expressed that lithium may interact with magnesium-dependent processes,[341] and this has been supported from a theoretical chemical viewpoint.[360] Not only might lithium act directly, but, because of its tissue concentration, it might have a permissive effect by 'unmasking' sites previously inaccessible to magnesium or calcium.

Much effort has been directed towards studying the effects of lithium on neurotransmitter amines, the catecholamines and indoleamines, though presently the results are inconclusive.[349]

Erythrocyte Lithium Levels and Membrane Transport. It has been proposed that membrane transport of lithium in erythrocytes (red blood cells) might be used to indicate defects in general membrane function in patients suffering from periodic affective disorders. The erythocyte plasma ratio (Li_e/Li_p) might be a useful diagnostic tool in the prediction of those patients who will respond to lithium therapy.[361] This claim has been vigorously contested (see ref. 341). A familial trait for high Li_e/Li_p has been reported in a test *in vitro*, though this was seen in patients who were periodic psychotics and also in their normal relatives.[362] Recent studies of cultured neuroblastoma cells suggest that the mechanisms of lithium transport differ in nerve cells from those of the erythrocyte.[363]

Studies of the membrane properties of erythrocytes from patients and controls have indicated that the main mechanisms of lithium transport are (*a*) a ouabain-sensitive component, the sodium pump; (*b*) a phloretin-sensitive component, lithium counter-transport exchange for other cations; and (*c*) leak diffusion across the membrane.[364–368] The role of these various mechanisms has been reviewed and an attempt made to explain the changes in membrane ATPases seen in manic depressive patients in different mood states and after treatment with lithium.[339, 367, 369]

Lithium is therefore of widespread clinical significance, though presently its mode of action is unknown. This brief summary of some aspects of the recent literature may be supplemented by review papers in the symposia and compendia already, more specifically, cited.[338, 359, 360]

Aluminium.—*Dialysis dementia.* Patients with renal failure who are undergoing chronic haemodialysis commonly consume about 4 g of aluminium hydroxide

[360] R. J. P. Williams, in 'Lithium: its Role in Psychiatric Research and Treatment', ed. S. Gershon and B. Shopsin, Plenum Press, New York, 1973, p. 15; J. J. R. Frausto da Silva and R. J. P. Williams, *Nature*, 1976, **263**, 237; R. J. P. Williams, in 'Neurobiology of Lithium', *Neuroscience Research Program Bulletin*, ed. W. E. Bunney and D. L. Murphy, NSRPB, Boston, 1976, **14**, 145; G. Eisenman, *ibid.*, p. 154; M. Eigen, *ibid.*, p. 142.

[361] J. Mendels and A. Frazer, *J. Psychiatr. Res.*, 1973, **10**, 9.

[362] G. N. Pandey, D. G. Ostrow, M. Haas, E. Dorus, R. C. Casper, J. M. Davis, and D. C. Tosteson, *Proc. Nat. Acad. Sci. U.S.A.*, 1977, **74**, 3607.

[363] E. Richelson, *Science*, 1977, **196**, 1001.

[364] J. Duhm, F. Eisenried, B. F. Becker, and W. Greil, *Pfugers Archiv.*, 1976, **364**, 147.

[365] J. Duhm and B. F. Becker, *Pfugers Archiv.*, 1977, **368**, 203.

[366] W. Greil, F. Eisenried, B. F. Becker, and J. Duhm, *Psychopharmacol.*, 1977, **53**, 19.

[367] W. Greil and F. Eisenried, in 'Lithium in Medical Practice', ed. F. N. Johnson and S. Johnson, M.T.P. Press, Lancaster, 1978, p. 415.

[368] M. Haas, J. Schooler, and D. C. Tosteson, *Nature*, 1975, **258**, 425.

[369] G. J. Naylor, D. A. T. Dick and E. G. Dick, *Psychol. Med.*, 1976, **6**, 257; G. J. Naylor, A. Smith, L. J. Boardman, D. A. T. Dick, E. G. Dick, and P. Dick., *ibid.*, 1977, **7**, 229.

gel per day, to mop up phosphate. Aluminium administered in this way is generally assumed not to be absorbed, but to be excreted in the faeces as insoluble aluminium phosphate. It is now recognized that patients who develop dialysis encephelopathy may have whole-brain Al levels up to 12 times that of controls,[370] resulting from gastrointestinal absorption of 100—568 mg of aluminium hydroxide per day.[371] Disabling dementia is also found in Alzeheimer's disease, where elevated levels of aluminium in the brain have also been demonstrated.[372] A specific role for serum parathyroid hormone in both the uptake and distribution of Al has been suggested.[373] A fierce medical debate on whether the use of aluminium hydroxide gels should be abandoned is currently in progress.

Platinum.—*Clinical Trials of Anti-tumour Agents.* A sharp increase in interest in *cis*-[Pt(NH$_3$)$_2$Cl$_2$] (11) has been noted amongst clinicians since the National Cancer Institute (U.S.A.) began sponsoring the drug in 1971. In August 1976, for example, 25 000 vials containing 10 mg were supplied, and 207 investigations were in progress by the Spring of 1977.[374] The drug is now established as being useful for certain tumour types, namely genito-urinary and those of the head and neck, and it holds promise for several others. Typical response rates are 26.5—40% for patients with far advanced adenocarcinoma of the ovary, treated with (11) alone,[375, 376] and 72% complete and 28% partial remissions in disseminated testicular cancer, treated with (11) + vinblastine + bleomycin in combination therapy.[377] Combination with bleomycin appears to be yielding substantially better results than have been seen before in treating epidermoid carcinomas of the head and neck.[378]

$$H_3N \diagdown \diagup Cl$$
$$Pt$$
$$H_3N \diagup \diagdown Cl$$

(11)

Combination therapy is a major advance for Pt, but so too is the finding that renal toxicity (kidney damage) can be suppressed by the induction of diuresis with mannitol, which enables doses of up to 3 mg per kg (by intravenous injection) to be tolerated.[378]

Complexes that are More Active than cis-[Pt(NH$_3$)$_2$Cl$_2$]. Clinicians seem to prefer complexes that are soluble in water [at least as soluble as (11), *i.e.* 9 mmol l^{-1}] and which have a wide spectrum of activity against a variety of tumours. Good anti-tumour activity in animal tests has been observed for *cis*-[Pt(A)$_2$Cl$_2$] com-

370 A. C. Alfrey, G. R. LeGendre, W. D. Kaehny, *N. Engl. J. Med.*, 1976, **294**, 184.

371 E. M. Clarkson and V. A. Luck, *Clin. Sci.*, 1972, **43**, 519.

372 D. R. Crapper, S. S. Krishnan, and A. J. Dalton, *Science*, 1973, **180**, 511.

373 G. H. Mayor, J. A. Keiser, D. Makdani, and P. K. Ku, *Science*, 1977, **197**, 1187.

374 M. Rozencweig, D. D. Von Hoff, J. S. Penta, and F. M. Muggia, *J. Clin. Hematol. Oncol.*, 1977, **7**, 672.

375 E. Wiltshaw and T. Kroner, *Cancer Treatment Reports*, 1976, **60**, 55.

376 H. H. Bruckner, C. J. Cohen, S. B. Gusberg, R. C. Wallach, E. M. Greenzum, and J. F. Holland, *Proc. Asco*, 1976, **17**, 287.

377 L. H. Einhorn and B. Furnas, *J. Clin. Hematol. Oncol.*, 1977, **7**, 662.

378 R. E. Wittes, E. Cvitkovic, I. H. Krakoff, and E. W. Strong, *J. Clin. Hematol. Oncol.*, 1977, **7**, 711.

15

plexes, where A is an alkyl, alicyclic, or heterocyclic amine,[379] *e.g.* as shown in Table 7, but their solubility in water is too low for clinical use (intravenous injection). The methylamine complex is far less potent than the ammonia complex, although this improves in the C_2, C_3, and C_4 straight-chain analogues, with a marked decrease in toxicity, especially for C_6 and C_7. Very high therapeutic indices were observed for the cyclopentylamine and cyclohexylamine complexes (low toxicities) against the ADJ/PC6A tumour, but the former is inactive against all other tumour systems, whilst the latter has a very low solubility (13 μmol l^{-1}). More promising is the isopropylamine complex, with 50% cures against the L1210 leukaemia, an LD$_{50}$ of 50 mg per kg, and activity against other tumours. The incorporation of hydrophilic *trans*-hydroxy-groups into Pt^{IV} com-

Table 7 *Anti-tumour activity of the compounds* [Pt^{II}(amine)$_2$Cl$_2$] *and some* Pt^{IV} *complexes against ADJ/PC6A mouse tumour*

Pt^{II} complexes

(amine)	LD$_{50}$/mg kg^{-1}	ID$_{90}$/mg kg^{-1}	TI^a
NH$_3$	13.0	1.6	8.1
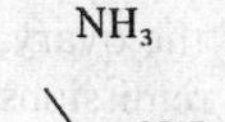 (CH$_3$)$_2$CH–NH$_2$ (isopropylamine)	33.5	0.9	37.2
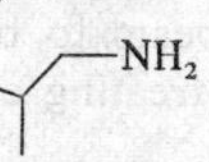 CH$_3$CH$_2$CH$_2$–NH$_2$	83	6.2	13.4
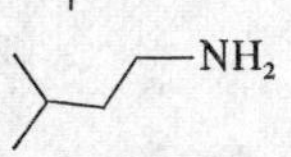 (butyl)–NH$_2$	1150	5.8	198
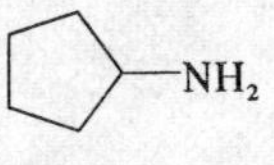 cyclopentyl–NH$_2$	480	2.4	200
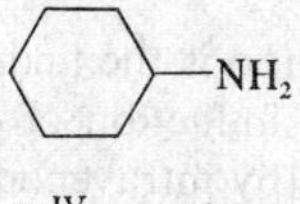 cyclohexyl–NH$_2$	> 3200	12	> 267

Pt^{IV} complexes

structure	LD$_{50}$/mg kg^{-1}	ID$_{90}$/mg kg^{-1}	TI^a
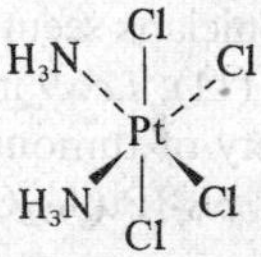 [Pt(NH$_3$)$_2$Cl$_4$] (Cl axial)	16	0.9	17
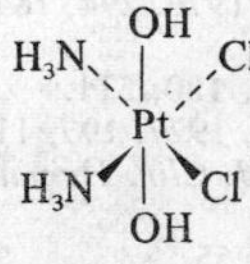 [Pt(NH$_3$)$_2$Cl$_2$(OH)$_2$] (OH axial)	135	< 12	> 11

(a) TI = therapeutic index = LD$_{50}$/ID$_{90}$, where LD$_{50}$ is the dose which kills half the animals, and ID$_{90}$ the dose which causes inhibition of tumour growth in 90% of the animals. The best drugs have high TI values, *i.e.* low toxicity and small effective doses.

[379] M. L. Tobe and A. R. Khokhar, *J. Clin. Hematol. Oncol.*, 1977, **7**, 114

plexes of the type *cis,trans,cis*-[Pt(A)$_2$(OH)$_2$Cl$_2$] improves the solubility in water and in some cases maintains activity as well. The NH$_3$, isopropylamine, and cyclopentylamine derivatives may well provide clinical alternatives to (11). Platinum(IV) is thought to be reduced to Pt^{II} *in vivo*; as yet, there is no example of an active Pt^{IV} complex with an active dichloro-platinum(II) analogue, and, despite the large number of compounds examined, no clear-cut patterns have emerged which allow the prediction of which complex will have a broad spectrum of activity and a low toxicity.

Sulphato-1,2-diaminocyclohexaneplatinum(II) (12) has been described as the most exciting discovery since (11). It is soluble in water (0.5 mol l^{-1}), active (i.p. or i.v.) against at least three animal tumours (L1210, Shay, and Gardner lymphoma OG), and at the optimal dose of 1 mg per kg has achieved *100% cures* in the L1210 system.[380, 381] It is about 30 times as potent as (11). A variety of other anions can replace sulphate and retain (*e.g.* hydroxide, tetraborate, ethylsulphate, monofluoroacetate) or increase (monobromoacetate) this activity, although the cyanate, maleate, and oxalate complexes are inactive.

(12)   (13)

Unidentate oxygen ligands are not expected to bind very tightly to Pt^{II} in solution, and there is a lot of work still to be done to clarify the structures of these complexes. Interestingly, and in contrast to most of the other complexes, significant concentrations of the malonato-complex (13) seem to reach the brain (40 p.p.m. of Pt in dog, seven days after a dose of 15.5 mg per kg).[381] Even more promising results have been obtained by increasing the ring size from cyclohexane to seven- or eight-membered. The bis(monobromoacetato)-1,2-diaminocyclo-octaneplatinum(II) complex is reported to be over 100 times as active as (11). Incidentally, the Pd^{II}, Cu^{II}, Cd^{II}, Ni^{II}, Zn^{II}, and Mn^{II} sulphato-1,2-diaminocyclohexane complexes are reported to be inactive (no characterization of the complexes given).[381]

This work has provided impetus for more fundamental studies of Pt chemistry, including the blue Pt complexes first described by Hoffman and Bugge in 1912 which were neglected until their anti-tumour activity and low toxicity were demonstrated.[382]

Platinum Blues. The starting material for formation of platinum blue is usually *cis*-[Pt(NH$_3$)$_2$(H$_2$O)$_2$]$^{2+}$, but at pH 6 this readily forms hydroxy-bridged dimers

[380] S. J. Meischen, G. R. Gale, M. Lanny, C. J. Frangakis, M. G. Rosenblum, E. M. Walker, L. M. Atkins, and A. B. Smith, *J. Nat. Cancer Inst.*, 1976, **57**, 841.

[381] (a) H. J. Ridgeway, R. J. Speer, L. M. Hall, D. P. Stewart, A. D. Newman, and J. M. Hill, *J. Clin. Hematol. Oncol.*, 1977, 7, 220, 231; (b) R. J. Speer, H. Ridgeway, D. P. Steward, L. M. Hall, A. Zapata, and J. M. Hill, *ibid.*, p. 210.

[382] J. P. Davidson, P. J. Faber, R. G. Fischer, Jr., S. Manscy, H. J. Peresie, B. Rosenberg, and L. Van Camp, *Cancer Chemother. Rep.*, 1975, **59**, 287

and trimers,[383] rather than the monomeric $[Pt(NH_3)_2(OH)(NO_3)]$. The blues, or purples,[384] formed by reactions of these aquo(hydroxo)-complexes with amides are thought to be oligomeric mixtures and are difficult to crystallize, but may give unusual (for Pt) e.p.r. spectra,[385] implying the presence of Pt^{III}. Indeed, in the crystal structure of *cis*-diamminoplatinum α-pyrimidone blue, every one in four atoms is effectively Pt^{III}, the rest being Pt^{II}.[386] It has been suggested that their biological activity arises *via* a breakdown into monomeric units, based on a study of lysozyme binding.[387]

Although previous data suggested that Pt pyrimidine blues may be used for the selective identification of tumorigenic cells,[388] more recent data do not confirm this, although the blues are still useful EM stains for the nucleus, ribosomes, and the cell surface.[389]

DNA Binding. The three-dimensional structures of the $[Pt(en)(guanosine)_2]^{2+}$ and $[Pt(en)(inosine\ 5'-monophosphate)_2]^{2-}$ complexes show Pt bound through N-7 of the purine rings, and in $[Pt(en)(5'-cytosine\ monophosphate)_2]$ the nucleotide is co-ordinated to two different Pt atoms, through N-3 of the pyrimidine ring and through a phosphate oxygen atom;[390] see cytidine (14) and guanosine (15). Heavy G–C involvement has been implicated in studies *in vitro* of the binding of Pt to DNA.[391, 392] However, the proposed chelation of *cis*-Pt^{II} to N-7 and O-6 of guanosine[393] as the basis for the selective anti-tumour action of the *cis*-isomer is open to doubt.[394] Some form of bifunctional attack on DNA would seem to be effective in modifying DNA replication, but this may be between bases on the same strand of DNA, or between appropriately adjacent sites on a particular base, or even between a strand of DNA and an associated protein molecule. Inter-strand cross-links alone are insufficient to account for cell killing, and the analogies with organic difunctional alkylating agents have clearly been misleading.

(14)   (15)

[383] F. Faggiani, B. Lippert, C. J. L. Lock, and B. Rosenberg, *Inorg. Chem.*, 1977, **16**, 1192.
[384] A. J. Thomson, I. A. G. Roos, and R. D. Graham, *J. Clin Haematol. Oncol.*, 1977, **7**, 242.
[385] B. Lippert, *J. Clin. Hematol. Oncol.*, 1977, 7, 26.
[386] J. K. Barton, H. N. Rabinowitz, D. J. Szalda, and S. J. Lippard, *J. Amer. Chem. Soc.*, 1977, **99**, 2827.
[387] C. C. F. Blake, S. J. Oatley, and R. J. P. Williams, *J.C.S. Chem. Comm.*, 1976, 1043.
[388] S. K. Aggarwal, R. W. Wagner, P. J. McAllister, and B. Rosenberg, *Proc. Nat. Acad. Sci. U.S.A.*, 1975, **73**, 928.
[389] P. K. McAllister, B. Rosenberg, S. K. Aggarwal, and R. W. Wagner, *J. Clin. Hematol. Oncol.*, 1977, 7, 717.
[390] R. Bau, R. W. Gellert, S. M. Lehovec, and S. Lowe, *J. Clin. Hematol. Oncol.*, 1977, 7, 51.
[391] I. A. G. Roos and M. C. Arnold, *J. Clin. Hematol. Oncol.*, 1977, **7**, 374.
[392] A. D. Kelman, H.-J. Peresie, and P. J. Stone, *J. Clin. Hematol. Oncol.*, 1977, **7**, 440.
[393] J. P. Macquet and T. Theophanides, *Bioinorg. Chem.*, 1975, **5**, 59.
[394] H. C. Harder and R. G. Smith, *J. Clin. Hematol. Oncol.*, 1977, 7, 401.

Roberts[395] has demonstrated that platinations of DNA of Chinese hamster cells in tissue culture can be circumvented by a post-replication repair process which is inhibited by caffeine. HeLa cells have a decreased capacity to carry out post-replication repair and are more sensitive to the toxic effects of (11), and differences in the response of various animals and human tumours to Pt complexes may be accounted for by differences in this repair pathway. Pt-resistant L1210 cells have been described.[396] Although resistant to (11), they respond to (12) or (13), which is an observation of potential clinical use if patients also become resistant to (11). Perhaps even chemically related Pt complexes can be cytotoxic by different mechanisms.

Radiation Sensitization. An interesting new development concerns the ability of Pt complexes to sensitize cells to irradiation.[397-399] Sensitization can occur both in the presence and absence of oxygen, and chemotherapy with (11) followed by radiation of a mouse mammary adrenocarcinoma produced a 40% increase in mean survival time, and complete tumour control in 5 of 12 rats with intra-cerebral brain tumours.[400, 401] It is not known whether the Pt complex scavenges the hydrated electrons produced by the ionizing radiation in the same manner as Ag^+ or $[Co(NH_3)_6]^{3+}$.

Animal Distribution Studies. The most useful radioisotopes for studies *in vivo* of the kinetics of pharmacologically active agents are γ-emitters, because no further work-up of the material is required before counting. Many of the Pt isotopes are not useful because they decay to interfering γ-emitting Ir isotopes, but the most promising is ^{193m}Pt ($T_{\frac{1}{2}} = 4.3$ days).[402, 403] Tissues which have a high percentage of labelled injected (8) in rats bearing the Walker 256 carcinosarcoma are the liver and kidneys: 24 hours after injection, 2.5% is in the blood, 3.5% in the liver, 2.8% in the kidneys, and 50% has been excreted in the urine. No significant concentration was observed in the tumour. Two blood-clearance phases were observed: one fast ($T_{\frac{1}{2}} = 16$ min) and the second very slow ($T_{\frac{1}{2}} = 7$ days). A major storage compartment for the slow phase appears to be the *skin*, and a patient has been observed to have 13% of the total administered dose concentrated in the skin 18 days post administration. Appreciable amounts of the drug can also be found in muscle and bone. The complex (11) binds very tightly to the proximal tubular cell membrane in the kidney, and it is not possible to protect by thiol reagents such as cysteamine, penicillamine, or N-acetylcysteine,[404] which are often useful for reducing metal-induced kidney damage.

It has been predicted that dynamic imaging may allow a determination of whether or not a patient is likely to be responsive to Pt therapy.[402] The kinetics

395 H. W. van der Berg, H. N. A. Fraval, and J. J. Roberts, *J. Clin. Hematol. Oncol.*, 1977, 7, 349.
396 J. H. Burcheval, K. Kalaher, T. O'Toole, and J. Chisholm, *Cancer Res.*, 1977, 37, 3455.
397 R. C. Richmond and E. L. Powers, *J. Clin. Hematol. Oncol.*, 1977, 7, 580.
398 A. H. W. Nias and I. I. Szumiel, *J. Clin. Hematol. Oncol.*, 1977, 7, 562.
399 I. I. Szumiel and A. H. W. Nias, *Chem. Biol. Interact.*, 1976, 14, 217
400 E. B. Douple, R. C. Richmond, and M. E. Logan, *J. Clin. Hematol. Oncol.*, 1977, 7, 585.
401 R. C. Richmond and E. L. Powers, *Radiat. Res.*, 1976, 68, 251.
402 W. Wolf, R. C. Manaka, and F. K. V. Leh, *J. Clin. Hematol. Oncol.*, 1977, 7, 741.
403 W. Wolf and R. C. Manaka, *J. Clin. Hematol. Oncol.*, 1977, 7, 79.
404 T. F. Slater, M. Ahmed, and S. A. Ibrahim, *J. Clin. Hematol. Oncol.*, 1977, 7, 534.

of urinary Pt excretion from an anaesthetized rat have already been determined, and the bladder, kidneys, skin envelope, and liver are readily visualized in Auger camera scans of the animal.[402] These distribution studies have been confirmed by atomic absorption analysis, and the biphasic plasma Pt decay occurs in species as widely different as dog, rat, and shark.[405]

Gold.—Gold compounds have been used as anti-inflammatory drugs for the treatment of rheumatoid arthritis for over 40 years ('chrysotherapy'), even though they were originally introduced on the mistaken assumption that the condition is related to tuberculosis. The effectiveness of gold(I) thiomalate (16) ('Myocrisin') and gold(I) thioglucose (Solganol') (17) was confirmed by controlled clinical trials in 1960; however, since they are not orally absorbed, they require parenteral injection (usually intramuscular; *ca.* 25 mg of Au per week and a total of 0.6—3.5 g are commonly administered). Excessive accumulation and retention of gold in the body may result, together with tissue damage; toxic dermatitis and blood dyscrasias are particularly prevalent. The incidence of fatal reactions (*ca.* 1.6 per 100 000 prescriptions) is about five times that of any other drug.[406] In spite of this, recent medical opinion[407] suggests that Myocrisin is as effective as any other drug for the treatment of rheumatoid arthritis, and one of the few that can alter the course of the disease (arrest the destruction of bone and cartilage).

$$Au—S—CHCO_2^- \ Na^+$$
$$|$$
$$CH_2CO_2^- \ Na^+$$

(16)

(17)

In patients receiving chrysotherapy, deposition of gold tends to be abundant in the synovial lining cells and the macrophages just below this layer in the affected joint,[408] but is not confined to inflamed joint tissue. It occurs in macrophages of many organs, including the renal tubular epithelium (kidney) and, after recent therapy, in seminiferous tubules, hepatocytes, and adrenal cortical cells. Gold has persisted in synovial and other tissues for up to 23 years after chrysotherapy had stopped. Little attention had been paid to the co-ordination chemistry of these drugs until it was emphasized[409, 410] that Au^I will adopt a co-ordination number of two, perhaps three, in solution and that the 1 : 1 complex has considerable affinity for further thiol (including glutathione, which is 2—3 mmol l^{-1} in human erythrocytes, and cysteine derivatives), giving species with empirical formula $Au_4(SR)_{6 \ or \ 7}$ which can themselves undergo thiol exchange.[411] Thio-

[405] C. L. Litterst, I. J. Torres, and A. M. Guarino, *J. Clin. Hematol. Oncol.*, 1977, **7**, 169.
[406] A. G. L. Kay, *Br. Med. J.*, 1976 (22 May), 1266.
[407] T. J. Constable, A. P. Crockson, R. A. Crockson, and B. McConkey, *Lancet*, 1975, ii, 1176.
[408] B. Vernon-Roberts, J. L. Dore, J. D. Jessop, and W. J. Henderson, *Ann. Rheum. Dis.*, 1976, **35**, 477.
[409] P. J. Sadler, *Struct. Bonding (Berlin)*, 1976, **29**, 171.
[410] P. J. Sadler, *Gold Bull.*, 1976, **9**, 110.
[411] P. J. Sadler and A. A. Isab, *J.C.S. Chem. Comm.*, 1976, 1051.

malate and thioglucose can be readily displaced by these reactions. In view of the abundance of thiols (including proteins) *in vivo*, it is perhaps not so surprising that gold reaches many sites.

Atomic absorption and neutron activation analyses both give satisfactory results for the estimation of the total gold in the plasma of rheumatoid patients, but only activation analysis is reported[412] to be suitable for determining the gold in various fractions (apart from albumin). Typically, more than 90% of gold is bound to albumin: albumin 2.5—38 μg atom of Au per litre, globulins 0.7— 3.5 μg atom of Au per litre, and free (low mol. wt.) 0.05—1.5 μg atom of Au per litre. Kamel *et al.*[413] have determined gold in protein fractions from patients who had been undergoing therapy for at least six months and who had serum gold levels of 1—3 μg ml^{-1}, using carbon furnace atomic absorption, and electro-phoresis and gel chromatography to separate the proteins. Appreciable gold levels were found in α_1-lipoprotein (or β_1-globulin), IgA and IgG γ-globulins; typical ranges here were: albumin 45—76% of total Au, α_1 2—4%, α_2 12—25%, β 5—14%, and γ 4.12%. There may be appreciable Cu–Au substitution of caeruloplasmin, which has elevated levels in rheumatoid arthritis. The gold level in nails (*ca.* 3 μg per gram) appears to correlate with that in serum (*ca.* 2 μg per ml) and is higher than that in the hair (*ca.* 0.6 μg per gram) of these patients.[414]

Significant reductions in the Cu and Zn levels of red blood cells from mice injected with aurothiomalate have been detected by analytical electron micro-scopy.[415] Levels of K and P were elevated but that of Na was lowered.

It has been suggested[416] that administration of penicillamine, to remove Au, or a combination of Au and penicillamine therapies, could produce toxic gold(III) complexes, but at present there is no evidence for the oxidation of Au^I to Au^{III} *in vivo*.

Isolated rheumatoid cells in monolayer culture produce large amounts of collagenase, which may be involved in joint destruction *in vivo*, but its production is not inhibited by aurothiomalate.[417] Aurothiomalate and aurothioglucose both dramatically inhibit the synthesis of prostaglandin $F_{2\alpha}$ and stimulate that of PGE$_2$ in an *in vitro* prostaglandin synthetase assay.[418] It is notable that the direction of prostaglandin synthesis by this enzyme can be controlled by copper (probably Cu^I, usually added as Cu^{II} plus excess glutathione). Most organic anti-inflammatory drugs inhibited synthesis of prostaglandins E_2 and $F_{2\alpha}$ equally.

In mycoplasma-induced arthritis the effectiveness of gold drugs is not due to antimycoplasmal activity.[419]

[412] R. J. Ward, C. J. Danpure, and D. A. Fyfe, in ref. 1, p. 301.
[413] H. Kamel, D. H. Brown, J. M. Ottaway, and W. E. Smith, *Analyst*, 1977, **102**, 645.
[414] H. Kamel, D. H. Brown, J. M. Ottaway, and W. E. Smith, *Talanta*, 1977, **24**, 309.
[415] T. C. Appleton, *Acta Pharmacol. Toxicol.*, 1977, **41**, 12.
[416] D. H. Brown, G. C. McKinlay, and W. E. Smith, *J.C.S. Dalton*, 1978, 199.
[417] J. M. Dayer, S. M. Krane, R. G. G. Russell, and D. R. Robinson, *Proc. Nat. Acad. Sci. U.S.A.*, 1976, **73**, 945.
[418] K. J. Stone, S. J. Mather, and P. P. Gibson, *Prostaglandins*, 1975, **10**, 241.
[419] P. C. T. Hannah, *J. Med. Microbiol.*, 1977, **10**, 87.

A significant advance in the design of anti-arthritic gold drugs has recently occurred. In 1972 Sutton *et al.*[420] described a series of Au^I phosphines of the type X—Au—PR_3 (where X was a halide or thiolate) which exhibited anti-arthritic activity after *oral* administration to adjuvant arthritic rats. One of the most active compounds in this series was (2,3,4,6-tetra-*O*-acetyl-1-thio-*β*-D-glucopyranosato-*S*)gold (18), 'Auranofin' (SKF D-39162). On a mg per kg basis, it was demonstrated that the oral potency of Auranofin was equal to, or better than, parenterally administered aurothiomalate,[421] and moreover it was less toxic. The compound inhibits antibody production, lysosomal release from phagocytizing leucocytes, and mediator release in immediate hypersensitivity reactions.[422] In contrast to aurothiomalate, it did not appear to inhibit sulphydryl group activity (the SH S–S interchange reaction between serum SH and dithionit-robenzoic acid). At 1—10 μmol l^{-1} levels it produces a dose-dependent reduction in the extracellular levels of lysosomal enzymes (*β*-glucuronidase and lysozyme) released from rat leucocytes during phagocytosis of zymosan particles; neither the ligands themselves (the phosphine oxide was tested) nor aurothiomalate were potent inhibitors.

CH$_2$OAc

O   S —Au —PEt$_3$

OAc

OAc

OAc

(18)

The effect of Auranofin and aurothiomalate on DNA and protein synthesis of phytohaemaglutinin-stimulated lymphocytes has been studied with a view to explaining the increased synthetic rate of IgG immunoglobulins (78 mg per kg per day) compared to normal controls (31 mg per kg).[423] Only Auranofin (1 μg per ml) inhibited synthesis, probably through interference with the transport system of the lymphocyte membrane.

A six-month clinical evaluation of Auranofin has been made on eight patients,[424] who received 3 mg orally twice a day for three months, or 3 mg twice a day for three weeks followed by 3 mg three times a day for nine weeks. The drug was absorbed and well tolerated; IgG levels dropped by the third week and clinical improvement was apparent after five weeks. Albumin levels increased, and Auranofin evidently protects against the enhanced degradation which occurs in RA patients. α_2-Globulin and rheumatoid titres decreased. After 12 weeks, blood Au levels reached 1.5 μg ml^{-1}. This general picture has been confirmed in another trial.[425]

[420] B. M. Sutton, E. McGusty, D. T. Walz, and M. J. DiMartino, *J. Med. Chem.*, 1972, **15**, 1095.
[421] D. T. Walz, M. J. DiMartino, L. W. Chakrin, B. M. Sutton, and A. Misher, *J. Pharmacol. Exp. Ther.*, 1976, **197**, 142.
[422] M. J. DiMartino and D. T. Walz, *Inflammation*, 1977, **2**, 131.
[423] A. E. Finkelstein, O. R. Burrone, D. T. Walz, and A. Misher, *J. Rheumatol.*, 1977, **4**, 245.
[424] A. E. Finkelstein, D. T. Walz, V. Batista, M. Mizraji, F. Roisman, and A. Misher, *Ann. Rheum. Dis.*, 1976, **35**, 251.
[425] F.-E. Berglof, K. Berglof, and D. T. Walz, *J. Rheumatol.*, 1978, **5**, 68.

Author Index

Corrigendum: After the text of this book had been printed, it was discovered that Figure 8 of Chapter 8 (on page 342) has been misattributed, and that it was taken from *Biochemistry*, 1977, **16**, 5228. We apologise to the publishers of *Biochemistry* for this error.